W9-CLI-003

No
THE TRVSTEES OF THE
PVBLIC LIBRARY
LVX
OMNIVM
CIVIVM
OF THE CITY OF
BOSTON
1852 1878

NEW ROADS TO YESTERDAY

New Roads to Yesterday

ESSAYS IN ARCHAEOLOGY

Edited by **JOSEPH R. CALDWELL**

ARTICLES FROM *SCIENCE*

Basic Books, Inc., Publishers • New York

Library of Congress Catalog Card Number: 65–25225

Manufactured in the United States of America

Designed by *Sophie Adler*

Contributors

Robert M. Adams is Director of the Oriental Institute, University of Chicago.

Y. Aharoni is an instructor in the Department of Archaeology, Hebrew University, Jerusalem.

François Bordes is Professor of Prehistory, University of Bordeaux.

Robert J. Braidwood is Professor in the Oriental Institute and Department of Anthropology, University of Chicago, and Field Director of the Oriental Institute's Iraq-Jarmo project.

Karl W. Butzer is Associate Professor of Geography, University of Wisconsin.

Joseph R. Caldwell is Head Curator of Anthropology, Illinois State Museum.

Michael D. Coe is Associate Professor of Anthropology, Yale University, New Haven.

Robert H. Dyson, Jr., is Field Director of the Hasanlu project and Assistant Curator of the Near Eastern Section, University of Pennsylvania Museum.

M. Evenari is Professor of Botany, Hebrew University, Jerusalem.

Edwin N. Ferdon, Jr., is Associate Director, Arizona State Museum, University of Arizona.

Kent T. Flannery is Associate Curator of Archaeology, United States National Museum, Smithsonian Institution, Washington, D.C.

Walton C. Galinat is a Research Fellow, Bussey Institution, Harvard University.

James B. Griffin is Director of the Museum of Anthropology, University of Michigan, Ann Arbor.

F. Clark Howell is Professor of Anthropology, University of Chicago.

Vernon J. Hurst is Chairman of the Department of Geology, University of Georgia, Athens.

Thorkild Jacobsen is Professor in the Oriental Institute, University of Chicago.

Arthur R. Kelly is Professor of Anthropology, University of Georgia, Athens.

William S. Laughlin is Professor of Anthropology, University of Wisconsin, Madison.

Richard MacNeish is Director of the Tehuacán project, Robert S. Peabody Foundation for Archaeology, Andover, Massachusetts.

PAUL C. MANGELSDORF is Fisher Professor of Natural History, Harvard University.

CHARLES A. REED is Associate Professor of Biology and Curator of Mammals and Reptiles, Peabody Museum, Yale University, Director of the Yale Prehistoric Expedition to Nubia, and zoologist to the Iraq-Jarmo and Iranian prehistoric projects of the Oriental Institute, University of Chicago.

IRVING ROUSE is Professor of Anthropology, Yale University.

L. SHANAN is a hydrological engineer in Tel Aviv.

RALPH S. SOLECKI is Associate Professor of Anthropology, Columbia University.

DENISE DE SONNEVILLE-BORDES is Master of Research, Centre National de la Recherche Scientifique, Laboratoire de Préhistoire de la Faculté des Sciences de Bordeaux.

N. TADMOR is a plant ecologist with the Ministry of Agriculture of Israel.

GORDON R. WILLEY is Bowditch Professor of Mexican and Central American Archaeology and Ethnology, Peabody Museum of Archaeology and Ethnology, Harvard University.

Contents

NEW ROADS TO YESTERDAY

Introduction

New Roads to Yesterday

Joseph R. Caldwell

In recent years archaeologists have been gaining knowledge of man's past in hitherto undreamed abundance. This newfound success is due to new techniques, such as radiocarbon dating methods, contributed by the physical sciences, the development of new, meticulous field methods, and novel concepts and theories. The excitement and ferment over the past decade, when archaeology's horizons were so rapidly expanding, found a ready-made historian in *Science* magazine, which originally published the essays in this volume. Many of these are landmarks that should excite the general reader as well as the professional archaeologist, for they are written by the innovators themselves.

Our arrangement of these contributions proceeds according to a chronological scheme: the development of man and cultures in the Old World; investigations of the settlement of the New World and cultural development there; the achievement of cities and civilizations in both hemispheres. The concluding chapter by Vernon J. Hurst and Arthur R. Kelly illustrates how heavily today's archaeology relies on the collaboration of the physical sciences.

In the hundred years or so since Charles Darwin taught us how to trace man's genealogy, many discoveries of our primitive forebears and relatives have been made. Skeletal parts of several types of ancient man are now known in detail, and we can be sure that additional evidence will be found. Yet, the certain fact of human evolution satisfies us no more than it did Darwin. Today, we are still learning more of the events, selective pressures, and genetic changes that brought us step by step to our present biological condition. We are still only approaching a knowledge of the association of events and pressures, not to mention genes, with the particular forms in which human evolution was expressed.

Current investigations in human palaeontology, therefore, reflect a constant effort to gather more information: to know the various paths human evolution has taken means, essentially, to fill the gaps in the fossil record of man and to establish varying degrees of relationship or descent. To understand the selective pressures that affected human evolution requires investigation of the specific events and conditions—geological, geographical, climatic, zoological, and cultural—relevant to this evolution at successive times in the past. Finally, since there was a time when man's ancestors were not human, we need to know how manlike individuals living in virtually cultureless groups could begin to modify the effects of the savage world around them, could develop unusual kinds of social behavior, and could build cultural traditions, which in turn influenced biological evolution.

This problem of culture influencing evolution is complex. Although its solution ultimately depends on archaeology, some of the most promising leads as to how we ought to interpret archaeological evidence come from studies of the abilities and sociality of apes and monkeys, our nearest living kin. It would seem that the mutual help and cooperation practiced by primates and other orders is one way to modify the external selective pressures exerted by environment and competing species. Male baboons will unite to repel a larger and stronger enemy and, one might wonder parenthetically, if man's failure to develop specialized biological weapons for fighting might not be related to the social cooperation of his primate ancestors. (There is many an observer, no doubt, who would assert that man's ferocity is best expressed through the group.)

Group living, moreover, can create evolutionary pressures on its own account—internal pressures, much different in quality from the external pressures of the physical environment. Observing the incessant sociality of macaques and baboons in the London and Basle zoos, M. R. A. Chance and H. Kummer suggest that their frequent, and sometimes dangerous, status-charged encounters ought to put a selective premium on alertness and, perhaps, intelligence. "The strain of this type of social system," they write, "must tax the discriminating power of the brain to breaking point. Consequently there is good reason to believe that the ability to adapt to social requirements could have played a more potent part in the evolution of our mental powers than has been evident heretofore."[1]

Recently, the continent of Africa has yielded startling evidence that human origins are older than was previously supposed, and that our hominid ancestors achieved some cultural features long before the development of the human form and brain. The implication is that early cultural elements played a decisive role in modifying external and internal evolutionary pressures and setting man on the course of his own distinctively human evolution.

In the first chapter of this volume, F. Clark Howell's "The Villafranchian and Human Origins," new material from Africa is used, along with a broad selection of other evidence, to propose events and situations relevant to the evolution of man during the earliest part of the Pleistocene epoch—probably, according to a recent dating by the potassium-argon method, considerably more than one million years ago. The term "Villafranchian" is applied to assemblages of mammalian fauna similar to those first recognized in the locality of Villafranca d'Asti in the upper Po drainage of Italy. Such assemblages, found in various other parts of the world, mark the beginning of the Pleistocene; they are notable for the first appearance of such modern genera as *Elephas, Bos,* and *Equus.* These genera are still distinctively associated with late Tertiary species and can be employed to distinguish an early Pleistocene time interval. Howell uses this interval as a means of inferring some of the events and situations involved in man's evolution.

Surveying all known associations of fossil hominids or their attributable stone implements with Villafranchian assemblages, Howell finds that, except for two possible occurrences in southwest Asia, the best evidence comes from Africa. The ancient sites of Fouarat in Morocco and Ain Hanech in Tunisia yield genuine stone artifacts. In Africa, south of the Sahara, are quantities of pebble tools and various other artifacts associated with the australopithecines, who, as now seems certain, must have made them. Howell concludes that Africa was the primary center for human origins and dispersals.

The primitive populations ancestral to man appear to have been drawn from the genus *Australopithecus,* with the events of the African Villafranchian playing a key role in their development. That australopithecine morphology was tending toward human is indicated by a pelvic structure with lumbar and lower limb modifications implying an upright structure and an efficient bipedal gait. Dentition, facial structure, and hands are also primitively human in varying degrees. The new find "*Homo habilis*" continued this trend.

Australopithecine finds in southern Africa tell us something about the adaptation of these creatures to the Villafranchian world. The sites are former caves containing skeletal parts of other species in about the same proportions as occur in carnivore kill sites in the open. It is therefore a reasonable inference that our australopithecines brought to the caves whatever they could secure from such kill sites. In addition, smashed baboon skulls, some with depressed or radiating fractures, and longitudinally split long bones imply that the australopithecine was a carnivorous predator himself. There is evidence from the Taungs site that crabs, eggs, turtles, birds, and rodents also formed a substantial proportion of his diet.

These erect-walking, skull-crushing creatures are, in all respects, worthy of a place in man's ancestry. The Villafranchian world, Howell tells us, entailed a period of trial in a new and exploitable environment. This was a crucial time for the development of man, for the higher primates, and for mammalian evolution in general. Perhaps, he adds, when we know more about such things as the radiations of the Villafranchian fauna and the desiccations that resulted in the present restricted habitats of the African apes, we shall also know more about this earliest history of mankind.

A consequence of the discovery that the australopithecines were making sharp-edged pebble tools is that we must decide whether to define their endeavors as a cultural tradition. These tools from a time previous to the appearance of *Homo*'s distinguishing skeletal features form the basis for the present tendency to emphasize ancient social and cultural factors in the evolution of the human form and brain. These implements are seeming proof of the old idea that man is a self-domesticated animal.

It will, of course, be some time before we can reasonably assess the role of culture in the australopithecine situation. We lack the evidence of perishable items long decayed, and of usages and understandings now vanished. Moreover, these creatures are so remote from us that cultural inferences become difficult. None of our recent so-called "primitive" cultures, which are the product of millennia of social development, offer any analogies. Linguists are fond of telling us that there are no longer any primitive languages; now the australopithecines pop up to remind us that there are no longer any really primitive cultures either. Far better to turn to the apes to find analogies illuminating the primitive foundations of human sociality and culture. (In this regard, the interested reader may profitably turn to a volume of essays edited by Sherwood L. Washburn.[2])

If perhaps we should not speak too assuredly of australopithecine "cultures," we may at least regard pebble tools as an early stage in the accumulation of artifacts and usages that eventually produced mankind. Our next glimpse of human evolution shows us individuals of our own genus *Homo* widely distributed in Eurasia as well as Africa, with divergent tool traditions that already imply the emergence of different kinds of culture. Aside from possibly late-Villafranchian discoveries in the Jordan and Orontes valleys of southwest Asia, which Howell mentions, the earliest archaeological finds in the northern continents are the massive Heidelberg mandible of Germany and some primitive hand axes from the highest terrace of the Somme River in France. These probably date back to a time before the Mindel glaciation of the Alpine sequence, estimated at about 500,000 years ago.

Two varieties of fossil man from Java, *Meganthropus* and *Pithecanthropus,* may be products of the earlier and later parts of the Mindel glaciation, respectively, while *Sinanthropus,* in China, may have lived during the following interglacial period. In Africa, human remains found along with evidence of a hand ax and cleaver industry at Ternafine, Algeria, are possibly of pre-Mindel age. In the same time range, hand axes and flake tools are characteristic of Africa, Europe, probably southwest Asia, and certainly most of India. In other parts of southern and eastern Asia, except Indonesia, hand axes are either rare or absent. These regions yield "chopping tools," possibly derived from more ancient pebble tools, which at Pekin may be associated with *Sinanthropus.*

Even in those remote ages, mankind had spread far across the Old World. Furthermore, geographical separations, it would seem, already had promoted cultural differences. The differing distributions of particular material-culture elements should become increasingly apparent as work in the field continues.

The improved workmanship of stone tools during this time range indictates a greater manual dexterity, probably attributable to an improved hand structure and the development of an opposable thumb. A startling increase in brain capacity has been described by the anthropologist Sherwood L. Washburn as most notable in the cortical areas concerned with hand, thumb, memory, foresight, and language. He proposes also that the obstetrical dilemma resulting from increasingly larger brains influenced a trend to delivery of the fetus at an earlier stage of development, with a consequent increase in infantile dependency on the parent.

> But this was possible only because the mother, already bipedal and with hands free of locomotor necessities, could hold the helpless, immature infant. The small-brained man-ape probably developed in the uterus as much as the ape does; the human type of mother–child relation must have evolved by the time of the large-brained, fully bipedal humans. . . . The slow-moving mother, carrying the baby, could not hunt, and the combination of the woman's obligation to care for slow-developing babies and man's occupation of hunting imposed a fundamental pattern on the social organization of the human species.[3]

The fossil remains from this time are, as we have seen, exceedingly sparse. *Sinanthropus* and *Pithecanthropus* are by far the best represented. These show, together with the finds from Ternafine in Algeria and probably the new cranium from Tanganyika mentioned by Howell, similarities that cause some prehistorians to group them as *Homo erectus,* a palaeospecies of man that presently seems to provide the available populations from which our own species might have developed. As C. B. M. McBurney has observed, this goes far to reinstate the view of earlier investigators who regarded *Pithecanthropus* and *Sinanthropus* as representative of a widespread evolutionary stage in the emergence of man.[4] Nevertheless, these specimens must have belonged to distinct, geographically separated lineages. Our current task is to learn which among such lineages of this stage contributed most substantially to the genetic inheritance of man, and whether to some degree all of them did. Eventually, archaeology and physical anthropology should be able to trace each lineage and appraise how each might have diverged from or coalesced with the main stream of human evolution.

One facet of this problem is to untangle the ancestry—or ancestries—of our own species from that of Neanderthal man, whom many accord a separate species rank. For this, we need more fossil evidence from the time of the beginning of the Riss glaciation, more than 200,000 years ago, to the beginning of the Würm glaciation, more than 100,000 years ago. We need to know if our *sapiens* ancestry derived from a pre-Neanderthal stage as it apparently also derived from the older *erectus* stage. The skeletal differences among the lineages involved are not so great as to make the prehistorian's task an easy one.

Even the classic, and perhaps most distinctive, Neanderthals who lived in western Europe are not so different from modern man as has been supposed. François Bordes, in "Mousterian Cultures in France," maintains that reconstructions of the Neanderthals as brutish creatures are erroneous. Part of Bordes's contribution to this volume is to show that they are close to us in many ways. From the perspective of the duration of prehistory, they are close to us in time. The last Neanderthals disappeared in France no earlier than 35,000 years ago and may have persisted longer in the Alps and in Africa. Nor is there such an abrupt difference between the cultures of the Mousterian period in France—formerly supposed to coincide exactly with the Neanderthal physical type—and the first Upper Palaeolithic cultures, once believed to have been entirely introduced by *sapiens.* Some supposedly Upper Palaeolithic innovations actually occurred first in Mousterian and even earlier Acheulian times; these include flint blades, backed knives, and end scrapers. In fact, it is difficult to distinguish between a very late Mousterian assemblage of Acheulian tradition and a succeeding Lower Périgordian I Upper Palaeolithic assemblage. The forms of stone tools are the same; only the percentages of occurrence of each form are different. This development, writes Bordes, poses a nice problem for those anthropologists who believe that all the peoples of the Mousterian period in France were of the western Neanderthal variety.

Parenthetically, it is not impossible that the specialized, collective hunting practices that reached a high development during the Upper Palaeolithic may also have had Mousterian beginnings. One type of assemblage, called "Denticulate Mousterian," because of an abundance of toothed flint flakes, perhaps scrapers, represents a people who apparently favored horses as game. Their debris is everywhere rich in horse bones. At Cave 2, Pech de l'Azé, a Denticulate Mousterian level, separated from a Typical Mousterian level, contained bones that were mainly horse. Bones in the Typical Mousterian were red deer and wild oxen.

Bordes's account of these distinctive assemblages is of interest. He regards them as representing separate cultures, and he has distinguished five so far: Mousterian of Acheulian tradition, Denticulate Mousterian, Typical Mousterian, Quina-type Mousterian, and Ferrassie-type Mousterian. From this start, he attempts by comparison to discover the individual cultural ancestries of these assemblages. He is also able to look for particular relationships to the succeeding assemblages of Upper Palaeolithic times. This sound procedure brings forth an appreciation of the cultural complexity that existed in France in those times.

So far so good, but to one who has never worked in France the stratigraphic sequences of these assemblages are astonishing. Denis Peyrony, Bordes, and others had shown that the old scheme of Mousterian development was untenable. Peyrony found layers of Mousterian of Acheulian tradition *between* layers of Typical Mousterian, and Bordes and others have had similar experiences (see Fig. 2–6). This is curious, for in normal stratigraphic sequences of later time spans, one never sees cultural assemblages, once replaced, reappearing in an unmodified condition. Contacts with other groups in the meantime will have assured the development of differences. History does not repeat itself in exact replicas. How then can this recurrence of assemblages in long stratigraphic sequences in France be explained?

Is it possible that the archaeologists are failing to distinguish those particular features that might disclose change? I think not. The implication is that the various separate Mousterian societies must have influenced one another very little. If true, this may indicate much about the relative isolation of each Mousterian group and of the limitations of opportunities for innovation. Perhaps it says something interesting and unusual about the quality of their cultures. On the other hand, if the Mousterian assemblages should each represent the tools used in the hunting and exploiting of particular kinds of game, there would have been less reason to modify these assemblages than if they did not.

Bordes's essay is limited to the Mousterian of France and adjacent Spain, where far more is known of the cultures than in any other region. Ralph S. Solecki, in "Prehistory in Shanidar Valley, Northern Iraq," indicates that the Mediterranean was central to Mousterian cultures. Altogether, we may say that these sites extend north and east through

Germany, central Europe, and as far as Teshik-Tash in the Uzbek Republic and then down to southwest Asia. They are also widespread in North Africa, but virtually absent in the eastern, central, and southern parts of that continent where, it appears, contemporary cultural developments had emerged less distinctively from an older Acheulian base.

Besides discussing various finds in southwest Asia, Solecki affords some observations about the Mousterian deposits in Shanidar Cave. He suggests that the people were "relatively stagnant culturally," estimating that 2,000 generations contributed to the 60,000 years of accumulation attributed to this deposit. Except for a brief vogue of a particular type of flint point (Emiran), there was no change in the tool inventory. This does sound like the kind of people who could have produced those changeless, recurrent assemblages of Bordes. The terminal date for the Mousterian at Shanidar is 46,000 years before the present. Subsequently, the site was deserted until 35,000 years ago, when it was reoccupied by Upper Palaeolithic men (Baradostian).

The introductory passages of Robert J. Braidwood's "Near Eastern Prehistory" remind us not to conclude this discussion without reference to the famous discoveries of Dorothy Garrod at Mount Carmel. Here, Mousterian levels, emphasizing the tool-preparation habits called Levalloisian, showed Neanderthaloid individuals with some *sapiens* characteristics. Still earlier Acheulian levels yielded some blades of a type that later are everywhere associated with modern man.

These finds, according to Solecki, have somewhat later dates than originally expected. Nevertheless, they are of the greatest importance for our understanding of the origins of modern man. Long before we shall be able to reach any firm conclusions, however, many more excavations will have to be made within the Mousterian geographical range. Bordes, we recall, cited French evidence of a transition from Mousterian of Acheulian tradition to a presumably Lower Périgordian I Upper Palaeolithic. There are also indications in western Iran of another cultural transition from Mousterian to Upper Palaeolithic.[5]

Unfortunately, in problems of cultural or morphological origins, a few transitional situations are insufficient evidence. Except in cases of sudden replacements of peoples, archaeological sequences nearly always assume the form of transitions, whether or not they represent the originators or merely the receivers of change. All that the evidence tells us at present is that the complex events leading to the emergence of modern man, whether viewed in the context of a number of Upper Palaeolithic cultural features or not, happened over an enormous area and during a vast time span. Some regions, such as the Levant, may have been ahead of others in this long continuing process. Knowing, however, that human interactions were taking place on a large scale by Mousterian times and given the long period that was involved, we can be sure that a great many breeding isolates and their cultures were drawn into the situation before it was over.

We should probably think in terms of genetic mutation and exchange systems, as well as cultural innovation and exchange systems, operating over millions of square miles. I do not think that we shall discover any one place where modern man emerged. Yet probably we will be borne out in our suspicions that certain regions, such as northwestern Europe, northwestern Asia, and South Africa, participated in the interactions that produced mankind to a distinctively lesser degree.

In Europe, we first find men of *sapiens* type in the latter part of the Würm glaciations, when flint industries, notably of blades, burins, and end scrapers, hallmarks of Upper Palaeolithic times, have become predominant. Denise de Sonneville-Bordes's "Upper Palaeolithic Cultures in Western Europe" indicates the cultural diversity that existed in those times. The older textbook view emphasized the sequence of periods of Upper Palaeolithic culture and the particular innovations that distinguished each from the others. The result was generally to ignore regional variations, differing regional sequences, and the great uncertainties in correlating one region with another.

Now with the kind of investigation de Sonneville-Bordes describes, we become conscious of such cultural boundaries as the Rhine River and of such regional climaxes as occurred in the area between the Loire River and the Pyrenees. A glance at Table 4–1 suggests that it was only in southwestern France and northern Spain that the classic French sequence prevailed. Other sequences in Italy, Belgium, Great Britain, western Germany, and Switzerland differ in important particulars or are incomplete. As we continue to collect radiocarbon dates and observe whether particular regions changed in concert with others, we shall be able to make many culture-historical inferences within the enormous complexity of Upper Palaeolithic Europe.

It may not be supposed that the complexities of the eastward regions will necessarily be less. A generalized horizon called Gravettian, of which the French Solutrian is presumably a regional variant, extended from Mediterranean lands through European Russia.[6] Regional differences may have become the basis for subsequent developments in Russia, Italy, the Balkans, and the western Sahara. Besides the great European plain in Russia, two other Upper Palaeolithic regions have been distinguished: the southern nonglaciated area of central Asia, Crimea, and the Caucasus, which would probably include Solecki's Baradostian levels at Shanidar; and a region in Siberia, which is of slightly later origin than the others.

Of particular interest is Madame de Sonneville-Bordes's observation that the greatest expansions of Upper Palaeolithic cultures occurred during the Aurignacian and Magdalenian periods, which were times of considerable cold. Considering the difficulty of the assessments involved here, it is, of course, possible that there will not be universal agreement on her position. Nevertheless, her view complements present opinion that the main adaptation of Upper Palaeolithic man was to cold climates and cold-loving fauna and that the decline of Upper Palaeolithic cultures was connected with the retreat of the last glaciation and the onset of warmer times.

Following on this, successions of new forests developed in northern Europe. These are correlated by pollen, varve, and radiocarbon dates with specific climates, transgressions of the Baltic Sea, and remarkable organic finds preserved in the anaerobic bogs of Germany and Denmark. The trend is toward composite tools, small flint microliths being set in series in handles of wood and bone to make various cutting, sawing, and reaping tools, as well as weapons. These methods were adopted with some rapidity over Europe, northern Africa, and western Asia. In India, the age of the numerous assemblages with microliths is uncertain. In Siberia, such assemblages clearly come later.

A parallel economic trend was a shift from specialized hunting of large and migratory game (the subsistence pattern of the best known Upper Palaeolithic societies) to a reliance on a variety of animal and plant resources. Continuing extinction of Pleistocene fauna and the sporadic development of new forest fauna must have meant that, as a rule, the prey was smaller than in Upper Palaeolithic times. There was a new emphasis on wild plant foods and, shortly thereafter, on fish and shellfish. These had been eaten since time immemorial, certainly, but were now becoming staples in some regions.

In northern Europe these post-Palaeolithic developments belong to the Mesolithic period and have been explained as a result of the new forest environments created by warming climates. Post-Palaeolithic developments elsewhere in the Old World are sometimes also called Mesolithic, not least because the microliths in the assemblages are a convenient time-marker. Yet the archaeologist Robert J. Braidwood has frequently inveighed against this usage. Although climates were changing in other geographic areas, it was not always in the same way. Cultural development in each region has to be understood in its own terms. Changing climates must indeed be considered in the widespread occurrence of food-collecting societies at this time, but such societies will also be explained in terms of normal cultural processes, accelerated perhaps by the depletion of many species of economically useful fauna. Moreover, those societies that adapted to a wider variety of food resources probably prospered and expanded their territories at the expense of those that did not.

The new emphasis on gathering edible wild plants was prerequisite, of course, to the eventual domestication of some plants. Speaking most generally of the origins of plant and animal husbandry, the evidence is clear that the first developments took place in societies that had a sustained practical interest in wild plants and a detailed knowledge of their properties. These were also the societies that, as a result of collecting wild foods, had achieved some degree of sedentary existence. In such circumstances, it is not surprising that there eventually came about a discovery that wild plants could be propagated, cultivated, and selected for desirable characteristics. We shall see that it happened in more than one part of the world during postglacial times. The necessary conditions must have occurred rather often.

Robert J. Braidwood's "Near Eastern Prehistory" and Charles A. Reed's "Animal Domestication in the Prehistoric Near East" had their genesis in the Iraq-Jarmo project. They are, however, complete surveys of their respective fields. The Iraq-Jarmo project was a planned attempt to combine the efforts of archaeologists and natural scientists to understand "the great swing from the food-gathering to the food-producing stage" in southwest Asia. There had been previous examples of such cooperation, but seldom on this scale or with such substantial results. Braidwood, who conceived and directed the project, has added some comments not in his original article in *Science* magazine. Reed also has made extensive additions to his essay in order to include new evidence.

The results of the project show immense sophistication and a healthy regard for evidence, and, perhaps, an additional advantage of having had natural scientists present on the scene. We learn that they question the case for extreme climatic changes in postglacial times in southwest Asia (but compare Solecki's view, pp. 100–103), even though such changes apparently took place in the Sahara. These investigators also discredit C. E. P. Brooks's "propinquity" theory, which explains the first domestications as a result of concentrations of men, plants, and animals in river valleys and oases following postglacial desiccation. If the climate of southwest Asia had become hotter and drier, they tell us, plants and animals would not have moved downward, but upward to cooler and moister slopes.

After these views had been expressed, Kathleen Kenyon made the alarming discovery of a very early settlement, with domesticated goats, in the oasis of Jericho. This stimulated the lively exchange of divergent opinions that is discussed in Braidwood's "Near Eastern Prehistory." However, perhaps the best reason for discarding the propinquity theory, at least as a general hypothesis, was Braidwood's discovery of the site of Jarmo on the hilly flanks of the Zagros Mountains in Iraqi Kurdistan. In this more elevated region, the presumed wild forms of early plant and animal domesticates are found today.

Jarmo was an early farming village with domesticated sheep, goats, pigs, wild cattle, and wheats and barleys (both wild and cultivated). Nevertheless, the western and southern flanks of the Zagros and Taurus mountains, which Braidwood and his co-workers regarded as the essential area of early domestications, now appears, according to Braidwood, too restrictive. Parallel but different developments may have occurred at lower elevations in southwestern Iran and in Palestine, while on the Anatolian plateau, James Mellaart's recently discovered site of Çatal Hüyük has yielded a wealth of still incompletely studied plant and animal remains. Whatever the limits of the region of the earliest domesticates may turn out to be, the situation has become too complex to be explained by the propinquity theory. Braidwood and Reed conceive of these developments, not as emanating from any particular locality, such as Jarmo, but as a result of the efforts and interactions of many societies over a large area. Very soon, as foreshadowed in the present essays, investigators will be describing what was done in particular regions. They will begin to speak of the culture-historical processes involved in the formation of viable complexes of domesticates as the various regions influenced one another.

The Iraq-Jarmo project has had an effect on archaeology beyond its immediate problems. It has made "problem-oriented archaeology," which indeed predates Heinrich Schliemann, particularly respectable in areas as far away as North America, which had not seen much of it in the preceding years. Some of the concepts being used in southwest Asia, such as "era of incipient cultivation," remind students with similar problems elsewhere of the long periods of trial and experiment that preceded cultivation in each case, which are now practically recognizable from archaeological evidence. The idea of a "primary farming community" can be used both as a historical datum and as an economic type, whereas "introgressive hybridization" suggests a process taking place when plants are moved outward from their natural habitat zones to less friendly environments. Even a characteristic form of illustration, the "droop chart," has come from the fertile minds of Braidwood and his colleagues to express the nuclearity of some regions vis-à-vis marginal areas at successive times (see figs. 5–5 and 5–6).

What knowledge we have of the specific times and places of the first plant domestications is derived from the present distributions of ancestral wild forms and, as at Jarmo, from archaeological discoveries of the seeds themselves. Essentially the same methods are used to determine the first domestications of animals: historic distributions of the wild forms together with examination of the bones from archaeological sites. The order in which various animals were domesticated in southwest Asia, as given in Charles A. Reed's chapter, begins apparently with the sheep. As a zoologist, however, Reed has gone deeply into a number of other relevant problems that generally do not occur to archaeologists and culture historians.

Evidence suggests a number of centers of plant domestication, of which the most important were western Asia in the Old World and Mesoamerica in the New. In the former area, plant domestications were apparently well under way by the eighth millennium B.C. and in Mesoamerica by the sixth millennium B.C.

Although these regions were absolutely independent, they show significant parallel developments. Both seem to have begun with societies that were already collecting wild seeds and that doubtless had a considerable pragmatic knowledge of plants and their properties. A degree of sedentary life must have been obtained by that time. In both regions, we see multiple local centers of plant domestication, from each of which particular plants were added to the spreading complexes of domesticables. We shall probably be able to describe adaptive variations in the environments to which the plants were introduced. There is already evidence of extensive hybridization.

In both hemispheres, areas in the hinterlands can be expected to show long periods of cultural resistance to this new and uncertain technology. Acceptance would, in the beginning, always have been a gradual affair with limited investments of time and energy, perhaps increasingly justified on those occasions when the season's hunting and collecting had been poor. Man seldom suffers change gladly, and, in these cases, the resistant nonagricultural traditions would be reinforced by social organizations already functioning in terms of the earlier economy. One supposes that it must have been the distaff side of the division of labor in most societies that, already concerned with plants and gathering, would have been first to experiment with the new techniques. We need not expect the menfolk to have easily resigned the status and perquisites of hunter-warriors for the doubtful joys of farming. Nor did they, in eastern North America, for example.[7] It is not known whether in western Asia the first cultivators were women, but, if so, and if their importance increased with the growth of agriculture, this would have some bearing on the question of ancient Old World matrilineal societies, which has been debated in Europe for a hundred years. In any case, field agriculture, as opposed to gardening, was in the hands of males by late prehistoric times. Some social and religious practices may have been perpetuated after the decline of woman's central role in the new economy, if central it ever was.

Previous to the development of agriculture in western Asia, there had been especially favorable ecological niches in which wild plants and animals were readily secured, either because of local abundance or because of the juxtaposition of a number of microenvironments. The appearance of food production now meant that some populations might expand into regions of particular suitability for rainfall agriculture. Other populations inhabiting such regions as the Nile and Indus valleys or the lush Caspian foreshore might be among those that for a time resisted the uncertainties of crop cultivation.

Questions of prehistoric settlement and economy obviously require the sort of combined approach by archaeologists and natural scientists so well illustrated in the Iraq-Jarmo project. And, in "Archaeology and Geology in Ancient Egypt," Karl W. Butzer shows some of the contributions that a geomorphologist can make.

In the Nile Valley, from the beginning of the fifth millennium to the middle of the third millennium B.C., rainfall (perhaps winter rainfall) was heavier than today. There was, at this time, an expansion of agricultural populations into parts of the desert that have not been inhabited since that time. Through analysis of land forms associated with existing Predynastic sites, Butzer is able to state some of the criteria by which new settlement areas were undoubtedly selected. With the exception of the Fayum, settlements are on low desert in the outer margin of the Nile flood plain. These locations were flood-free, though adjacent to agricultural land with a water supply. The problem is complicated because there has been a steady rise of the Nile flood plain, and most sites are covered by several meters of silt. It is not surprising then that, although hundreds of Predynastic cemeteries are known from the more outlying areas, corresponding settlement sites have not been found. They must exist, writes Butzer, under the Nile alluvium. These are facts, he adds, that must be taken into account, not only for the location of sites, but for any estimates of populations. Calculations based on discovered sites would be altogether too low. His estimate of Egyptian population during later Predynastic times is from 100,000 to 200,000 inhabitants.

Turning now to a remoter part of the Old World, southeast Asia and the Pacific, knowledge becomes so imperfect that one must rely heavily on historical reconstructions from present linguistic, ethnological, and racial distributions. In "Polynesian Origins," Edwin N. Ferdon finds little trustworthy evidence of origins in the southeast Asian region from which the Polynesians are believed to have set out across the Pacific some time prior to the Christian Era.

Linguistically, the Polynesians are placed in a Malayo-Polynesian family, which is distributed throughout Melanesia and Indonesia across the Indian Ocean to Madagascar. There are differing views about their origins, racial affiliations, whether they were originally as physically homogeneous as they seem today, when they began their migrations, and the nature of their cultural history after they reached the islands. Subsurface archaeology, until recently, was neglected in this search because, as the author tells us, a misconception lingered that "sites are shallow, refuse is sparse, and there seems to have been relatively little cultural change through time." But excavations are beginning to throw light on the Polynesians.

Prior to Thor Heyerdahl's famous voyage from Peru to Raratoria Atoll, the widely accepted view was that the Polynesians were one composite race that broke away from Indonesia and migrated to the Pacific in one or more waves. Cultural differences among existing Polynesian groups were rarely explained as a result of outside influences, but rather as a result of splinter migrations, which separated at various times and modified, in isolation, the particular cultural inventories they took with them.

It was this archaeologically unsupported conception of a single group origin with multiple independent developments that Heyerdahl challenged with his *Kon-Tiki* journey in support of his theory of an American Indian origin of the Polynesians. The older view, writes Ferdon, hid the full complexity of the problems, but we must not let the present East versus West controversy have the same effect. We now have archaeological information coming from several areas, and we must also consider more fully Pacific currents, prevailing winds, the variable effects of planned and unplanned voyages, as well as the numerous factors that must have governed the acceptance or rejection of objects and ideas transported from island to island in this manner. Voyagers might also unknowingly have passed by many an island and transported themselves a thousand miles deeper into the Pacific. Because of these peculiarities, Polynesia must differ from most culture areas on land. The possible sources of cultural influences on any Polynesian island are not necessarily from adjacent areas.

Turning to the New World, recent work has done nothing to change long-held opinion that the first migrants to the Americas, the ancestors of the American Indians, were physically modern men coming in small bands by way of Siberia, gradually filtering down through North and South America. Recalling that the first appearance of *sapiens* in the Old World had been about 40,000 years ago, the earliest dated discoveries in the Americas seem quite late. The earliest, a radiocarbon date of about 22,000 B.C. from Tule Springs, Nevada, has been discredited. Uncertainty surrounds a chipped stone assemblage near Sandia, New Mexico, though its date may be prior to 15000 B.C. The most conservative estimate based on radiocarbon dates establishes the presence of the earliest settlers well before 12000 B.C. Because of lowered sea levels, it would have been possible virtually to walk from Siberia to Alaska.

Alaska, surprisingly enough, has not supplied the earliest dated remains. In "Eskimos and Aleuts: Their Origins and Evolution," William S. Laughlin describes the site of Anangula on the shore of Bering Sea. This has three radiocarbon dates of around 6000 B.C. associated with a rather Upper-Palaeolithic-appearing core and blade industry that resembles materials in Japan and Siberia of 7000 to 11000 B.C. Anangula and a later site, Chaluka, probably represent ancestral Eskimo-Aleuts as distinguished from the less Mongoloid American Indians proper. Slightly later cores and blades of the Denbigh Flint Complex are believed by most workers to be in the ancestry of the proto-Eskimo Dorset culture of the east Canadian Arctic and Greenland. It is significant that there appear to be no early core and blade industries south of the Eskimo and Aleut territories.

Laughlin is interested in the Eskimos and Aleuts as a single basic stock of humanity that has been in an unusual region for some thousands of years, adapting to a diverse series of environments. The resulting developments can be investigated by archaeology and allied disciplines providing "a unique opportunity for studying 'microevolution,' population history from the standpoint of genetics, and biological and cultural adaptation." He discusses linguistic and archaeological evidence, the relation of the availability of particular kinds of faunal resource to economic productivity and population profiles, blood-group evidence of relationships, longevity, adaptation to cold and glare, and the evolutionary consequences of the fact that these people are divided into "breeding isolates," generally choosing mates within their local groups. Judging from the earliest known skeletons, the Eskimos and Aleuts have not become demonstrably more similar, but have undergone some parallel changes. The most notable of these is a shift from longer to shorter heads, which has also taken place among American Indian groups as well as in various parts of the Old World. A similar but smaller change, writes Laughlin, occurred in east Greenland where there can be no question of external admixture. The question of local evolutionary changes must be given more attention.

Laughlin's view of the earliest migrations to the New World is based on the now submerged Bering platform, on cultural connections between the site of Anangula and Japan, on the probable linguistic relationship between the Eskimo and Aleut, on one hand, and the Siberian Chukchi, Koryak and Kamchadal, on the other, as well as on the physical similarities among these groups. He believes that some 15,000 to 10,000 years ago the Bering platform was inhabited by contiguous isolates, which had the effect of a cultural linkage all the way from present Hokkaido to what is now Umnak Island. Later, when the platform was gradually inundated, populations withdrew to their present locations. Inasmuch as early American Indian groups were established in the two continents well before 10000 B.C., their separation from Eskimo-Aleut-Chukchi groups had probably been ensured by differences in economic adaptation and therefore by differences in routes of migration. Since the Bering platform was more than 1,000 miles wide, "the ancestral Indians, with their land-based economy, could have crossed often, following big game, without coming into contact with the Mongoloids [that is, the more Mongoloid Eskimo and others—EDITOR], who worked their way along the coastal edge of the reduced Bering Sea."

James B. Griffin's chapter, "Some Prehistoric Connections between Siberia and America," offers a summary of Upper Palaeolithic and later developments in Siberia compared to contemporary events in neighboring North America from about 12000 B.C. He finds a number of similarities and indications of connections between these areas. The earliest American materials are relatively late, but Siberia, far out on the limb of Old World developments, had also been laggard in the appearance of Upper Palaeolithic styles and innovations. Laughlin's opinion that, after Bering Strait became a channel, major migrations ended, does not mean that cultural diffusions ended as well. Abundant ethnological evidence of boreal connections between the Old World and the New indicates quite plainly that they did not. The Arctic Denbigh Complex, which Griffin has no hesitation in bringing from Siberia, and the Norton Pottery Complex of Alaska, which he derives from the Lena River area of Siberia, are prehistoric cases in point. The long Archaic period in America east of the Rocky Mountains, from about 8000 to 1500 B.C., witnessed changes and adaptations that were primarily developments of the native American populations. From 1500 to 500 B.C., however, there are some new culture traits which seem to Griffin best explained as a result either of diffusion from Asia and to some degree also by population movement. Chief among these is pottery, perhaps brought to eastern North America by "stimulus diffusion," in which knowledge of a particular technology is diffused, but not specific examples of the technology. Other features that Griffin regards as new in eastern North America at this time, and conceivably derived from Eurasia, are burial mounds and a roundheaded (brachycephalic) physical type. (In regard to the latter, compare Laughlin's views.)

Despite the difficulties of tracing boreal connections between Asia and the Americas, more survey, excavation, and utilization of the kind of approach outlined by Griffin will bring more information. The question of transpacific contributions to prehistoric American cultures is more difficult, whereas the possibility of transatlantic crossings prior to the Norsemen is not seriously considered by many Americanists today. Gordon R. Willey in "New World Prehistory" is unconvinced that even the Pacific contacts, which certainly occurred, had any appreciable effect.

A few other students, notably Robert Heine-Geldern, have taken a different view. Ferdon's chapter points up how much more we need to know of southeast Asia and Polynesia to make even a partial appraisal of this matter.

Aside from boreal connections and more questionable influences from other areas, American culture history was clearly independent of the Old World. This enhances our interest in the prehistoric Americas for the study of culture processes. We have already seen some parallels between the independent agricultures of the Old World and the New; other parallels will be described below.

Gordon R. Willey's chapter on prehistory in the New World is a synthesis of prehistoric developments in all the Americas. By dividing the subject into eras or stages of subsistence technology, and by describing as "nuclear" the regions from Mexico to Peru that advanced earlier and farther than the others, he shows the cultural interconnections of the two continents in a way not yet achieved for the Old World. Except for his citation of the since discredited early date from Tule Springs, Nevada, no important changes needed to be made in his article as originally published.

Following a survey of scattered finds of rough percussion-shaped flint implements, which may or may not represent early food-gathering societies, Willey considers the first solid body of evidence in the Americas. This pertains to specialized hunting societies, whose distinctive lanceolate flint points have sometimes been found with the remains of extinct and often migratory varieties of large animals. The beginnings of this adaptation, which is also discussed by Griffin, must be prior to 10000 B.C. In North America, it is represented by fluted lanceolate points of the Folsom and Clovis types, while later the unfluted Yuma and Plainville types are the hallmarks. Similar adaptations to the south are represented by the Iztapan and Lerma remains in central and northwestern Mexico, the El Jobo points of Venezuela, the Aympitín industry of the Andes, and the Magellan I culture of the straits. A few camp sites and animal-kill sites of these early hunters have been investigated, but we still have no clear picture of their life. Yet the analogy to the Upper Palaeolithic specialized hunting societies of the Old World is patent. The same warming climates and animal extinctions that caused the disappearance of the latter apparently had a similar effect on the specialized hunting societies in the New World. We may add that the latest distinctive points of the hunters of the North American plains are found far to the north in Canada at a date of 2800 B.C.[8] This may indicate their northward movement following the retreating ice, a supposition often made in the case of their Old World counterparts.

It is certain that other economic adaptations existed before and during the time of the hunting societies. The latter have been overstressed in the literature because the remains are recognizable as pertaining to an economy of this particular type. But long before the disappearance of hunting societies other groups, using a greater variety of plant and animal resources, arose. In eastern North America these are assigned to the "Eastern Archaic Stage," in western North America they belong to the "Desert Culture." Interpreting such developments in my chapter, "The New American Archaeology," I mention the idea of various regional "efficiencies," each representing adaptation to a different environment, with differing potentialities for exploitation and cultural development, and with necessarily differing receptivity to cultural elements from other areas. To take one example, the "primary forest efficiency" of eastern North America can be seen developmentally as a series of events that ought to be different from those discovered in the histories of plains, mountain, desert, and maritime efficiencies. It is similar in some respects, however, to the independent developments in the forested area of Mesolithic Europe.

Both areas are peripheral to the centers of cultural development in their respective hemispheres. In both regions, after the retreat of the glaciers we find the development of hunting-gathering economies geared to a forest existence. In eastern North America, we know that some of these economies gradually developed a pattern of seasonal movements to particular sources of food and industrial materials, and it would indeed be surprising if this did not happen in Europe as well.[9]

Hunting and gathering were well established in both areas before it was discovered that shellfish could be made a dietary staple, which seems to have happened at the same time in both regions. Perhaps this discovery provided a greater degree of residential stability, for in both areas the earliest crude pottery is found in the upper levels of shell heaps. We are referring to the fiber-tempered pottery of the southeastern United States and the pottery of the Ertebølle shell heaps in northern Europe. Whether the latter might not be independent of earlier potteries, first made in the Mediterranean area and southwest Asia, is not important. What is important is that a sufficient degree of sedentism had been obtained in northern Europe for fragile pottery to be practical to use. Both areas later showed less dependence on shellfish. Also, in both areas, the pattern of forest efficiency, coupled with varying degrees of exploitation of maritime resources, prevented reliance on agriculture long after the technology of agrarian economies had become familiar. A similar situation seems to have been true in Europe beyond the range of the Danubian agricultural developments and through the forest belt of Russia. In eastern North America, agricultural practices had only a limited ascendancy even by historic times; forest resources still provided much of the diet.

Peoples who, after the appearance of effective food production, continued to live by hunting, gathering, or fishing in outlying regions are sometimes described as "marginal." In the past there have been many more such peoples than there are today. In "Prehistory of the West Indies," Irving Rouse describes some of the events that took place as marginal peoples gradually gave way to agriculturalists. This region, where Columbus first set eyes on the New World, has seen a fortunate blending of archaeological results with the theoretical guidelines set out by such ethnologists as Julian Steward.[10]

The first inhabitants of the West Indies were hunting and fishing peoples. Later, the area was partly occupied by "Tropical Forest" agriculturalists from South America whose economies were heavily based on bitter manioc and other root crops. On this agricultural basis cultures of "Circum-Caribbean" type developed, distinguished by a greater complexity of social and religious life. This did not happen in the West Indies alone. Other Circum-Caribbean cultures emerged in northern Amazonia and in the mountainous area between Mesoamerica and the central Andes. In the latter region, however, they developed on a seed agriculture with Mesoamerican connections rather than on an agriculture of the Tropical Forest type.

Rouse's reconstruction is aided by an imposing series of radiocarbon dates essential for the understanding of this large and geographically fissiparous region. It appears that the West Indies were uninhabited until about 2000 B.C. The irregular distribution of the earliest sites suggests to Rouse that the first settlements may have been a result of boats having accidentally been blown out to the Antilles. During the first millennium B.C. certain Tropical Forest agriculturalists pushed out of the lower Orinoco Valley and migrated to Venezuela, where they may have learned seafaring. In any event, they were able to move out into the West Indies at about the time of Christ. They introduced pottery and agriculture as far as Puerto Rico. During the latter part of the first millennium A.D. they continued to push back the earlier marginal populations, eventually to the southwestern tip of Haiti and the western tip of Cuba. Those who reached Cuba continued to live there until the time of Columbus.

Michael D. Coe and Kent V. Flannery, in "Microenvironments and Mesoamerican Prehistory," discuss one aspect of hunting and collecting that continued in most areas after the advent of crop cultivation. Within most geographical regions are a number of microenvironments: ". . . the immediate surroundings of an ancient archaeological site itself, a nearby stream bank, or a distant patch of forest." The Tehuacán Valley of highland Mexico, investigated by Richard S. MacNeish, shows that the development of maize agriculture there did not entirely supplant the hunting-gathering utilizations of Tehuacán microenvironments until after the introduction of irrigation. In lowland Guatemala, on the other hand, the very old site of Ocos shows a greater reliance on maize. Several microenvironments, which could have provided game and wild plant foods, were hardly exploited.

One of the characteristics of ancient hunting-gathering societies, in contrast to societies of specialized hunting, was a considerable reliance on wild plant resources. It is readily demonstrable in both Old and New World agricultural beginnings that societies already making use of wild plant foods provided the contexts in which plant domestication methods were slowly learned. We have already seen that, at Jarmo in Iraqi Kurdistan, both cultivated and wild wheats and barley were used; and we shall see shortly that the use of wild and cultivated plants together was also the case in the New World situations of early plant domestications. Willey's aforementioned chapter cites the desert food-collecting context at Bat Cave, New Mexico, where primitive domesticated corn appeared between 3500 and 2500 B.C. Even more illuminating is his description of MacNeish's outstanding work in Tamaulipas, northeastern Mexico, where, 4,000 years after the first traces of domesticated plants, the composition of food refuse shows that subsistence was still primarily from wild sources: 76 per cent wild plants, 15 per cent animals, and only 9 per cent cultigens. Probably Tamaulipas was excessively slow in this respect, but it appears that we must always distinguish between the ability to domesticate plants and the ability to make of the domesticates the basis for full-scale food production. There are different social, historical, and botanical factors involved in each kind of achievement, and the most likely microenvironments for the initial domestications are not necessarily the best environments for full-scale agriculture.

Plant domestications were not limited to the abilities of hunting-gathering societies of a single type, such as the desert food-collectors of Bat Cave and Tamaulipas. Many other desert food-collectors did not domesticate potentially domesticable plants. Initial plant domestications were achieved by a number of societies inhabiting regions of different environment. A number of distinct hearths of early plant cultivation, which developed before the spread of maize agriculture from Mesoamerica, are described by Willey. In the tropical forests of the Amazon there was a tradition of cultivating tropical root crops that we have already encountered in Rouse's chapter: bitter manioc, sweet manioc, and a variety of yam. Apparently this began well before 1000 B.C. On the Peruvian coast another series of cultivated crops appeared before 2200 B.C.: squash, peppers, gourds, cotton, achira tubers, and Canavalia beans. In the northern Mississippi Valley, plants that were cultivated before 1000 B.C., and prior to the appearance of maize, were probably gourds, pumpkins, squash, Chenopodium, sunflower, and possibly marsh elder and an amaranth. In Tamaulipas, northern Mexico, to which we have already referred, squash, peppers, gourds, and small beans were domesticated much earlier, between 7000 and 5000 B.C. After 3000 B C., a primitive small-eared maize appeared.

In the chapter "Domestication of Corn," by Paul C. Mangelsdorf, Richard S. MacNeish, and Walton C. Galinat, there is a discussion of recent investigations in the Tehuacán Valley of Puebla, southern Mexico.[10] At a date prior to 3400 B.C., there existed an early cultivated maize, two species of squash, tepary beans, chili peppers, amaranths, avocados, and sapotes. At this time wild maize and other plants were still being collected and eaten, and, as we have seen in the chapter by Coe and Flannery, which also cites the Tehuacán evidence, wild plants formed a considerable proportion of diet until the advent of irrigation there in the first millennium B.C.

The studies at Tehuacán were a result of MacNeish's search, first begun at Tamaulipas, for the origins of domesticated maize. The authors tell us that five major dry caves in the Tehuacán Valley have yielded culturally stratified situations of the oldest well-preserved cobs for botanical analysis, the oldest cobs probably being those of a wild maize used well before 5200 B.C. This confirms the view that the ancestor of maize is maize, and neither the related teosinte nor Tripsacum. This maize also appears to be the ancestor to two previously recognized races of Mexican corn, Nal-Tel of southern Mexico and Chapalote of the northwestern region. The earliest cultivated form in Tehuacán, probably prior to 3400 B.C. and called "early cultivated," differs little from its wild ancestor, quantities of which continued to be used for some thousands of years.

Perhaps the early cultivated variety was slightly modified in the beginning simply by its growing in an environment improved by man's removal of competing vegetation. Appearing earlier than 1500 B.C. is another form of domesticated corn called "early tripsacoid," which is suspected to be a product of hybridization between maize and one of its two relatives. Neither teosinte nor Tripsacum is known in the Tehuacán Valley today; tripsacoid maize is believed to be an introduction from the Balsas River region of the adjacent state of Guerrero, the nearest place where teosinte and Tripsacum are still common. The introduced tripsacoid evidently hybridized with both the wild and domesticated corn at Tehuacán to produce hybrids with characteristics intermediate between those of the parents. These back-crossed to produce great variability in both wild and cultivated populations. Both the "early cultivated" and tripsacoid corns became virtually extinct, but the result of their crossing was the Nal-Tel–Chapalote complex. Concerning this the authors tell us:

> It was this corn, more than any other, which initiated the rapid expansion of agriculture that was accompanied by the development of, first, large villages and later, secular cities, the practice of irrigation, and the establishment of a complex religion. If it is too much to say that this corn was responsible for these revolutionary developments, it can at least be said that they probably would not have occurred without it. Perhaps it is not surprising that present-day Mexican Indians have a certain reverence for these ancient races of corn, Nal-Tel and Chapalote, and continue to grow them although they now have more productive races at their command.

Just as hunting-gathering societies formed the context for the achievement of food production, so too did food production make possible larger settlements and eventually cities and civilizations. The processes that slowly led to urban formations in the Americas may not be unlike those that resulted in the first cities of the Old World. In Willey's chapter, we find an orderly presentation of the New World evidence. The spreading village-farming pattern permitted in different areas the development of two kinds of town. These gave rise respectively to the great dispersed lowland Maya centers, such as Tikal and Palenque, and the concentrated urban agglomerations of Chanchan in Peru, Aztec Tenochtitlán, and probably the older Teotihuacán.

> The concentrated city adheres more to the concept of the city in the western European definition of the term. It was a truly urban agglomeration. Its traditions were heterogenetic, and its power extended over a relatively large territorial domain. The city was, in effect, the capital of an empire . . . Although the cities and civilizations which developed in Middle America and Peru in the first millennium A.D. were unique and distinct entities in their own right, it is obvious that they also drew from the common heritage of culture which had begun to be shared by all of Nuclear America at the level of village farming life. This heritage was apparently built up over the centuries, through bonds of interchange and contact, direct and indirect . . . During the era of city life these relationships continued, so that a kind of cosmopolitanism, resulting from trade, was just beginning to appear in Nuclear America in the last few centuries before Columbus.

Thus we are mindful of a close relationship between cities and civilization. Both ambiguous conceptions partly treat the same realities, and it is a reasonable assumption that the origins of cities and civilizations are inseparable. The bonds of interchange and contact, of which Willey writes, offer some explanation for these phenomena. For prehistoric highland Mexico, William Sanders has also written of a process of "symbiosis," arguing along the same lines. Because of great diversity of geographic resources in that area, regional cultures could continually exchange products and stimulate one another.[11]

Indeed the cultural role of modern cities is best expressed in terms of networks of exchange and communication. Each city is a link in innumerable exchange networks. These provide its lifeblood and reason for existence. The particular networks in which it is a link, taken together with its traditions, some of which stem from older or even defunct networks, go a long way toward giving any city its special character.

Another aspect of interchange and contact is reflected in the view of civilization held by the late Robert Redfield that a civilization consists of a number of little (local) traditions joined together in a Great Tradition.[12] The bonds of interchange and contact of the Great Tradition can help explain why, as compared to the relatively isolated primitive cultures, civilizations show such a great mass and prolixity of cultural forms. As a form of interaction, the Great Tradition serves to keep little traditions in communication over significant periods of time and provides the exchanges of materials, techniques, and ideas that foster innovation.[13]

In another recent article, Gordon R. Willey has described something that may be Redfield's Great Tradition in a more primitive form.[14] At the very dawn of the prehistoric civilizations in Mexico and Peru, a number of distinct societies are found to be in communication. "Great Art Styles" extending over large areas, Olmec in Mexico and Chavin in Peru, indicate in each case the spread of a distinctive religious symbolism linking various areas within each region. And Willey asks, as historian Arnold Toynbee had asked before him, whether religion might have been a causal factor in the development of civilizations. This seems not at all unlikely. But a more general point of view would encompass many systems of communication and interchange, of which religion is only one.

In western Asia the precociously extensive Anatolian town of Çatal Hüyük has yielded much evidence of having been a religious center.[15] Still earlier, fortified Jericho[16] might be more explicable if we knew to which exchange networks it belonged. Or perhaps its urbanity simply derived from the capacity of its surroundings to support a large number of people. In this century we do not as yet have archaeology so much by the tail that our generalities can explain everything.

Most towns and cities in western Asia, nonetheless, are probably explicable in terms of the picture we have of increasing interaction, the same bonds of communication and exchange that Willey and Sanders have in mind. We can think of this both in terms of long-distance trade and the increasing size of continuous areas of communication, *oikoumenē*, which the archaeologist sees as sharing significant stylistic features which they would not share without communication.

Trade had begun in western Asia at least as early as Upper Palaeolithic times with traffic in ornamental shells from the Red Sea and Persian Gulf. Later there was an extensive carrying-trade in obsidian from the region of Lake Van in present Armenia. Before 4500 B.C. there was a trade in native copper from Anatolia and the Iranian Plateau. With regard to areas of communication, Robert J. Braidwood and Bruce Howe call attention to a "Ubaidian *oikoumenē*" with close contacts among constituent societies linked from the Mediterranean to distant reaches in Iran.[17] One may think that more innovations ought to have appeared at this time as a result of increased contacts, perhaps the temples at Eridu in Iraq, smelting and casting of copper, stamp seals intimately bound up with ownership and commerce, perhaps mold-made bricks, and possibly the potter's wheel.

It may be that the succeeding Warka period in Mesopotamia was a time of decreased communication. At least Ann Perkins found that northern and southern Iraq became less alike.[18] Moreover, some regions of Iran where traditions of painted pottery persisted strongly are now less like Warkan Mesopotamia than they had appeared earlier, when Iranian Buff Ware was not, after all, so different from Ubaid pottery.

The next period, sometimes called Protoliterate, is clearly another time of increased interactions among many far-flung peoples and societies. As in the Ubaid period earlier, we can regard such interactions as fostering idea-exchange and hence an increased tempo of innovation without being relieved of the necessity of examining each specific innovation in terms of the particular events that brought it into being. The greater proportion of traits that Ann Perkins listed as linking northern and southern Iraq were innovations belonging to this time span. More narrowly within this time Seton Lloyd defined a "late Uruk" interval corresponding to Level VI of the long stratigraphic sequence at the site of Warka in southern Mesopotamia. To this he ascribed such innovations as earliest temple buildings in square-section bricks (*riemchen*), cone mosaic ornament, and the first use of cylinder seals.[19] Cuneiform writing on clay tablets comes into use slightly later at about 3000 B.C. but it is not at all clear that the apparently archaic Sumerian script of the oldest clay tablets in southern Mesopotamia (in Warka Level IV) is necessarily older than the archaic Protoelamite script of adjacent southeastern Iran.[20]

Some elements have an immense geographical spread. Cylinder seals are now found from Iran to Egypt. The curious and distinctive "beveled rim bowls" occur from the Mediterranean littoral to Tal-i-Iblis some 1,500 miles away in southeastern Iran.[21] This period of interaction also may have had more distant effects. Mesopotamian elements reaching Egypt at this time, including writing, must have hastened the development of Egyptian civilization. What is more, on the eastern side of the nexus new discoveries in Pakistan and Afghanistan are carrying the antecedents of Indus Valley civilization back to a point where they are almost old enough to have been stimulated in like manner.

The archaeology of civilizations is in some ways more complex and illuminating than the archaeology of primitive societies. The chapters by Dyson; Adams; Jacobsen and Adams; and Evenari, Shanan, Tadmor, and Aharoni, indicate that many specialists may be needed to do justice to the material of this time range. In addition to all those whose collaboration is desirable at a prehistoric site, the essays published here make clear that we need such specialists as historians, epigraphers, and numismatists. More complex technologies mean that the archaeologist must also be prepared to call on experts in glass, metals, glazes, dyes, and so on.

The great networks of communication that brought so many things and ideas to the various capitals and provincial centers also had less happy results. Robert M. Adams in "Agriculture and Urban Life in Early Southwestern Iran" is struck by the constant wars, destruction, and economic disasters that are recorded from the time ancient Elam became involved in Mesopotamian campaigns. These catastrophes continued through the history of Khuzistan almost to the present day. It seems to Adams that man's well-being may depend more often on the nature of his social institutions than on the presence or absence of particular items of material equipment. And to this picture may be added the knowledge that in western Asia, after the advent of civilization, most regions were not only subject to the disasters of their own military adventures, but sooner or later became the victims of each of their neighbors and of all the world empires and their policies.

The chapter by Robert Dyson, Jr., "The Hasanlu Project," documents the situation by offering a list of some twenty-odd conquests of the present province of Azerbaijan in northwestern Iran, beginning with the Assyrians in 850 B.C. and continuing through the Pahlavis in A.D. 1925. His report, however, is concerned with the investigation of a provincial center during the time span of 1200 B.C. to about 500 B.C. His main objective is "elucidation of the chaotic and mysterious period which corresponds to the first appearance of the Medes and the Persians" in the first half of the first millennium B.C. The historical records of the neighboring Assyrian and Urartian kingdoms indicate that by 850 B.C. the inhabitants of the region were Mannaeans, speaking Hurrian. After about 600 B.C., writes Dyson, they were completely absorbed by the Medes, and a short time later the region became part of the Achaemenian Empire.

In his discussion, which describes the investigation of a single site, Dyson gives an illuminating account of fundamental archaeological procedures. He describes the establishment of a stratigraphic sequence, relative and absolute chronologies, the characteristics of the local population, local economy, and the connections established through foreign trade. The project is continuing, and we may judge that Dyson's objective will be achieved by more excavations of exactly this kind.

If Dyson's work is a model of present archaeological excavation, Adams' aforementioned work is a model for what may be learned in a region without digging. Provided some kind of sequence is known from previous excavations, historical records, or both, a surface survey can be very illuminating. In this one, many sites were identified by period on the basis of pottery fragments lying on the ground. Inferences were made by comparing the distribution of sites of each period with physical situations, elevation, position in regard to ancient and modern watercourses and irrigation canals, position with regard to present and presumably past patterns of rainfall, and the results compared with statements in historical documents. Thus he is able to say that the earliest agricultural villages in the Susiana region of Khuzistan are found on the upper plains, which now have the highest annual rainfall, and to infer that at that time agriculture depended mainly or exclusively on rainfall.

From about 5000 to 3000 B.C. the number of sites greatly increases, extending farther south along the margins of old streams with more numerous channels than exist today. Here then is reason to think that irrigation practices had been introduced. Subsequently, some sites (in addition to the great town of Susa) begin to stand out as small towns, while the total number of sites within the survey area decreases by two-thirds. The newly emerging pattern, writes Adams, consists in part of drawing the population together into larger and more defensible political units. Some of these, on the analogy of the better known Mesopotamian materials—for much of Khuzistan is an extension of the great Mesopotamian plain—begin to attain urban status and to be surrounded with defensive walls.

This trend continued after Elam became involved in Mesopotamian wars and political alignments. Similar observations are offered in this chapter for the entire later history of the region, from its destruction by the Assyrians, through incorporation into the Achaemenian, Greek, Parthian, Sassanian and Moslem empires. The effect of such events on agriculture and urbanization is discussed in a fruitful union of historical and archaeological evidence. Here in Khuzistan is further documentation of the conditions of well-being that prevailed under the Sassanians; we are beginning to understand the cultural influence they exerted all the way to western Europe and the Far East. "For Iran, at least," writes Adams, "we are justified in regarding the Sassanian administrators as the spiritual ancestors of the modern teams of developers, and in hoping that the latter are as successful by contemporary standards as the Sassanians must have seemed in their own time."

Great weirs constructed across the major rivers are still locally identified as "Roman," doubtless because of the role in their construction played by the 70,000 legionaries captured with the Emperor Valerian by Shapur I. Aerial photographs and ground reconnaissance enabled Adams to locate the major branches of the Sassanian canal system. It apparently had been designed and executed under a series of comprehensive plans, and it contrasts favorably with recent small-scale private irrigation projects in the Near East, where the lack of central planning is immediately evident. From Moslem times, the region exhibited a gradual agricultural and municipal decay as shown by tax records and the disappearance of towns. Great tracts of marginal lands were put for a time into operation as money-making schemes, while better lands, on the other hand, were probably held by an "intrenched peasantry," which had not been unduly disturbed by the conquest. Prosperity did not begin to return to Khuzistan until the opening of the Karun River to steamship navigation and until the development of oil resources after World War I. Now with government-sponsored agricultural improvement, and projects such as the Diz (now Pahlavi) Dam and the reestablishment of the sugar cane industry, the region is entering a new period of prosperity.

In "Salt and Silt in Ancient Mesopotamian Agriculture," Thorkild Jacobsen and Robert M. Adams examine the problems of soil salinity and sedimentation that contributed to the decline of agriculture in Iraq and the Iranian province of Khuzistan. Soils readily become saline in regions where low rainfall is combined with poor drainage and existent quantities of salt. Often the problem is brought about by agricultural practices, so that in a good many parts of western Asia and the Sahara, areas that were once cultivated will no longer produce a fair crop. In fact, our authors point out that modern American and European methods of cultivation when introduced to this area sometimes *increase* soil salinity. For the successful future development of agriculture in this region, an understanding of the 6,000 year record of its irrigation agriculture is indispensable.

From 2400 B.C., the farm lands of southern Iraq have become increasingly salty. This is shown by records of ancient surveyors—comparisons show a continuous decrease in the proportions of wheat acreage to the acreage of more salt-resistant barley—and by a serious decline in the region's crop yields. Jacobsen and Adams think that growing salinity must have played an important part in the decline of the Sumerian city-states and the shift of political power to Babylon in the eighteenth century B.C.

Sedimentation is inevitably associated with Iraqi agriculture and, in turn, brings its own problems. Canals require constant cleaning. The silt removed is eventually carried by wind and rain to the surrrounding fields, while additional silt is deposited by the irrigation water itself. The deposition of silt is a massive, continuing process; the authors estimate that perhaps ten meters have accumulated over the northern end of the alluvium during the past 5,000 years.

Archaeological survey of the Diyala region, using large-scale maps and aerial photographs, succeeded in locating many irrigation watercourses attributable to various times. The later canals were still indicated by spoil ridges from their repeated cleaning. The location of the older watercourses could be inferred from the patterns of the settlements; settlements of every period always described networks of lines representing the approximate location of those streams that were so necessary for settled agricultural life.

On the basis of these maps and photographs, it was possible to distinguish two successive phases of settlement and irrigation. The earlier, beginning about 4000 B.C., shows a linear pattern of settlement largely confined to the banks of the major natural watercourses. Water for the fields was drawn more or less directly from the watercourses, and accumulation of silt was not then the major problem it later became. These natural watercourses followed nature's regimes, and the system persisted for more than three millennia, in spite of periods of abandonment. The second phase, a vastly enlarged and complex system of brachiating branch canals, continued from Achaemenian times into the Islamic period. This was a far more complete exploitation of available land and water, resulting in a much greater population.

Indeed, the first true cities in the area were introduced during this phase by Alexander the Great's followers. Our authors tell us that this was "a whole new conception of irrigation which undertook boldly to reshape the physical environment at a cost which could be met only with the full resources of a powerful and highly centralized state." Unfortunately, however, the system required this same great administrative agency to maintain it. The long branch canals tended to fill rapidly with silt, and the massive weirs needed periodic reconstruction and continuous maintenance. During social unrest and military adventures, the maintenance of the system could only fall back on local communities that were unable to handle it. Eventually, natural factors such as the raising of the level of fields by silt deposition, coupled with decreasing capacity for more than rudimentary maintenance, brought about the virtual desertion of the area. We learn that Hulagu Khan's invading Mongol horsemen must have surveyed an already deserted and desolate scene, for which they have hitherto been unjustly blamed as the authors.

In "Ancient Agriculture in the Negev," M. Evenari, L. Shanan, N.

Tadmor, and Y. Aharoni combine techniques of botany, hydrology, ecology, and archaeology to investigate ancient ways of obtaining water for agriculture in the desert. Rainfall in the barren Negev of southern Israel is only about 100 millimeters a year, yet this region once contained towns and flourishing farms. Potsherds associated with the ruins there indicate settlements at least as early as the fourth millennium B.C. There is evidence of considerable prosperity during Nabatean, Roman, and Byzantine times. Just as in southwestern Iran, the Moslem conquest seems to have initiated a slow decline of agriculture.

Water utilization techniques included chain-well systems, such as the *qanat*, apparently used in Iran at least as early as the time of Alexander. A chain-well system in the Arava Valley of the Negev could not be dated closely and might have belonged anywhere from Middle Bronze Age to Roman-Byzantine times. Two other kinds of water utilization technique were large watershed runoff systems and small watershed runoff farms. Both kinds may be as old as Middle Bronze Age I and were certainly in use from the period of the Judean Kingdom until the end of the agricultural history of the Negev.

In large watershed systems (up to 10,000 hectares) rainfall runoff is brought by spillways and canals to irrigated terraces. The hydrology of these differs considerably from that of the small runoff farms. Over large areas a greater amount of rain is required to produce runoff, for relative to their acreage, large areas produce less of this kind of water. Moreover, their flash-flood flows can destroy the strongest of engineering structures. Such systems suffer additional instability in that the wadis cut ever deepening gullies, which require the water to be periodically raised by new diversion canals; at the same time the elevation of the terraces continually rises through silting. In several instances, by comparing capacities of spillways, canals and drop structures, the authors succeeded in determining successive modifications of such systems. When large watershed complexes eventually became useless, parts of them were converted into the smaller and simpler runoff farms.

In the small watershed runoff farms (up to 100 hectares), water from the hillsides was directed by low terrace walls to cultivated valleys. In a typical instance, about 70 hectares of watershed supplied 2 hectares of cultivated land.

The authors reasoned that since the Negev had once supported a considerable agriculture it could be made to support agriculture today. Two ancient runoff farms were therefore reconstructed. In one of the farms 250 fruit tree saplings showed rapid growth during the severe drought years 1959–1960. On the other farm, 500 kilograms of barley were produced in 1960 while elsewhere in the country crops failed utterly.

It is appropriate to conclude our preview of the essays in this volume with studies showing that we have much of value to learn from man's past. We could also add a few comments about what is being learned from archaeology, in general.

The development of new conceptions, indicated throughout these essays, should interest both archaeologists and the general reader, for in the usefulness of its ideas, archaeology is able to transcend its particular subject matter. As in all other disciplines, the tools of archaeology are not only the shovels, microscopes, and computers, but consist also of the resources of the mind. How the archaeologist thinks about the evidence brought up from the ground—the conceptions he brings to the task, and the questions he is prepared to ask—ultimately determines his understanding of what he sees before him.

The new evidence from prehistoric Africa, that man is not just the producer of culture but is, in fact, its product, carries far-reaching implications. This means, according to the anthropologist Clifford Geertz, that man's nervous system does not merely enable him to acquire culture but positively demands that he do so if he is going to function at all: "better to be a thoroughgoing ape than half a man," as Ernest Hooton once put it. And Geertz predicts a fundamental revision in the theory of culture itself, that " . . . we are going in the next few decades, to look at culture patterns less and less in terms of the way they constrain human nature and more and more in the way in which, for better or for worse, they actualize it; less and less as an accumulation of ingenious devices to extend preexisting innate capacities, and more and more as part of those capacities themselves . . . . The tension between the view of man as a talented animal and the view of him as an unaccountably unique one should evaporate, along with the theoretical misconceptions which gave rise to it."[22]

Thought can be no more precise than its words and definitions, and archaeologists are improving some of these. Even such well-worn terms as "town" and "city" inadequately express the nature of many an ancient settlement that differs from our western European understanding of what towns and cities should be like (see Chapter 11). Other ideas and categories long known to be inadequate or ambiguous—civilization, religion, economy, warfare, and so on—are often fused in reality in curious and unexpected ways, causing endless difficulties for archaeologists who try to use them. In this volume, the reader will find investigators sometimes at a loss when there are no conceptions to express adequately the nature of findings. Here and there he will read of new conceptions which seem to fit the evidence better. Some of these might perchance become useful in other humanistic studies as well.

The conceptions come from a variety of sources. Not a few have been invented on the spot, as it were, in the need of finding some way to grasp an archaeological situation. Among those expressing some of the situations of human biological development are the ideas of *micro-evolution* (Chapter 9), *lineages* and *palaeospecies* (Chapter 1), *external and internal evolutionary pressures* (Introduction), *genetic mutation and exchange systems* (Introduction), and *continuous and discontinuous traits with reference to breeding isolates* (Chapter 9).

Ideas about culture and cultural development include a concept of culture in archaeology (Chapter 12), a view of *cultural "facts"* as derived from inferences (Chapter 12), the use of *types* in archaeology (Chapter 9, Chapter 12), other conditions of inquiry (Chapter 12), *cultural process* (Chapter 12), *cultural innovation and exchange systems* (Introduction), *cultural evolution* outstripping *diffusion* (Chapter 5), *stimulus diffusion* (Chapter 10), *shock stimulus* (Chapter 3), *diffusion* outstripping *cultural evolution* (Chapter 12), *continuous intra-areal diffusions* of cultural forms (Chapter 12), *primary and secondary diffusion* (Chapter 11), *multilinear cultural evolution* (Chapter 12), *area tradition* (Chapter 11, Chapter 12), *traditions or "hearths"* of plant cultivation (Chapter 11), *nuclear and nonnuclear groups* with respect to general cultural development (Chapter 5, Chapter 12), *innovations* confined to particular aspects of culture (Chapter 12), *nuclear and nonnuclear groups* with respect to the development of particular aspects of culture (Chapter 5, Chapter 11, Chapter 12), *rates and magnitudes of change* of cultural forms (Introduction, Chapter 2, Chapter 3).

Terms for interaction situations are *area co-tradition* (Introduction), *symbiosis* (Introduction), *interaction sphere* (Chapter 12), *stages* of *growth and decline* of *communication* (Chapter 12), *oikoumenē* (Introduction, Chapter 5), *Great Tradition* and *little tradition* (Introduction), *Great Art Styles* (Introduction). Ideas representing adaptive relations to the physical environment include the physical settings of archaeological sites (Chapter 7), *microenvironment* (Chapter 13), forest, plains, and maritime *efficiencies* (Chapter 5, Chapter 12), *food-gathering economies* (Chapter 5, Chapter 11), *food-collecting economies* (Chapter 5, Chapter 11), *big-game hunting economies* (Chapter 11), *incipient cultivation* (Chapter 5, Chapter 11), *primary* or *village farming community* (Chapter 5), *root crop* as contrasted to *seed crop cultivation* (Chapter 11, Chapter 15), and trends or phases of *settlement patterns* (Chapter 18).

In conclusion, the development of new and interesting social and historical conceptions is not the only promise of an archaeology that has already extended our notions of human history by a million years. Archaeology is a historical pursuit that lends itself to generalization. The reader will find that these chapters disclose astonishing parallels in the development of the independent histories of the Old World and the

New (Introduction, Chapter 11). There is here no discussion of historical "laws," although something is said about principles (Chapter 12). In general, however, we are probably still only approaching that level of analysis at which principles are discovered. Archaeologists everywhere write in very much the same terms and are increasingly aware of similar cultural and historical processes manifested in their respective areas. Perhaps there is only a finite number of social and historical processes behind the events of history. If so, to know them must be to make the events more intelligible, and in every case to enhance our appreciation of their unique qualities.

NOTES

1. Article in *Social Life of Early Man*, S. L. Washburn, Ed. *Viking Fund Publ. in Anthrop., No. 31* (1961).

2. *Loc. cit.*

3. S. L. Washburn, "Tools and Human Evolution," *Scientific Amer.*, Sept. 1960.

4. C. B. M. McBurney, *The Stone Age of Northern Africa* (Penguin Books, London, 1960), p. 101.

5. K. V. Flannery, personal communication.

6. McBurney, *op. cit.*, pp. 38, 56, 225.

7. J. R. Caldwell, "Eastern North America" in *Courses toward Urban Life*, R. J. Braidwood and G. R. Willey, Eds. *Viking Fund Publ. in Anthropol., No. 32* (1962).

8. R. S. MacNeish, "A Speculative Framework of Northern North American Prehistory as of April, 1959," *Anthropologia*, 1, 7–21.

9. J. R. Caldwell, "Trend and Tradition in the Eastern United States," *Amer. Anthropol. Assoc., Mem. No. 88* (1958).

10. J. H. Steward, "American Culture History in the Light of South America," *Southwestern J. of Anthropol.* (1947), 3, 85–107.

11. W. T. Sanders, article in *Prehistoric Settlement Patterns in the New World*, G. R. Willey, Ed. *Viking Fund Publ. in Anthrop. No. 23* (1956).

12. R. Redfield, "The Social Organization of Tradition," *Far Eastern Quart., Vol. XV, No. 1*, Nov., 1955, pp. 13–21.

13. J. R. Caldwell, "The Origins of Civilizations," *Fulbright Series 1* (mimeographed, United States Comm. for Cult. Exch. with Iran, Tehran, 1963).

14. G. R. Willey, "The Early Great Art Styles and the Rise of Pre-Columbian Civilizations," *Amer. Anthropol.*, Vol. 64, No. 1, part 1 (1962).

15. J. Mellaart, "A Neolithic City in Turkey," *Scientific Amer.*, Apr., 1964.

16. K. Kenyon, *Antiquity*, **30**, 184 (1956).

17. R. J. Braidwood and B. Howe, in "Southwestern Asia beyond the Lands of the Mediterranean Littoral," *Courses toward Urban Life*, R. J. Braidwood and G. R. Willey, Eds. *Viking Fund Publ. in Anthropol. No. 32* (1962).

18. A. L. Perkins, "The Comparative Archeology of Early Mesopotamia." *Orient. Inst. Studies in Ancient Orient. Civilizations No. 25*, pp. 195–6 (1957).

19. S. Lloyd, "Uruk Pottery," *Sumer*, IV (1948).

20. J. R. Caldwell, "The Ceramic Sequence at Tall-i-Ghazir, Khuzistan, Iran." Ms.

21. J. R. Caldwell, "Archeological Reconnaissance near Kerman, Iran." Ms.

22. C. Geertz, "The Transition to Humanity," in *Horizons in Anthropology*, S. Tax, Ed. Chicago Univ. Press, Chicago, 1964, pp. 37–48.

Part I

OLD WORLD BEGINNINGS

Chapter 1

The Villafranchian and Human Origins

F. Clark Howell

Few people lack curiosity about their ancestors, their genealogy, and the ways of the world in the past. This curiosity extends to the ancestry and history of all mankind. In the last hundred years, theological explanations have—especially in countries of Western cultural tradition—tended to be replaced by a more rational approach to the matter. This approach is the concern of human palaeontology, and its inception is closely linked with the name of Darwin.

A century ago, only a few early human skeletal remains were known, and these only from western Europe. These represented either the approximately 35,000-year-old Cro-Magnon people or the (then scarcely recognized) 50,000-year-old "Classic" Neanderthal folk. Much understanding of the course of human evolution has been gained since Darwin's time, and especially within the past quarter century.[1] There are still some extraordinary gaps in the fossil record of the family Hominidae with respect both to specific ranges of Pleistocene time and particular geographic areas. Some such gaps, as the Middle Pleistocene range in Mediterranean Africa have become known only in the last decade, as investigations have been vigorously pursued in the field. However, fossil hominid specimens are still almost entirely unknown from this range of time in western Asia and in sub-Saharan Africa. The discovery in 1960, of a hominid cranium, whose age is close to half a million years according to potassium-argon (K-Ar) measurements on associated sediments,[2] in upper Bed 2 at Olduvai Gorge, Tanganyika, helps fill the sub-Saharan African gap. The specimen was associated with an African early Chelles-Acheul stone-tool industry and represents, in the writer's opinion, a variety of *Homo erectus*. The most striking hiatus, particularly since stone artifacts are far from uncommon, is the complete absence of hominid skeletal remains from the Pleistocene in the Indian subcontinent.

In spite of these deficiencies in knowledge, a good deal is known of the major evolutionary stages in hominid phylogeny for the latter half of the Pleistocene. Several distinct, largely geographically restricted lineages are recognized to have existed during the Middle, and to have persisted into the early Upper, Pleistocene; these probably represented palaeospecies,[3] although some workers consider them to have been generically distinct.[4] The latter interpretation is quite unsatisfactory. It is based on two suppositions: (1) that (some) populations of *Homo erectus* were, broadly, contemporaneous with human populations already referable to another species, probably an early phase of *Homo sapiens*, and (2) that descendants of the former persisted, in some sort of isolation (in southeast Asia), as a distinct lineage, unlike and separate from other human populations (in western Eurasia). Substantial evidence, from empirical field data and also from evolutionary theory, militates against this view. Hence, for the present, it seems most appropriate to regard *Homo erectus* as a species some of whose populations were ancestral to *Homo sapiens*. The transformation occurred, perhaps, sometime within the Lower Middle Pleistocene time range.

From a portion of one such (*not* Asian) lineage, anatomically modern man (*Homo sapiens*) evolved, but the details of this transformation are still largely obscure.[5] This best documented aspect of human palaeontological knowledge encompasses only the later phases of man's evolution. It is well known largely because of its recency and because there are apparently more abundant traces of human occupation in datable Pleistocene contexts, especially since these can often be linked with effects of the extensive continental glaciations of the northern hemisphere. This record thus begins, broadly, during the time of the first of these great continental ice sheets, variously named in Germany,[6] Britain,[7] Poland,[8] and European Russia.[9]

The main lines of hominid evolution were set during an earlier segment of the Pleistocene. The mammalian fauna of that time range, in Europe, Asia, and Africa, reveal the first appearance of new and modern genera in an otherwise often archaic assemblage. This was an extended period of fluctuating, but generally cooler and more temperate, climates compared with the late Tertiary (Pliocene), leading to mountain glaciation and changing biotopes and biota. It was a time marked by extensive mountain building, faulting, and upwarping to the extent of several thousand meters, as seen in the Alps, Pyrenees, Caucasus, Atlas, and Himalayan ranges.[10] This long interval, representing the Lower Pleistocene and often termed the Villafranchian after its characteristic fauna, encompasses a time probably as long as the whole of the Middle and

Upper Pleistocene. Half a million years is perhaps a modest estimate.

When this essay was written (early 1959) the writer drew the Lower (including the Villafranchian)–Middle Pleistocene boundary at the top of the so-called Cromerian interglacial stage (see figs. 1–1 and 1–2). A stronger case could now be made, on mammalian palaeontological and palaeofloristical grounds, for drawing the boundary *below* the Cromerian stage. (And in agreement with some authors, B. Kurtén and K. D. Adam for example, a full stage earlier, below the "Günz glacial."[11]) The writer believes that the traditional Lower, Middle and Upper divisions of the Pleistocene have lost their meaning since the decision was made[12] to include certain marine and continental rock- and bio-stratigraphic units, often regarded as "Pliocene," within the Pleistocene. A fourfold system of subdivisions, already used by some German (and other) authors, is Basal (= Villafranchian), Lower (= Cromerian plus Mindel glacial), Middle (= Great interglacial plus Riss glacial), and Upper (= Last interglacial plus Last glacial). Pleistocene is more meaningful and far less confusing, at least in Europe.

My estimate for the duration of the Pleistocene was the not very original million years, and hence the half-million-year "modest estimate" for Basal Pleistocene. However, more recent research would suggest a two-, even three-million-year duration for the Pleistocene. In the terminology used above, it is not unlikely that the Lower–Middle–Upper divisions alone totaled about a million years.

The recent work by Cesare Emiliani,[13] who employed oxygen isotopes in the analysis of climatic change from ocean-bottom cores, gives some promise of providing an "absolute chronology" for the Pleistocene. However, the published climatic curves appear to encompass only the latter half of the Pleistocene—that is, the three major continental glaciations and the intervening Great and Last interglacial stages, the stratigraphy of which closely parallels the core profiles.

The Lower Pleistocene represents perhaps the most crucial period for future research in human palaeontology. Such efforts promise results of great significance to the understanding of formative phases of hominid phylogeny and the elucidation of those distinctive associated patterns of behavior that differentiated the first hominids from other higher primate (pongid) antecedents and collaterals. I have attempted in this chapter to indicate a part of what is known of the Villafranchian stage from traces of hominids of this period.

Villafranchian: Europe and Asia

The base of the Pleistocene is best defined by three lines of evidence: tectonics, climatic deterioration, and the appearance and distribution of new forms of animal life. In the majority of stratigraphic sections, an unconformity, representing an interval of uplift and erosion, separates terminal Pliocene deposits from overlying marine or continental (Villafranchian) sediments of the Basal Pleistocene. The first signs of marked cooling are evident in the appearance of north temperate or arctic forms (such as *Cyprina islandica* in Mediterranean waters) in marine invertebrate faunas, and in vegetation changes demonstrated by palynology or by the particular conditions of sedimentation. Continental deposits of Basal Pleistocene age contain mammalian assemblages referred to as Villafranchian, from the type locality of Villafranca d'Asti in the upper Po River drainage basin.[14] Such faunas are characterized by the first appearance of the modern genera *Elephas* (*Archidiskodon*), *Bos* (*Leptobos*), and *Equus* (and, in some areas, *Camelus*), but in association with a number of other typically late Tertiary species.[15] The term "Villafranchian" has also often been extended to apply to this Basal Pleistocene interval as well as to the fauna. The termination of the interval is ill defined; it is probably best taken as the base of the Cromerian "interglacial" stage[16] or its marine equivalent in the Mediterranean basin, the Sicilian.[17] This stage immediately precedes the first major continental glaciation. However, the base of the Günz glacial is probably the best boundary.

In western and southwestern Europe,[18] evidences of fluctuating sea levels, both transgressive and regressive, and of variably cooler climates, unlike the preceding Pliocene,[19] are recognized at a number of localities. The tilted deep-water Plaisancian and brackish Astian sediments of the (later) Pliocene sea, which flooded many areas of lowland southern Europe that are continental at present, are uncomformably overlain in a number of regions (in southern France and Italy) by the marine deposits of the Basal Pleistocene Calabrian sea.[20] The Calabrian, a transgressive sea that was subsequently regressive (compare the Po Valley; also the Emilian stage of *Emilia*[21]), has as its continental equivalent, developed in unsubmerged and emergent uplands, a series of fluviolacustrine sands and gravels with a markedly cool temperate flora and a characteristic Villafranchian mammal fauna. The contrast with the subtropical vegetation and archaic mammal faunas of the Pliocene is striking and clearly delineated.

The comparable successions afforded by the North Sea Basin, in East Anglia[22] and the Netherlands,[23] and by the polliniferous clays and lignites of the lacustrine basin of Leffe (Bergamo, Tuscany),[24] illustrate the main pattern of climatic change during the Villafranchian in western and southern Europe (Fig. 1–1). The evidence would seem to indicate at least two major colder stages, the latter double, prior to the well-defined Cromerian interglacial stage. These two colder stages are separated by the still inadequately known Tiglian "interglacial" stage.[25]

This sequence is paralleled in the Rome region by the Acquatraversan and Cassian phases of colder climate which precede the transgressive Sicilian sea and the extensive eruptions of the Sabatino volcanoes.[26] A considerable body of evidence indicates several phases of fairly extensive mountain glaciation during this interval, probably including the Donau[27] and Günz stages in the northern Alps. The first Himalayan glaciation is perhaps the Asian equivalent. However, there is a dearth of evidence of direct correlation of these subalpine stages either with continental Villafranchian deposits in the lowlands or with equivalent marine horizons.

The European Villafranchian lacks any trace of higher primates, and there are no stone implements testifying to occupation of the continent by hominids.[28] In the past few years, an increasing number of reports were made from Europe of lithic assemblages, reputedly artificially worked, and hence hominid-produced, and of earlier Pleistocene antiquity. All those cases with which the author is familiar from the literature are suspect either on archaeological and/or geological (contextual) grounds. These occurrences surely need to be approached more critically both in view of the effects of factors in nature producing certain forms of fracture and with regard to indubitable hominid-fashioned lithic artifacts in controvertible contexts, now well-documented in Africa. An example is the description[29] of (derived) artifacts from *recent* alluvial deposits in the Oltenia region of sub-Carpathian Romania. These are attributed a pre-Mindel (Lower Pleistocene) antiquity (which may be true, but on available evidence practically impossible to prove), and are regarded as a very early human industry (although the artifacts, in substantial part at least, are on technological and typological grounds compatible with an early phase, for example, Abbevillian, of a human lithic industry of about Mindel age). However, the recent discoveries in Vallonet Cave (Alpes-Maritimes) certainly indicate terminal Basal Pleistocene occupation by hominids in Europe.[30]

This was probably a consequence of two factors: (1) Europe was not a primary center in the original hominid radiation, and (2) the extent of the Mediterranean and Black seas,[31] not only in the Pliocene but also during the Calabrian and Sicilian transgressions of the early Pleistocene, created impassable water gaps that effectively isolated Europe. The first evidence of hominid occupation of the European continent is well along in the earlier Middle Pleistocene. It corresponds to the time of the Romanian regression[32] of the Mediterranean, when eustatic lowering of sea level, attendant upon the first major continental glaciation, evidently permitted expansion into Europe of those peoples (probably represented only by the Mauer, or Heidelberg, mandible) responsible for the Abbevillian hand-ax industry.[33]

In eastern Asia, the continental Villafranchian is best known in northern China and in the southern foothills of the Himalayas.[34] In northern China,[35] the Pliocene, a period of dry subtropical-to-tropical climate with extensive lakes between stretches of desertic country, was terminated by diastrophic movements resulting in extensive erosion (Fenho erosion interval). These formerly widespread lakes were consequently displaced, and rejuvenated rivers and streams greatly enhanced their fluviatile activities. In the synclinal basins of Nihowan (Hopei), Taiku, and Yûshe (Shansi), to mention only the best known, this erosion surface underlies a torrential lacustrine series of basal conglomerates overlain by sands, marls, and clays (lower Samenian series); the clays include a plant bed with cool dry flora (Taiku Basin) and the series is capped with sands and silts that yield a characteristic Asiatic Villafranchian fauna.[36] The entire series was tilted and is separated unconformably (by the Huang-shui erosion interval) from the early Middle Pleistocene red loams (upper Sanmenian series). No evidence of hominid fossils or of tool-making activities has been recorded from the earlier, Villafranchian, beds. The first evidence of hominid occupation is revealed at locality 13, Choukoutien, correlative with the upper Sanmenian series on faunal grounds, which has yielded a single small chert chopping-tool but no human skeletal remains. In the past few years a mandible and incomplete skull of *Homo erectus* have been recovered near Lantien (Shensi) from such a horizon,

The entire Pliocene and Lower Pleistocene succession is magnificently represented in sub-Himalayan northwest India, both in the Potwar region (Punjab) to the northwest and the Siwalik Hills to the east.[37,38] The Pliocene, exposed in the middle Siwaliks series, is represented by fresh-water sandstones and shales, 6,500 to 10,000 feet in thickness, deposited under tropical to subtropical climates with a trend toward increased aridity. The entire series was tilted during a major phase of mountain building followed by an interval of severe erosion at the end of the Pliocene. These movements, either through the development of anticlines or as a result of faulting, created a series of depressions trending northeast–southwest. The depressions, which were filled during the Lower Pleistocene with great thicknesses of alluvial sediments (upper Siwaliks series), derived from the adjacent uplands. The earlier Tatrot zone of this series is characterized by a thick (100-foot) basal conglomerate overlain by nearly a thousand feet of coarse sandstone; interspersed silty or conglomeratic horizons contain a still inadequately known fauna which is characteristically Villafranchian.[39, 40]

Reference 40, which states that *Ramapithecus brevirostris* occurs in the Tatrot zone, is incorrect. G. E. Lewis[41] first regarded the specimen as either from the end-Middle Siwaliks (= Dhok Pathan zone) or initial-

Upper Siwaliks (= Tatrot zone). However, he subsequently referred to it as the Nagri zone.[42] Elwyn Simons has recently made a careful restudy of the original specimen.[43] This genus is apparently also represented in the Mio-Pliocene of eastern Africa, as shown by the recent discovery of a partial maxilla (with teeth) and lower molar at Fort Ternan, Kenya; it has been referred by L. S. B. Leakey[44] to a distinct genus, *Kenyapithecus wickeri.*

The overlying Pinjor zone, some 1,000 to 1,500 feet of laminated silts and sands, clearly distinguished lithologically from the Tatrot zone, appears to have been deposited under warmer and more temperate conditions by sluggishly meandering streams, in contrast to the great alluvial activity that accompanied the deposition of the preceding Tatrot. A rich Villafranchian fauna is represented in the Pinjor zone.[45, 46] However, neither the Tatrot nor the Pinjor provide hominid remains, nor is there evidence of stone tools testifying to hominid occupation. Such tools first appear, as the "Pre-Soan" (Punjab flake industry) chopper–chopping-tool assemblage,[47, 48] in the overlying boulder conglomerate, attributed to the second Himalayan glaciation of early Middle Pleistocene age.

Discussion of the southeast Asia area of the Sunda shelf, an extension of the mainland in Middle and late Pleistocene time, is less pertinent here because of the extensive Pliocene submergence. Only isolated mountain peaks were emergent along the southern (Zuider Mountains) and northern (Kendeng Hills) coasts in eastern Java, and some few uplands in the western region.[49] A Villafranchian fauna is known from fresh-water sandstones and coarse conglomerates where these overlie Pliocene marine beds at several localities.[50] However, skeletal remains of two distinct hominids (*Meganthropus, Homo modjokertensis*) first appear in the early Middle Pleistocene.[51] Sundaland, including Sumatra, Java, Borneo, and much of the present ocean floor had become largely continental by that time as a result of further uplift and marine regression consequent upon continental glaciation in the Northern Hemisphere. There is no evidence to suggest hominid occupation of the Sunda shelf during the Villafranchian stage.

The present evidence appears to indicate that continental Eurasia was not occupied by hominids during the Lower Pleistocene—that is, the Villafranchian stage. In the summer of 1959, the locality of 'Ubeidya, on the western rim of the central Jordan Valley, provided the first indications of a rich Pleistocene mammal fauna, with associated stone implements and hominid skeletal fragments (vault fragments and two teeth), from exposures of tilted fluvio-lacustrine sediments of the so-called *Melanopsis* stage.[52] These beds, which overlie faulted Neogene fresh-water and brackish sediments with basaltic intrusions and which are overlain by Pleistocene basalts, were previously regarded by L. Picard[53] as of Upper Pliocene age, but subsequently as of Lower Pleistocene age. The extremely rich vertebrate faunas, which include fish, frogs, a variety of reptiles (turtles, lizards, snakes), various birds, and a rich assemblage of both small and large mammals, has been attributed to the Villafranchian stage, but an early Cromerian age is not wholly excluded. The richness of the faunas will eventually permit detailed comparison with both southwest European Villafranchian faunas and the now well-known Cromerian fissure and cave faunas of southeastern Europe, especially Hungary. The stone-tool assemblage includes chopping tools on pebbles, flakes, multifaceted spheroids ("missile stones"), as well as bifacially pointed pebbles. The preliminary description suggests affinities with stone-tool assemblages from the upper limits of Bed I, or even lower Bed II, at Olduvai Gorge, Tanganyika.

The discovery of *Elephas* (*Archidisicodon*) *meridionalis* in ancient high-level (30–35m.) gravels of the Orontes Valley (near Hama, Syria), together with a few stone artifacts (including flakes and polyhedral spheroids), suggests such deposits may be of broadly comparable age to those farther south in the Jordan Valley.[54]

These important discoveries surely indicate hominid occupation of western Asia by Cromerian times, if not even in the later Villafranchian. The evidence is admittedly and necessarily negative, and further field investigations, particularly in the sub-Himalayan Siwaliks, are sorely needed. The important point is that the available evidence does not bear out the opinions of certain earlier workers[55] that Asia, particularly central Asia, was a primary center for hominid (or higher primate) origin and dispersal. Such a conclusion is contradicted not only by our understanding of higher primate relationships, based on comparative anatomical and palaeontological studies, but also by biogeographical and palaeogeographical conditions. On the other hand, a variety of evidence shows that the Villafranchian stage in Africa was crucial in the earlier phases of hominid evolution.

African Villafranchian

MEDITERRANEAN AFRICA

Several Lower Pleistocene localities with the Villafranchian fauna are known in northwestern Africa[56] (Fig. 1–2). One of the best-stratified localities is Fouarat near Port Lyautey, Morocco, along the southern border of the Rharb Plain. A Villafranchian fauna is present here in coarse sands and sandstones representing a littoral facies of Calabrian gulf, which in places filled depressions in Pliocene marine sediments.[57] Hominid occupation of the area is perhaps first recorded at a slightly later stage, probably corresponding to the regressive Emilian. Flaked pebbles, thought by some to represent primitive pebble tools, have been collected from the Arbaoua conglomerates, of Basal Pleistocene age.[58] These deposits, like the reddened Marmora sandy loams, mantle this region and (broadly) represent the continental equivalent of the Calabrian (Maghrebian) transgression.[59] However, the artificial nature of the specimens is difficult to confirm when they are discovered in this gravel context. P. Biberson[60] has published a very full discussion of these and other lithic assemblages from Morocco, along with a fine appraisal of their stratigraphic contexts and temporal relationships. Many, though not all, of these assemblages comprise specimens that were surely fashioned by hominids.

His subdivisions of the Basal Pleistocene differ from those in this chapter (see column Morocco, Fig. 1–2) in that he treats the Maghrebian stage as (regressive) end-Pliocene. Thus he would equate the marine stages Messaoudian and Maarifian with the Mediterranean, Calabrian, and Sicilian stages, respectively. Hence, his age attributes are a full stage *earlier* than those proposed by the writer (who generally followed the interpretations of G. Choubert). The writer has not personally examined the field relationships in Morocco and, consequently, is unable to evaluate properly the merit of either interpretation. Biberson recognizes four stages of Pre-Chelles-Acheul stone industry ("Pebble-Culture"). Stages 2 and 4 are related, respectively, to the aforementioned marine cycles; Stage 1 to the continental Moulouyian (pre-Messaoudian) cycle; and Stage 3 to the continental Saletian (pre-Maarifian) cycle.

The Villafranchian fauna of North Africa is more adequately known in Tunisia (Garaet Ichkeul), and particularly from old lake basins in northern Algeria (Bel Hacel; St. Arnaud). The base of Lake Ichkeul, near Ferryville, comprises southerly tilted deep-water Plaisancian and lagoonal Astian marine sediments which are overlain by fresh-water lake beds of Basal Pleistocene age. The lower sands and gravels of the latter, separated from the Pliocene sediments by a thin conglomerate, are richly fossiliferous;[61] a mild temperate flora, but with substantial boreal elements, occurs in intercalated, more or less sandy, clays.[62] However, the full succession at Ichkeul is still poorly known, and the faunal assemblage is incomplete, since the beds are only partially exposed during times of low water, on the northern foreshore of the present lake. At Bel Hacel,[63] emergent dune sandstones, which concordantly overlie transgressive Plaisancian-Astian sediments, are eroded and filled with alluvial, weathered and reddened, conglomerates which contain terrestrial and fresh-water molluscs and a Villafranchian fauna. The more than 100-meter-high Sicilian beach rests horizontally and unconformably on the tilted and compressed series.

On the Constantine-Setifian Plateau near St. Arnaud, the earlier Pleistocene is exposed in deep ravines dissecting thick marls and other calcareous clayey sediments, with intercalated gravel and conglomeratic horizons, which fill old marshy or lacustrine depressions. In the Oued Boucherit,[64] two horizons, separated by a meter of sterile brown clay, contain a Villafranchian mammal fauna. The lower horizon (Ain Boucherit) is a coarse calcareous conglomerate, and the upper horizon (Ain Hanech), a cracked clay, rather sandy or with light gravels at the base. A quantity of undoubtedly primitive stone implements has been recovered here, largely from the upper horizon.[65]

The Ain Hanech stone-tool assemblage may very well be comparable in a broad way with that from the upper Bed I/lower Bed II at Olduvai Gorge, Tanganyika. (L. S. B. Leakey[66] has suggested that it may even correspond to the lower limits of Bed II.) The problem should be clarified when the respective mammal faunas have been fully studied and described. These specimens (Fig. 1–3) are fashioned from naturally worn dolomitic limestone pebbles, exhibiting fresh, concave flaking scars and ranging in size from that of a tangerine to that of a good-sized orange. They are either battered over most of the surface to a multifaceted polyhedral form (boules) or are flaked along a margin, unilaterally or bilaterally, to produce an irregular sinuous edge characteristic of choppers or chopping tools. Unfortunately, the Ain Hanech site has been worked only briefly, since both political circumstances and the proximity of a Moslem cemetery have prevented extensive excavations.

No hominid skeletal remains have yet come to light in the formations of the North African Villafranchian. An incomplete hominid skull (cranio-facial portion), referable to *Australopithecus* sp., is known from fluviolacustrine sediments of the northeastern sector (Bahr El Gazal depression) of the Chad Basin, in association with possibly Lower Villafranchian type mammal fauna.[67] There is unquestionable evidence of hominid occupation of this region during the terminal phases of the Villafranchian, and a site such as Ain Hanech might one day provide fossilized remains of the creatures themselves.

SUB-SAHARAN AFRICA

The central African Pleistocene was initiated by prolonged and extensive uplift, 1,000 to 1,500 feet at least, accompanied by downwarping in adjacent areas and by fracturing along ancient troughs. Extraordinary volcanic explosions, especially of tuffs and ashes, also took place along the eastern Rift Valley. As a consequence, the mid-Tertiary surface of erosion, whose relief was relatively gentle even during the end-Tertiary phases of valley incision, was severely deformed, and drainage systems were disrupted and even reversed.[68,69] Extraordinary depths of lacustrine deposits and subaerial, partially volcanic-derived, sediments were accumulated. These are particularly well preserved in the western (Albertine) and eastern (Gregory) rift valleys and represent good exposures of the continental Villafranchian (Fig. 1–4).

The Villafranchian, with an overlying Middle Pleistocene series (Rawe beds), is also represented along the southern shores of a minor rift valley trending east–west, now Kavirondo Gulf, an eastern embayment of Lake Victoria.[70] At Kanam, along the north and northeast slopes of the extinct volcano which forms Homa Mountain, a great thickness (more than 100 feet) of lacustrine brown clays, with intercalated fine, laminated stony tuffs from intermittent volcanic explosions, has provided a good Villafranchian assemblage. A few pebble tools have also been recovered from such horizons. A small fragment of a hominid mandible is also known,[71] although it was once believed to date back to the Lower Pleistocene, it is now known to be of considerably more recent age.

Philip V. Tobias[72] has demonstrated in a detailed study the extent to which pathology has resulted in deformation of the symphysial region of the Kanam mandible fragment. Also, its association with the indubitable Villafranchian mammal fauna from Kanam is not confirmed by radiometric (uranium) determinations.

In the Albertine rift valley, an extensive lake existed in the Basal Pleistocene, as indicated by the massive (at least 1,500 to 2,100 feet, to judge from borings) tilted and contorted beds of the Kaiso series.[73] These are particularly well exposed along the margins of Lake Albert and its southern tributary, the Semliki River, as well as along the northern shores of Lake Edward[74] and in the region south of the latter, adjacent to the volcanic highlands north of Lake Kivu. The beds seem to attain their maximum exposed thickness in the southerly reaches of Lake Albert and at the adjacent mouth of the Semliki Valley. Thinner exposures in the upper Semliki and the northern reaches of Lake Edward may merely reflect earlier, more prolonged and intense subsidence of the rift floor in this region.

The Kaiso series is complex, and three main stages of sedimentation have been distinguished.[75] Only the basal stages are generally regarded as Lower Pleistocene; the upper stage is thought to be earlier Middle Pleistocene. The lower stage (100 to 120 feet thick in the northern reaches of the valley) is largely silty, with some minor gravel horizons. In some localities, this stage is found to overlie a basal ironstone horizon capped by unstratified sands; the ironstone seems to represent a laterite capping the down-faulted peneplain surface and provides an important datum point. This earlier stage is essentially nonfossiliferous. The middle stage (300 to 600 feet thick in the lower reaches of the valley) is predominantly clayey, with selenite evaporites and zones of gypsum. It is characterized by fine sands and sandstones and by discontinuous ironstone horizons and limonite lenses thought to represent desiccated pools. These horizons occasionally provide silicified wood and have yielded a typical, though small (thirteen species), Villafranchian mammal fauna. Several such ironstone "bone-beds" appear to be present, though they are restricted largely to the middle Kaiso stage.[76] However, another, perhaps lower, bed is also present in the lower Semliki Valley. The site of Kanyatsi,[77] along the northern shore of Lake Edward just east of the Semliki outflow, has yielded traces of worked stone implements adjacent to an ancient subaerial soil horizon within the middle Kaiso stage. The specimens represent fresh flakes of quartz and quartzite, but the cores from which these were struck are apparently absent, and no pebble choppers or chopping tools have yet been found.

The central region of the Gregory rift valley has failed to provide certain evidence of Lower Pleistocene formations. This is perhaps because there was extensive uplift during the later Pleistocene, or because there was relatively little deposition but considerable volcanic activity. The northern and southern reaches of the valley do afford such exposures. The tilted and step-faulted beds of the Omo series, just north and northwest of Lake Rudolph, testify to an ancient and extensive Lower Pleistocene lake. A rich Villafranchian fauna occurs in extraordinary profusion in sandstone horizons intercalated in a massive succession of lacustrine volcanic clayey tuffs.[78] Neither hominid skeletal remains nor primitive stone implements have been recovered from the Omo series, which, since their discovery by the Bourg de Bozas expedition in 1902–3, have only been worked during one relatively brief field season. Further investigations in this most inaccessible region are surely warranted, especially since in 1959 I recovered Pre-Chelles-Acheul artifacts from the eroded sandstones.

In the southern reaches of this rift valley, the Lower Pleistocene is exposed along the tributary Eyasi trough and the adjacent region to the west of the Crater Highlands. At Olduvai Gorge,[79] which extends from Lake Lagarja on the Serengeti plains some thirty-five miles to the western boundary faults of the Balbal depression, the magnificent lacustrine series of the Middle Pleistocene is underlain by a flow of olivine basalt, which buries or obscures earlier horizons. The extensive new investigations carried out since 1959 by L. S. B. Leakey and his collaborators at Olduvai Gorge have provided a wealth of new data bearing on the Pleistocene stratigraphy, mammalian palaeontology, and palaeoanthropology of this extraordinary locality.

Geological investigations by Richard L. Hay[80] have demonstrated the essential soundness of the fourfold succession of main beds recognized years ago by Hans Reck. Bed I, now demonstrably of Basal Pleistocene age on the basis of its contained mammal fauna (Villafranchian affinities) is from 15 to about 40 meters thick. It comprises a conformable series of various sedimentary deposits (largely subaerial tuffs and hardened clays with more fluviatile deposited and reworked tuffs and conglomerates in the western and eastern reaches of the Gorge) and, in the easterly reaches as far as around the junction of the Side Gorge, two successive olivine basalt lava flows intercalated in the subaerial and fluviatile sedimentary sequence. The top of the bed is often well defined by a prominent Marker Bed (B), a flaggy indurated sandy tuff. A succession of K-Ar determinations indicates that the basalts are approximately 1.86–1.88 million years old, with the overlying Bed I sediments extending in age up to over a million years.[81] Bed II, conformably overlying Bed I, and up to some 30 meters thick, comprises largely lacustrine clays (Main Gorge) deposited in a small basin under saline or alkaline conditions, with laterally equivalent (west and east reaches of the Gorge) fluviatile, aeolian, and pyroclastic deposits. K-Ar determinations indicate that the base of this bed somewhat exceeds a million years in age and that the top is under half a million years in age.

Slightly to the south, in the Vogel River region on the northwest escarpment high above the Eyasi graben, are the Laetolil beds, a series of upfaulted subaerially deposited tuffs.[82] These appear to be earlier than, as well as in part contemporaneous with, the Olduvai series. The upper Laetolil beds have yielded a rich Villafranchian assemblage of mammals, including some microfauna.[83] The lowest horizons are also fossiliferous, but the fauna is very poorly known. Pebble tools and a hominid maxilla fragment,[84] the latter believed to resemble the australopithecine (*Australopithecus*),[85] attributed to these beds may in fact be of later Pleistocene age. The basal bed (Bed I) at Olduvai Gorge is known to contain a Villafranchian faunal assemblage comparable to that from Omo, rather than of Middle Pleistocene affinity. Pre-Chelles-Acheul stone implements of the Oldowan industry were first found here nearly thirty years ago. In the 1960–1961 season, Dr. and Mrs. L. S. B. Leakey recovered a beautifully preserved skull of a new form of australopithecine from this bed in association with an occupation surface rich in such artifacts and with the bones of small game taken by this creature (announced at the fourth Pan-African Congress on Prehistory, Léopoldville, August, 1959).

These were the first demonstrably australopithecine remains recovered from Bed I. They are referable to the genus *Australopithecus*, and more precisely to the form *Paranthropus* (= *Australopithecus robustus*). A tibia and fibula and other fragmentary hominid skeletal remains of another individual have subsequently been recovered from this site (FLK Main) and probably represent "*Homo habilis.*"

Central and eastern Africa afford some of the richest Villafranchian faunal localities in the world, coupled with an excellent Pleistocene succession. The faunas from these sites differ somewhat in composition, that from the Laetolil beds being probably the youngest, overlapping basal Olduvai, and that from Kaiso being perhaps the oldest. The Omo fauna overlaps both Laetolil and Kaiso, and that from Kanam is probably broadly equivalent.[86,87] This interpretation of the relative age relationships of these faunas is probably incorrect. The Kanam fauna appears to be the oldest and is perhaps overlapped by the younger Kaiso faunas in part. The Olduvai Bed I fauna, described by L. S. B. Leakey and his collaborators,[88] provides a superb picture of the Upper Villafranchian fauna. The fauna from Laetolil is apparently only a local facies of that represented in Bed I, and the deposits may be shown to be laterally broadly equivalent. It is not unlikely now that the well-described Omo fauna is no older than that of Upper Bed I, or probably Basal Bed II, at Olduvai Gorge.[89]

As yet it is impossible to determine the magnitude of the climatic change that occurred during the Villafranchian of central Africa because of the effects of tectonics in this unstable region. At present, only a broad block correlation with other areas of the Old World, on the basis of faunal content, can be made. Future detailed investigations of the sediments in these basins, particularly of nonsequences and old soil horizons, coupled with palynological research, promise to throw some light on this problem.

At three such localities, Kanyatsi (Lake Edward), Kanam (Kavirondo, Kenya), and Olduvai Gorge, there is clear evidence of hominids, sometimes in the form of skeletal remains and, in all cases, in the form of deliberately fashioned stone tools. Few open habitation sites of the hominids themselves have yet come to light; however, this is largely a result of too little field work in difficult regions. There is every indication that such sites will be forthcoming in the future with concentrated work by prehistorians and Pleistocene geologists.

Olduvai Gorge has afforded a number of hominid occupation places in Bed I. These are, from top (latest) to bottom (oldest): FLK N I, rich faunal associations with vertically dispersed hominid occupation in clays underlying Marker Bed B; FLK Main, fairly extensive hominid occupation surface, on swamp clays sealed in by tuffs, with abundant small mammal fauna, other small vertebrates, stone artifacts of Oldowan industry, and *Australopithecus robustus* and *"Homo habilis"* skeletal remains; FLK NNI, occupation place, with probable carnivore kill-scavenge disturbance, with lower vertebrate and mammal remains, Oldowan stone (and bone) artifacts, and cranial (jaw and vault portions) and postcranial (hand, foot, clavicle, rib, and pelvic portions) skeletal remains referable to *"Homo habilis"*; MK, hominid occupation site with abundant associated fauna and Oldowan stone artifacts, and "*Homo habilis*" teeth; DK, patterned dispersal of emplaced stones. The FLK Main and NNI sites are demonstrated by repeated K-Ar determinations to be approximately 1.75 million years of age, with the latter older than the former by an unknown number of years (presumably at least some thousands of years). The "*Homo habilis*" remains from Bed I and lower Bed II seem to represent a transitional form between a more australopithecine condition (below) and an early *Homo erectus* condition (above). Ample evidence of such early hominid forms, including abundant skeletal remains, is afforded by the australopithecines of southern Africa.[90]

Australopithecine Sites

The australopithecines rank among the most numerous and best known of all Pleistocene hominids. Usually classified as a distinctive sub-family (Australopithecinae)[91] of the Hominidae, but quite probably representing merely a distinct genus, *Australopithecus*,[92] the group contains two probably subgenerically distinct forms, *Australopithecus* and *Paranthropus*. On the basis of the associated faunal assemblages, it appears that the australopithecines are all probably late Villafranchian.[93] However, there is a possibility that the younger form (*Paranthropus*) may have persisted into the early Middle Pleistocene.[94]

Australopithecines are known from five sites at three localities (Fig. 1–5) in southern Africa.[95] With one exception (Taungs) they occur in fossiliferous breccia, composed of calcite-cemented dolomite soil, which infills former caves, formed by solution or subsidence along ancient fracture planes in dolomitic limestones of the Transvaal system. Studies of the mode of formation and the sequence of infillings of the caverns have demonstrated the significance of the degree of communication of the cavern with the outside in the accumulation of sediments.[96] Travertines and intercalated bands of thin gray marly breccia, representing residual calcified material from dissolution of the dolomite, accumulate prior to the formation of any substantial opening to the surface. Subsequently, as the opening becomes progressively enlarged, surface-derived material collects in sufficient quantity to represent a state of equilibrium with outside conditions. Such breccia accumulations may serve as climatic indicators, through analysis of the sand fractions of breccia residues, minus the carbonate cement derived from roof drip, and comparison with modern dolomite soils in regions of differing rainfall in southern Africa. Very satisfactory results have been thus achieved for the Sterkfontein, Swartkrans, and Kromdraai (site A) sites; the method is not directly applicable to the Limeworks Cave site, Makapansgat, where alluviation or slope-wash, in the higher levels, is a complicating factor. In general, the climate appears to have been somewhat drier in this region when *Australopithecus* lived and somewhat, or even considerably, wetter when *Paranthropus* lived than it is now. C. K. Brain[97] refers the three older Transvaal sites (Sterkfontein, Limeworks, Swartkrans) to a major dry interpluvial stage (with at least three separate peaks) and the youngest site, Kromdraai, to a succeeding, wetter pluvial stage.

The Taungs site, long since destroyed by quarrying activities, was a cave formed by solution in the capping carapace of a massive Basal Pleistocene travertine banked up against a dolomite limestone cliff (Campbell Rand series).[98] The filling of the cavity was calcified sandy breccia overlain by contaminated travertine with sandy lenses from which the type australopithecine remains were most probably recovered.

The fauna associated with the australopithecines is not only varied but differs from site to site. In general it comprises other primates (both rare monkeys and abundant baboons), numerous rodents, insectivores, hyracoids, lagomorphs, numerous carnivores, including hyaenids and sabretooths, suids, an extinct sivathere, equids, and numerous antelopes. The following frequencies were obtained from counts of over 7,000 bone fragments, out of a much larger number, from remnants of the gray marly breccia at Limeworks Cave:[99,100] 92 per cent, antelope (293 individuals; more than two-thirds medium to small, the remainder large or very small varieties); 4 per cent, other ungulates (four zebras, six chalicotheres, five rhinoceroses, one hippopotamus, 20 pigs, six giraffids); 1.6 per cent, carnivores (17 hyenas, one leopard, one jackal, one wild dog, one sabretooth, and nine other small and medium species); 1.7 per cent, baboons (45 individuals), rare rodents (hares and porcupines), and very rare birds and reptiles (tortoise, water turtle). Five australopithecine individuals represent only 0.26 per cent of the total assemblage.

The various segments of the skeleton are very unequally represented at this site. The frequency of cranial fragments is particularly high among the nonbovid ungulates (88 per cent), the carnivores (75 per cent), the rodents (100 per cent), and the primates, including the australopithecines (95 per cent). The proportions are lower (34 per cent) among the antelopes (of all sizes); and there are interesting differences in frequencies of antelope postcranial elements: cervical vertebrae (7 per cent), other vertebrae (5 per cent), ribs, and so on (5 per cent), scapulae (9 per cent), innominates (8 per cent), forelimbs (37 per cent), hindlimbs (20 per cent), feet (6 per cent); these figures do not apply to very small antelopes, which are represented exclusively by cranial elements.

Any consideration of the diet and life habits of the australopithecines must take into account the associated fauna and frequencies of preserved skeletal segments. Hence, such figures are important and are much needed from the other sites. Several possibilities exist as to the manner in which the australopithecines and the associated fauna came to be incorporated into the breccias. The sites might have been (1) natural crevices into which animals fell; (2) crevices into which bone accumulations were swept by natural agencies; (3) carnivore lairs into which prey or scavenged carcasses were carried; (4) rubbish heaps; or (5) actual occupation sites of the australopithecines. There is no evidence at any of the sites to support (1) or (2), although the gravelly breccia at Limeworks Cave was partly fluviatile in origin; there is also evidence of stratification in the upper brownish breccia at Swartkrans, probably due to deposition in isolated pools, but certainly not the consequence of stream activity. It will always be difficult to decide between (4) and (5), but the important and still unsettled question is whether the sites were occupied and the bones were accumulated by carnivores, in particular sabretooths or hyenas, or both, or by carnivorous australopithecines.

It is necessary to bear in mind that these sites are known because of commercial lime-quarrying activities. In most cases such efforts were directed toward the basal travertines, formed largely when the caves were still solution cavities. In the case of Taungs, the cave was discovered as a consequence of such mining in the massive cliff-forming travertines at Buxton, in one of which the cave happened to be situated. All the fossil mammals at the Limeworks Cave, Makapansgat, have been obtained by sorting through the extensive dump heaps left behind by the miners. Only in the case of Swartkrans[101] (in part) and some of the excavations at Sterkfontein have investigations been carried out which would permit some comprehension of the fossiliferous breccias as they existed *in situ*. However, the extreme consolidation of the fossil cave earths, a consequence of calcareous cementation, necessitates the use of explosives, so results are definitely limited with respect to details of the pattern of association and the arrangements of bones in the deposits. In the present state of knowledge, it is indeed doubtful if the matter can be definitely settled until a new site can be excavated, with every attempt made not to disturb the stratigraphy and fossil associations.

Raymond A. Dart[102,103] has repeatedly maintained that the extraordinary accumulations of mammalian skeletal remains in the fossiliferous breccias are a direct consequence of the predatory and carnivorous habits of australopithecines. Such remains represent, in his opinion, not only slaughtered prey but also scavenged carnivore kills; many of the bones were useful as tools and weapons for pounding, cleaving, scraping, stabbing, and slicing. On the basis of the preserved remains inventoried at Limeworks Cave, specialized functions have been attributed to specific bones and portions of animal skeletons that were put to use by those "flesh-eating, skull-cracking and bone-breaking, cave-dwelling apes."

There are in fact two distinct issues involved here. The question as to whether the bones were employed as implements and weapons by australopithecines presupposes that these creatures were carnivorous and were therefore responsible for the fossiliferous accumulations. The use of these bones is extremely difficult to verify, since none of the sites yield any trace of specimens which have been deliberately worked or shaped. There is no doubt that the jaws, teeth, horns, and shattered or damaged limb bones which Dart attributes to an "osteodontokeratic culture" might be employed in the fashions he has so exhaustively and imaginatively outlined. However, as in the case of the so-called bone and antler industry from the Choukoutien locality in northern China, attributed by H. Breuil[104] to the Middle Pleistocene hominid found there, this is extremely difficult to confirm scientifically, even though both claims may prove entirely valid.

The question of the carnivorous habits of australopithecines is a separate matter and one which should be resolved from existing evidence. The parts and proportions of the animal skeletons preserved do coincide closely with remains at carnivore kill sites in the open, even after the usual scavengers have been at work.[105] Moreover, although contrary claims have been made,[106] both brown and spotted hyenas may eat and accumulate bones in and about their lairs, at least at times.[107] Nonetheless, this possibility does not account for the enormous concentration of bones at the sites. Also, there are discrepancies between the proportion of cranial and postcranial elements of the antelopes compared with the other ungulates, the carnivores, and the primates.

Two main points are important in connection with the dietary habits of these creatures: (1) the evidence from Taungs, Sterkfontein, and Limeworks Cave of baboon skulls bearing evident signs of localized depressed or radiating fractures, smashed-in walls or tops of the cranial vault, openings in the vault or base, and twisted facial skeletons all testify to predatory activities which are those of a hominid rather than any hyaenid or felid carnivore; and (2) the substantial quantity of various antelope and other long bones, which are not only broken and smashed, but also split longitudinally, and which usually fail to reveal any traces of carnivore gnawing, is further testimony to hominid habits.

Such evidence has been convincing not only to me, but also to other workers[108] who have examined the specimens in question. Moreover, it seems very likely on the basis of the Taungs evidence that eggs, crabs, turtles, birds, rodents, and smaller antelopes were a substantial part of australopithecine diet. Eggs and crabs are easily collected, and it is not particularly difficult to kill other species of smaller mammals. The australopithecines were probably carnivorous predators as well as scavengers of the kills of other carnivores (of which there was then an abundance of forms long since extinct). The marked disproportion between the bovid and nonbovid ungulates and the relatively few carnivores in the Limeworks Cave inventory may merely be a reflection of this latter fact and of the limited hunting capabilities of such creatures. Such a conclusion does not preclude the possibility that carnivores also, at least periodically, occupied such sites and contributed to the bone accumulations. This can hardly be denied until careful excavations have been carried out which prove the situation to have been otherwise.

Until recently, none of the australopithecine sites were known to contain artifacts. Consequently, many workers asserted that such primitive creatures, although admittedly hominids, were incapable of making, and perhaps even of using, tools. Quite possibly this lack of stone implements has also convinced some workers that bone, horn, and teeth of other animals were used by australopithecines as weapons and implements. A few split and flaked dolomite pebbles from the calcified stony and sandy fluviatile horizon, which overlies the pink and gray breccias at Limeworks Cave,[109] suggest but do not afford conclusive proof of tool-making activities. However, there is no doubt about the validity of the implements (Fig. 1–6), referred to a pebble-tool (Pre-Chelles-Acheul) industry, recovered recently from the Sterkfontein locality.[110]

A full description of the outcome of this work has been published by J. T. Robinson and R. J. Mason.[111] A total of 286 specimens was recovered from the Extension Site breccia, of which only about a third are truly worked as artifacts; the bulk of the series represents natural unmodified pebbles (32) and split or otherwise fractured pebbles (156). There is also a single split and polished bone implement. Fragmentary skeletal remains (teeth) referred to *Australopithecus africanus* were associated in the same breccia. The significance of the discovery has been variously interpreted; that of the previously mentioned authors is that the tools were not the work of *Australopithecus*, but represents a post-Oldowan industry and the tool-making of a hominid referable to genus *Homo* (ex-*Telanthropus capensis*). The writer thinks the evidence may be interpreted otherwise, especially in view of the recent discoveries in Bed I/II at Olduvai Gorge, and adheres to the statements made by him in this essay.

The specimens derive from a reddish-brown breccia at the extension site first thought to be broadly contemporary with the basal pink australopithecine-bearing breccia of the type site. Robinson's[112] more recent investigations indicate, however, that these breccias are separated unconformably, as a consequence of subsidence, and that the latter pink breccia underlies the reddish-brown breccia of the extension site. The latter contains the foreign and worked stones as well as some fauna (including *Equus*, absent in the type site basal pink breccia) and some remains (isolated permanent teeth, a juvenile maxilla fragment with several teeth) referred to *Australopithecus.*

The specimens recovered from Sterkfontein include pebble and core choppers (8), a chopper-hammerstone (1), and rough retouched end-struck flakes (2). Quartz, quartzite, chert, and diabase pebbles foreign to the deposit were also found; about half (24) of these were plain, and the other half (23) exhibited evidence of fracture from use. The small flakes struck from the choppers and cores are missing and indicate that the specimens were collected and worked elsewhere, near streams where raw material was available, prior to being carried to the site for use. Some of the specimens show extensive battering rather than careful flake removal, suggesting either hard use or, according to J. Desmond Clark,[113] poor workmanship. The implements are fresh and unweathered and cannot have been washed in from the outside, since the breccias must have accumulated under an overhanging roof, and since the breccias fail to reveal such conditions of deposition. The artifacts seem to be concentrated at the western end of the site near the original entrance to the cave and were undoubtedly left behind by the hominids who occupied this end of the cavern.

The recovery of pebble tools in association with australopithecines is a momentous discovery. There now seems little doubt that these primitive creatures were already capable of using and manufacturing implements of stone and, presumably, of other nonpreserved materials as well. The extraordinary concentration of other mammalian bones would indicate that these creatures were capable of killing the moderate- and smaller-sized species; probably they also scavenged carnivore kills. There is no indication that they had the ability, the equipment, or the organization necessary for killing very large mammals, in contrast to Middle Pleistocene peoples. Such carnivorous habits would have required some sharpened stone implements, such as flakes and chopping tools, for cutting open the hide of kills to obtain meat. There is no trace of the use of fire at this early time.

Clark[114] has recently suggested that the availability of water in the cave systems may have been an important factor in their having been occupied. Except for small and seasonally dry streams, the cave and fissure systems and the springs related to them provide the best source of water in such limestone country. Such sources of water would have attracted game and australopithecines alike and would have provided ideal conditions for the latter to prey on antelopes, pigs, baboons, and other animals which came to drink there. This also readily accounts for the profusion of animal bones accumulated in the cave and for the presence of implements necessary to butcher the slaughtered game.

Australopithecine Morphology

Hominids and pongids (apes) are generally regarded as closely related higher primate groups (hominoids). There are obvious and significant morphological and behavioral differences between the living representatives of the two families. Acceptance of the fact of evolution and the reality of such close affinity indicates, however, that such divergences were fewer and less sharply delineated in the remote past. The *primary* adaptation of the hominid radiation required transformation of the locomotor skeleton to permit fully upright posture and an efficient bipedal gait.[115] This mode of terrestrial locomotion contrasts markedly with the arboreal or terrestrial quadrupedalism of the lower catarrhine monkeys or with the arboreal brachiation of the pongids, coupled, in the larger pongine (anthropoid ape) species, with semi-pronograde quadrupedalism. This basic locomotor adaptation of the hominids was doubtless preadaptive for subsequent evolutionary changes that affected the skull—reduction of the facial skeleton and extraordinary enlargement of the cerebral hemispheres and cranial vault. The latter has been linked with enhanced cultural capacities, although the manifold interrelationships between structure and function, and their behavioral significance, are still very largely obscure.

The pelves of four australopithecine individuals are now known. The innominate bone of an additional individual from Limeworks Cave brings this total to five. Three represent *Australopithecus* and were found at Limeworks Cave[116] and Sterkfontein.[117] (At the latter site, much of an associated vertebral column was also found.) *Paranthropus* is represented by one incomplete innominate bone from Swartkrans.[118] These fortunate discoveries demonstrate conclusively that the pelvic structure of these creatures was that characteristic of primitive bipedal hominids. They also greatly add to our understanding of the hominid locomotor transformation, which involved a complex of interrelated structural modifications. These constituted a basic reorientation of the pelvis in relation to the trunk, interrelated changes that permitted an erect trunk and full extension of the lower limbs in stable upright posture (and in the female, maintenance of the bony birth canal).

Such changes involve (1) expansion of the iliac blade, especially the auricular area, coupled with sacral rotation and accentuated lumbar lordosis; (2) shortening and anterior rotation or "twisting" of the ilium, with attendant development of a sigmoid curvature of the iliac crest; (3) thickening of the outer bony table above the acetabulum (to aid in balance and weight support); (4) development of an iliac cristal tubercle, in line with the strengthened supra-acetabular region (related to differentiation and expansion of the iliotibial tract as an aid in stabilization of the hip and knee joints in standing and walking erect); (5) enlargement and approximation to the acetabulum of the anterior inferior iliac spine (related to the size of the straight head of the *rectus femoris* muscle, as part of the general enlargement of the *quadriceps femoris* muscle group in bipeds); (6) shortening of the ischium and altered form of the ischial tuberosity (the full significance of this is obscure, but it is apparently related to the position of the extensor, or hamstring muscle, lever arm); (7) enlargement and displacement of the *gluteus maximus* muscle as a powerful extensor (rather than a lateral rotator as in apes and monkeys); and (8) altered function of *gluteus medius* (and *minimus*) muscles as abductors (to maintain lateral stability in walking erect).

There are accompanying interrelated modifications in the proximal head of the femur. These include: (1) enlargement of the femoral head; (2) development of the lesser trochanter (related to an altered disposition of the *psoas major* muscle); (3) development of the anterior segment of the greater trochanter and the intertrochanteric line (related to the attachment of the Y-shaped ligament of Bigelow and the joint capsule); (4) development of the *linea aspera*; (5) shift of *gluteus maximus* to a posterior rather than a lateral insertion, as in apes, in the place of *adductor minimus* and expansion of *vastus intermedius*; and (6) notable reduction of the *quadratus femoris* muscle. Other modifications of the distal end of the femur, including the obliquity of the shaft, the marked depth of the patellar surface, the configuration of the intercondylar notch, and the enlargement of the lateral condyle, were apparently associated in large part with enhanced stability of the knee joint in orthograde progression.

The basic morphological pattern characteristic of hominid bipedalism is apparent in the australopithecine lower limb skeleton so far as it is known.[119] There are a number of minor differences—for example, in the form of the ischial region—from the pelvic morphology of *Homo sapiens*, a not unexpected finding in a primitive Lower Pleistocene hominid. The australopithecine lower leg and foot, except for a talus of *Paranthropus*, is still largely unknown.

Much of the foot skeleton (minus phalanges) of "*Homo habilis*," very *Australopithecus*-like in teeth and jaw morphology, is known now from the FLK NNI site at Olduvai Gorge. The foot is neither functionally nor morphologically quite "modern human," but exhibits well developed arches, a stout and non-divergent hallux, but weight-transmission rather laterally, as in apes, judging from the orientation of the ankle joint.[120]

The hand of the same form differs from modern man more markedly, in a number of features of the digits, although the overall structure is clearly of the hominid type. J. R. Napier[121] succinctly describes it as "a short powerful hand with strong, curved digits, surmounted by broad, flat nails and held in marked flexion," the thumb "strong and opposable though possibly rather short," with a powerful power grasp capability and probably a somewhat limited effectiveness of precision grip (pulp-to-pulp contact between thumb and fingers).

The morphological pattern of the australopithecine dentition is also hominid rather than pongid.[122] In the deciduous dentition this is evident in the evenly curved dental arch, lacking diastemata; the small milk incisors; the small, nonprojecting spatulate milk canines; the quadricuspid upper first milk molar; and the nonsectorial, quinticuspid lower first milk molar with well-developed anterior fovea and cusps of approximately equal height. In the permanent dentition this is evident in the evenly curved (parabolic) dental arch; the small incisors; the small, nonprojecting spatulate canines, lacking a talonid and with the internal cingulum forming a basal tubercle; the double-rooted upper first premolar; the nonsectorial, bicuspid lower first premolar; and the replacement sequence, in which both the permanent canine and medial incisor tend to erupt relatively early.

The related structure of the facial skeleton is also primitively hominid, and this is paralleled in a number of structural details of the cranial base and the occiput. The brain, in proportion to body size, in certain aspects of its form and proportions, and in its tendency toward delayed maturation, approaches a primitive hominid rather than a pongid condition.

There are consistent morphological distinctions between the earlier form *Australopithecus* and the younger form *Paranthropus*.[123] These are evident not only in the deciduous and permanent dentitions and the facial skeleton, but also in the structure of the cranial base and vault, as well as in the known portions of the postcranial skeleton. *Australopithecus* was a small, gracile bipedal creature, weighing certainly no more than 75 to 85 pounds in the larger males. *Paranthropus,* on the other hand, was a far more robust and massive creature of probably half again that body weight. Such, probably subgeneric, differences indicate a pronounced bifurcation within a primary australopithecine radiation, at least in the Basal Villafranchian and possibly even in the later Pliocene. Unfortunately, the general absence of fossiliferous Pliocene horizons in sub-Saharan Africa has thwarted investigation of the earlier evolutionary phases and primitive hominoid antecedents of the australopithecine group.

At Swartkrans, another distinct hominid, designated *Telanthropus*, is found in direct association with the australopithecine *Paranthropus*.[124] This form, which is still only inadequately known, differs markedly from *Paranthropus* in dental and mandibular morphology and in certain features of the maxilla (in particular in the structure of the nasal floor). In all such characteristics, some of which reveal some resemblances to the earlier australopithecine, *Australopithecus*, this hominid is further evolved morphologically than any known australopithecine and approaches therefore the phylogenetic status of certain earlier Middle Pleistocene forms attributed to the genus *Homo*. The full implications of this conclusion cannot be properly evaluated until additional, more complete, specimens have been discovered.

Tertiary Hominoids and Villafranchian Hominids

It is significant that the African Villafranchian hominids differ considerably in morphology from later Tertiary hominoids of eastern Africa and Eurasia. Four hominoid genera are recognized from the earlier Miocene of eastern Africa, three and possibly four genera[125] are recognized from the Middle and Late Miocene and the Lower Pliocene of peri-Alpine Europe, and four genera are recognized from the Upper Lower, and Middle Siwaliks of Asia (Fig. 1–7).

In both Africa and Europe, primitive hylobatids (gibbons) are already evolved in the lower half of the Miocene. Most of the basic cranial and dental morphology of the group is already established in these forms; in fact the dentition is already basically hominoid in the Oligocene form *Propliopithecus*. However, in some features of postcranial structure and in limb proportions these forms differ significantly from their living representatives.[126] Similarly, the basic dryopithecine (pongid) dental characteristics are manifest in the earlier Miocene hominoids of eastern Africa, although cranial and facial morphology is distinct from that of evidently specialized living varieties of the group.[127] Moreover, the fundamental locomotor pattern of the recent large-bodied brachiator is evident in the morphology of shoulder and elbow joints, whereas the skeleton of the hand is distinctly primitive cercopithecoid or monkey-like.[128] The later Miocene and Pliocene dryopithecine hominoids of Europe and their Pliocene counterparts of Asia, known nearly exclusively from jaws and teeth, are typically pongid in mandibular and dental structure and therein do not differ fundamentally from the living gibbons and great apes.

Oreopithecus, known for over three quarters of a century, which has received much publicity and been much discussed, is a notable exception to this statement. This primate occurs in the Pontian (Lower Pliocene) lignites of Tuscany.[129] It is surely significant that *Oreopithecus* reveals features of dental morphology that are not *typically* pongid, although there can be no question that it is fully hominoid rather than primitive cercopithecoid (cercopithecid).[130] A majority of workers tend to agree that the other hominoids of the later Tertiary are typical pongids, with all the attendant dental specializations that would effectively exclude such creatures from hominid ancestry. J. Hürzeler's[131] painstaking reexamination of all the *Oreopithecus* material proves, however, that *all* later Tertiary hominoids were not typically pongid.[132] Hence, the longstanding argument over the "cercopithecid" (Old World monkey) or "pongid" (anthropoid ape) origin of hominids can be clarified, since many of the more "generalized" hominid features, which some workers have regarded as indicative of lower catarrhine affinity, are present in either primitive or nonpongid hominoids. Consequently, an oreopithecine hominoid group might well provide the ancestral stage from which orthograde bipedal hominids were subsequently to evolve. It is still premature to assert the correctness of such a hypothesis to the exclusion of all others.

It seems clear that *Oreopithecus* must represent a distinct lineage within the radiation of the Hominoidea. The postcranial skeleton leaves no doubt that this creature was hominoid and not cercopithecoid (simian). On the other hand, there are numerous and profound structural peculiarities in the dentition that are quite unparalleled among pongids or hominids, living or extinct. This important fact, plus the detailed dental resemblances with the Lower Oligocene primate *Apidium,* now increasingly well-known from recent discoveries by Elwyn Simons in the Fayum of Egypt, would suggest an early differentiation of an oreopithecid lineage. This conclusion has been supported by other writers in recent years, and was impressed on me during examinations of the now abundant skeletal remains of *Oreopithecus* in 1959 and again in 1960. [I am deeply grateful to Dr. J. Hürzeler for his kindness in permitting me to examine these unique specimens.]

It should be recalled that the whole of the African Pliocene, a span of over ten million years, is still largely unknown.[133] Late Tertiary hominoids surely occupied the more central area of the continent, although fossiliferous deposits of this time range are still unknown. Until some evidences of hominoid varieties are forthcoming from the upper Neogene of sub-Saharan Africa, any hypotheses of hominid origins will lack support. There is widespread evidence of extensive sub-Saharan desiccation, between 1°N and 20°S, from the late Miocene throughout the Pliocene. This was a period of desertification during which great distributions of fine, light, unstratified aeolian sands of the Upper Kalahari system occurred.[134] Conditions appear to have been such that, except for rare, and probably small, basins of sedimentation (proto-rift valleys), which are usually obscured by volcanic lavas and other deposits, the preservation of mammalian fossils was literally precluded.

This was certainly a crucial time for mammalian evolution in general, as well as for higher primate evolution in particular. Thus, the origins of the incredibly rich eastern-central African grasslands fauna is unknown; yet this long interval was probably vital for such a radiation, since the Villafranchian antelopes appear largely referable to existing genera. For the pongid hominoids, whose ecological requirements and habits of locomotion were apparently becoming those of forest-dwelling, vegetarian (or frugivorous) brachiators, such desiccation had profound effects on distribution. These effects are fully evident today in the sparse and restricted habitats of the African apes. For those hominoids that were preadapted toward terrestrial bipedalism, it was a period of trial in a new and exploitable environment. The Villafranchian stage, as well as the later Pleistocene phases of hominid history, testifies to the achievement of this primary radiation.

NOTES

1. The best single, well-illustrated reference work to the hominid (and primate) fossil record is the recent work by J. Piveteau, *Traité de Paléontologie, VII: Primates; Paléontologie humaine* (Masson, Paris, 1957).

2. J. F. Evernden and G. H. Curtis, The potassium-argon dating of Late Cenozoic rocks in East Africa and Italy, *Current Anthropol.*

3. F. C. Howell, *European and Northwest African Middle Pleistocene Hominids*, in press.

4. W. E. Le Gros Clark, *The Fossil Evidence for Human Evolution* (Univ. of Chicago Press, Chicago, 1955).

5. F. C. Howell, *Am. J. Phys. Anthropol.* **9**, 379 (1951); *Quart. Rev. Biol.* **32**, 330 (1957); in, *Hundert Jahre Neanderthaler 1856–1956* (Kemink en Zoon, Utrecht, 1958), p. 185.

6. P. Woldstedt, *Norddeutschland und angrenzende Gebiete im Eiszeitalter* (Koehler, Stuttgart, 1950).

7. D. F. W. Baden-Powell, *Geol. Mag.*, **85**, 279 (1948); W. B. R. King, *Quart. J. geol. Soc. London*, **111**, 187 (1955); R. G. West, *Quaternaria*, **2**, 45 (1955); R. G. West, *J. Glaciol.*, **3**, 211 (1958).

8. W. Szafer, *Ann. soc. géol. Pologne*, **22**, 1 (1953).

9. H. Spreitzer, *Quartär*, **3**, 1 (1941); V. Gromov, *Am. J. Sci.*, **243**, 492 (1945).

10. See the summary list of uplifted highlands in R. F. Flint, *Glacial and Pleistocene Geology* (Wiley, New York, 1957).

11. "Mammal migrations, Cenozoic stratigraphy and the age of Pekin Man and the Australopithecines," *J. Paleontol.*, **31**, 215–27 (1957).

12. *Intern. geol. Congr. London., 18th Congr.*, 1948.

13. C. Emiliani, *Science*, **123**, 924 (1956); *J. Geol.*, **63**, 538 (1955); *ibid.*, **66**, 264 (1958).

14. L. Pareto, *Bull. soc. géol. France*, **22**, 210 (1865).

15. É. Haug, *Traité de Géologie* (Colin, Paris, 1911); A. T. Hopwood, *Proc. Geologists, Assoc. Engl.* **46**, 46 (1935); *ibid.*, **51**, 79 (1940). Although several mammals are included as new arrivals in the Villafranchian, all are not *necessary* to prove a lower Pleistocene age; thus, *Equus* never reached Java and *Leptobos* is not present in sub-Saharan Africa.

16. See C. Reid, "The Geology of the Country around Cromer," *Mem. Geol. Survey G. Brit. Engl. Wales* (1882).

17. M. Gignoux, *Compt. rend.*, **147**, 1497 (1908); A. C. Blanc, *Geol. Meere Binnengewässer*, **5**, 137 (1942).

18. In France: Sâone basin (Bresse-Dombes region), Puy-en-Velay (Haute-Loire), Allier Valley (Perrier, Mont Coupet, Senèze localities), and Rhone Valley (St. Vallier, Drôme); in Italy: Po Valley (Piemonte, Emilia), upper Arno Valley (Tuscany), Monte Mario (Rome), Puglia and Calabria, and Sicily. For an excellent general review of the evidence in southern Europe, see H. L. Movius, Jr., *J. Geol.*, **57**, 380 (1949); also various papers in *Rept. Intern. Geol. Congr. London, 18th Congr., 1948* (1950), part 9, sect. H, and A. C. Blanc et al., *Compt. rend. congr. géol. intern., 19e Congr., Algiers, 1952* (1954), fasc. 15, p. 215.

19. For useful summaries of European Pliocene floras see G. Depape, *Ann. soc. Sci. Bruxelles*, **48**, 39 (1928); also, W. Szafer, *Inst. Geologiczny, Prace (Warsaw, Wydawnictwa Geol.)*, **11** (1954).

20. M. Gignoux, *Ann. géogr.*, **18**, 141 (1909); *Ann. Univ. Lyon*, **1**, No. 36 (1913); *Bull. soc. géol. France*, **14**, 324 (1915); *Ann. Univ. Grénoble*, **28**, 9 (1916).

21. R. Ruggieri and R. Selli, *Rept. Intern. Geol. Congr. London, 18th Congr., 1948* (1950), part 4, sect. H, p. 85.

22. C. Reid, "The Pliocene Deposits of Britain," *Mem. Geol. Survey Gt. Brit. Engl. Wales* (1890); F. W. Harmer, *Proc. Geologists' Assoc. Engl.*, **17**, 416 (1902); P. G. H. Boswell, *ibid.*, **63**, 301 (1952); C. P. Chatwin, *East Anglia and Adjoining Areas* (Geological Survey and Museum, London, 1954); for general discussion, see W. B. R. King, *Quart. J. geol. Soc. London*, **111**, 187 (1955).

23. I. M. van der Vlerk, *Proc. Koninkl. Ned. Akad. Wetenschap*, **B56**, 34 (1953); I. M. van der Vlerk and F. Florschütz, *Verhandel. Koninkl. Ned. Akad. Wetenschap. Afdel. Natuurk.*, **20**, 1 (1953); R. Lagaaij, *Mededel. Geol. Sticht., Ser. C, No. 5* (1952); for palaeogeographic maps, see J. H. van Voorthuysen, *Geol. en Mijnbouw*, **19**, 263 (1957); for general summary, see A. J. Pannekoek et al., *Geological History of the Netherlands* (Government Printing Office, s'Gravenhage, 1956).

24. S. Venzo and F. Lona, *Atti soc. ital. sci. nat. e museo civico storia nat. Milano*, **89**, 43 (1950); S. Venzo, *Geol. Bavarica*, **19**, 74 (1953); *Actes congr. assoc. intern. étude quatern., 4e Congr., Rome-Pise, 1953* (1956), vol. 1, p. 65.

25. W. H. Zagwijn, *Geol. en Mijnbouw*, **19**, 233 (1957); A. Schreuder, *Arch. néerl. zool.*, **7**, 153 (1945).

26. A. C. Blanc, *Quaternaria*, **4**, 95 (1957).

27. B. Eberl, *Die Eiszeitenfolge im nördlichen Alpenvorlande* (Filser, Augsburg, 1930); I. Schafer, *Geol. Bavarica No. 19* (1953), p. 13.

28. Supposedly artificially flawed specimens have been reported from the East Anglian "Crags" [see especially J. R. Moir, *The Antiquity of Man in East Anglia* (Cambridge Univ. Press, Cambridge, 1927)]; however, most workers are now convinced that these are natural specimens. See A. S. Barnes, *Anthropol.*, **48**, 217 (1938); *Am. Anthropol.*, **41**, 99 (1939); S. H. Warren, *Southeast Nat. Antiq.*, **53**, 48 (1948).

29. C. S. Nicolăeuscu-Plopsor and I. N. Moroșan, *Dacia*, n.s., **3**, 9–33 (1959).

30. H. de Lumbey, S. Gragnière, L. Bavoral, and R. Pascal, *Bull. du Musée d'Anthrop. prehist. de Monaco*, **10**, 5–20 (1963).

31. M. Pfannenstiel, *Geol. Rundschau*, **34**, 341 (1944); *Bonner Geogr. Abhandl.*, **6**, 1 (1950); S. Erinc, *Rev. Geogr. Inst. Univ. Istanbul.*, **1**, 85 (1954).

32. A. C. Blanc, *Geol. Meere Binnengewässer*, **5**, 137 (1942).

33. It has been maintained that the Abbevillian, at least in the Somme valley where it is best known, is of "first interglacial" (= Günz-Mindel = Cromerian) age. See H. Breuil, *Proc. Prehist. Soc.*, **5**, 33 (1939); H. Breuil and L. Koslowski, *Anthropol.*, **41**, 449 (1931); *ibid.*, **42**, 27, 291 (1932); *ibid.*, **44**, 249 (1934). However, a variety of both geological and palaeontological evidence indicates that assignment of a younger age is fully warranted, and that this was probably an ameliorative phase within or toward the end of the first major continental glaciation.

34. See the excellent general summary by H. L. Movius, Jr., *Papers Peabody Museum Am. Archeol. Ethnol. Harvard Univ.*, **19**, 1 (1944); also H. de Terra, *Proc. Am. Phil. Soc.*, **77**, 289 (1937).

35. P. Teilhard de Chardin, *Bull. Geol. Soc. China*, **16**, 195 (1937); *ibid.*, **17**, 169, (1937); *Compt. rend. soc. géol. France, No. 17* (1938), p. 325; *Inst. Géo-Biol., Pekin, No. 7* (1941); P. Teilhard de Chardin and C. C. Young, *Bull. Geol. Soc. China*, **12**, 207 (1933); G. B. Barbour, *ibid.*, **10**, 71 (1931); H. de Terra, *Inst. Géo-Biol., Pekin, No. 6* (1941).

36. See faunal list in P. Teilhard de Chardin and P. Leroy, *Inst. Géo-Biol., Pekin, No. 8* (1942).

37. H. de Terra, *Proc. Am. Phil. Soc.*, **77**, 289 (1937); H. de Terra and P. Teilhard de Chardin, *ibid.*, **76**, 791 (1936); G. E. Pilgrim, *Records Geol. Survey India*, **40**, 185 (1910); W. D. Gill, *Quart. J. Geol. Soc. London*, **107**, 375 (1952).

38. H. de Terra and T. T. Paterson, *Carnegie Inst. Wash. Publ. No. 493* (1939), p. 1.

39. E. H. Colbert, *Trans. Am. Phil. Soc.*, **26**, 1 (1935).

40. The interesting and little known pongid *Ramapithecus* is the last Siwalik primate to occur in the Tatrot zone.

41. G. E. Lewis, *Am. J. Sci.*, **227**, 161 (1934).

42. G. E. Lewis, *Am. J. Sci.*, **234**, 139 (1937).

43. E. Simons, *Postilla No. 57* (1961), pp. 1–9.

44. L. S. B. Leakey, *Annals and Mag. Nat. Hist., Ser. 13*, **4**, 689 (1962).

45. De Terra and Paterson, *loc. cit.*

46. D. A. Hooijer and E. H. Colbert, *Am. J. Sci.*, **249**, 533 (1951); also D. A. Hooijer, *Proc. Koninkl. Ned. Akad. Wetenschap.*, **B55**, 436 (1952).

47. De Terra and Paterson, *loc. cit.*

48. H. L. Movius, *Trans. Am. Phil. Soc.*, **38**, 329 (1948); *Cahiers hist. mond.*, **2**, 257, 520 (1955).

49. H. de Terra, *Trans. Am. Phil. Soc.*, **32**, 437 (1943); R. W. van Bemelen, *The Geology of Indonesia* (Nijhoff, The Hague, 1949).

50. G. H. R. von Koenigswald, *Proc. Koninkl. Ned. Akad. Wetenschap.*, **B38**, 188 (1935); *Quartär*, **2**, 28 (1939).

51. D. A. Hooijer, *Am. J. Phys. Anthropol.*, **9**, 265 (1951); *Proc. Koninkl. Ned. Akad. Wetenschap.*, **B55**, 436 (1952); *ibid.*, **B60**, 1 (1957); *Quaternaria*, 3, 5 (1956). A contrary opinion is held by G. H. R. von Koenigswald, *Proc. Koninkl. Ned. Akad. Wetenschap.*, **B59**, 204 (1956); *Actes congr. assoc. intern. étude quatern., 4ᵉ Congr., Rome-Pise, 1953* (1956), vol. 1, p. 5.

52. M. Stekelis et al., *Bull. Res. Council Israel, Sect. G*, **9**, 175 (1960); also N. Schulman, *ibid.*, **8-G**, 63–90 (1959).

53. L. Picard, *Bull. Geol. Dept. Hebrew Univ.*, **4**, 1 (1943).

54. W. J. van Liére and D. A. Hooijer, *Ann. archéol. de Syrie*, **11**, 165 (1961).

55. D. Black, *Bull. Gel. Soc. China*, **4**, 133 (1925); W. D. Matthew, *Climate and Evolution* (N.Y. Acad. Sci., New York, ed. 2, 1939).

56. C. Arambourg, *Bull. soc. géol. France*, **19**, 195 (1949); "La Paléontologie des vertebrés en Afrique du Nord française," *Congr. géol. intern., 19ᵉ Congr., Algiers*, 1952, *Monogr.* Arambourg distinguishes Lower (Ichkeul, Boucherit, Fouarat) and Upper (Bel Hacel, Ain Hanech) Villafranchian faunas; the former is more archaic and includes *El. africanavus, Anancus, Stylohipparion*, and *Libytherium*, whereas the latter includes predominantly *El. meridionalis* and *Equus*, with *Libytherium* and, rarely, *Stylohipparion*.

57. G. Choubert et al., *Notes et mém. serv. géol. Maroc, No. 71* (1948), p. 15; G. Choubert, *Notes et mém. serv. géol. Maroc No. 76* (1950), p. 13.

58. P. Biberson et al., *Compt. rend.*, **245**, 938 (1957).

59. G. Choubert and R. Ambroggi, *Notes et mém. serv. géol. Maroc, No. 117* (1953), p. 3; G. Choubert et al., *Compt. rend.*, **243**, 504 (1956); G. Choubert, *ibid.*, **245**, 1066 (1957); *Actes congr. assoc. intern. étude quatern., 5ᵉ Congr. Madrid-Barcelona, 1957* (in press).

60. P. Biberson, *Publ. Serv. Antiquities du Maroc*, Nos. 16–17 (1961).

61. C. Arambourgh and M. Arnould, *Bull. Soc. nat. Tunisie*, **2**, 149 (1949).

62. C. Arambourg et al., *Compt. rend.*, **234**, 128 (1952); *Arch. Mus. nat. hist. natl. Paris, sér.* 7 (1953), vol. 2, p. 1.

63. M. Dalloni, *Bull soc. géogr. archéol. prov. Oran*, **61**, 1 (1940); *Actes congr. assoc. intern. étude quatern., 4ᵉ Congr., Rome-Pise, 1953* (1956), vol. 1, p. 19; *Bull. soc. hist. nat. Afrique nord.*, **45**, 134 (1954).

64. C. Arambourg, *Bull. soc. géol. France*, **19**, 195 (1949); *Compt. rend.*, **229**, 66 (1949).

65. C. Arambourg, *Compt. rend. soc. géol. France, No. 7* (1949), p. 120; *Bull. soc. préhist. franç.*, **47**, 348 (1950); *Compt. rend.*, **236**, 2419 (1953); C. Arambourg and L. Balout, *Bull. soc. hist. nat. Afrique nord*, **43**, 152 (1952). Crude bifaces, referable to a Basal Chellean industrial stage, were discovered in 1952, derived from a higher level at the Ain Hanech site.

66. L. S. B. Leakey, personal communication.

67. Y. Coppens, *C. R. Acad. Sci. Paris*, **252**, 3851–3852 (1961).

68. F. Dixey, *Quart. J. Geol. Soc. London*, **102**, 339 (1946); "The East African Rift Valley," *Colonial Geol. and Mineral Resources Suppl. Ser.* (London, 1956). See especially the recent excellent critical appraisal by H. B. S. Cooke (see note 69, p. 60).

69. H. B. S. Cooke, *Annex. Trans. Geol. Soc. S. Africa* (1958).

70. P. E. Kent, *Geol. Mag.*, **79**, 117 (1942); E. P. Saggerson, *Geology of the Kisumu District* (Geological Survey, Kenya, Nairobi, 1952).

71. L. S. B. Leakey, *The Stone Age Races of Kenya* (Oxford Univ. Press, London, 1935); compare M. F. Ashley Montagu, *Am. Anthropol.*, **59**, 335 (1957).

72. P. V. Tobias, *Actes 4e Congr. Panafricain de Préhist. et d l'étude du Quarternaire*, Léopoldville, 1959 (1962), p. 341; also, P. V. Tobias and K. P. Oakley, *Nature*, **185**, 944 (1960).

73. This series is named after the village of Kaiso, the type locality, on the eastern shore of Lake Albert. See E. J. Wayland et al., *Occasional Papers Geol. Survey Uganda*, **2**, 1 (1926); also, K. A. Davies, *Geol. Mag.*, **88**, 377 (1951); N. Harris, J. W. Pallister, J. M. Brown, *Geol. Survey Uganda Mem. No. 9* (1956).

74. V. E. Fuchs, *Geol. Mag.*, **71**, 97 (1934).

75. J. Lepersonne, *Ann. soc. géol. Belge*, **72**, 1 (1949).

76. The particular horizon of the type Kaiso mammal fauna is not clear. Possibly it corresponds to the middle or even the upper stage; but this problem may be settled by careful study and comparison of the molluscan faunas within the several stages of the Kaiso series. W. W. Bishop (Geological Survey of Uganda, Entebbe) is now carrying out systematic investigations in the Western Rift, both at Kaiso and in the Lake George-Lake Edward region (see "Proc. Assoc. African Geol. Services, Léopoldville, Belgian Congo, Meeting, July 1958," in press).

77. J. de Heinzelin de Braucourt, "Le fossé tectonique sous le parallèle d'Ishango," *Publ. Inst. Natl. Parcs Nat., Bruxelles* (1950); also, *Proc. Intern. Sci. Sol, 5e Congr. Léopoldville, 1954* (1954), vol. 4, p. 435.

78. C. Arambourg, *Mission scientifique de l'Omo, 1932–1933, I* (Museum Natl. Hist. Nat., Paris, 1935; 1943), vols. 1, 3.

79. L. S. B. Leakey, *Olduvai Gorge* (Cambridge Univ. Press, Cambridge, 1951); *Nature*, **181**, 1099 (1958).

80. R. L. Hay, *Science*, **139**, 829-833 (1963).

81. J. F. Evernden and G. H. Curtis, personal communication.

82. P. E. Kent, *Geol. Mag.*, **78**, 173 (1941); also, L. Kohl-Larsen, *Auf den Spuren des Vormenschen* (Strecker and Schroeder, Stuttgart, 1943).

83. W. O. Dietrich, *Palaeontographica*, **A94**, 43 (1942); *ibid.*, **A99**, 1 (1950).

84. A. Remane, Z. *Morphol. Anthropol.*, **42**, 311 (1951); *Am. J. Phys. Anthropol.*, **12**, 123 (1954).

85. J. T. Robinson, *Am. J. Phys. Anthropol.*, **11**, 1 (1953).

86. Cooke, *op. cit.*

87. See faunal lists in L. S. B. Leakey, *Olduvai Gorge*, note 74.

88. L. S. B. Leakey, *Olduvai Gorge, 1951–1961* (Cambridge Univ. Press, 1960).

89. L. S. B. Leakey, personal communication.

90. Other riverine situations have provided *selected* assemblages of pebble tools referred to the Pre-Chelles-Acheul industries. The Kafuan industry, generally regarded as the oldest such manifestation, has been recovered from very high-level gravels of

the Kafu and Kagera rivers in western and southeastern Uganda, respectively [C. van Riet Lowe, *Mem. Geol. Survey Uganda No. 6* (1952); *Proc. Pan-African Congr. Prehist., 3rd Congr., Livingstone, 1955* (1957), p. 207]. The industry, in a more evolved facies it seems, is probably also represented in the Katanga, southeastern Belgian Congo [G. Mortelmans, "Mélanges en hommage au Prof. Hamal Nandrin," *Publ. soc. roy. Belge anthropol. préhist.* (1952), p. 150; *Actes congr. panafr. préhist., 2e Congr., Alger, 1952* (1955), p. 295; *Proc. Pan-African Congr. Prehist., 3rd Congr., Livingstone, 1955* (1957), p. 214] and in the calcified basal older gravels of the Vaal River in southern Africa [C. van Riet Lowe, *S. African Archaeol. Bull.*, **8**, 27 (1953)]. There is little question that some of the selected specimens are indeed artifacts of a very primitive stone industry although this is likely *not* true of those from the Kafu River; until larger assemblages are found *in situ*, preferably on old land surfaces, in clays or other fine sediments, there will always be a considerable measure of doubt concerning their full authenticity.

91. J. T. Robinson, *Am. J. Phys. Anthropol.*, **12**, 181 (1954).

92. S. L. Washburn and B. Patterson, *Nature*, **167**, 650 (1951).

93. H. B. S. Cooke, *S. African Archaeol. Bull.*, **7**, 59 (1951); F. C. Howell, *Am. J. Phys. Anthropol.*, **13**, 635 (1955); R. F. Ewer, *S. African Archaeol. Bull.*, **9**, 41 (1956); *Proc. Pan-African Congr. Prehist., 3rd Congr., Livingstone, 1955* (1957), p. 135.

94. K. P. Oakley, *Am. J. Phys. Anthropol.*, **12**, 9 (1954); *Proc. Pan-African Congr. Prehist., 3rd Congr., Livingstone, 1955* (1957), p. 155.

95. Sterkfontein, Swartkrans, and Kromdraai are situated in a shallow valley about six miles northwest of Krugersdorp, southern Transvaal; Limeworks Cave is located in the Makapansgat Valley, several miles northeast of Potgietersrust, central Transvaal; the limeworks at Buxton, near Taungs, is just west of the Transvaal border, adjacent to the Kaap escarpment bordering the Harts River, in Bechuanaland (Cape Province).

96. C. K. Brain, *Proc. Pan-African Congr. Prehist., 3rd Congr., Livingstone, 1955* (1957), p. 143; and *Transvaal Museum, Pretoria, Mem. No. 11* (1958), p. 1; also, H. B. S. Cooke, *S. African J. Sci.*, **35**, 204 (1938); S. H. Haughton, *Trans. Geol. Soc. S. Africa*, **50**, 55 (1947).

97. C. K. Brain, see note 95.

98. F. E. Peabody, *Bull. Geol. Soc. Am.*, **65**, 671 (1954); also, R. B. Young, *Trans. Geol. Soc. S. Africa*, **28**, 55 (1925).

99. R. A. Dart, *Transvaal Museum, Pretoria, Mem. No. 10* (1957), p. 1; *Proc. Pan-African Congr. Prehist., 3rd Congr., Livingstone, 1955* (1957), p. 161; *Am. Anthropol.*, **60**, 923 (1958).

100. The account by W. I. Eitzman [*S. African J. Sci.*, **54**, 177 (1958)] of the original constitution of the Limeworks Cave and the mining activities which largely destroyed it suggests that there were probably three separate fossiliferous breccias, separated by travertines in the lower portion of this great cavern. He states that the lower and middle of these were densely packed with bone, whereas the upper horizon was more discontinuous and far less fossiliferous. It is clear that only a very small proportion of the bone accumulations has been salvaged from the mine dumps and that the great bulk of the fossil mammals (and australopithecines) was consumed in the lime kilns.

101. J. T. Robinson, *Ann. Transvaal Museum, Pretoria*, **22**, 1 (1952).

102. R. A. Dart, see note 75.

103. R. A. Dart, *Intern. Anthropol. Ling. Rev.*, **1**, 201 (1953); *Smithsonian Inst. Ann. Rept. 1955* (1956), p. 317; *Am. Anthropol.*, **60**, 715 (1958).

104. H. Breuil, *Palaeontol. Sinica, Peking,* **D6**, 1 (1939).

105. S. L. Washburn, *Am. Anthropol.*, **59**, 612 (1957).

106. A. R. Hughes, *S. African J. Sci.*, **51**, 156 (1954); *Am. J. Phys. Anthropol.*, **12**, 467 (1954); R. A. Dart, *Am. Anthropol.*, **58**, 40 (1956).

107. J. Stevenson-Hamilton, *Wild Life in South Africa* (Cassell, London, 1954).

108. R. A. Dart, *Nat. Hist.*, **26**, 315 (1926); *Am. J. Phys. Anthropol.*, **7**, (1949); see also, K. P. Oakley, in *Appraisal of Anthropology Today*, S. Tax et al., Eds. (Univ. of Chicago Press, Chicago, 1953), pp. 28–30.

109. C. K. Brain et al., *Nature*, **175**, 16 (1955); see K. P. Oakley, *Antiquity*, **30**, 4 (1956). This horizon has also yielded a fragment of an australopithecine maxilla [R. A. Dart, *Nature*, **176**, 170 (1955)].

110. J. T. Robinson and R. J. Mason, *Nature*, **180**, 521 (1957); see K. P. Oakley, *Antiquity*, **31**, 199 (1957). Foreign pebbles were first reported from this region (at Kromdraai) by S. H. Haughton, *S. African Archaeol. Bull.*, **2**, 59 (1947).

111. J. T. Robinson and R. J. Mason, *So. African Archaeol. Bull.*, **17**, 87–125 (1962).

112. J. T. Robinson, *Leech*, **28**, 94 (1958).

113. J. D. Clark, personal communication.

114. *Ibid.*

115. F. Weidenreich, *Anat. Anz.*, **44**, 497 (1913); A. Keith, *Brit. Med. J.*, **1**, 451, 499, 545, 587, 624, 669 (1923); S. L. Washburn, *Cold Spring Harbor Symposia Quant. Biol.*, **15**, 67 (1950); L. W. Mednick, *Am. J. Phys. Anthropol.*, **13**, 203 (1955).

116. R. A. Dart, *Am. J. Phys. Anthropol.*, **7**, 301 (1949).

117. R. Broom et al., *Transvaal Museum, Pretoria, Mem. No. 4* (1950), p. 1; R. Broom and J. T. Robinson, *Am. J. Phys. Anthropol.*, **8**, 489 (1950).

118. R. Broom et al., *Transvaal Museum, Pretoria, Mem. No. 6* (1952), p. 1.

119. W. E. Le Gros Clark, *Am. J. Phys. Anthropol.*, **13**, 19 (1955).

120. J. R. Napier, personal communication.

121. J. R. Napier, *Nature*, **196**, 409–411 (1962).

122. W. E. Le Gros Clark, *J. Roy. Anthropol. Inst.*, **80**, 37 (1952); J. T. Robinson, *J. Dental Assoc. S. Africa*, **7**, 1 (1952); *Transvaal Museum, Pretoria, Mem. No. 9* (1956), p. 1.

123. W. E. Le Gros Clark, *Evolution*, **8**, 324 (1954); *Z. Morphol. Anthropol.*, **46**, 269 (1954); also, see the excellent summary by G. Heberer, *Primatologia, I*, **4**, (1956), p. 379.

124. W. E. Le Gros Clark, *Am. J. Phys. Anthropol.*, **11**, 445 (1953).

125. See G. H. R. von Koenigswald, *Proc. Koninkl. Ned. Akad. Wetenschap.*, **B59**, 318 (1956). This author would distinguish *Paidopithex rhenanus*, of which a complete right femur is known, from the Pontian of Rheinhesse and the Schwäbian lignites, giving it as a separate genus (rather than as a species, as I have done here).

126. W. E. Le Gros Clark and D. P. Thomas, *Fossil Mammals of Africa* (British Museum of Natural History, London, 1951) vol. 3, p. 1; J. Hürzeler, *Ann. paléontol.*, **40**, 1 (1954).

127. W. E. Le Gros Clark, *Quart. J. Geol. Soc. London*, **105**, 225 (1950); W. E. Le Gros Clark and L. S. B. Leakey, *Fossil Mammals of Africa* (British Museum of Natural History, London, 1951), vol. 1, p. 1.

128. W. E. Le Gros Clark, *Proc. Zool. Soc. London*, **122**, 273 (1952); J. Napier, *Fossil Mammals of Africa* (British Museum of Natural History, London, in press).

129. H. de Terra, *Science*, **124**, 1282 (1956).

130. See A. Remane, *Abhl. Akad. Wiss. Lit. Mainz, Math.-Naturw. Kl., No. 2* (1955).

131. J. Hürzeler, *Schweiz. Palaontol. Abhandl.*, **66**, 1 (1949); Eclogae Geol. Helv., **44**, 404 (1951); *Verhandl. naturforsch Ges. Basel*, **65**, 88 (1954); *ibid.*, **69**, 1 (1958).

132. See W. L. Strauss, Jr., *Quart. Rev. Biol.*, **24**, 200 (1949).

133. In Mediterranean Africa the Pontian is little known in southern and northern Tunisia and practically not at all in Morocco. The richest locality, except for primates, is Qued el Hamman, Oran (Algeria). Findings from the later Pliocene are extremely rare; this period is known largely from the Constantine (Algeria) sites of Ain el Bey and Ain el Hadj Baba. A fairly substantial faunal assemblage, probably Middle Pliocene, from the Gart el Moluk, Wadi Natrun, in Egypt, is known. The situation is even more discouraging south of the Sahara where a Pliocene fauna is known only from diamond diggings in Little Namaqualand, southwest Africa.

134. L. Cahen and J. Lepersonne, *Mem. soc. géol., paléontol., Hydrol. Belg. ser. 8, No. 4* (1952), p. 1; J. de Heinzelin, "Sols. paléosols et désertifications anciens," *Publ. Inst. Natl. Etude Agron. Congo Belge, Bruxelles* (1952).

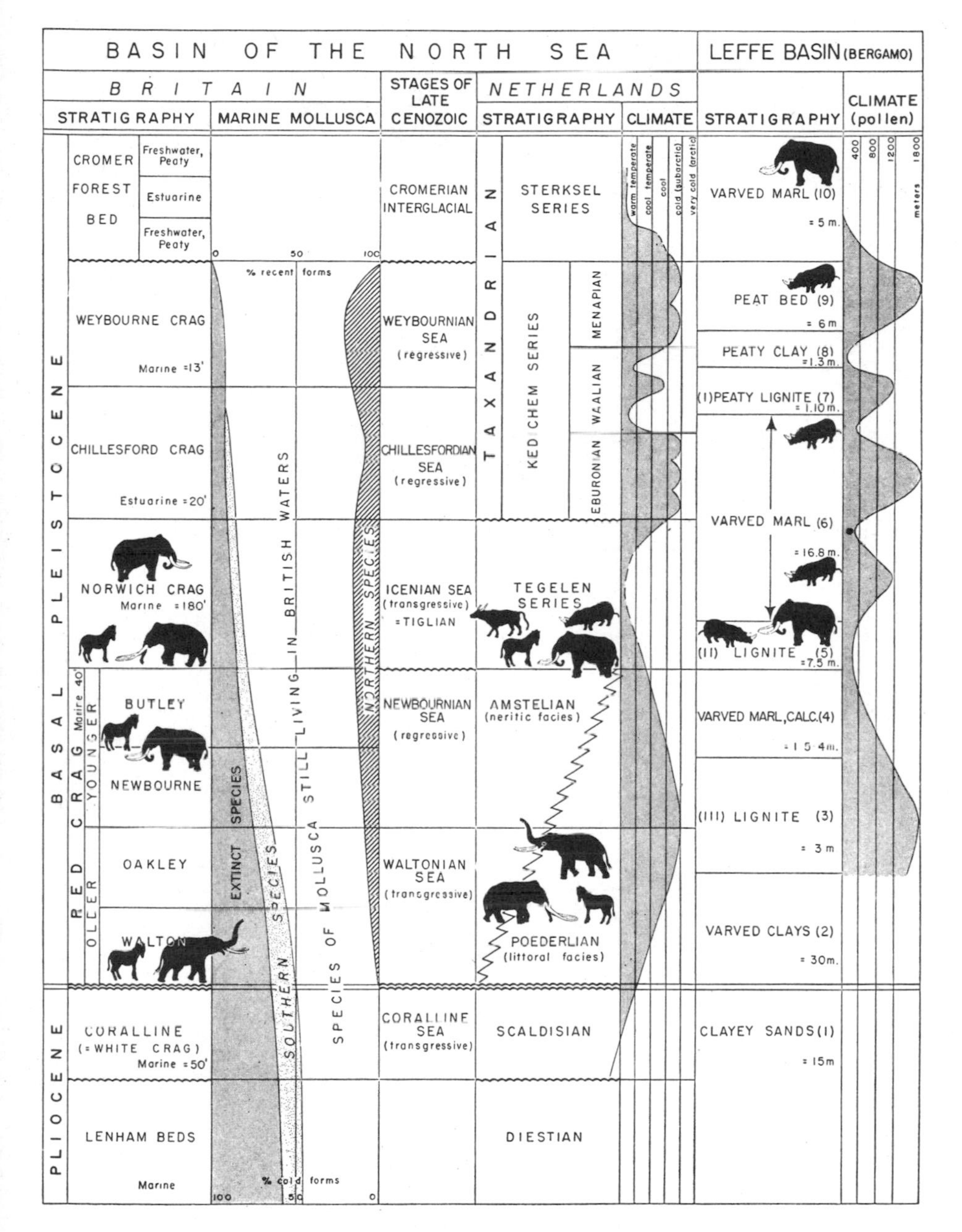

Fig. 1–1. Villafranchian stratigraphy and climatic change in the North Sea Basin and the Leffe Lake Basin (Tuscany). The silhouettes indicate some of the characteristic species, in particular faunal assemblages. The Leffe climate curve, based on palynological evidence, is expressed in terms of present altitude corresponding to vegetation; Leffe is 400 meters above sea level. (Compiled from various sources; see text references.)

	EASTERN ATLANTIC		WESTERN MEDITERRANEAN	
	MOROCCO	STAGES OF LATE CENOZOIC	ALGERIA	TUNISIA
	STRATIGRAPHY		STRATIGRAPHY	STRATIGRAPHY
PLEISTOCENE	Marine conglomerates, sandstones. +90 –100 m. (Sidi Messaoud, Cap Cantin, Mazagan)	SICILIAN SEA	Marine conglomerates, sandstones +90 –100 m. (Sahel, Oran, Arzew, Oued Isser) (overlies tilted Bel Hacel Villafranchian)	High-level (+100 m.) conglomerates and gravels, calcrete-cemented, in oueds. (no marine Sicilian)
	Mouloyian Pluvial. Cooler. +150 m. terrace Calcareous lacustrine deposits in southern basins erosion	MARINE REGRESSION	Lacustrine depressions in Chélif and Oranian Sahel and Algerian Sahel. Bel Hacel, Ain Hanech Setifian Lake, St. Arnauld: Deep series (+100 m.) of marls and lacustrine clays and silts, with intercalated fossiliferous sandstones, gravelly sands, and calcareous conglomerates.	Ichkeul Lake Top not observed
	uplift Reddened loams (Marmora) 50 m. and Conglomerates (Rharb, Arbaoua) 100 m.			Sandy-clays with thin sandstones (4)
BASAL	uplift shelly sandstones Moghrebian Transgression +300 m. (warm, impoverished fauna; 50% Med. species) reddish shelly gravels and conglomerates FOUARAT	CALABRIAN SEA	Traces of marine sediments +200 - 300 m., in Oran, Mostaganem, Chélif plain. Ain Boucherit	No marine Calabrian. +50 m. Tilted 70° south Fossiliferous sands and gravels; clay lens; clay horizon; clay horizon Basal conglomerate. 30 cms.
PLIOCENE	Plaisancian-Astian sediments marls; sands and gravels.	PLIOCENE SEA	Plaisancian-Astian sediments marls; sands and gravels. (Chélif)	Plaisancian-Astian sediments marls; sands and gravels.

Fig. 1–2. Villafranchian stratigraphy in northwestern Africa. The silhouettes indicate some of the characteristic species, in particular faunal assemblages. (Compiled from various sources; see references.)

Fig. 1–3. Pre-Chelles-Acheul implements from the site of Ain Hanech (Algeria). (× ½) (Courtesy C. Arambourg.)

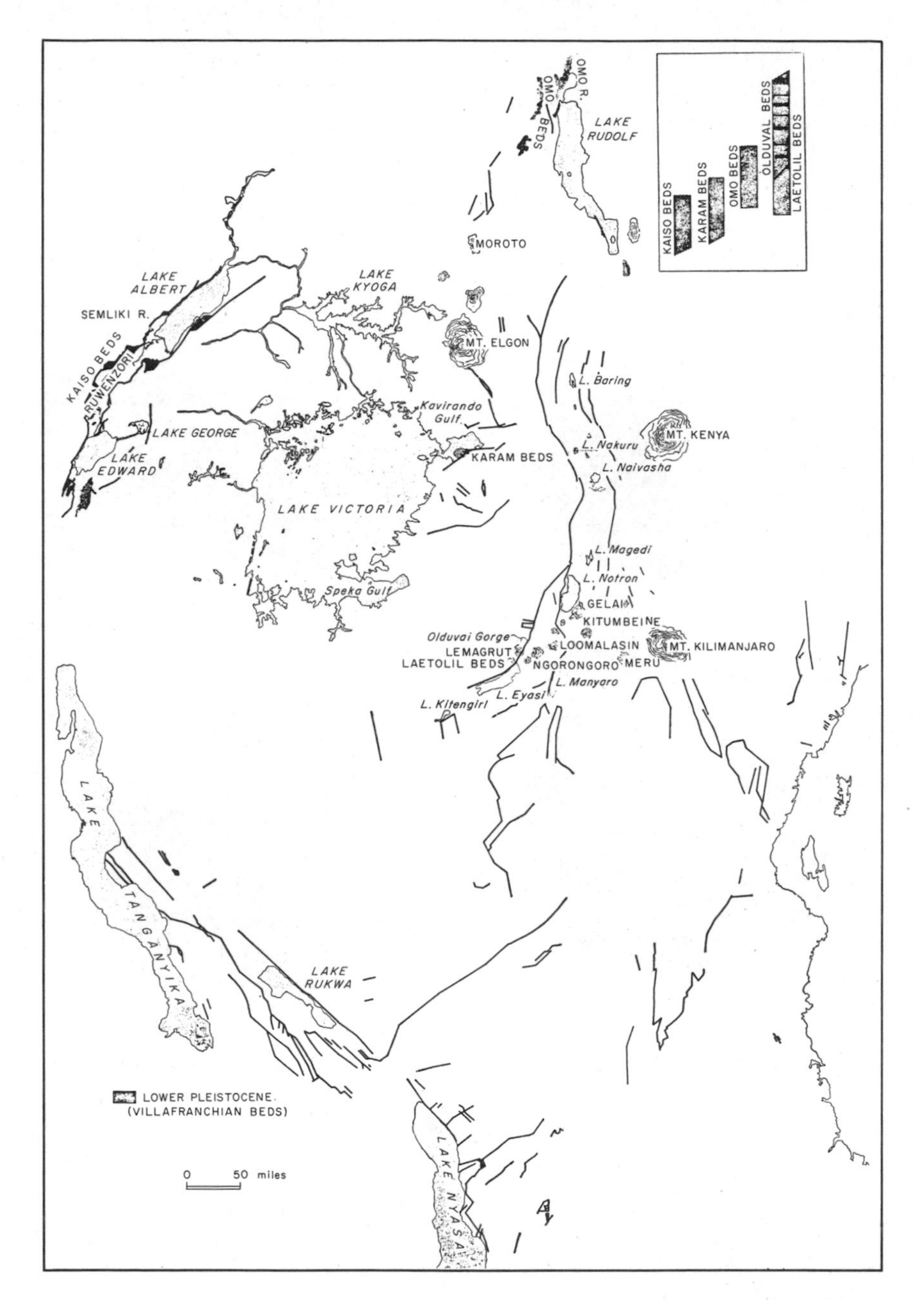

Fig. 1–4. The Rift Valley system and Villafranchian localities in eastern Africa. (Base map and major faults redrawn from the geological map of East Africa prepared on behalf of the Inter-Territorial Geographical Conference, 1952; stratigraphical inset from H. B. S. Cooke, *Annex. Trans. Geol. Soc. S. Africa* [1958].)

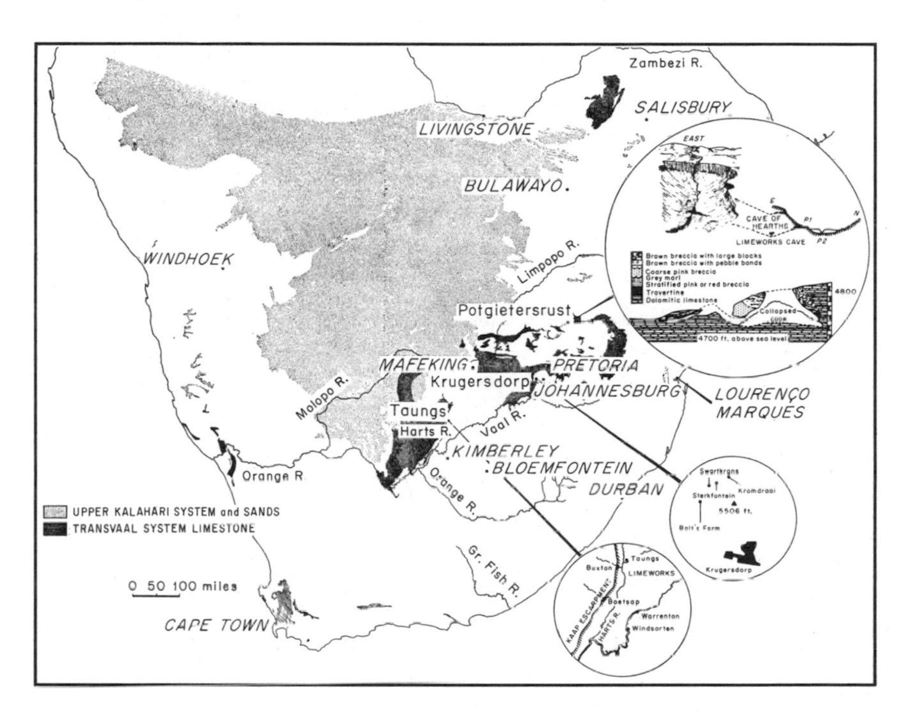

Fig. 1–5. Australopithecine localities in southern Africa. (Base map redrawn from the geological map of southern Africa in A. L. du Toit, *The Geology of South Africa,* 3rd ed., 1954; insets adapted from G. B. Barbour, H. B. S. Cooke, J. T. Robinson, and F. E. Peabody.)

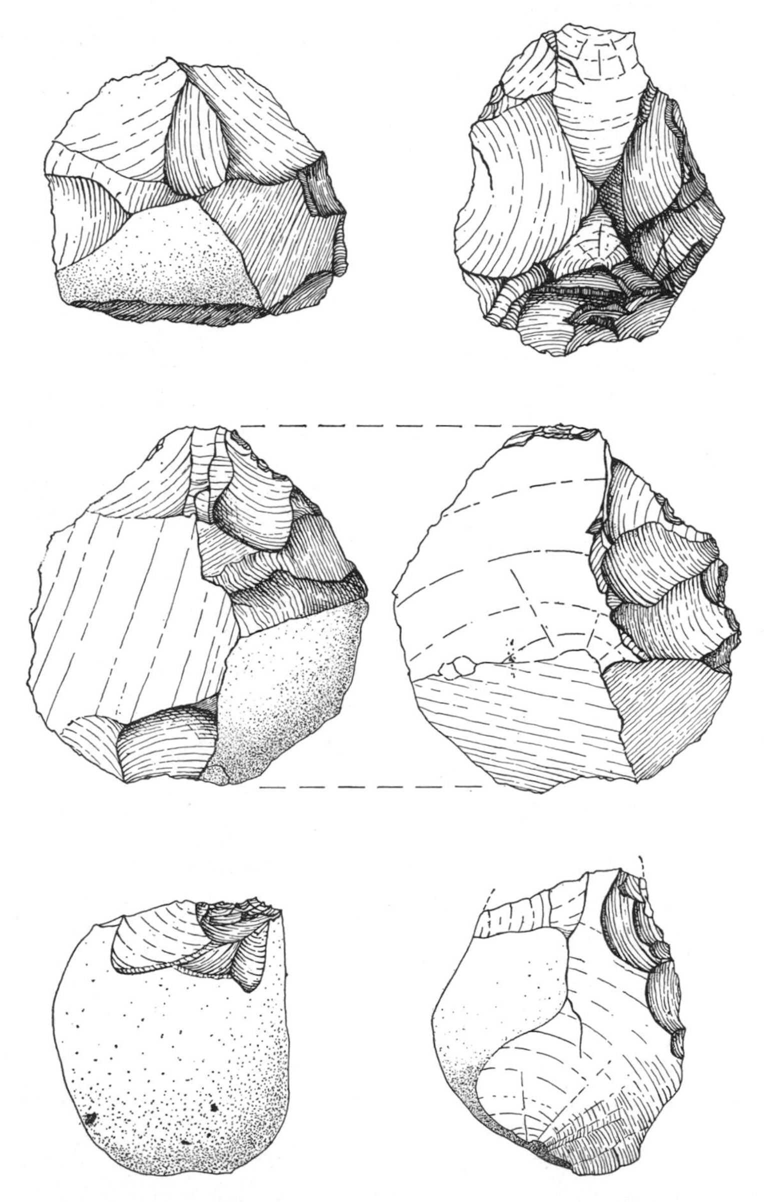

Fig. 1–6. Pre-Chelles-Acheul implements from the australopithecine-bearing breccias at Sterkfontein (Transvaal). (After J. T. Robinson and R. J. Mason, *Nature,* **180**, 521 [1957].)

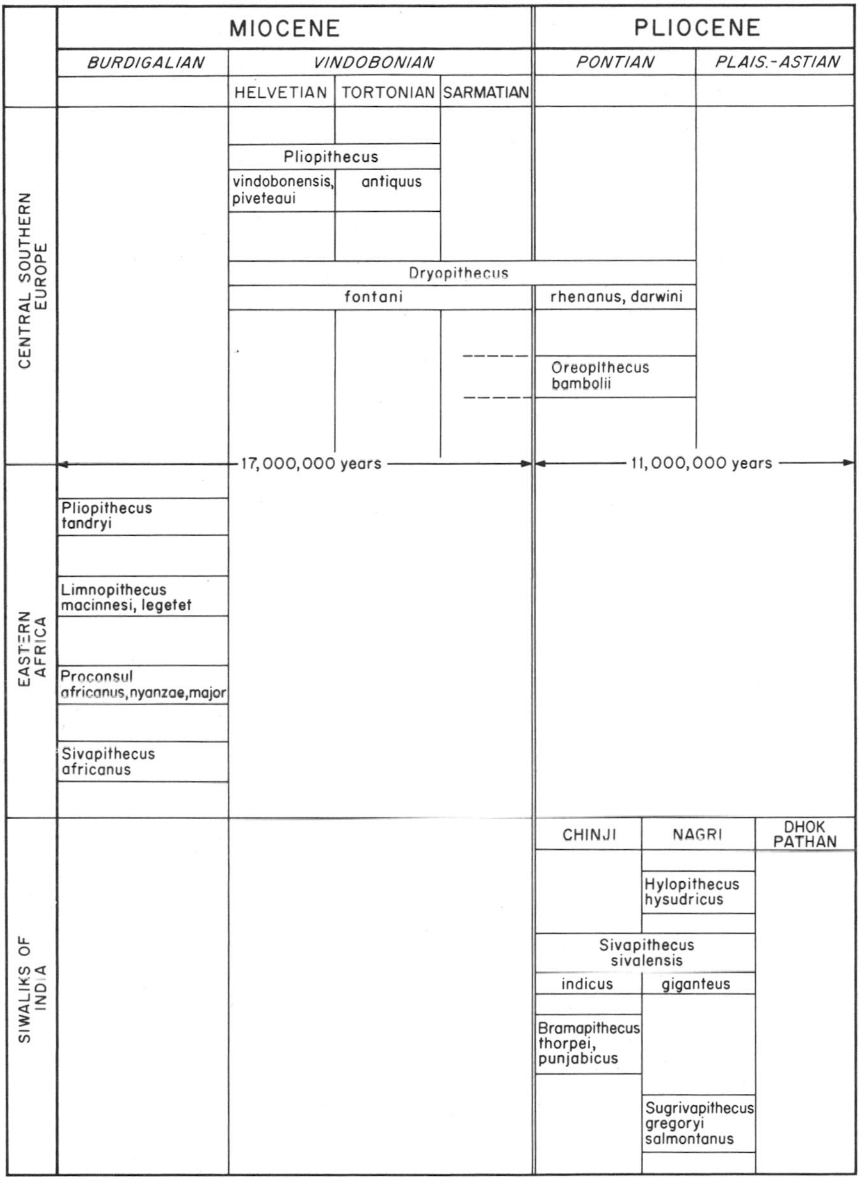

	MIOCENE				PLIOCENE	
	BURDIGALIAN	*VINDOBONIAN*			*PONTIAN*	*PLAIS.-ASTIAN*
		HELVETIAN	TORTONIAN	SARMATIAN		
CENTRAL SOUTHERN EUROPE		Pliopithecus				
		vindobonensis, piveteaui	antiquus			
		Dryopithecus				
		fontani			rhenanus, darwini	
				- - - - -	Oreopithecus bambolii	
	← 17,000,000 years →				← 11,000,000 years →	
EASTERN AFRICA	Pliopithecus tandryi					
	Limnopithecus macinnesi, legetet					
	Proconsul africanus, nyanzae, major					
	Sivapithecus africanus					

SIWALIKS OF INDIA	CHINJI	NAGRI	DHOK PATHAN
		Hylopithecus hysudricus	
	Sivapithecus sivalensis		
	indicus	giganteus	
	Bramapithecus thorpei, punjabicus		
		Sugrivapithecus gregoryi salmontanus	

Fig. 1–7. Temporal distribution of hominoids in the Neogene of Africa, Asia, and Europe.

Chapter 2

Mousterian Cultures in France

François Bordes

One of the stages in the history of mankind of much popular interest is that linked with Neanderthal man. The imagination is caught by this being, close enough to *Homo sapiens* to be called a man, but distant enough, in shape as well as in time, to appear in a way an "alien" in the sense in which the word is used by science-fiction writers. Hence, many misconceptions are to be found in popular books, even textbooks, the most common being the one about the "brutish Neanderthal." Reconstructions show him as only a little better off than the big apes, and his tools (Mousterian) are described as "crude" by people who would not, to save their lives, be able to make them. The truth is, indeed, quite different.

In the long story of man's evolution, the people of the Mousterian are very near us. If we suppose that the first to make tools, thus bridging the gap between apes and man, were the Australopithecinae, we must put them very early—at least 600,000, perhaps 1,500,000 years ago. On the other hand, the last of the Mousterian peoples lived about 35,000 to 40,000 years ago (in France) and probably much later in Africa. Their culture is also much closer to that of the first Upper Palaeolithic men than is generally believed, and the "gap" between Middle and Upper Palaeolithic does not exist. It was merely the result of insufficient data.

First, let us have a look at the geological, chronological, and ecological setting of Mousterian man. The Pleistocene period witnessed, four times at least, the advance of huge sheets of ice in northern Europe and in the mountains of central Europe. These glaciations are known in Europe under the names of Günz, Mindel, Riss, and Würm (from the oldest to the latest). The exact correspondence with the North American glaciations is yet to be worked out. Within each of these glaciations were periods of less severe climate, the interstadials. Between the glaciations were long periods of fairly warm conditions, the interglacials. There are no special names for these European interglacials, as there are for interglacials in the United States; we identify them in terms of the glacial period which preceded and of that which followed them.

The roots of some Mousterian cultures can be found during the Riss glaciation in the Clactonian complex, an assemblage of cultures without hand axes. The roots of others are found in the Acheulean complex. During the Last Interglacial (Riss/Würm), there are already some assemblages which can be called Mousterian, but it is in the Last Glacial period, the Würm, that the true Mousterian evolved. This Last Glacial period is divided into substages, the number of which is given variously by different authors. Some authors give two subdivisions, others three, but now, in France, most Pleistocene geologists agree that there were four: Würm I, II, III, and IV. However, in central Europe only three are usually recognized. Hence, Würm I and II of the French classification correspond to Würm I of the German classification. We need not bother about who is right, for the question has little bearing on the subject of this discussion.

Sites and Climate

As a general rule, Mousterian peoples in France lived in two different environments—caves and rock shelters in the valleys of southern and eastern France and open-air sites on the great loessic plains of the northwest and the plateaus of the south. This led to differences in their ways of life and had certain influences on assemblages of tools. Except, possibly, during the interstadial, Mousterian man lived under cold climatic conditions. The climate at the beginning of the Last Glacial was probably more damp than truly cold, but in due course the average yearly temperature fell far lower than that of today in the same areas. (However, we must not make the mistake of imagining the Mousterian environment to have been the barren ground or even the tundra of present-day northern countries. The latitude was the same as today, and the summers must have been fairly long and warm.) As a consequence of the severe cold, the limestone of the shelters flaked off; the cultural remains were, accordingly, covered with congelifracts when the shelter was deserted by man, so the different layers are separated and give a good basis for the determination of stratigraphy.

The fauna was a "cold" fauna—reindeer, arctic fox, and arctic hare—but there was a great abundance of big game: bison, wild oxen, horses, several kinds of deer, mammoth, rhinoceros, and so on. The carnivores included bears, lions, panthers, hyenas, and wolves.

Typological Subdivisions

For a long time it was thought that the Mousterian was a single culture which evolved in the following fashion. First, logically following the Acheulean, came the Mousterian of Acheulean tradition, with many hand axes. Then came the Typical Mousterian, without hand axes. Then came the Evolved Mousterian, without hand axes and with special types of scrapers (as in La Quina, Charente, southwestern France). Doubt was first cast on the validity of this straightforward scheme by the great French prehistorian Denis Peyrony. In the lower shelter at the classic site of Le Moustier (Dordogne, southwestern France), he found several layers of Mousterian of Acheulean tradition *between* two layers of Typical Mousterian. The picture has been further complicated as a result of my own studies of Mousterian assemblages, made by statistical methods. It is now known that the term *Mousterian* encompasses a complex of cultural groups, some closely related, others not so closely related or even of different origin. The following different groups are recognized.

1. *Mousterian of Acheulean tradition.* Contrary to what is often stated, this is not always an "old" Mousterian, and it can be found in Würm II as well as in Würm I. But an evolution is clearly to be seen. First there is a phase (type A) in which there are numerous hand axes (from 8 to 40 per cent of the artifacts), among which are triangular forms (Fig. 2–1, No. 14), together with heart-shaped axes (cordiforms) (No. 12). These are associated with fairly numerous side scrapers (from 20 to 40 per cent) (No. 2), denticulate tools (about 10 to 15 per cent) (No. 7), some points (No. 1), and knives made on a flake on which one of the edges has been blunted by abrupt retouch (No. 5). The side scrapers, generally flat, are of several types. The backed knives, together with the hand axes, are typical of this Mousterian of Acheulean tradition. There are also other tools, some of Upper Palaeolithic type, such as end scrapers (Nos. 3 and 9), gravers (No. 8), borers (Nos. 10 and 11), and truncated flakes or blades (No. 6).

After the Würm I/Würm II interstadial, this Mousterian of Acheulean tradition passes into an evolved phase (type B). The hand axes are far less numerous than in the type A phase (seldom more than 4 or 5 per cent), and none is triangular. There is a corresponding drop in the number of side scrapers (down to 4 to 10 per cent). The number of knives increases (sometimes up to 20 per cent), and these are more elongated, being made on blades as well as on flakes (Fig. 2–1, No. 4) and suggesting the Chatelperron knife of the early Upper Palaeolithic

(Périgordian I). There is a similar increase in the number of denticulate tools (up 25 per cent and more). The flaking technique is more lamellar, and even little blades appear, as is shown by the appearance of bladelet cores (No. 13).

It is, in fact, very difficult to distinguish between a very late Mousterian of Acheulean tradition and an early Périgordian I. From the point of view of tools in the assemblage there is really no distinction between the two; only the percentages are different.

At the beginning, Périgordian I has a basic tool kit of Mousterian tools, with, however, more blades, more burins and end scrapers, and a special type of backed knife, the Chatelperron knife, which, even if foreshadowed by the Mousterian backed knives, is generally more elongated. Mousterian types of backed knives do survive. Very quickly, this Périgordian I evolves toward the Upper Périgordian (or Gravettian), losing the Mousterian types of tools, developing end scrapers and many types of burins; and the backs of the knives, instead of being curved as in the Mousterian or the Chatelperron types, tend to be straight. At the same time, bone tools are more numerous and better made.

2. *Typical Mousterian* (Fig. 2–2, Nos. 1–7). The Typical Mousterian at first seems very much like the Mousterian of Acheulean tradition (type A). There are side scrapers (rather flat) (25 to 55 per cent), some denticulates and notched tools, and well-made points. But there are few or no hand axes or backed knives (0.5 per cent at most). Some thick side scrapers of La Quina type (see below) are to be found (as in the Mousterian of Acheulean tradition), but these are so few (at most, 1 per cent) as probably not to be truly significant. (When one undertakes to make a side scraper on a thick flake, the odds are good that he will make a Quina-like tool without trying to.)

So, the main difference between Mousterian of Acheulean tradition (type A) and Typical Mousterian lies in the fact that hand axes and knives are found in the former and not in the latter. This is a truly significant difference.

3. *Denticulate Mousterian* (Fig. 2–2, Nos. 8–17). In this group there are no hand axes (at least typical ones) or backed knives. There are few or no points and very few scrapers. If you are willing to call any flake with some working on the edge a "scraper," the figure may be as high as 13 per cent, but if you insist on a narrow definition of the scraper, it is very unlikely that the figure will be more than 3 to 7 per cent. But there are notched tools and denticulate tools galore. In some layers the two types, taken together, comprise nearly 80 per cent of the assemblage; side scrapers, end scrapers, burins, borers, and so on, constitute the remainder.

4. *Quina-type Mousterian* (Fig. 2–3). Here the picture is definitely

different. Few or no hand axes and backed knives are found, but there are many more side scrapers (up to 75 per cent or more), often magnificent ones. And among them, side by side with the ordinary types, are special ones—scrapers of the Quina type. These are made on thick flakes, usually have a convex working edge, and have a special type of retouch, like the overlapping scales of a fish (Fig. 2–3, No. 1). Such scrapers may be either side scrapers or scrapers of the transverse variety, with the edge opposite the butt of the flake (No. 8). There are also bifacial scrapers (No. 3), not to be mistaken for hand axes. On one side the flaking is very shallow and flat; on the other there is scalar retouch. One edge is often left unworked, or is more crudely worked than the other, but sometimes the two edges are retouched equally well, and then it is easy to mistake these tools for hand axes. There are also some denticulates, some notched tools, and some burins, borers, and end scrapers. The end scrapers may be either of the carinate or of the nosed types (Fig. 2–3, No. 7). Another tool found in relative abundance is the "limace" (shaped like a slug) (No. 4). This one is not unknown, however, in the Typical Mousterian.

The relative proportions of Quina-type scrapers (with respect to scrapers in general) in the different Mousterian groups are summarized in Table 2–1.

Levallois Technique

Such are the four main typological subdivisions of the Mousterian. Cutting across this typological division is a technical division. In the Middle Palaeolithic there was a special method of producing flakes, called the Levallois technique after a suburb of Paris where this special type of flaking was first found, in the last century. To make a Levallois flake you take a flint nodule (Fig. 2–4, No. 1) and flake it off all around the margins (No. 2). Then, using each of these flake scars in turn as a striking platform, you flake away the upper surface of the nodule (nos. 3 and 4). Then you prepare a special striking platform (either plain or faceted) at one end and you strike off a large flat flake (No. 5); the shape of the flake is predetermined by the previous shaping of the core (No. 6).

This technique, developed in Europe by the peoples of the Middle Acheulean, was used until the end of the Mousterian and even later. It was not used by every Mousterian tribe, however. For instance, there is a Mousterian of Acheulean tradition with Levallois flaking (as in Le Moustier) and there is another with little such flaking (as in the Pech de l'Azé Cave, Dordogne). The same is true for the Typical Mousterian and for the Denticulate Mousterian. But in the case of the Quina-type

Mousterian, things are a little more complicated. Assemblages of the Quina-type Mousterian that show use of the Levallois technique are known as Ferrassie-type Mousterian (from La Ferrassie, Dordogne). The use of the Levallois technique has the effect of lowering the percentage of Quina-type scrapers, since Levallois flakes are flat, and Quina-type scrapers can only be made on thick flakes (Table 2–1). Also,

TABLE 2–1.

Relative proportions of Quina-type scrapers (with respect to scrapers in general) in the various cultural groups of the Mousterian.

CULTURAL GROUP	PROPORTION (%)
Mousterian of Acheulean tradition	0 to 0.9
Denticulate Mousterian	0 to 0.1
Typical Mousterian	0 to 1
Quina-type Mousterian	15 to 25
Ferrassie-type Mousterian	6 to 9

in the Ferrassie-type Mousterian there are fewer transverse scrapers than in the Quina-type Mousterian, for Levallois flakes, being rather elongate, are not suitable for use as transverse scrapers.

The existence of a special "culture," the Levalloisian, characterized by the existence of numerous unretouched Levallois flakes, has been proposed by some prehistorians. Actually, such findings represent only a facies of various Mousterian groups, found in places where flint was plentiful and readily available—mainly in open-air sites where a nomadic way of life prevailed.

The question arises, of course, whether the different types of Mousterian represent distinctive cultures, or whether, like the Levalloisian, they are merely facies of the same culture. They might be the result of cultural evolution or only of seasonal variations. This question probably can be clearly answered in the present stage of our knowledge.

Hypothesis of Seasonal Variation

The hypothesis of seasonal variation seems unacceptable for several reasons. It is difficult to accept the idea that the peoples of the Mousterian changed their tool assemblage four times a year, according to season. Moreover, the thickness of occupation layers in the caves and shelters argues against a one-season stay. Each layer indicates a stay of considerable length. One might suppose that there were spring, summer, autumn, and winter caves, occupied only at a particular time of year, but it is difficult to imagine the existence of a kind of convention among all the Mousterian tribes, governing the use of a cave, assuring that a given cave would be kept as a "spring cave" and that no summer or winter cultural material would be mixed with the spring tools. Moreover, we have very good reason to think that these caves were occupied all year round. It is possible to tell, from a study of a reindeer's antlers and teeth, how old the animal was when it was killed, and as a consequence, since the typical birth season of the reindeer is known, we can tell at what time of year a reindeer was killed. It appears that reindeer were killed at all times of the year by the occupants of these caves—proof that man occupied the caves all year round.

Hypothesis of Cultural Evolution

As I said earlier, doubt was cast long ago on the validity of the hypothesis of cultural evolution when Denis Peyrony excavated the lower shelter at Le Moustier and found three layers of Mousterian of Acheulean tradition between two layers of Typical Mousterian. Actually, the sequence at Le Moustier is even more complex than this (see Fig. 2–5). The sequence is substantiated and the evidence is supplemented by findings from other caves and shelters—for instance, the cave at Combe-Grenal, near Domme (Dordogne), in which I have been making excavations since 1953.

The cultural sequence in the Combe-Grenal Cave, from top to bottom, is as follows (Fig. 2–6): (1) Mousterian of Acheulean tradition (here in terminal position); (2) several layers of unidentifiable Mousterian (*A*2, *A*3, *B*1, *B*2); (3) two layers of Typical Mousterian with Levallois flaking (*B*3 and *B*4); (4) a poor layer (*C*), which is probably Denticulate Mousterian; (5) layer *D*1, probably Denticulate Mousterian with Levallois flaking; (6) layer *D*2, probably Typical Mousterian with Levallois flaking; (7) two layers (*E*1 and *E*2) of Denticulate Mousterian, the first with Levallois flaking, the other with less Levallois flaking; (8) three layers of Denticulate Mousterian (*F*, *G*, *H*1) without Levallois flaking; (9) layer *H*2, also Denticulate Mousterian but with Levallois flaking; (10) layer *I*, Quina-type Mousterian; (11) layer *J*, Denticulate Mousterian without Levallois flaking; (12) layers *K*, *L*, *M*, *N*, *N*1, Quina-type Mousterian; (13) layer *P*, Ferrassie-type Mousterian; (14) layers *Q*, *R*, *R*1, Typical Mousterian with Levallois flaking; (15) layers *U*1, *U*2, *W*, *X*, *Y*, all Ferraissie-type Mousterian; (16) layer *Z*, Typical Mousterian with Levallois flaking; and (17) layer α, Denticulate Mousterian with Levallois flaking. (It should be understood that "without Levallois" flaking does not mean that there are no Levallois flakes in the assemblage but means only that there are very few.) Under layer α are several other layers only now being excavated. Most of them belong to the Typical Mousterian, except the lowermost which can be late Acheulean.

So, at Combe-Grenal we can clearly see, interstratified, almost all types of Mousterian. Elsewhere in the Dordogne, as at Combe-Capelle (lower site), Quina-type Mousterian lies below Ferrassie-type Mousterian, not above it as at Combe-Grenal. At Combe-Capelle, too, the Mousterian of Acheulean tradition is in a terminal position. However, in the upper shelter at Le Moustier the Quina-type Mousterian lies above the Mousterian of Acheulean tradition.

Another hypothesis links these variations in tool assemblages to the environment. But it is easy to show that Mousterian people who lived under very different environmental conditions had the same type of tool assemblage. The site of Aïn Meterchem, in Tunisia, yields a Mousterian assemblage which is very close to that found at La Ferrassie. On the other hand, in the same geological layer, and thus representative of people living under very similar climatic conditions, one sometimes finds two very different cultural horizons, as at Pech de l'Azé, Cave II (Dordogne).

So, the existence of different cultures within the Mousterian complex appears to be an established fact. The question then arises, what were the origins of these cultures?

Cultural Origins

The Mousterian of Acheulean tradition poses no serious problem. It is logically derived from the Upper Acheulean. The only question is whether it passed through an intermediate Micoquian stage or whether it evolved directly, the Micoquian being in part contemporaneous with the Old Mousterian of Acheulean tradition. The latter theory would not necessarily exclude the possibility of a Micoquian stage as well. The Quina-type Mousterian has a possible antecedent in the so-called Tayacian assemblage of layer 3 (of Riss glacial age) at La Micoque near Les Eyzies (Dordogne) or in the High Lodge type of Clactonian industry in England. The Ferrassie-type Mousterian might have its roots in layer 4 (Riss glacial age) of La Micoque, or in cultures like that found at Ehringsdorf (near Weimar, Germany), which dates from the Last Interglacial.

But when one comes to consider the origin of Denticulate Mousterian or Typical Mousterian, one is almost at a loss. Of course, some cultural horizons at La Micoque might be antecedents of the former, but unhappily these layers have been so thoroughly crushed by frost heaving (cryoturbation) that it is difficult to differentiate between true denticulate tools and other artifacts. As yet, the origins of the Denticulate Mousterian are not known, but Denticulate Mousterian is present at the very beginning of the Last Glacial and extends to the end of the Mousterian.

The Typical Mousterian is also a problem. It closely resembles the Mousterian of Acheulean tradition (type A) but lacks hand axes and backed knives. However, since some of the Mousterian of Acheulean tradition is fairly poor in hand axes and knives, it is possible that Typical Mousterian does derive from it. There is some indication that the Rissian levels found by Dr. E. Bonifay at the cave of Rigabe, southeastern France, could be counted among the ancestors of Typical Mousterian.

What happened to these different Mousterian cultures in Würm III? There is little question that the Mousterian of Acheulean tradition must have developed into the Périgordian I. This poses a nice problem for those anthropologists who believe that all the peoples of the Mousterian in France were of the "western Neanderthal" variety, and hence that these strains were dead ends from the standpoint of evolution. Actually, very little is known about the peoples who made the Mousterian hand axes. The skull of a young child from Pech de l'Azé Cave (a child so young that the skull tells us little) and some cranial fragments from other sites are all that we have. However, there is no other possible origin, culturally speaking, for the early Périgordian, and physical anthropologists must accept this fact. The early Périgordian is a Western culture and is unknown outside of France and Spain (and perhaps Germany). Distribution maps of the early Périgordian and of the Mousterian of Acheulean tradition are very similar except for a small zone of the latter in the Middle East (Palestine).

It was pointed out long ago that there are resemblances between the Quina-type Mousterian and another stage of the early Upper Palaeolithic, the Aurignacian. In fact most of the tool types characteristic of the Aurignacian are in a way foreshadowed in the Quina-type Mousterian, just as the Périgordian I tools are foreshadowed in the Mousterian of Acheulean tradition. Carinate and nosed scrapers (Fig. 2–3, Nos. 2 and 7) exist in the Quina-type Mousterian. Actually, such tools are present as early as the "Tayacian" in the layers at La Micoque. The special Quina retouch (see Fig. 2–3) is very close to the kind of retouch used by the Aurignacians on the side of their blades, and some double side scrapers on blade-like flakes might be regarded as forerunners of the more elongate "Aurignacian blade" (Fig. 2–3, No. 9).

But these affinities are less marked than those between the Mousterian of Acheulean tradition and the Lower Périgordian. Distribution for the Aurignacian is much wider than for the Lower Périgordian. The evidence suggests that the peoples of the Aurignacian invaded the West, bringing with them a well-developed culture. If there is a bridge between the Quina and Ferrassie types of Mousterian and the Aurignacian, it is to be found outside of France.

There is some slight evidence that Ferrassie-type Mousterian survived for a short time into Würm III in Provence. No trace has been found in Upper Palaeolithic cultures of derivatives from either Typical Mousterian or Denticulate Mousterian.

Ways of Life

What do we know about possible differences in the ways of life of these several groups within the Mousterian? As yet, very little. Until recently, little attention has been given, in excavating, to such matters as the horizontal distribution of tool types in a shelter. Modern excavations, in France and elsewhere, are not yet numerous enough to give any clear answers. In any case, to obtain such information is no easy task. Mousterian layers are usually very rich and contain a great mixture of broken bones, tools, chips, and flakes, together with congelifracts, pebbles, ashes, and sometimes charcoal. Even the most careful excavation will not always reveal, for instance, whether some beautiful scrapers in a given grid square *A*3 are exactly contemporary with a hand ax in grid square *E*5. In fact, we can seldom achieve this degree of certainty. And to make distribution graphs for layers more than one or two centimeters thick would lead us exactly nowhere. In a thick layer, the only appropriate parts for such studies are the bottom and the top; the bottom is better if the layer overlies another layer that is compact and sterile.

However, fireplaces can often be found, and rough distribution maps for some tools can be worked out. In Pech de l'Azé (Cave I), for instance, we found that the hand axes and scrapers were more numerous near the fireplaces. There seems to have been a special place outside for the making of hand axes, for most of the finds there are of flakes such as would be struck off in that process. In the same site, the lower (and richest) layer ended against a low stone wall, just a little outside the cave and the adjacent shelter.

At Combe-Grenal we found a posthole in layer *G*, extending down to layer *K*. A cast made of this hole shows quite clearly that the post must have been a pointed wooden shaft driven into the ground, and that the tip mushroomed against a stone in layer *K* (Fig. 2–7). This shaft was perhaps one of a row of shafts used to support skins or woven branches to close the cave.

For half a century it has been known that the Mousterian peoples buried at least some of their dead. Most of these burials seem to be associated with the Quina-type and Ferrassie-type Mousterian cultures;

however, some are from the Typical Mousterian. We also know that in every Mousterian context mineral color (manganese dioxide, red ocher) was used. Some bits have been scratched to make colored powder; others are pointed, like pencils (Fig. 2–8). These colors presumably served for body painting; there is no cave painting that can be traced back to the Mousterian.

Some hunting habits can also be deduced from the animal remains in the layers. Thus, it seems that peoples of the Denticulate Mousterian favored horses as game. Layers from this culture are everywhere rich in horse bones. At Pech de l'Azé (Cave II), in the same geological layer and very close together, there are two cultural levels; one is Denticulate Mousterian, and the other is Typical Mousterian. There is no indication of a long lapse of time or of any significant climatic change between the two. The animal remains from the former are mainly horse, and those from the latter are red deer and wild oxen.

Conclusions

All this gives a picture of life in Mousterian times that is rather different from the picture, too often given, of brutish half-men, crouching in caves, terrified and not very clever. It seems that Mousterian tribes may, at times, have numbered some thirty to fifty individuals. These people had weapons to combat cave lions and cave bears, whose remains are sometimes found in the occupation layers. They used paints, were clever flint workers, and buried their dead. It is obvious that they did not lack inventive powers.

Most of the stone tools which were developed in Upper Palaeolithic times by *Homo sapiens* were invented by Mousterian or even by Acheulean peoples. The blade (that is, a blade made through a special technique of *débitage* and not the result of a flaking accident) goes back at least to the end of the Acheulean, and in some Mousterian assemblages blades comprise up to 40 per cent of the *débitage*. End scrapers and burins were known in the Middle Acheulean. The backed knife is an Acheulean invention also. But if all these tools already existed in the Acheulean, they were further developed and diversified in the Mousterian. Even the multiple tool is found in the Mousterian (Fig. 2–1, No. 8); some complex tools—for instance, a burin combined with an end scraper—are also found, but rarely.

The peoples of the Mousterian also experimented with bone tools, but there they fall very short of the achievements of men in the Upper Palaeolithic. They never did more than make some bone spear points, and in the main they used only bone splinters, shaping them crudely. But in this respect the first people of the Upper Palaeolithic (Périgordian I) appear not to have done much better.

To conclude, it does not seem that, culturally at least, there is any great gap between the Mousterian cultures and the early Upper Palaeolithic cultures that followed. At least one of the latter has its roots quite clearly in the Mousterian of Acheulean tradition. And even if some anthropologists deny to Neanderthal man (*sensu strictu*) the right to be counted among our direct ancestors, one thing is sure: these ancestors of ours were at a cultural level very like that of the Mousterian peoples. So we come uncomfortably close to the old joke: It was not William Shakespeare who wrote *Hamlet* but another man who lived at the same time and whose name was also William Shakespeare!

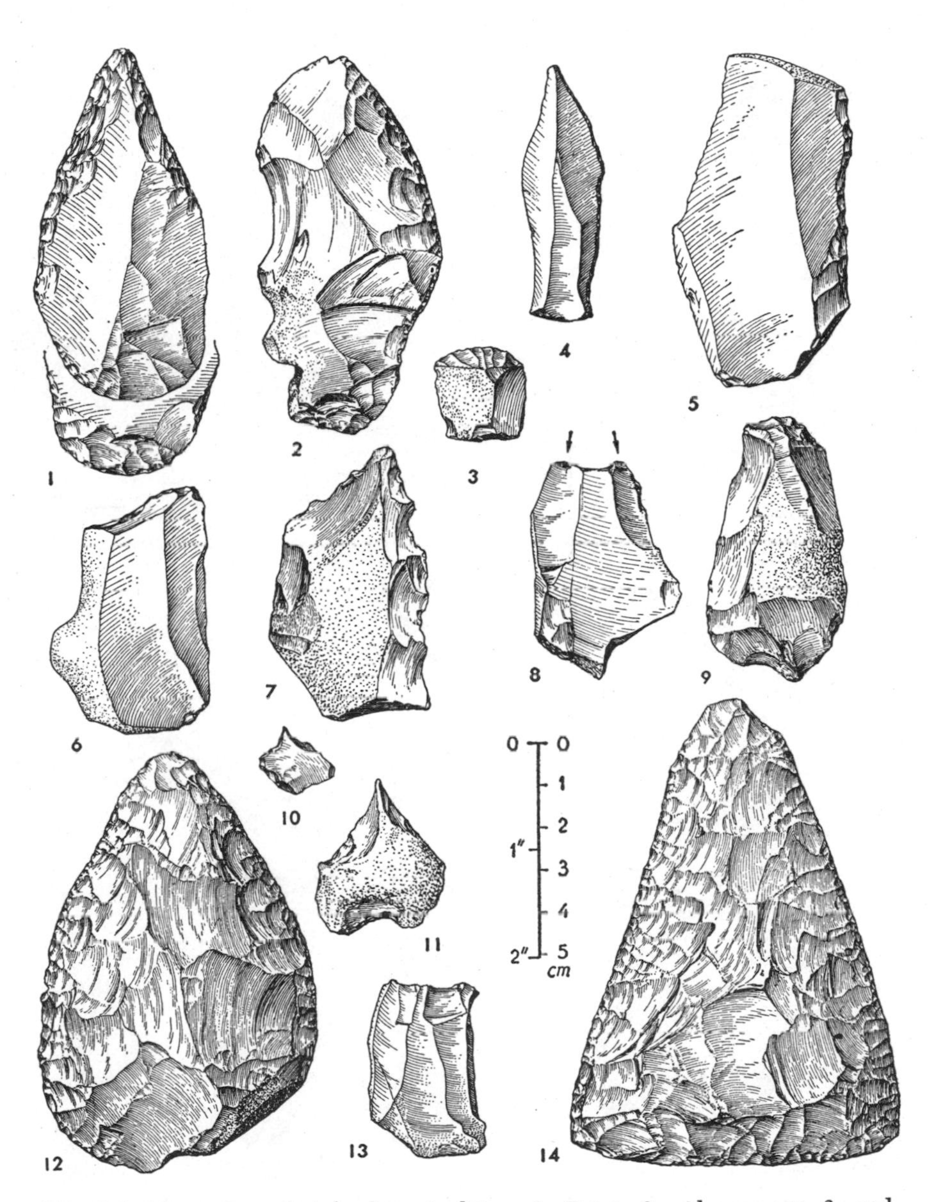

Fig. 2–1. Mousterian of Acheulean tradition. 1, Point; 2, side scraper; 3, end scraper; 5, backed knife; 7, denticulate tool; 10, 11, borers; 12, cordiform hand ax; 14, triangular hand ax. Evolved Mousterian of Acheulean tradition: 4, backed knife; 6, truncated flake; 8, double burin; 9, end scraper; 13, bladelet core.

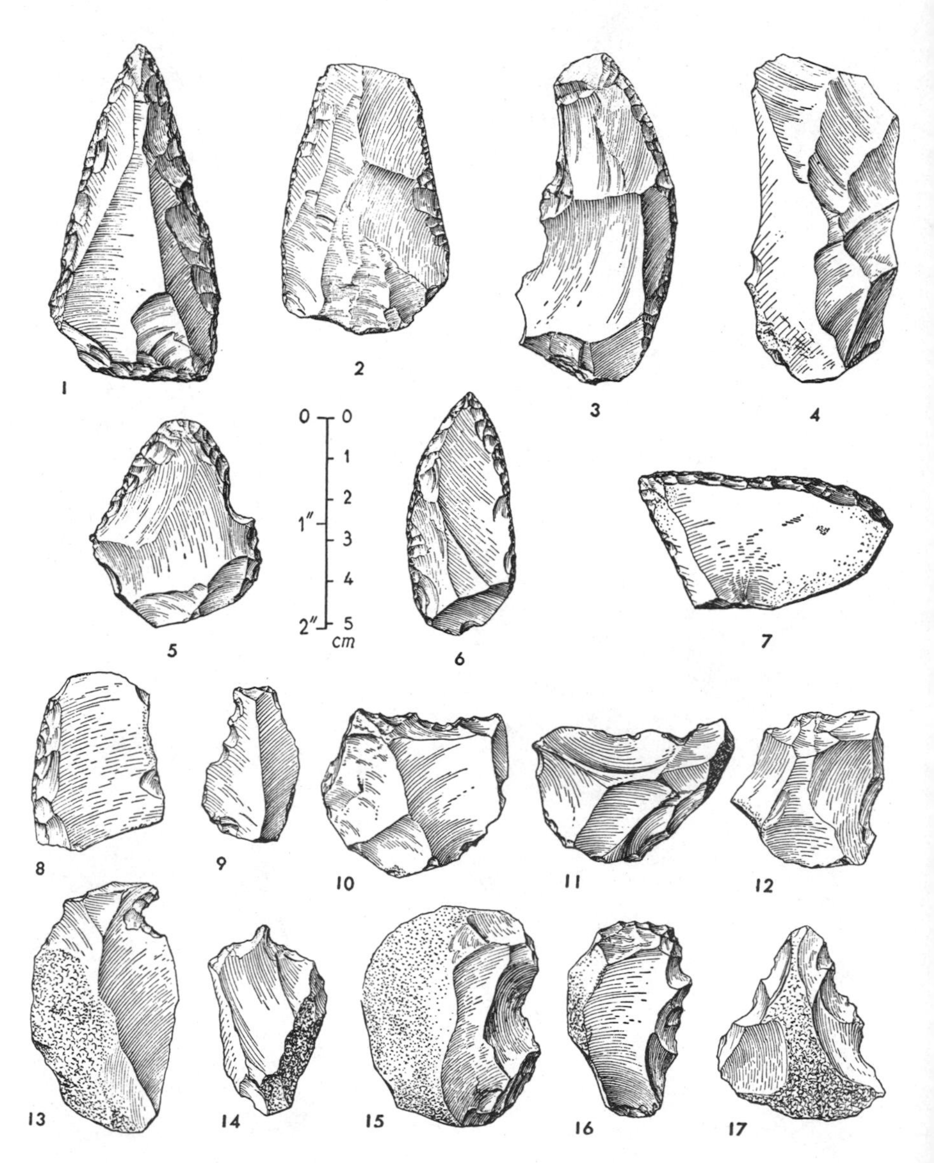

Fig. 2–2. Typical Mousterian. 1, Point; 2, double side scraper; 3, side scraper; 4, Levallois flake; 5, end scraper on a flake; 6, point; 7, transverse scraper. Denticulate Mousterian: 8, side scraper; 9, 10, denticulate tools; 11–13, notches; 14, borer; 15, notch; 16, 17, denticulate tools.

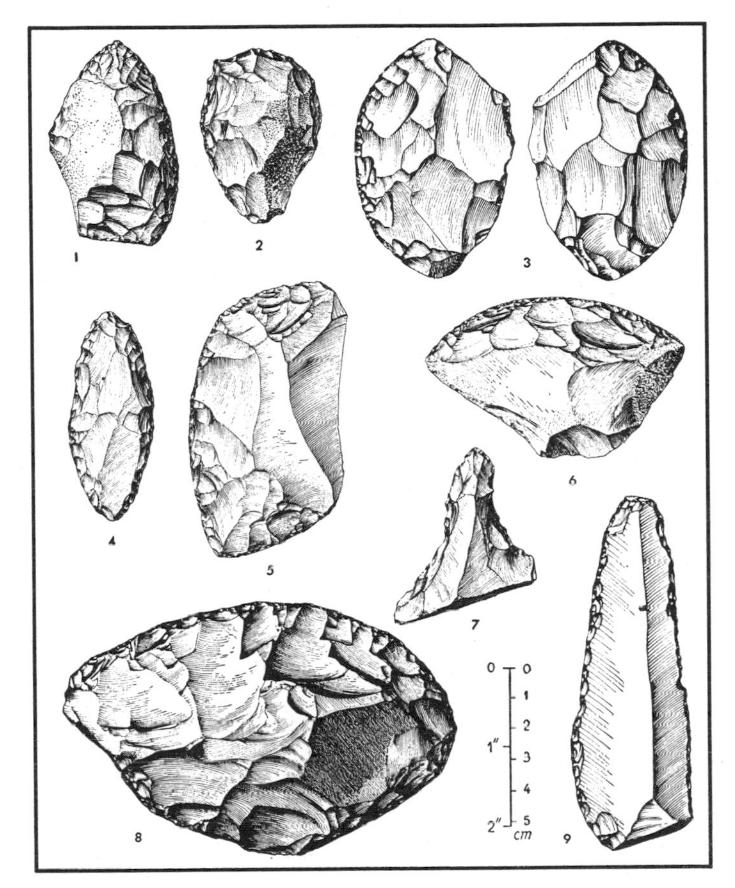

Fig. 2–3. Quina-type Mousterian. 1, Side scraper, Quina type; 2, end scraper; 3, bifacial scraper; 4, "limace"; 5, side scraper; 6, transverse scraper, Quina type; 7, nosed end scraper (Aurignacian type); 8, transverse scraper, Quina type; 9, retouched blade (Aurignacian type).

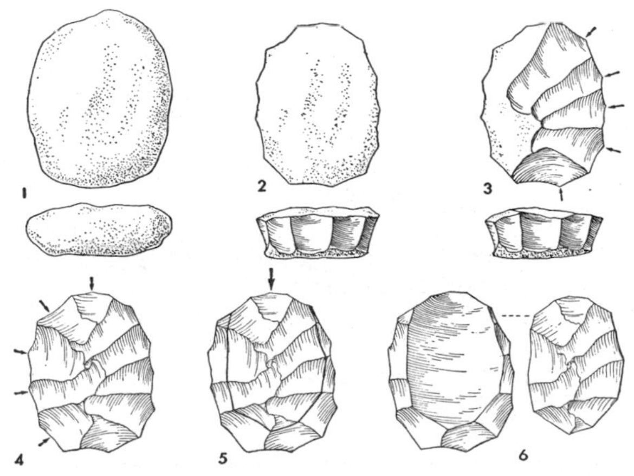

Fig. 2–4. Steps in the making of a Levallois flake.

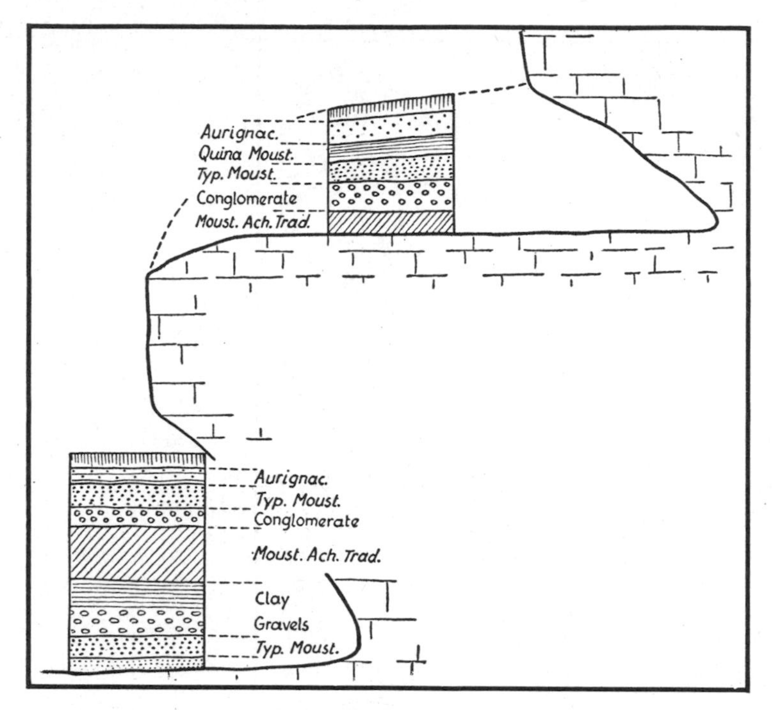

Fig. 2–5. Schematic section at Le Moustier (Dordogne), upper and lower shelters, showing the interstratification of several types of Mousterian.

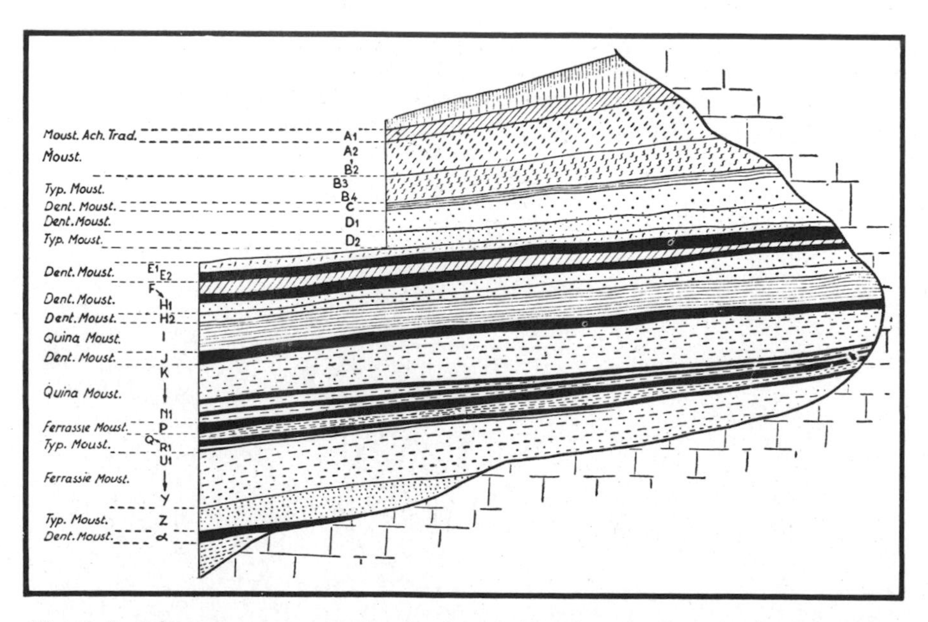

Fig. 2–6. Schematic section at Combe-Grenal (Dordogne), showing the interstratification of several types of Mousterian.

Fig. 2–7. Cast of a posthole at Combe-Grenal, layer *G* (length of the cast, 21 centimeters).

Fig. 2–8. Fragments of manganese dioxide from the Mousterian-of-Acheulean-tradition level of Pech de l'Azé (Dordogne). (Top row, from left) unworked lump, triangular pencil, trapezoidal pencil, rounded pencil; (bottom row) scratched lumps.

Chapter 3

Prehistory in Shanidar Valley, Northern Iraq

Ralph S. Solecki

The archaeological investigations of two sites in Shanidar Valley, northern Iraq (figs. 3–1 and 3–2), have been made more significant through the use of interdisciplinary studies. The combined information provides concrete data regarding man and his environment in this region from the Middle Palaeolithic age (perhaps 100,000 years ago) to the present.

The significance of the Shanidar Valley investigations is that here, in this one locality, there is an almost continuous sequence of human history dating from the time of the Neanderthals. The information derived from these investigations contributes to biological, palaeontological, climatological, and geological studies, as well as archaeological and anthropological ones—the major concerns of the project. The Shanidar data do much to elucidate man's history in a most interesting period of his existence—the time of the Neanderthals and the replacement of this long dominant people by *Homo sapiens.*

The project is of further special interest because Shanidar lies within the area where domesticated plants and animals—the basis for the great Neolithic economic, social, and cultural revolution—appear to have been first developed. The Shanidar excavations provide data reflecting the effect on the people in this remote valley of the introduction of the new mode of living, which was dependent on the products of the fields and on tamed animals rather than exclusively on the hunt. The great alternations of climate and temperature that mark the Pleistocene, a recent geological period of the ice ages dating back more than a million years, are reflected in the cultural history.

The sites, Shanidar Cave and the nearby village site of Zawi Chemi Shanidar, have given us a long preface to Mesopotamian history. Thus far, the cultural sequence for Shanidar Valley is outlined on a relatively firm basis by carbon-14 dates from about 50,000 years ago, and by "guess dates" for periods before that. Palaeoclimatological inferences have been made on the basis of pollen remains and of trace elements in soil studies. Osteological materials from seven Neanderthals and 28 representatives of Post-Pleistocene *Homo sapiens* have been found. There is also a wealth of faunal data. The presence of domesticated animals in Shanidar Valley at the relatively early date of 8900 B.C. seems likely.[1]

The Sites

Shanidar Cave (Fig. 3–3) is situated at longitude 44°13′E, latitude 36°50′N, about 400 kilometers due north of Baghdad, within the outer folds of the Zagros Mountains. The cave, of limestone-solution origin, is about 2.5 kilometers from the Greater Zab River, a major tributary of the Tigris River. The precipitous mountains there reach an elevation greater than 1,900 meters. The region is relatively well wooded. There is still some wild game to be seen in the area.

The cave lies at a measured elevation of 765 meters, facing south. The mouth is about 25 meters wide and 8 meters high, and the cave extends about 40 meters to the rear, with a maximum width of about 53 meters. Its earthen floor is about 1,200 square meters in area. The cave is inhabited by several families of Kurdish shepherds during the winter months.

During the four seasons of excavation,[2] a series of cultural deposits nearly 14 meters deep were explored down to bedrock. The deposits consist of an easily dug loamy soil and material indicative of at least five major rockfalls. These rockfalls were very effective man-traps—apparently they caused the death of most of the Neanderthals so far found in the cave. From top to bottom the occupation sequence includes four major layers, arbitrarily labeled layers A, B, C, and D (Fig. 3–4). Layer B was divided subsequently into two parts, B1 and B2. There are cultural, stratigraphical, and chronological breaks between each of these layers, so far as can now be determined.

Layer A consists of extensive, multicolored, dry, and dusty ash beds, hearths, and black organic-stained soil. It includes remains of modern, historic, and Neolithic age. As deduced from observations of contemporary Kurdish herdsmen occupation at Shanidar Cave and neighboring caves, much of the heavy organic staining must be due to the droppings of livestock herded in the interior of the cave.

Layer B is somewhat thinner and markedly less heavily stained with organic matter than layer A. The two divisions of this layer, B1 and B2, are distinguishable from each other by soil coloration, artifact content, and carbon-14 dates. The upper part, B1, is Proto-Neolithic[3] and is dated at about 8650 B.C.[4] The lower part, B2, is Mesolithic (or very late Upper Palaeolithic) and is dated at about 10000 B.C.[5]

Shanidar B1 is contemporary with the basal layer of the Zawi Chemi Shanidar village site, which has a carbon-14 date of about 8900 B.C.[6] The artifact contents of cave and village layers are quite similar. In addition to bone artifacts and chipped stone implements, larger tools of ground stone, such as querns, mortars, and hand rubbers, were found. These indicate that some sort of vegetal foods, possibly acorns or even cereal grains, were prepared as part of the diet. The find, in the cave, of fragments of matting or basketry, the oldest yet known, suggests that collecting-baskets may have been used. Twenty-eight skeletons were associated with the B1 layer in Shanidar Cave; of these, twenty-six were found in a cemetery group. Associated with the cemetery were platforms of stones[7] and an arclike alignment of flat stones (Fig. 3–5).

There is evidence that the Proto-Neolithic people ranged far for manufacturing materials. Obsidian was brought in from the north, probably from the Lake Van region. A material which looks very much like bitumen was used as an adhesive;[8] bitumen is found more than one hundred miles to the south.

Several pieces of evidence suggest that the B1 peoples had a more assured food supply than their predecessors in the valley. First, a number of pits, which may have been food storage pits, were found intruding into the B2 layer (Fig. 3–6). Second, a number of "luxury" items, such as beads, pendants, and inscribed slates, are found for the first time in the cave in layer B1—items not strictly related to the onerous and time-consuming business of securing a living (Fig. 3–7).

The B2 layer, in contrast to the overlying B1 layer, contained no grinding stones. Several pits that were noted in this layer could have served as storage pits for food that was not stone-ground (of course, perishable material such as wood could have been used for preparing food). The artifact assemblage is different from that found in layer B1. It includes a large number of microliths of the "geometric" type, carefully made and reflecting an expert and sophisticated flint-chipping industry (Fig. 3–8). It is inferred from the technological and cultural level of these people that they were more oriented to the hunt than their followers a thousand years later. We may also assume that they had a more complex technology and economy than their predecessors at Shanidar.

Layer C is easily distinguishable from layer B on the basis of stratigraphy and artifact remains (Fig. 3–9). The top part of this layer has been dated by the radiocarbon method at about 26700 B.C.;[9] the bottom part, at about 33100 B.C.[10] Thus far in the excavations we have been unable to find remains linking layers B and C. There is an abrupt change of industries, from a blade-tool type reminiscent of the Upper Palaeolithic "Aurignacian" (here called "Baradostian")[11] in layer C, to the more highly evolved microlithic industry of basal layer B. The stone equipment of layer C indicates a high degree of skill in the woodworking crafts (only a few worked bones were found). Especially numerous are the burins, which are characterized by several types of working-bit. No human skeletal remains have been found in this layer, but it is assumed that these people were a variety of true *Homo sapiens.*

Layer D (Fig. 3–10) is the thickest layer in the cave (about 8.5 m.). There is evidence of heavy occupational concentration toward the middle of the layer. The Mousterian artifacts include rather typical points, scrapers, and knives made on unifacial flakes (Fig. 3–11). Seven Neanderthal skeletons (six adults, numbered from I to VI, and one child, unnumbered) were found in the upper third of the layer.[12] Since only about one-tenth of the cave has been excavated thus far, more human remains are likely to be found in future seasons. Shanidar Neanderthals I and V were found near the top of layer D, in the level dated approximately 44000 B.C.[13] T. Dale Stewart, of the U.S. National Museum, is studying the adult skeletons.[14, 15] The late Muzaffer Senyürek, of the University of Ankara, studied the skeleton of the Shanidar child.[16]

Investigation

Three broad avenues of investigation are discussed here. The first is the establishment of a chronological framework to serve as support for the study (Fig. 3–12). Comparative studies of the Shanidar sites and of other sites of the same age in the same broad geographical zone also are to be made. The second approach is the study of populations that lived in the valley, from perhaps 100,000 years ago, to find what can be learned of them and their movements from their ancient leavings. The third is investigation to find where the Shanidar Valley fits in the great food-production revolution that supposedly took place in southwestern Asia.

Chronological Framework

Archaeology without the backdrop of a time scale has little meaning. The cultural-temporal positioning of the occupations on the basis of the artifact typology was accomplished first. The chronology was fixed by sixteen carbon-14 dates from all four layers at Shanidar Cave and by one from Zawi Chemi Shanidar. The samples were dated by four different laboratories in studies of which several were duplicate checks.[17] The dates range from about A.D. 1750 for layer A to about 48000 B.C. for layer D. Several obsidian samples from layers B and C were also dated.[18]

On the framework of this chronological scale, the climatological data obtained from studies of noncultural materials in the deposits were arranged. We enlisted the aid of a palynologist, Arlette Leroi-Gourhan of Paris, whose findings (nine samples) were independently corroborated, with one exception, by trace-element analyses of the soils (five samples), made by Bruno E. Sabels, of the University of Nevada.[19] These analyses indicated marked fluctuations of climate in the late Pleistocene (Fig. 3–13), bearing out the geological observations made elsewhere in Kurdistan by Herbert E. Wright, Jr., of the University of Minnesota.[20, 21]

Data for the 8.6-meter level (the lowest for which there is information), well into the Mousterian layer, indicate a climate much warmer than that in the area today and growth of the date palms (*Phoenix dactylifera*) not far away.[22] Data for the 7.5-meter level show a reversal to an exceedingly cool climate and growth of fir trees (*Abies*) in the area. Pollens near the top of layer D, at depths of 4.25 and 4.35 meters, suggest a return to warm climate about 44000 B.C. Findings for layer C suggest a change from a dry, steppe environment near the start of the Upper Palaeolithic Baradostian occupation, at about 34000 B.C., to a wet and cold climate near its end, at about 25000 B.C. Data are lacking for the next 15,000 years, to the base of layer B. Howeevr, findings for the two parts (B1, B2) of layer B indicate a relatively cool climate changing to a warmer one similar to the present climate. A culture horizon comparable to B1 at Zawi Chemi Shanidar also provided evidence of warmer conditions.

We are painfully aware that the minimum four conditions postulated by Edward S. Deevey, Jr.,[23] as requisite for the application of pollen analysis to the problems of prehistory, are only half satisfied at Shanidar. Lacking are a "standard pollen sequence" and a "knowledge of the regional plant ecology." But a start has been made toward a climatological sequence for Shanidar Valley.

The question at this point in the studies is this: If the suggestions of climate changes are correct, where do the alternations at Shanidar fit into the Pleistocene climate sequence? The dated part of the cave chronology, to about 48000 B.C., can be checked against generalized curves. Beyond that, there is some doubt. An attempt to date the Shanidar Cave deposits below the oldest carbon-14 determinations can be made by rough extrapolation on the basis of guessing the rate of accumulation of cultural deposit in feet per 1,000 years.[24-27] Assuming a constant rate of cultural deposition of about 1.25 feet per 1,000 years, we guess that the Shanidar Cave deposits began accumulating close to 100,000 years ago. The chronological fit of the projected curve with Flint and Brandtner's interpretations of climate change since the Last Interglacial[28] is better than the fit with Zeuner's or Emiliani's interpretations[29] (Fig. 3–13). The deepest pollen sample (8.6 m.) and the corresponding trace-element sample (8.3 m.), which reflect a very warm climate, could correspond with Flint and Brandtner's "Eem," or Last Interglacial. The next Shanidar curve position could be interpreted as corresponding with their Early Würm Stadial. The much disputed Göttweig Interstadial[30] in their analysis corresponds with the climatological evidence from the lower part of layer C and the upper part of layer D: a colder climate about 20,000 years ago is indicated at Shanidar. The gap between layers B and C falls in what is called the Würm Maximum in the European Alpine sequence. The Zagros Mountain glaciers advanced down the slopes during this interval, causing in all probability a lowering of temperatures and a retreat of the flora and fauna.[31] Man could not tolerate such an icebox very long (the glaciers came down to elevations of about 1,500 m.), and he sensibly left for the Florida of his time. According to the inferred Shanidar climatological data, the climate had improved greatly (from man's standpoint) by 10000 B.C., and by about 8000 B.C. had changed to a warmer, postglacial climate very much like that of today.[30, 33]

Unfortunately, no other pollen or trace-element climatological studies have been made for this range of prehistorical time in the Near East, and thus there are no data with which the Shanidar data can be compared.[34] Changes in climate have been established on faunal evidence; such evidence, however, even in natural (noncultural) contexts is recognized as being of secondary reliability.[35]

One of the best-known examples of climate sequence based on faunal evidence from a Near Eastern archaeological site is at Mount Carmel.[36] Primarily involved are two faunas, a cool-wet-loving deer and a warm-dry-loving gazelle. F. E. Zeuner[37] adapted the Mount Carmel faunal-frequency chart in his correlation with the European late Pleistocene sequence. This touched off a debate which is still alive as fresh data are gathered.[38, 39, 40, 41] Especially illuminating are D. A. Hooijer's[42] investigations of the deer and gazelle frequencies from Ksâr 'Akil, in the Lebanon. These throw some doubt on the climate interpretations from the Mount Carmel sequence.

Another late Pleistocene climate sequence, based on faunal evidence, primarily the large bovines, has been offered for the North African and Mediterranean area.[43, 44]

Concerning faunal curves, Hooijer's admonition[45] is pertinent: "What the vertebrate palaeontologist does rather more than anything else when studying the 'fauna' of an occupation site is sampling the history of the menu of the local population of prehistoric man." Today, you cannot find a Kurd at Shanidar who will eat snails or the flesh of boars, even though both abound. These are restatements of the observation that the fauna of an area, as identified in archaeological contexts, is passed through the filter of human occupation. Cultural selection of fauna in a particular region, however, is obviously dependent upon the existing faunal inventory, which must have first passed through the screen of natural environment.

Of import to the Shanidar study is a comment by Charles A. Reed of Yale University, who has studied Shanidar faunal data from three excavation seasons. He says that the remarkable thing about the fauna of Shanidar Cave is that outwardly all of the bones look to be of the same age and of recent date, having the appearance of a "single-age, post-Pleistocene fauna."[46] The bones or faunal evidence, therefore, suggest that the climate at Shanidar did not change a great deal from Middle Palaeolithic times on. Yet the primary climatological data, the available pollens, show otherwise. At present there is no ready explanation of this contradiction. Broadly speaking, animals are less sensitive to climate change than plants. Furthermore, some animals, such as sheep, are less sensitive to change than others.[47, 48] Could this be the root of the problem at Shanidar? Or is it that here, as in the Mediterranean area,[49, 50] the hunters had access to a large region with a very wide range of environmental conditions from which to draw their game animals?

We must wait until a regional sequence of pollen data has been obtained from the Shanidar area before we can say that the climate-change yardstick can be applied in the Near East with precision. Nevertheless, in the two faunal curves and the admittedly incomplete pollen curve of Shanidar (Fig. 3–13), some broad correspondence can be seen, with one notable exception. The plot for Tabūn B in the Mount Carmel diagram, indicating a wet and cold climate, is not in accord with the other curves. It occurs at about the time when the Mousterian cultures in this part of the Near East were dwindling.

Human Populations of Shanidar Valley

Study of human populations of the Shanidar Valley is a complex process, its complexity compounded by incompleteness of the Shanidar and related investigations and by a lack of data. However, some general observations can be made on the basis of the available Shanidar data.

The Mousterian layer seems to have been built up by a series of

Neanderthals who were relatively stagnant culturally; there were 2,000 generations of them in the perhaps 60,000 years of its accumulation. Preliminary analysis of the tool types, from bottom to top, indicates that, except for the brief vogue of what looks like the "Emiran" type point,[51] there were no changes in the deposit. These basally inverse retouched points were found in the middle of a heavy occupational zone at a depth of about 8.5 meters—a zone which, as noted earlier, is probably evidence of a climate warmer than that of today. Neither fauna nor culture seems to have been much affected by the change to a very cool climate that is indicated by remains at a depth of about 7.5 meters. About this time (about 60,000 years ago), three of the six adult Neanderthals (Nos. II, IV, and VI) whose skeletons have been recovered were killed, all crushed by rocks (Fig. 3–14). Shanidar III, found at a depth of 5.4 meters, lived in a warmer climate perhaps 50,000 years ago. Shanidar I (figs. 3–15 and 3–16) and Shanidar V were found at a depth of about 4.3 meters in a horizon dated about 44000 B.C., also in a warm environment. Physical violence, compassion for the living,[52] and a certain regard for the dead are reflected in the skeletal finds. T. Dale Stewart[53, 54] has shown that the Shanidar Neanderthals have morphological features similar to the Tabūn skeletons of Mount Carmel, which postdate the Shanidar Neanderthals by at least 5,000 years. Stewart has also shown, on the basis of his studies of the skulls of Shanidar I and II, that an almost classic Neanderthal skull form was retained over a period of at least 15,000 years. The form was seemingly unaffected by the climatic changes inferred here. The carbon-14 dates recently obtained for Tabūn B (about 39,500 years ago) and Tabūn C (about 41,000 years ago)[55, 56] suggest that the Mount Carmel sequence falls almost entirely within the Last glaciation (Würm in the Alpine sequence).

There are certain differences in tool types between the Mousterian cultures of Mount Carmel and Shanidar—for example, the absence of the Levallois core technique at Shanidar[57] and evidence of the technique at Mount Carmel. This undoubtedly must have some basis in the difference in environment. The recently obtained dates for Mount Carmel point up the difficulties in making cultural correlations on the basis of industry typologies[58] and negate my conclusions from typological comparison of the Mount Carmel and the Shanidar Mousterian.[59] If the Mount Carmel dates were some 6,000 years older, they would make the cultural sequences, at least, match better with those of other sites in the Near East.

Viewed broadly, the Shanidar Mousterian is a reasonably good example of the Mousterian culture horizon which ranged from western Europe and North Africa to Uzbekistan[60, 61] and central Asia[62] in a rough ellipse around the Mediterranean, Black, Caspian, and Aral seas.[63] An interesting parallel can be drawn between the Shanidar and the Teshik-Tash Mousterian,[64, 65] where, in similar mountainous environments, the principal animals hunted were goats. In the Iraq-Iran area, an occupation closely related to the Shanidar Mousterian occurred at the shelter-cave sites of Hazar Merd, Babkhal, Spilik, Bisitun, and Warwasi, and occupations less closely connected, at the open sites of Tarjil, Serandur, and Telegraph Pole 26/22.[66, 67] Hazar Merd and Bisitun, at least, appear to be statistically related to Shanidar.[68]

The Mousterian lingered on longer at Mount Carmel than it did in the Zagros Mountains, a finding that upset former chronological estimates for Mount Carmel.[69] The terminal date for the Mousterian has been set at about 35,000 to 40,000 years ago in the Levant;[70] this is about the same as the date established for the beginning of the Upper Palaeolithic III Baradostian of Shanidar. The date for the end of the Mousterian in Libya has been set at about 35000 to 40000 B.C.[71] The Mousterian came to Haua Fteah and to Shanidar at about the same time, but for some reason it seems to have been late in appearing at Ksâr 'Akil and Mount Carmel. One must conclude that the Acheulean industries lingered on until quite late at Mount Carmel, if the inferred chronology is true (Fig. 3–13). The Mousterian was part of an abrupt introduction, possibly stemming from Africa.[72] But neither the spread nor the final extinction of this culture was uniformly smooth. The carbon-14 dates hint that the final Mousterian cultures occurred at the same time as initial Upper Palaeolithic occupations in the Near East.

What happened eventually to the last Shanidar Neanderthals is not known. It is hardly likely that unfavorable climate was a contributing cause of their departure. The possibility that they were eliminated by a prehistoric catastrophe cannot be ruled out. There is no evidence, but perhaps the Upper Palaeolithic true *Homo sapiens* contributed to this extinction at Shanidar. However, if the Neanderthals had been bested in combat (and this surely would have been an unequal fight), it might be expected that the newcomers would have taken over their homes.[73] There seems to be evidence of cultural intermixture of the Upper Palaeolithic and the Mousterian Middle Palaeolithic in a transitional industry (the Emiran) in the Levant,[74] and similarly, passage of the Mousterian culture into the Upper Palaeolithic Périgordian I in France.[75] But such was not the case at Shanidar. After, at most, a lapse of 10,000 years, the new occupants of Shanidar Cave, the Baradostian people, took over a wilderness restocked with game, with no one to dispute their hunting territory claims.

On the basis of their tool inventory, it appears that these Upper Palaeolithic peoples were most closely related to the southwest Asian and European blade- and burin-using populations.[76, 77, 78] The Baradostian industry is unique in Iraq, although it has been reported from Warwasi Cave in western Iran.[79] Nothing is known of cave or "home" art. It looks as though the Baradostian people had adapted their hunting methods to local conditions and pursued the same game animals (mainly goats) as their Neanderthal predecessors in Shanidar Valley. Probably they drove these gregarious herbivores over cliffs or trapped them in blind canyons nearby.

In the four known caves in the Zagros area which contain a Mousterian occupation overlaid by a Mesolithic one, only Shanidar and Warwasi caves have an intervening Upper Palaeolithic occupation. There are more numerous, related occupations to the west in the Levant. There the closest parallels to the Baradostian are at Yabrud (Shelter II, Layers 4 and 5) in Syria, at Abu Halka (Layer IVc) in Lebanon, and at Mount Carmel Wad E in Palestine—or what R. Neuville[80] calls Upper Palaeolithic III.[81] The Upper Palaeolithic had a longer and more complete cultural history in the Levant than in the mountain hinterland of the Zagros area. While it is evident that the Baradostian did not have the time spread of the Mousterian, a more likely explanation of the sparseness of distribution for the Zagros area is a lack of population. It is possible that this interior mountain environment necessitated a special economic adjustment.

About 26000 B.C., it is surmised, the climate became too cold for man at Shanidar and he left. The next occupation was not until about 15,000 years later. No barren soil layer was noted between layers B and C with which this hiatus can be correlated. This apparent desertion of Shanidar in the later Upper Palaeolithic is paralleled at other dated and undated sites in the Near East, from Kara Kamar[82] in Afghanistan to southern Turkey.[83] There may well have been a low population density during the peak of the Last glaciation, between about 13,000 and 23,000 years ago.[84] This is borne out by the dwindling number of sites even in the historically rich area of the Levant, where the Upper Palaeolithic Stage V of Neuville[85, 86] is clearly defined in only two sites, both in Palestine (Mount Carmel Wad C "Atlitian," and el Khiam E). The hiatuses occur between the period of blade and burin industries of the Upper Palaeolithic and the Mesolithic of the very late Pleistocene. In neighboring Soviet Asia we find a similar situation, with even longer hiatuses between the Mousterian and the Mesolithic.[87, 88] After the close of the last glaciation, about 9000 to 10000 B.C., there was a rash of Mesolithic settlements. As in the Near East, they blossomed over what is now Soviet Asia like desert flowers after a rain, taking advantage of an apparent cultural vacuum, meeting with little or no resistance. Surely something new must have been added to the economy, or new techniques and innovations must have broadened the economic base, contributing to this evident population spurt.[89]

The date of this movement, at least in this part of the Near East, was about 10000 B.C., and the movement probably lasted not more than one and a half millennia. There are ten known sites of this relatively brief culture horizon (generally belonging to the "Zarzian"[90]) in the Iraq-Iran Zagros Mountain area in contrast to the lone pair of known Baradostian sites, spanning a period about five times as long. The Mesolithic layer B2 of Shanidar Cave is one of these culturally related "Zarzian" sites. The others are components in the cave sites of Zarzi, Hazar Merd, Babkhal, Palegawra, Hajiyah, Barak, and Warwasi and in the open sites of Turkaka and Kowrikhan.[91, 92, 93, 94] Outside the Zagros Mountain area, on the Caspian Sea in Iran, are the Belt and Hotu caves,[95, 96, 97] which have occupations of comparable date and culture.

The origins of this Mesolithic culture are not definitely known. Certainly it did not stem from the Baradostian at Shanidar. It could have come from the Levant, but related cultures to the north beyond the Caucasus area may have been just as compelling.[98] This people marked the end of the true hunters and gatherers, analogous to the Azilians and Tardenoisians in Europe prior to the Neolithic cultures. The sudden abundance of snail shells, the marked discoloration of the soil (from vegetal stuffs?), and the suggested presence of pits and basins in the Mesolithic layer of Shanidar Cave suggest that these people were successfully launched on the road of experimentation with nontraditional food.

The cultural analog to the qualified "Zarzian" horizon in the Levant is the "Kebaran" of the Upper Palaeolithic Stage VI.[99, 100] In North Africa, the analog to this culture horizon is the Oranian, which appears to date from later than 15000 to 12000 B.C. in the Maghreb.[101, 102] Is it possible that Oranian-related cultures swung eastward along the Mediterranean and inland through the Zagros arc and on eastward? It is too soon to say, but we are on the threshold of knowing. If this were true, then the movement of the later Proto-Neolithic horizon would be an interesting phenomenon to trace. The matter of the relationship of "Zarzian" to the open-air sites of the Ukraine, a thousand miles away, is still not clear.[103]

The next culture horizon in Shanidar Valley is known from the Proto-Neolithic occupation at the cave (layer B1) and at the open site of Zawi Chemi Shanidar.[104] These sites were probably seasonally occupied. Related components are found at such sites as Karim Shahir, M'lefaat, and possibly Asiab.[105, 106, 107, 108, 109] The evidence from Zawi Chemi Shanidar shows that this culture, especially, was well on its way toward full food production; the domesticated sheep was already known—an innovation probably brought in from some other area.[110] In the Levant the analogous Proto-Neolithic culture seems to have been the Natufian of Palestine.[111] A more distant analog is the Caspian culture in North Africa (Maghreb), emergent there sometime after 9000 B.C.[112] The interrelationships among these widely dispersed cultures, like those for the previous Mesolithic horizon, are not yet positively shown, but the thread of cultural similarities cannot be dismissed.[113]

The contrast between the Mesolithic and the Proto-Neolithic cultures at Shanidar is very marked. The compelling problem is that of determining the origin of the Shanidar B1–Zawi Chemi Shanidar culture horizon (Proto-Neolithic), with its focus on economic change. It may be that the preceding Shanidar B2 type culture and its Mesolithic equivalents elsewhere were for some reason, possibly ecological, already heavily predisposed toward experimental food collection and preparation of such edibles as acorns, nuts, and wild grass seeds. Social changes must have accompanied the new mode of life, but inferences about this are somewhat more difficult to make. At any rate, it is an inescapable fact that a food-production revolution of a sort is evidenced at Shanidar Cave layer B1 and at Zawi Chemi Shanidar. It did not evolve directly out of Shanidar B2, but the change probably took place not very far away.

This great revolution seems to have occurred at just about the time of an abrupt world-wide rise in temperature.[114] Undoubtedly the same sort of climatic change had occurred before in man's history, but without a similar aftermath, so far as we know. Presumably, man did not have the right combination of mental, technological, and social attributes earlier in his development to search out and utilize radically new ways of getting a living, or else he was not in an area where the proper combination of ecological factors obtained. But given these, a kind of trigger was needed to make him depart from being a perpetual "lotus-eater," forever dependent upon hunting and gathering for his existence. In the area under discussion, the rise in temperature could have served as just such an indirect stimulus. The same sort of shock stimulus and subsequent concatenation of events was felt in the American Southwest at about the same time (about 8000 B.C.),[115, 116] where, paralleling developments in the Near East, there were shifts to an economic base more dependent upon food gathering, especially the gathering of vegetal foods, than on hunting. The hallmark of the so-called American Desert Culture was the flat milling stone, or quern, and the gathering-basket.[117] At Shanidar and related sites we find evidence of the introduction of querns and hand milling stones [also, at Shanidar Cave, baskets (?), and at Zawi Chemi Shanidar, possibly some kind of reaper (Fig. 3–17)][118] indicating vegetal foods. The wild goat, known since the first occupation at Shanidar Cave, was now of minor importance as compared to the sheep that was found domesticated in Shanidar Valley. The stage was set for a "mixed-farming" economy.

As for the Desert Culture of the American Southwest, it seems that food production did not take hold there as it did in the Near East. Lacking was the combination of potentially domesticable animals and wild cereal prototypes, which in the Near East were the touchstone to civilization.[119]

NOTES

1. C. A. Reed, *Z. Tierzüchtung Züchtungbiol.*, **76**, 1 (1961), pp. 31–38.

2. For the 1951 season, see R. S. Solecki, *Sumer*, **8**, 127 (1952); *ibid.*, **9**, 60 (1953). For 1953, see R. S. Solecki, *Smithsonian Inst. Ann. Rept.*, **1954**, 389 (1955) [reprinted in *Sumer*, **11**, 14 (1955)]; *Sumer*, **9**, 229 (1953). For 1956–57, see R. S. Solecki, *Smithsonian Inst. Ann. Rept.*, **1959**, 603 (1960); *Sumer*, **13**, 165 (1957); *ibid.*, **14**, 104 (1958). For 1960, see R. S. Solecki, *Trans. N.Y. Acad. Sci.*, **23**, 690 (1961).

3. The terms *Palaeolithic*, *Mesolithic*, and *Neolithic* are not precisely defined but serve as a useful nomenclature for establishing broad cultural perspectives in prehistory.

4. Sample W-667 is dated 10,600 ± 300 years ago [R. S. Solecki and M. Rubin, *Science*, **127**, 1446 (1958); M. Rubin and C. Alexander, *Am. J. Sci. Radiocarbon Suppl.*, **2**, 183 (1960)].

5. Sample W-179 is dated 12,000 ± 400 years ago [M. Rubin and H. E. Suess, *Science*, **121**, 481 (1955)].

6. Sample W-681 is dated 10,870 ± 300 years ago [R. S. Solecki and M. Rubin, *Science*, **127**, 1446 (1958); M. Rubin and C. Alexander, *Am. J. Sci. Radiocarbon Suppl.*, **2**, 184 (1960), date the same sample (W-681) 10,800 ± 300 years ago].

7. The platforms, associated with evidence of fire, were probably of ceremonial origin.

8. R. S. Solecki and R. L. Solecki, "Two bone hafts from the Proto-Neolithic horizon at Shanidar, Northern Iraq," in preparation.

9. Sample W-654 is dated 28,700 ± 700 years ago [M. Rubin and C. Alexander, *Am. J. Sci. Radiocarbon Suppl.*, **2**, 184 (1960)].

10. Sample GRO-259 is dated 35,080 ± 500 years ago [letter from H. DeVries (1 Aug. 1959)].

11. R. S. Solecki, thesis, Columbia University (1958).

12. R. S. Solecki, *Trans. N.Y. Acad. Sci.*, **23**, 690 (1961).

13. Sample GRO-2527 is dated 46,000 ± 1,500 years ago [R. S. Solecki, *Smithsonian Inst. Ann. Rept.*, **1959**, 629 (1960); letter from H. DeVries (1 Aug. 1959)].

14. T. D. Stewart, *Sumer*, **14**, 90 (1958) [reprinted in *Smithsonian Inst. Ann. Rept.*, **1958**, 473 (1959)]; *Yearbook Am. Phil. Soc.* (1958), pp. 274–278.

15. *Bibliog. Primatol.*, **1**, 130 (1962) (Adolph H. Schulz anniversary volume); *Sumer*, **14**, 104 (1958).

16. M. Senyürek, *Anatolia*, **2**, 49 (1957); *ibid.*, **2**, 111 (1957); "A Study of the Deciduous Teeth of the Fossil Shanidar Infant," *Publ. Univ. Ankara, No. 128* (1959).

17. The samples were dated at the Laboratory of Physics, University of Groningen, Groningen, Netherlands; the Lamont Geological Observatory, Palisades, New York; the U.S. Geological Survey Radiocarbon Laboratory, Washington, D.C.; and the Geochronological Laboratory, London Institute of Archaeology, London University, London, England.

18. I. Friedman, R. L. Smith, C. Evans, B. J. Meggers, *Am. Antiquity*, **25**, 476 (1960). I believe, on the basis of the evidence, that obsidian dates older than those of layer B are not reliable.

19. R. S. Solecki and A. Leroi-Gourhan, *Ann. N.Y. Acad. Sci.*, **95**, 729 (1961).

20. H. E. Wright, Jr., in *Prehistoric Investigations in Iraqui Kurdistan*, R. J. Braidwood and B. Howe, Eds. (Univ. of Chicago Press, Chicago, 1960), pp. 1–184.

21. H. E. Wright, Jr., *Eiszeitalter Gegenwart*, **12**, 131 (1961); *Ann. N.Y. Acad. Sci.*, **95**, 718 (1961).

22. Previously, date palms had not been found, in an archaeological context, in Mesopotamia before Sumerian times.

23. E. S. Deevey, Jr., *Am. Antiquity*, **10**, 135 (1944).

24. R. S. Solecki, unpublished manuscript.

25. C. B. M. McBurney has hit upon the same generalization from carbon-14 dates, using a more involved method. See notes 26 and 27.

26. C. B. M. McBurney, *Advan. Sci.*, **18**, 494 (1962).

27. C. B. M. McBurney, *Nature*, **192**, 685 (1961).

28. R. F. Flint and F. Brandtner, *Am. J. Sci.*, **259**, 321 (1961).

29. H. E. Wright, Jr., *Bull. Geol. Soc. Am.*, **72**, 933 (1961).

30. *Loc. cit.*

31. Wright, "Eiszeitalter Gegenwart," *loc. cit.*, *Am. N.Y. Acad. Sci. loc. cit.*

32. *Ibid.*

33. This would substantiate Dr. Wright's observation for climate change in Kurdistan.

34. Willem Van Zeist's "Preliminary palynological study of sediments from Lake Merivan, S. W. Iran" (unpublished manuscript, 1961), is not applicable here.

35. Wright, *Bull. Geol. Soc. Am.*, *loc. cit.*

36. D. A. E. Garrod and D. M. A. Bate, *The Stone Age of Mount Carmel* (Oxford Univ. Press, Oxford, 1937), pp. 1–240.

37. F. E. Zeuner, *Dating the Past* (Methuen, London, 1958), pp. 1–516.

38. Wright, *Prehistoric Investigations in Iraqui Kurdistan*, *loc. cit.*

39. D. A. Hooijer, *Advan. Sci.*, **18**, 485 (1962).

40. F. Clark Howell, *Proc. Am. Phil. Soc.*, **103**, 1 (1959).

41. C. B. M. McBurney, *The Stone Age of Northern Africa* (Penguin, London, 1960), pp. 1–288.

42. D. A. Hooijer, *Zool. Verhandel.*, **49**, 1 (1961).

43. E. S. Higgs, *Advan. Sci.*, **18**, 490 (1962); E. S. Higgs and D. R. Brothwell, *Man.*, **41**, 138 (1961).

44. E. S. Higgs, *Proc. Prehist. Soc.*, **27**, 144 (1961).

45. Hooijer, *Advan. Sci.*, *loc. cit.*

46. C. A. Reed and R. J. Braidwood, in *Prehistoric Investigations in Iraqi Kurdistan*, R. J. Braidwood and B. Howe, Eds. (Univ. of Chicago Press, Chicago, 1960), p. 165.

47. Higgs, *Proc. Prehist. Soc., loc. cit.*

48. K. Kowalski [*Folia Quarternaria*, **8**, 1 (1962)] finds that small rodents are very useful indicators of climate in cave sections. No such study has been made as yet at Shanidar Cave.

49. Hooijer, *Advan. Sci., loc. cit.*

50. Hooijer, *Zool. Verhandel, loc. cit.*

51. R. S. Solecki, *Smithsonian Inst. Ann. Rept.*, **1954**, 389 (1955).

52. Compassion for the living is indicated by evidence of care of the infirm and wounded (one individual, at death, was recovering from a stab wound in the rib) and by evidence of surgery (an arm, useless since birth, had been cut off above the elbow).

53. Stewart, *Sumer, loc. cit.*; *Yearbook Am. Phil. Soc., loc. cit.*

54. T. D. Stewart, *Science*, **131**, 1437 (1960); paper delivered at American Association of Physical Anthropologists meeting; *Smithsonian Inst. Ann. Rept.*, **1961**, 521 (1962).

55. K. P. Oakley, *Advan. Sci.*, **18**, 415 (1962).

56. I believe that the "guess date" of "as early as 37000 B.C." [D. R. Brothwell, *Proc. Prehist. Soc.*, **27**, 155 (1961)] for the Skhul finds of Mount Carmel is within the limits of credibility.

57. R. S. Solecki, *Smithsonian Inst. Ann. Rept.*, **1959**, 603 (1960).

58. F. Bordes, *Science*, **134**, 803 (1961); F. Bordes, in *Evolution of Man*, S. Tax, Ed. (Univ. of Chicago Press, Chicago, 1960), vol. 2, pp. 99–110.

59. R. S. Solecki, *Smithsonian Inst. Ann. Rept.*, **1959**, 603 (1960).

60. H. L. Movius, Jr., *Bull Am. School Prehist. Res. No. 17* (1953), pp. 11–71.

61. H. L. Movius, Jr., *Proc. Am. Phil. Soc.*, **97**, 383 (1953).

62. V. A. O. Ranov, *Akad. Nauk Tadjikstan SSR Stalinabad*, **1**, 89 (1961); V. A. O. Ranov, in *Noveishei etat geologicheskogo razvitiia territorii Tadzhikistana* Dushanbe, U.S.S.R., (1962), pp. 35–65.

63. F. E. Zeuner, in *A History of Technology*, C. Singer, E. J. Holmyard, A. R. Hall, Eds. (Oxford Univ. Press, Oxford, 1954), vol. 1, pp. xlviii–lix. The distribution can be extended around the Caspian Sea and eastward.

64. Movius, *Bull. Am. School Prehist. Res. No. 17, loc. cit.*

65. Movius, *Proc. Am. Phil. Soc., loc. cit.*

66. R. J. Braidwod and B. Howe, in *Prehistoric Investigations in Iraqi Kurdistan*, R. J. Braidwood and B. Howe, Eds. (Univ. of Chicago Press, Chicago, 1960).

67. R. J. Braidwood, *Advan. Sci.*, **16**, 214 (1960).

68. This conclusion is based on a rough statistical estimate for Shanidar Mousterian. It looks like a "typical" Mousterian.

69. R. S. Solecki, *Trans. N.Y. Acad. Sci.*, **24**, 712 (1959).

70. Oakley, *loc. cit.*

71. C. B. M. McBurney, *Advan. Sci.*, **18**, 494 (1962).

72. C. B. M. McBurney, in *Neanderthal Centenary*, G. H. R. von Koenigswald, Ed. (Drukkerij, Utrecht, Netherlands, 1958), pp. 1–325.

73. In the Levant area the Upper Palaeolithic had evolved through two stages (U.P. I, U.P. II) before the Baradostian appeared in the Zagros.

74. D. A. E. Garrod, *J. Roy. Anthropol. Inst.*, **81**, 121 (1951).

75. F. Bordes, in *Neanderthal Centenary*, G. H. R. von Koenigswald, Ed. (Drukkerij, Utrecht, Netherlands), pp. 1–325.

76. R. S. Solecki, thesis, Columbia University (1958).

77. D. A. E. Garrod, *J. World Hist.*, **1**, 13 (1953).

78. D. A. E. Garrod, *Bull. Soc. préhist. Franç.*, **54**, 439 (1957).

79. R. J. Braidwood, *Iranica Antiqua*, **1**, 2 (1961).

80. R. Neuville, *Arch. Inst. Paleontol. Humaine*, **24**, 1 (1951).

81. R. S. Solecki, thesis, Columbia University (1958).

82. C. S. Coon and E. K. Ralph, *Science*, **122**, 921 (1955); C. S. Coon, *The Seven Caves* (Knopf, New York, 1957), pp. 1–338.

83. I. K. Kökten, *Belleten*, **19**, 271 (1955); M. Senyürek and E. Bostanci, *ibid.*, **22**, 171 (1958); E. Bostanci, *Anatolia*, **4**, 129 (1959); E. Bostanci, *Belleten*, **26**, 233 (1962).

84. M. Rubin and H. E. Suess, *Science*, **123**, 442 (1955). Haua Fteah was evidently occupied during the height of the Last glaciation. (See note 26.)

85. Howell, *loc. cit.*

86. Neuville, *loc. cit.*

87. Movius, *Proc. Am. Phil. Soc.*, *loc. cit.*

88. H. L. Movius, Jr., *Actes Congr. Intern. Quaternaire, 4e* (1953), pp. 3–20.

89. P. N. Tretiakov and A. L. Mongait, *Contrib. Ancient Hist. U.S.S.R.* (Russian Translation Series of the Peabody Museum of Archaeology and Ethnology, Harvard Univ.), **1**, 1 (1961); A. Mongait, *Archaeology in the U.S.S.R.* (D. Skvirsky, trans.) (Foreign Language Publishing House, Moscow, 1959), pp. 1–429.

90. The name "Zarzian" was given these sites by R. J. Braidwood and B. Howe [*Prehistoric Investigations in Iraqi Kurdistan*, R. J. Braidwood and B. Howe, Eds. (Univ. of Chicago Press, Chicago, 1960), pp. 155, 180; *Science*, **127**, 1419 (1958)] after D. A. E. Garrod's site in Iraq [*Bull. Am. Prehist. Res.*, **6**, 13 (1930)].

91. R. J. Braidwood and B. Howe, in *Prehistoric Investigations in Iraqi Kurdistan*, R. J. Braidwood and B. Howe, Eds. (Univ. of Chicago Press, Chicago, 1960).

92. Braidwood, *Advan. Sci.*, *loc. cit.*

93. Braidwood, *Iranica Antiqua*, *loc. cit.*

94. R. J. Braidwood and B. Howe, in "Courses Toward Urban Life," *Viking Fund Publ. in Anthropol. No.* 32 (1962), pp. 132–146.

95. Howell, *loc. cit.*

96. R. S. Solecki, *Smithsonian Inst. Ann. Rept.*, **1954**, 389 (1955).

97. C. S. Coon, *Cave Explorations in Iran in 1949* (University Museum, Philadelphia, 1951), pp. 1–124; *Proc. Am. Phil Soc.*, **96**, 231 (1952).

98. Garrod, *J. World Hist.*, *loc. cit.*

99. *Ibid.*

100. Garrod, *Bull. Soc. Prehist. Fran.*, *loc. cit.*

101. McBurney, *Nature*, *loc. cit.*

102. McBurney, *The Stone Age of Northern Africa*, *loc. cit.*

103. Garrod, *J. World Hist., loc. cit.*

104. R. L. Solecki, "Zawi Chemi Shanidar, a Post-Pleistocene Village in Northern Iraq," in preparation.

105. R. J. Braidwood and B. Howe, in *Prehistoric Investigations in Iraqi Kurdistan*, R. J. Braidwood and B. Howe, Eds. (Univ. of Chicago Press, Chicago, 1960).

106. Braidwood, *Advan. Sci., loc. cit.*

107. Braidwood, *Iranica Antiqua, loc. cit.*

108. R. J. Braidwood and B. Howe, in *Courses toward Urban Life*, R. J. Braidwood and G. R. Willey, Eds. *Viking Fund. Publ. in Anthropol. No. 32* (1962), pp. 132–146.

109. The distinctiveness of Shanidar Cave layer B1 should correct the impression that, after the "Zarzian," caves were not inhabited by the Zawi Chemi Shanidar peoples [see Note 53 and R. J. Braidwood, *Science*, **127**, 1419 (1958)].

110. Reed, *Z. Tierzüchtung Züchtungbiol, loc. cit.*

111. D. A. E. Garrod, *Proc. Brit. Acad.*, **43**, 211 (1957); Garrod calls Natufian a "Mesolithic Culture." J. Perrot, in *Courses toward Urban Life*, R. J. Braidwood and G. R. Willey, Eds. *Viking Fund Publ. in Anthropol. No. 32* (1962), pp. 147–164.

112. McBurney, *The Stone Age of Northern Africa, loc. cit.*

113. There are rough correspondences in the timing, the basic tool kit, luxury goods, and certain economic traits which hint at a kind of unity, as yet undefined, and a widespread culture horizon in the African Mediterranean area and Near East.

114. W. S. Broeker, M. Ewing, B. C. Heezen, *Am. J. Sci.*, **258**, 429 (1960).

115. R. S. Solecki, "Clues to the Emergence of Food Production in the Near East," in preparation

116. J. Jennings, "Danger Cave," *Soc. Am. Archaeol. Mem. No. 14* (1957).

117. *Loc. cit.*

118. Solecki and Solecki, *loc. cit.*

119. The fieldwork discussed in this chapter was made possible by grants from the American Philosophical Society, Columbia University, the National Science Foundation, the Smithsonian Institution, and the Wenner-Gren Foundation and by the generous cooperation of the Government of Iraq and the Iraq Petroleum Company. I thank Rose L. Solecki and James H. Skinner for critical reading of the manuscript and Alan Mann for help in its preparation.

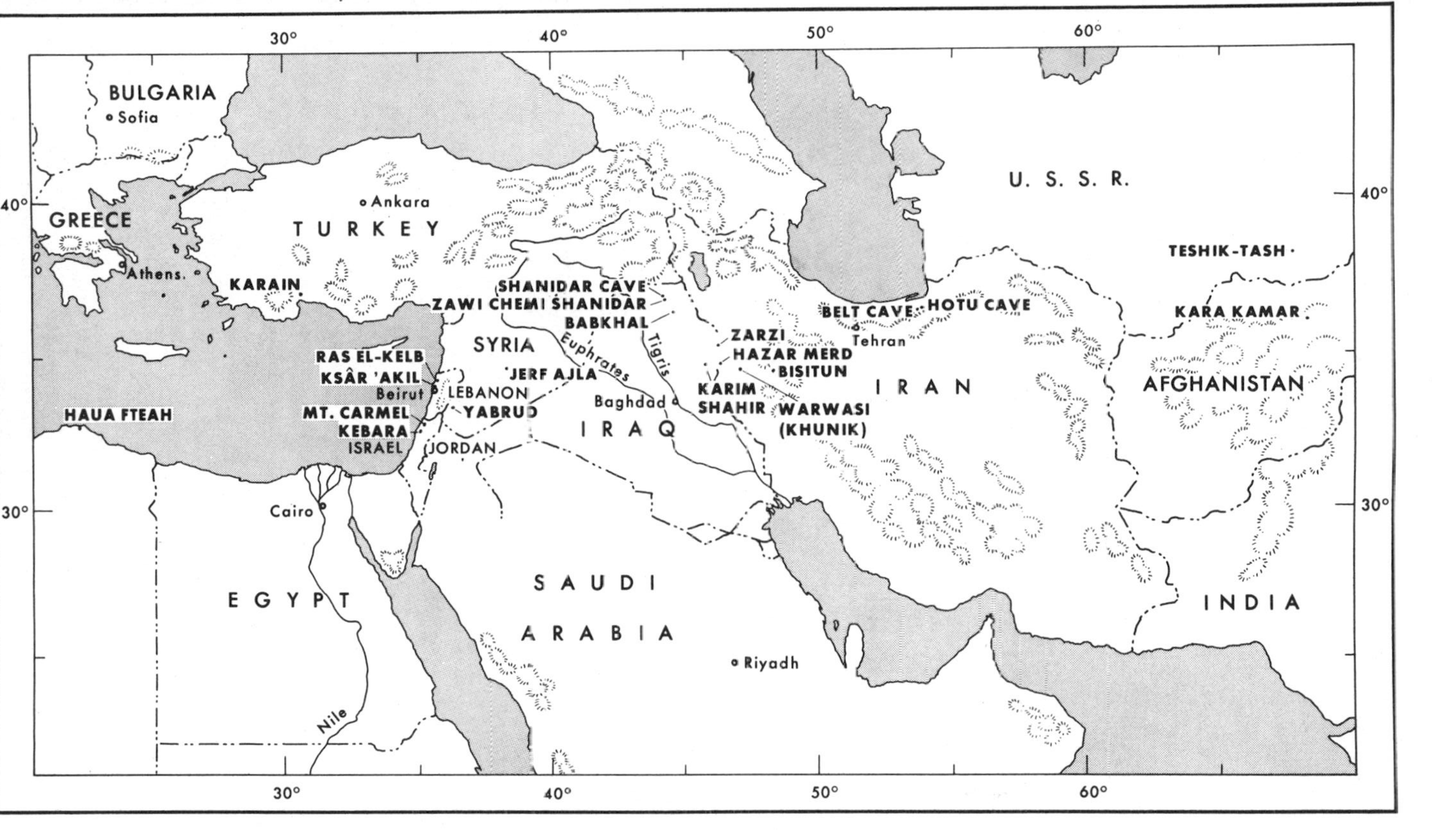

Fig. 3–1. Selected archaeological sites in northern Africa, the Near East, and the Middle East; Haua Fteah (Libya); Mount Carmel and Kebara (Israel), Ksâr 'Akil and Ras el-Kelb (Lebanon); Yabrud and Jerf Ajla (Syria); Shanidar Cave, Zawi Chemi Shanidar, Babkhal, Zarzi, Hazar Merd, and Karim Shahir (Iraq); Belt Cave, Hotu Cave, Bisitun, and Warwasi (Iran); Kara Kamar (Afghanistan); and Teshik-Tash (Uzbekistan).

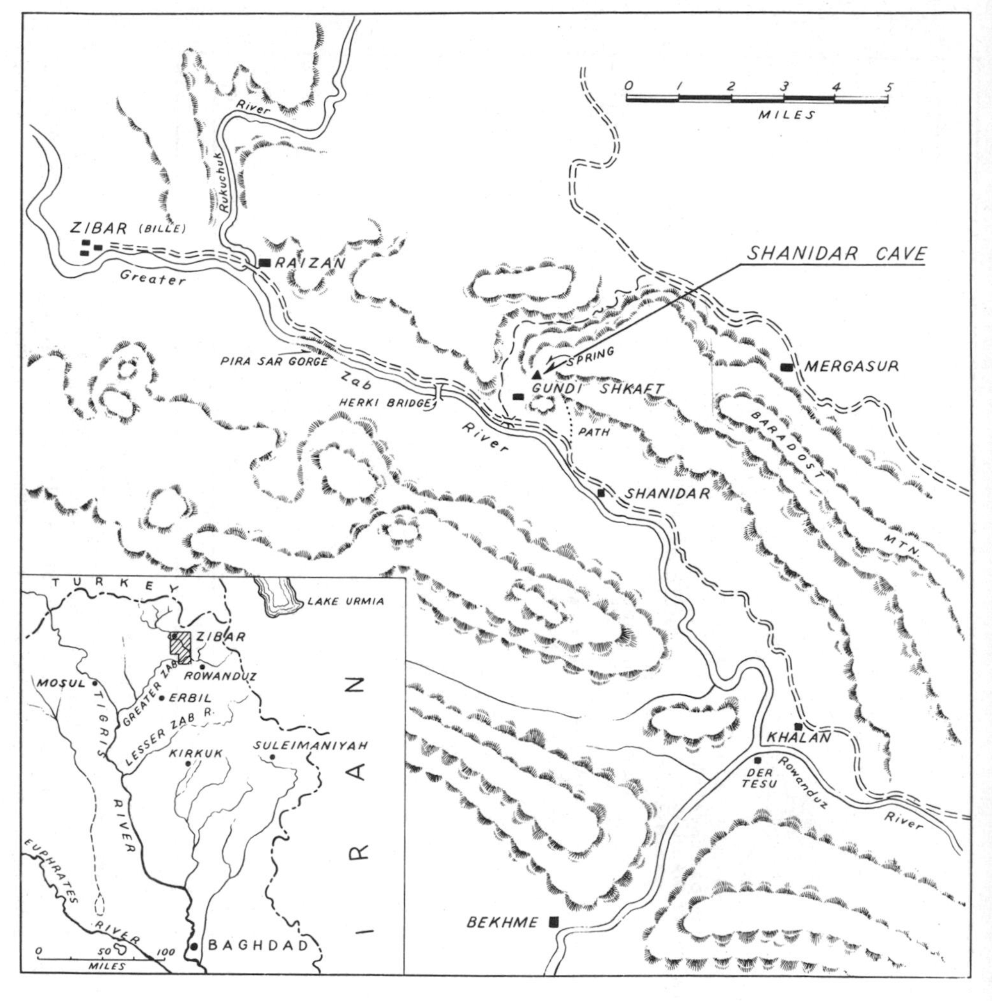

Fig. 3–2. Map showing the location of Shanidar Valley in northern Iraq.

Fig. 3–4. (Right) Schematic cross section of the Shanidar Cave excavation, showing the major cultural layers, the pertinent radiocarbon dates, and the relative positions of the Shanidar Neanderthals.

Fig. 3–3. The limestone cave of Shanidar, seen from the south. The swallow holes at the right enter into the cave. The long grass slope in front receives nourishment from the spilled human occupational debris.

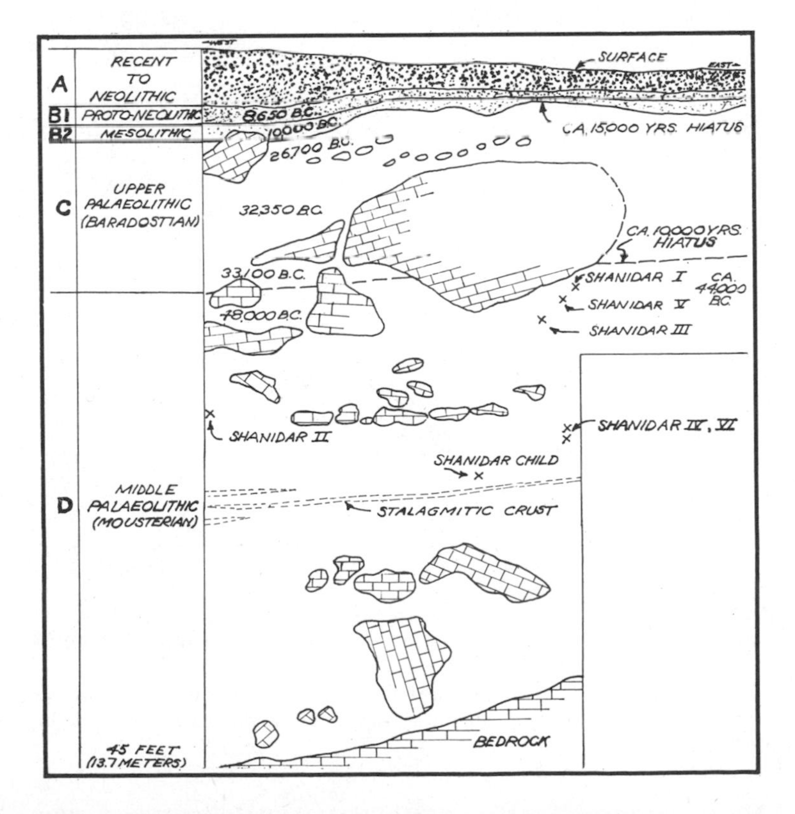

Fig. 3–5. Looking northeast over the cemetery and associated features in the Proto-Neolithic layer of Shanidar Cave. The stone wall and the rough pavements of stones may be part of a mortuary custom of this age. The light, broad horizontal streaks in the upper part of the section are ash lenses in layer A, the Recent-to-Neolithic layer.

Fig. 3–6. A stone-filled pit with four associated boulder querns and quern fragments in the Proto-Neolithic layer of Shanidar Cave. These indicate food grinding and probably food storage.

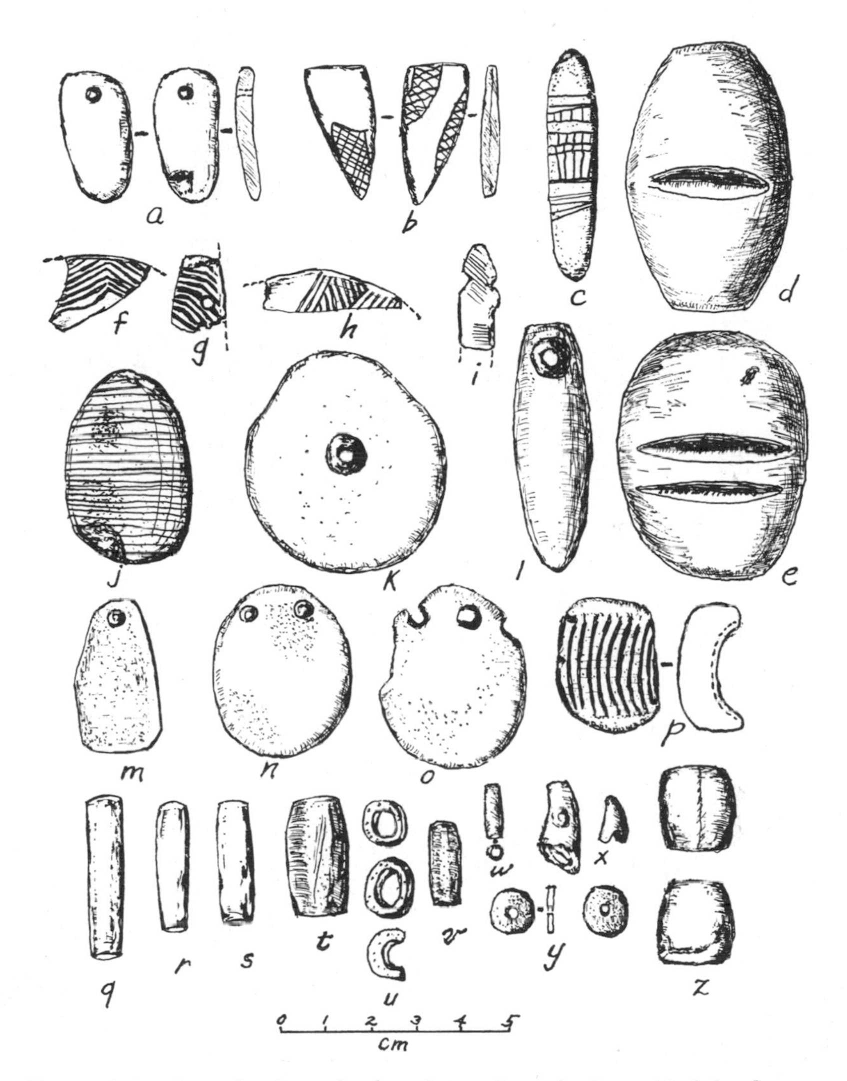

Fig. 3–7. Pendants, beads, and other objects from the Proto-Neolithic layer at Shanidar Cave (*a–d*) and Zawi Chemi Shanidar village (*e–z*). *a*, Shell pendant; *b*, incised slate tablet; *c*, incised slate pebble; *d*, single-grooved plano-convex steatite stone object; *e*, double-grooved steatite stone object; *f–h*, incised fragments of bone tools; *i*, carved fragment of bone tool; *j*, flat pebble bearing parallel incised scratches; *k*, flat single perforated pebble pendant; *l*, elongate single perforated pebble pendant; *m*, flat single perforated green stone pendant; *n*, *o*, double perforated limestone (marble?) pendants; *p*, small steatite object with single U-shaped groove in which there are nine deeply incised cuts; *q–s*, cut tubular bone beads; *t*, barrel-shaped steatite stone bead; *u*, three squat steatite stone beads; *v*, tubular limestone (marble?) bead; *w*, small tubular cut bone bead; *x*, perforated animal teeth (probably *Cervus elaphus*); *y*, two flat disk beads of indeterminate material; *z*, two broad bone beads.

Fig. 3–8 (*a–v*). Various forms of finely executed points and side blades of flint from the upper and lower subdivisions of layer B (Proto-Neolithic and Mesolithic horizons) at Shanidar Cave. They are pressure retouched. With the exception of *m*, a "Gravettian" type point, and *s*, a single-shoulder based point, all the specimens shown are blunted-back retouched on one side. Some of them were probably side blades for composite implements. From the points, at least, it is inferred that the bow and arrow were known—a great technological advance.

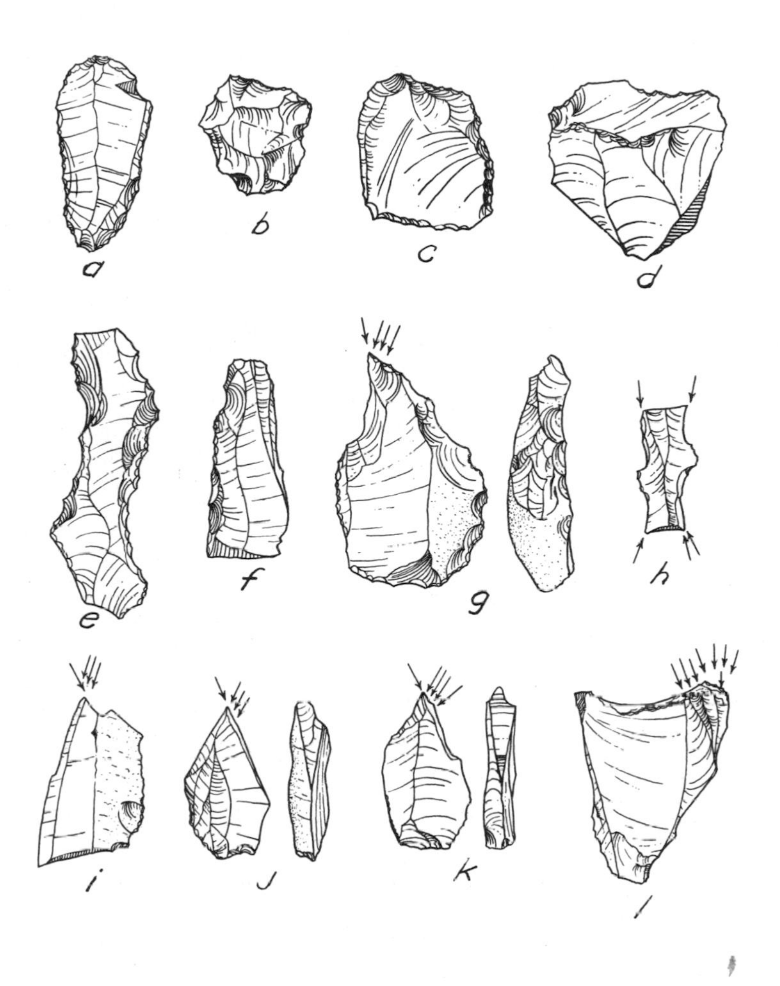

Fig. 3–9 (*a–l*). Flint artifacts from layer C (Upper Palaeolithic, Baradostian horizon) at Shanidar Cave. They were made by percussion striking and pressure retouching, principally on blades. These artifacts indicate a heavy preoccupation with wood-working (very few bone implements were found). A fireside activity requiring special talents is clearly shown by the diversification of the tool kit, indicating gouging, incising, cutting, shaving, and scraping arts, specialized forms being found within each group. *a*, End scraper; *b*, "circular" scraper; *c*, side and end scraper combination; *d*, "nosed" steep scraper on a blade core; *e*, notched or "strangled" blade; *f*, chisel-ended implement; *g*, combination nosed burin (or graver) and end scraper; *h*, multiple-ended burin; *i*, angle-struck burin; *j*, stepped-bit "bec de flute" type burin; *k*, nosed-bit "bec de flute" type burin; *l*, heavy bitted burin with polyhedric facets.

Fig. 3–10. The Shanidar Cave excavation, looking west toward the "find" spot of Shanidar II in the keyway pit. The deepest part of the excavation is at right.

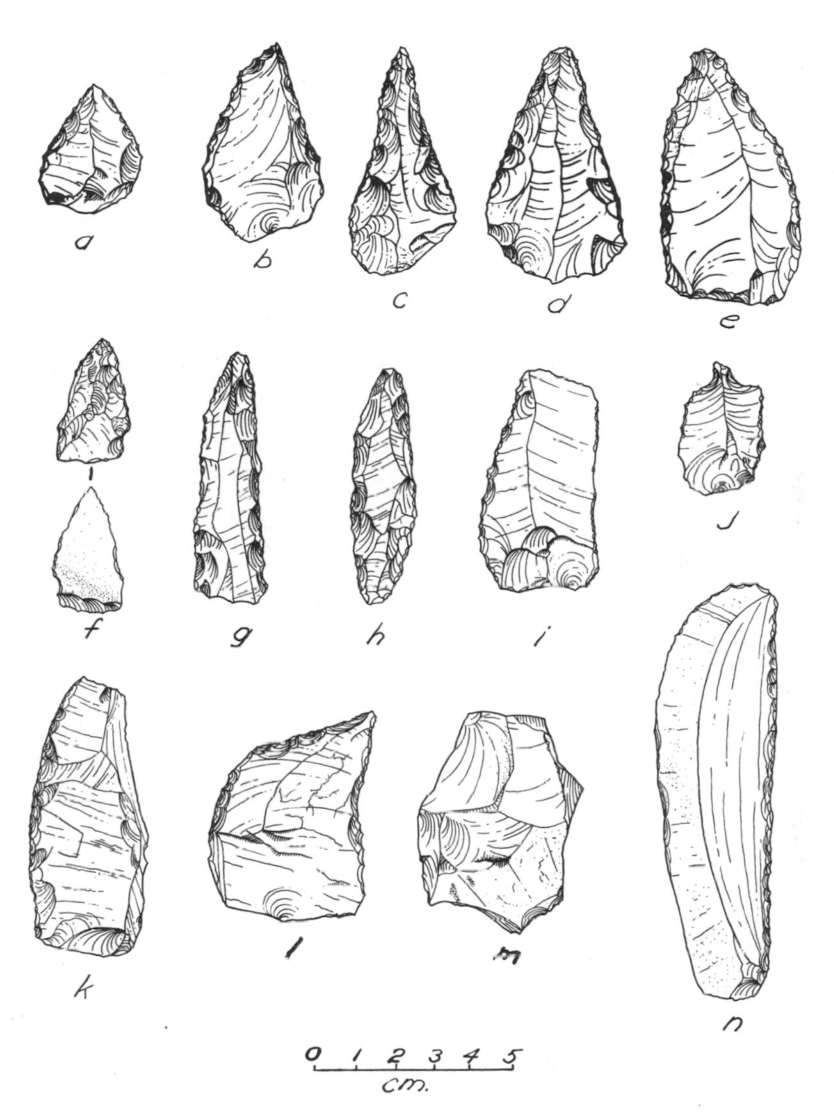

Fig. 3–11. Examples of artifacts from layer D (Middle Palaeolithic Mousterian horizon) at Shanidar Cave. They are percussion struck, and made on unifacial flakes of flint. These represent the simplest implements at Shanidar Cave, presumably used for tipping spears, as skinning knives, and as simple wood-working tools. No Levallois prepared cores were found, although many of the artifacts exhibit "facetted butt" preparation on their basal ends. *a–e*, Typical Mousterian points; *f*, "Emireh"-type point with basal inverse retouch; *g*, elongated Mousterian point; *h*, double-ended point; *i*, convex-edged side scraper; *j*, borer; *k*, convex-edged side scraper on a thick flake; *l*, assymetrically shaped point or "*déjeté*" type side scraper; *m*, flake core; *n*, unusually long side scraper and knife combination.

A TENTATIVE CHRONOLOGICAL CORRELATION OF SELECTED SITES IN THE NEAR AND MIDDLE EAST AND NORTH AFRICA*

TIME SCALE B.P.	LIBYA	PALESTINE	LEBANON	SYRIA	IRAQ	IRAN	AFGHAN.	GENERAL LEVEL OF CULTURE
	HAUA FTEAH	MOUNT CARMEL AND KEBARA	KSÂR 'AKIL AND RAS EL-KELB	YABRUD AND JERF AJLA	SHANIDAR CAVE, ZAWI CHEMI SHANIDAR, AND OTHER SITES IN NORTHERN IRAQ	CASPIAN CAVES, WARWASI AND BISITUN	KARA KAMAR	
10,000	LAYER XIII (10,600 B.P.)[1]	EL-WAD LAYER B; KEBARA LAYER B			SHANIDAR CAVE LAYER B-1 (10,600 B.P.)[6] (12,000 B.P.)[6] LAYER B-2; ZAWI CHEMI SHANIDAR LAYER B (10,800 B.P.)[6]; Open Sites: KARIM SHAHIR, M'LEFAAT, GIRD CHAI; Cave Sites: ZARZI, HAZAR MERD B, BABKHAL (U), PALEGAWRA	HOTU AND BELT (11,500 B.P.)[7]; WARWASI	TRENCH-A LEVEL 4 (10,580 B.P.)[7]	PROTO-NEOLITHIC (Simple food production equipment, "luxury items" and microliths)
15,000	LAYER XVII (18,400 B.P.)[1]	KEBARA LAYER C	KSÂR 'AKIL COMPLEX 1	YABRUD SHELTER III	↕ Hiatus			MESOLITHIC (Microlith and backed-blade industries)
20,000		EL-WAD LAYER C						
25,000								
		EL-WAD LAYER D; KEBARA LAYER D	KSÂR 'AKIL (28,500 B.P.)[3] (6.50-7.00m)		TOP LAYER C (28,000 B.P.)[6]			UPPER PALAEOLITHIC (Blade, burin, and end-scraper industries)
30,000	LAYER XXVI (34,000 B.P.)[1]	EL-WAD LAYER F; KEBARA LAYER F (34,700 B.P.)[2]		YABRUD SHELTER II LAYERS 2-5	BOTTOM LAYER C (35,000 B.P.)[6]	WARWASI	LOWER LOESS (>34,000 B.P.)[7]	
35,000		ET-TABŪN LAYER B (39,500 B.P.)[2]	KSÂR 'AKIL COMPLEX 2	YABRUD SHELTER II LAYER 6	↕ Hiatus			
40,000		(41,000 B.P.)[2] ET-TABŪN LAYER C	KSÂR 'AKIL COMPLEX 3 (44,000 B.P.)[4] (16m)	JERF AJLA TRENCH A (43,000 B.P.)[7]				
45,000	LAYER XXXII (46,000 B.P.)[1]				TOP LAYER D (46,000 B.P.)[6]; Cave Sites: SPILIK, HAZAR MERD C, BABKHAL (L)			MIDDLE PALAEOLITHIC (Flake, point, and side-scraper industries)
50,000			RAS EL-KELB BASAL LAYER (>52,000 B.P.)[5]	YABRUD SHELTER I	UPPER LAYER D (50,000 B.P.)[6]; Open Sites: TARJIL, SERANDUR, TELEGRAPH POLE 26/22	WARWASI; BISITUN BASAL LAYER		
55,000								
60,000					(Base of Shanidar Cave estimated 100,000 B.P.)			

Fig. 3–12. A tentative chronological correlation. The dates given for the various sites are round numbers based on carbon-14 determinations reported as follows: 1) C. B. M. McBurney, *Advan. Sci.*, **18**, 496 (1962); 2) K. P. Oakley, *ibid.*, **18**, 415 (1962); 3) J. Perrot, in R. J. Braidwood and G. R. Willey, "Courses toward Urban Life," *Viking Fund Publ. No. 32* (1962), p. 150; 4) J. Franklin and S. J. Ewing, personal communication (Sept. 1962); 5) D. A. E. Garrod and G. Henri-Martin, *Bull. Musée de Beyrouth* **16**, 4 (1961); 6) R. S. Solecki (see text); 7) C. S. Coon, *The Seven Caves* (Knopf, New York, 1957), pp. 210, 252, 253, 315. Layers without carbon-14 dates are given relative positions in their respective sequences.

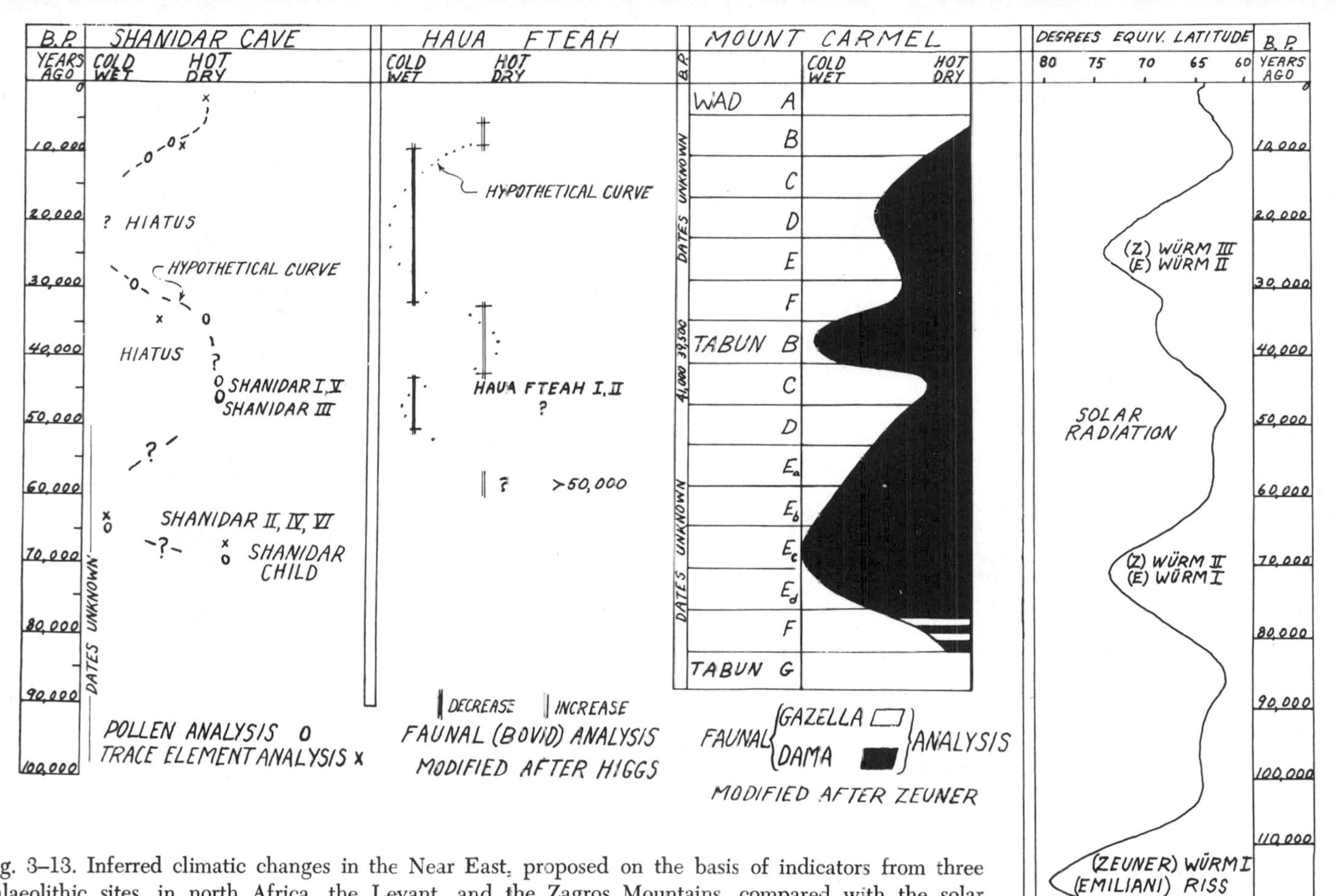

Fig. 3–13. Inferred climatic changes in the Near East, proposed on the basis of indicators from three Palaeolithic sites, in north Africa, the Levant, and the Zagros Mountains, compared with the solar radiation curve for the upper Pleistocene.

Fig. 3–14. The skull of Shanidar II, as it appeared when discovered, crushed under a rockfall. The skull is lying on its right side, face to the front. The stone to the left was found directly over the left temple. This individual lived in a cool climate.

Fig. 3–15. The skull of Shanidar I, dated about 44,000 B.C., as it looked in the Shanidar laboratory after removal from the cave. The vertebrae are in place. This individual lived in a warm climate.

Fig. 3–16. The skull, restored, of Shanidar I. (Smithsonian Institution)

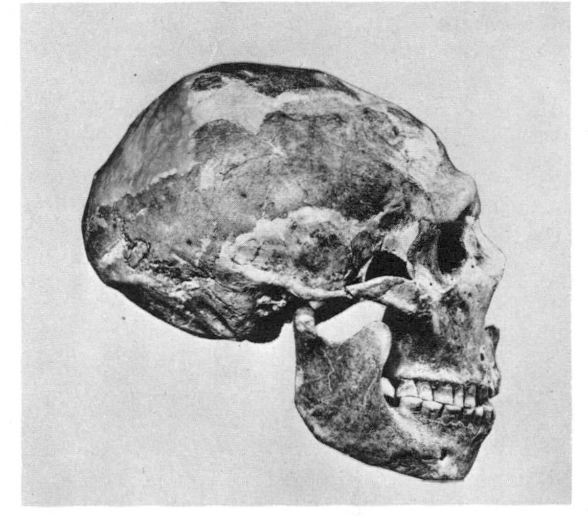

Fig. 3–17. (A) A knife, 20.9 centimeters long, made of a flint blade held with a tarry substance in a bone handle, found in the Proto-Neolithic cemetery at Shanidar Cave. (B) A laterally grooved bone handle, 21.7 centimeters long, which presumably held flint blades, from the Proto-Neolithic layer at the Zawi Chemi Shanidar village site. This was probably a sickle for cutting grasses.

Chapter 4

Upper Palaeolithic Cultures in Western Europe

Denise de Sonneville-Bordes

Western Europe had its first "great civilization" during Upper Palaeolithic times (sometimes called the Age of Reindeer). For about 25,000 years—from 35000 or 40000 to about 10000 B.C.—this area was inhabited by powerful tribes of *Homo sapiens* physically like us. Like the Mousterians, their predecessors and perhaps ancestors, they dwelled in caves, rock shelters, and open-air camp sites; and they lived by hunting and fishing, in a glacial environment. Like their predecessors they made tools and weapons by chipping hard stones, mainly flint, but because they used new techniques, they made more specialized and more complex implements than the Mousterians. They also made tools of bone, ivory, and antlers (mainly of reindeer and red deer). Above all, with these Upper Palaeolithic peoples, mankind took a decisive step forward in mental development, as the art objects (found for the first time in an archaeological context) and paintings on the walls of the caves, as well as the greatly increased number of human burials, bear witness.

This earliest culture of *Homo sapiens* in western Europe was not so closely knit as one would have expected among primitive peoples. The various cultures that the component peoples developed, successively or simultaneously, have only rather general characteristics in common. The greatest complexity is in southwestern France, between the Loire River and the Pyrenees. This region, the richest in sites and cultural remains, is rightly considered the classical region of definition of these cultures. Most of the names of Upper Palaeolithic cultures (the Périgordian, the Aurignacian, the Magdalenian, and the Azilian) are derived from type sites in this region, the only exception being the Solutrean. From the stratigraphical sequences and the typological comparisons, Henri Breuil and Denis Peyrony, the famous French prehistorians, established a general classification (Table 4–1). During Upper Palaeolithic times, local factors played a considerable role in the choice of sites for human occupation. Thus, there are hiatuses and regional differences outside the classical region.

Climate and Fauna

Climate, fauna, vegetation, and human life were still, at this time, under the direct influence of the great glaciations that were the main features of Pleistocene European prehistory. Coming down from the Scandinavian mountains, a thick and enormous ice sheet—an *inlandsis*—covered the lowlands of northern Europe and stretched out over the shallow North Sea. Near the borders of this huge glacier, in what is now England and Wales and the Belgian Ardennes, were areas where peoples lived (Fig. 4–1). The sites of southern Germany and Switzerland were under the influence of the great alpine glacier, which spread out to the north over the Swabian Jura and Bavarian plateau and to the west of the French Jura down to the Rhone River. The French Massif Central, less high, shows only localized glacial features, and human penetration of this area was possible along the main valleys. The Pyrenees, where glaciers formed only on the peaks, were never impassable, and cultures there spread through the lower valleys, on the Atlantic side between what are now the Basque countries and the Spanish Cantabrian Mountains, on the Mediterranean side between present-day Roussillon and Catalonia.

The Upper Palaeolithic cultures appeared during the Würm glaciation, the last of the four great glaciations in Europe. As in the glaciations that preceded it, there were minor advances and regressions during the Würm glaciation. The advances, or stadials, are called Würm I, II, III, and IV; the intervening regressions, or interstadials, are called Würm I/II, II/III, and III/IV. Following the Mousterian culture, which spanned Würm I and Würm II and ended during the Würm II/III interstadial, the Upper Palaeolithic began at the end of Würm II/III. It continued through Würm III and Würm IV and disappeared at the very beginning of postglacial times.

Thus, western Europe had, in general, a severe climate during these times, with variations depending on latitude and on position relative to the ice sheets. That the climate was severe can be seen from the fauna and the deposits in rock shelters. Under the roofs of the caves and shelters, angular congelifracts became detached from the wall by

TABLE 4–1

Upper Palaeolithic cultures in western Europe.

SOUTH-WESTERN FRANCE	NORTHERN SPAIN	ITALY	BELGIUM	GREAT BRITAIN	WEST GERMANY	SWITZERLAND
Azilian	Azilian		Epi-Magdalenian	Epi-Magdalenian	Epi-Magdalenian	Epi-Magdalenian
Final	Final		Final	Creswellian		Final
Upper (V–VI)	Upper (V–VI)		Upper		Upper	Upper (V–VI)
Middle (III–IV)	Middle (III–IV)		Magdalenian		Middle (?)	Middle
Lower (I–II)	Magdalenian				Magdalenian	Magdalenian
Magdalenian						
Upper	Upper					
Middle	Middle					
Lower	Solutrean					
Solutrean						
Final (V)						
Proto-Magdalenian		Grimaldian				
Final IV						
Evolved III	Evolved	Evolved	Evolved			
		Périgordian	Périgordian			
Middle II	II		II			
I	I		I			
Aurignacian	Aurignacian	Aurignacian	Aurignacian	Aurignacian (?)	Aurignacian	Aurignacian (?)
Lower	Lower					
Périgordian	Périgordian					
Mousterian	Mousterian	Mousterian	Mousterian	Mousterian	Mousterian	Mousterian

the action of the intense cold. When the climate was less severe and precipitation was greater, layers of sand and clay were often deposited by the water. These rarer periods of warm climate and greater humidity are marked by the formation of stalagmitic floors. From the succession of these different types of sediments in the numerous sites of southwestern France, Denis Peyrony and later François Bordes were able to formulate a general outline of the climate in this region during Upper Palaeolithic times.

The fauna of the Upper Palaeolithic is well known from the remains of game found in the archaeological layers and from the numerous examples of Upper Palaeolithic art. Like the Mousterian fauna, this is a cold fauna. It includes arctic animals such as reindeer, musk ox, wolverine, arctic fox, arctic hare; animals that have completely disappeared from our planet, such as mammoth, woolly rhinoceros, cave bear; mountain game, such as mountain goat, chamois; and great herds of horses, oxen, bison, and red deer. The Saiga antelope (now a dweller of the cold steppes of Central Asia), till then almost unknown in Atlantic Europe, made occasional appearances in numerous herds in the hills of southwestern France, but only during the Magdalenian. The cold fauna, with reindeer, penetrated at last into the meridional peninsulas of Spain and Italy.

Technical Innovations

It generally has been believed that blade *débitage* was a characteristically Upper Palaeolithic innovation and that, in the Upper Palaeolithic, blades were the basic stone form from which implements were manufactured. In truth, this invention dates back to the Acheulean, and blade *débitage* was widely used in some Mousterian cultures. Besides blades (Fig. 4–2, Nos. 4–7; Fig. 4–3, No. 1; Fig. 4–4, Nos. 1, 5; Fig. 4–5, No. 7; Fig. 4–6, Nos. 7, 13, 16) and bladelets (Fig. 4–2, No. 2; Fig. 4–3, No. 5; Fig. 4–4, No. 2; Fig. 4–6, Nos. 8–12), the men of Upper Palaeolithic times also made flakes (Fig. 4–2, Nos. 8, 10; Fig. 4–3, Nos. 2–4, 6, 7; Fig. 4–6, No. 14). At some stages of their cultures—for instance, in the Aurignacian—flake implements are more numerous than blades. So the Upper Palaeolithic is not exclusively a "blade culture," even though a good proportion of the tools, and, in some stages (such as the Upper Magdalenian) an overwhelming majority of them, are blade or bladelet tools.

These populations showed true originality in inventing new types of tools. However, the Mousterian tool types—for instance, side scrapers (Fig. 4–2, No. 9) or denticulate tools (No. 10), which were especially numerous in the early Périgordian—were made till the end of the

Palaeolithic. The typical tools of the Upper Palaeolithic, which had occurred occasionally in the Mousterian, were now the most important tools, with varied subtypes: end scrapers (Fig. 4–2, No. 8; Fig. 4–3, No. 1; Fig. 4–6, No. 2), burins (Fig. 4–3, Nos. 3, 4; Fig. 4–4, No. 1), and borers (Fig. 4–4, No. 3). Multiple tools were numerous: double scrapers (Fig. 4–5, No. 2), multiple burins (Fig. 4–2, No. 5; Fig. 4–4, No. 5; Fig. 4–6, No. 1), multiple borers (Fig. 4–6, Nos. 4–6). Composite implements were invented, and two or more different tools were made on the same piece of flint—for instance, end scraper and burin (Fig. 4–4, No. 4; Fig. 4–6, No. 13). This common tool assemblage gives a kind of uniformity to the Upper Palaeolithic cultures, even if there is a considerable range of variation in the proportions of the different types from culture to culture. In addition to the tools common to these cultures there were tools found only in one culture and characterized either by the technique of retouch or by the general morphology, or by both: in the Périgordian, backed points with abrupt retouch (Fig. 4–2, nos. 6, 7); in the Aurignacian, thick scrapers (Fig. 4–3, nos. 2, 6, 7); in the Solutrean, foliate tools with a very flat parallel retouch (Fig. 4–6, nos. 1, 5–7).

In making tools of antler, ivory, and bone, the Upper Palaeolithic men were true innovators. They used two different techniques of *débitage*. In the first or "wedge" technique, a beveled reindeer antler was used as a wedge to split another antler lengthwise. The second, and later, technique was the "groove and splinter" method: with a burin, two parallel grooves were made in the bone or antler, down to the marrow. The splinter of bone or antler thus detached was scraped, then polished on sandstone to form the tool. The splinters were made into various tools, some of them quite elaborate: bone points with split bases (Fig. 4–3, No. 8; Fig. 4–7, No. 2) or beveled bases (Fig. 4–3, No. 9); needles (Fig. 4–5, No. 4); half-cylindrical wands (Fig. 4–7, No. 7); harpoons (nos. 3–6); spear-throwers (No. 1); and thong-stroppers or shaft straighteners (No. 8). The tools were hollowed out, smoothed, and shaped to make them easier to grasp, bind on, and handle. Decorative patterns and even, during the Magdalenian, engraving and carving sometimes made these implements real works of art (Fig. 4–7, nos. 1, 8).

Stratigraphical Sequences

In southwestern France, the rich and numerous Upper Palaeolithic sites can be divided into two main zones (Fig. 4–1). The first comprises the limestone countries north of the Garonne River. Open to the Atlantic Ocean to the west, this zone forms a belt around the high and cold lands of the Massif Central and includes Poitou, Charentes, and

Périgord. The second zone includes the Pyrenees region and its piedmont, from the Adour Valley to the Ariège. To these southwestern areas may be added the valleys of the Saône and the Rhone, north-south passageways where cultural territories meet: the Solutrean does not cross the Rhone, nor does the Magdalenian cross the Durance. The Upper Palaeolithic in this region has been used as a reference base for the rest of Europe.

In the excavations made there, mainly by Denis Peyrony and Élie Peyrony near Les Eyzies (Dordogne) at La Ferrassie (Fig. 4–8, top) and Laugerie-Haute (Fig. 4–8, right) and later by François Bordes at Laugerie-Haute and by Hallam L. Movius at the Pataud Shelter, the general stratigraphic sequence is as follows. After the enormous Mousterian deposits, which represent the Würm I and the Würm II stadials, the Upper Palaeolithic begins, in the reddish and clayey layers of the Würm II/III interstadial, with a culture, the early Périgordian (La Ferrassie, layer E), whose lithic components are often crushed as a result of frost action (Fig. 4–2, No. 10). Deposits of thermoclastic *éboulis* with cold fauna, representative of the beginning of Würm III, contain the various levels of the Aurignacian, I–IV (La Ferrassie, layers F, H, H′, and H″). In a layer containing more sandy, small *éboulis*, formed when the climate had moderated slightly, are found remains of the evolved Périgordian (La Ferrassie, layers J, K, and L). Apparently this moderation did not last, for remains of the final Périgordian are found in a layer containing large thermoclastic *éboulis* (Laugerie-Haute, layers B and B′). Above it comes an exceptional culture, the Proto-Magdalenian (Laugerie-Haute, layer F), also found over the same Périgordian by Movius at Pataud. The Aurignacian V has been found, to date, only at Laugerie-Haute. The cold climate persisted into the Lower Solutrean, as shown by the fact that these levels, too, contain thermoclastic material (Laugerie-Haute, layer H′), but the Middle Solutrean (layer H″) and the Upper Solutrean (layer H″′) evolved during the period of climatic moderation of the Würm III/IV interstadial. In this same interstadial we find the remains of Lower Magdalenian culture; Phases I, II, and III are found in sandy and clayey layers, where the proportion of the angular *éboulis* produced by frost action on the walls of the shelter increases from the bottom to the top (Laugerie-Haute, layers I′, I″, and I″′).

In the earliest Würm IV deposits there is extensive evidence of the collapse of the roofs of shelters, caused by the erosion of frost action on the interior of the shelters. This is followed by deposition of layers made of loose, angular *éboulis*. In these layers are found materials of the Upper Magdalenian, classified in three phases—IV, V, and VI—at La Madeleine shelter. Heavier precipitation and more temperate conditions progressively brought the great glacial period to an end. The sandy and clayey layers (often rich in snail shells) of the Magdalenian and Azilian, last of the Palaeolithic cultures, are frequently covered by stalagmitic floors. The cold reindeer fauna is now replaced by a residual fauna, with red deer, wild boar, and lynx. Mesolithic times begin.

Some dates obtained by the radiocarbon technique by Hessel de Vries of the University of Groningen, the Netherlands, make it possible to determine the stratigraphical chronology with some precision. These dates are 26930 B.C. ± 250 years (GRO-1491) for the Aurignacian of the Caminade shelter (excavations by François Bordes, Denise de Sonneville-Bordes, and B. Mortureux); 19785 B.C. ± 250 years (GRO-1876) for the Proto-Magdalenian of Laugerie-Haute (excavations by F. Bordes) and about 18250 B.C. for the same culture at Pataud (excavations by Movius)—a rather surprising discrepancy, for these two layers seem to be strictly contemporaneous; and 18700 B.C. ± 300 years for the Lower Solutrean of Laugerie-Haute (excavations by François Bordes).

Typological Classification

From the study of tool assemblages we can trace the evolution of the cultures whose succession is established by the stratigraphical sequences.

The cycle of the Aurignaco-Périgordian cultures, which occupy the beginning of the Upper Palaeolithic, appears from the work of D. Peyrony to be more complex than H. Breuil had believed. Breuil had envisioned a linear development of a single civilization, the Aurignacian, which he subdivided into Lower Aurignacian, with Châtelperron points; Middle Aurignacian, with split-base bone points; and Upper Aurignacian, with Gravette points. From his excavations at La Ferrassie and Laugerie-Haute, Peyrony distinguished in this complex two different cultures: the Périgordian, which combines the inferior and the superior stages of Breuil's Aurignacian, and the Aurignacian (*sensu stricto*), which is Breuil's Middle Aurignacian. Recent excavations at Laugerie-Haute and Pataud have somewhat complicated Peyrony's schema.

The Périgordian has, in every one of its stages, points with backing by abrupt retouch—Châtelperron points with a curved back (Fig. 4–2, No. 6) in the lower stage and Gravette points with a straight back (No. 7) in the evolved stages. In the lower stage, the ordinary tools, often somewhat awkwardly made, are associated with Mousterian-like tools —side scrapers (No. 9) and denticulates (No. 10). In the later stages the proportion of these Mousterian-like tools diminishes, and the ordinary tools—mainly end scrapers (No. 8) and burins, often multiple ones (No. 5)—are better made. Not found at earlier levels, backed bladelets

(No. 2) make their appearance. Special tools, such as truncated blades (No. 4), tanged Font-Robert points (No. 3), and tiny multiple Noailles burins (No. 1), first appear in the evolved Périgordian. No art objects and very few bone tools occur in the Lower Périgordian. In the evolved Périgordian, where only a few types of bone tool occur, and these not numerous, little female statuettes called "Venuses" are found.

The Aurignacian is characterized by implements made with a special kind of wide, heavy retouch (Fig. 4–3, No. 1); by thick carinate scrapers (No. 6) and nosed scrapers (No. 2); by busked burins (No. 4); by bladelets with a semi-abrupt retouch (No. 5); and by a rich, elaborate set of bone tools. The evolution of this culture can be clearly followed at La Ferrassie: decrease in the proportion of implements with the special Aurignacian retouch, increase in the proportion of burins, and increase in the ratio of nosed scrapers to carinate scrapers. The split-base bone points (No. 9), typical of the very earliest stage, are replaced first by flat, lozenge-shaped points, then by lozenge-shaped points with an oval section, then by biconical bone points with a circular section. The very late Aurignacian V of Laugerie-Haute has dihedral burins (No. 3), thick denticulate scrapers (No. 7), and bone points with a single bevel on which the bone canals can be seen (No. 8)—a feature which occurs only at this level of this site during the Palaeolithic. The first animal paintings, in red or black, appear in the early Aurignacian, together with coarse, deeply grooved engravings.

Everywhere, in France, the Aurignacian and the Périgordian complexes are totally independent of each other, without reciprocal influences. The typical tools of the one are not found in the other, and the common tools occur in different proportions. The technical and typological continuity between the Lower and the Upper Périgordian seems to indicate that the Périgordian and the Aurignacian evolved contemporaneously.

The Proto-Magdalenian is known only from Laugerie-Haute and Pataud. The tools are especially beautiful—well-retouched blades; numerous burins, often double (Fig. 4–4, Nos. 1 and 5); end scrapers and burin-scrapers (No. 4); borers (No. 3); numerous backed bladelets (No. 2); and sundry bone tools. The relation of the Proto-Magdalenian to the other cultures is yet to be established.

All known subdivisions of the Solutrean occur only in southwestern France, and even there they are infrequent. All have tools with the flat Solutrean retouch, with parallel edges, and a very uniform distribution of the common tools—simple or double-end scrapers (Fig. 4–5, No. 2) and a few burins, borers, and composite tools. The special tools, always plentiful, change from one substage to the other. Among them are unifacial points (No. 5), laurel-leaf points (No. 6), shouldered points (No. 8), and willow-leaf points (No. 7), all with the characteristic Solutrean retouch. Bladelet tools are rare; Mousterian-like tools are fairly numerous. Despite the invention of the eyed needle (No. 4), the bone tools are mediocre and lack variety. This culture disappears after showing a remarkable mastery of flint work. The Solutreans also made beautiful bas-reliefs of animals.

The Magdalenian has an elaborate assemblage of bone tools, which are found in abundance and which Breuil used for purposes of classification. The lithic assemblages gathered by Peyrony increase the precision of classification. The Lower Magdalenian is characterized by several types of bone point (Fig. 4–7, No. 2), among them grooved points. Among the characteristic stone tools are small flakes with a very abrupt retouch [the "raclettes" (Fig. 4–6, No. 7)] and small triangles (No. 10), the first of numerous geometric forms to appear in the Palaeolithic. After the Magdalenian III, in which the first half-cylindrical wands appear (Fig. 4–7, No. 7), the relative proportions of the various types of stone tools remain the same. There is remarkable unity among the Magdalenian lithic assemblages. They include burins, often double (Fig. 4–6, No. 1) and always plentiful; end scrapers (No. 2), less numerous; sundry composite tools, mainly scraper-burins (No. 13), borers, *and* microborers (Nos. 4–6) of various types; and numerous backed bladelets, sometimes denticulate (No. 12). In the Upper Magdalenian, harpoons with one side barbed (Fig. 4–7, No. 4) and then with two sides barbed (No. 5) are added to this lithic assemblage. In the late Magdalenian some new types are added to the usual flint-tool kit: parrot-beaked burins (Fig. 4–6, No. 14), shouldered points (No. 11), and foliate points. Some sites also contain numerous geometrical microliths: triangles, semilunates, rectangles (No. 9), trapezoids (No. 8), and the so-called microburins (No. 3), the waste products of the fabrication of the other forms. All these geometrical microliths occur frequently in the Mesolithic cultures that follow. Of all the early cultures the Magdalenian is by far the richest in art forms, be they engraved or sculptured weapons, like the spear-throwers (Fig. 4–7, No. 1), or implements like the shaft straighteners (No. 8), or paintings and engravings on cave walls.

The Azilian, which marks the end of the Upper Palaeolithic, has a less varied lithic assemblage of rather awkward, smaller tools. They include mainly flint points with a curved back–Azilian points (Fig. 4–9, No. 3) –and very short end scrapers (No. 2). Since tools of these types are already present in most late Magdalenian assemblages, it is very likely that there was a connection between the two cultures. Wands, punches, and flat harpoons with or without a basal perforation shaped like a buttonhole (No. 1) were made of the antlers of red deer. Some pebbles engraved with geometric patterns or painted with red and black dots or bars are the only manifestation of artistic expression.

North of Provence, in the Rhone and the Saône valleys, Aurignacian is known and Upper Périgordian is well represented. Ardèche is rich in Lower Solutrean sites. At Solutré, near Mâcon, over the Upper Pèrigordian layers, in which thousands of horse bones occur, are Middle and Upper Solutrean levels, but no shouldered points occur in them. Throughout the Rhone Valley, no parrot-beaked burins occur in the Upper Magdalenian.

Extension, Geographical Gaps, and Facies

Outside this region of southwestern France, the Upper Palaeolithic is neither so complex nor so continuous (Table 4–1). Spain, more than any other country, belongs to the same cultural community, in spite of the barrier of the Pyrenees. The Atlantic coast was more widely used than the Mediterranean coast as a route of communication. The sites in Cantabrian Spain between the sea and the high Cantabrian Pyrenees are very numerous until one reaches the present-day provinces of Asturia, whereas Catalonia has very few Palaeolithic sites. Only traces of the early Périgordian are found in Catalonia. On the other hand, the Aurignacian, with split-base points, is widely found, mainly in Cantabrian Spain. Here the Upper Périgordian is less rich and less complex than it is in France. The Solutrean, which first occurs here in its middle stage, with laurel-leaf points, has some original tools, typically Spanish: laurel leaves with a concave base (Fig. 4–5, No. 1) and curved bone points with a median flat area striated by the fine oblique lines (No. 3). In the Asturias, associated with tools of classical type, are found tools of quartzite, often coarsely made. These include side scrapers, planes, thick-flake end scrapers, denticulate and notched tools, and large retouched blades. Despite changes in other aspects of the assemblage, these tools continue, with the same distribution of types, in the Magdalenian. The Magdalenian first occurs in Cantabrian Spain in its median phase (Magdalenian III), but it is mainly and magnificently represented by Stages V and VI, with special laterally perforated harpoons, typically Spanish (Fig. 4–7, No. 3). As in France, it is overlain by Azilian layers, which are numerous in the Basque country. In eastern Spain the cultures attributed to the Upper Palaeolithic, such as Solutrean and Magdalenian at the Parpallo site, near Valencia, are very special and seem to be partially related to the Mediterranean cultures.

Provence, which lies east of the Rhone and fronts on the Mediterranean, and Italy down to Mount Circeo, near Rome, belong to the same cultural complex as France at the beginning of the Upper Palaeolithic, yielding Aurignacian with split-base points and Upper Périgordian with Gravette points. But toward the end of Würm III, these Mediterranean areas, where neither the Solutrean nor the Magdalenian penetrated, separated from the continent, culturally speaking. A special culture, the Grimaldian, a kind of Périgordian, occurred early and continued to the end of glacial times. This is characterized by geometrical microliths, round scrapers, and microburins, and is poor in bone tools.

In Belgium, in the caves of the Ardennes, the Aurignacian with split-base points overlies important Mousterian levels, without any intercalation of Lower Périgordian. Here the Aurignacian is similar to the Aurignacian of France, though perhaps poorer in typical blades and in busked burins. The Upper Périgordian seems poorer and rarer. There is a long gap, and then evidence of a Palaeolithic occupation of the Upper or late Magdalenian phase, at the end of Würm IV. This differs from the late Magdalenian in France: there are no parrot-beak burins, there is an abundance of borers, and barbed harpoons are rare. There are some Azilian points. It differs from, and is certainly older than, the epi-Magdalenian of open-air sites, where Azilian points and tanged points are abundant. This Epi-Magdalenian of a slightly later date is very common on the plains of northern Europe. There are few works of art in this Belgian Magdalenian.

The caves of Great Britain have produced rather poor assemblages, the oldest of them being attributed to the Aurignacian, the more recent and more numerous, to a very late Magdalenian-like culture, the Creswellian. The bone-tool assemblage includes needles, arrow-shaft straighteners, and barbed harpoons, sometimes of Magdalenian type, sometimes of a later type—flat, with angular barbs. Curved-back points in the shape of large trapezoids (Fig. 4–6, No. 16), truncated or bitruncated, link these cultures to the Epi-Magdalenian of the Northern Plains, but the fauna is still of glacial type, with reindeer.

In western Germany, the Rhine Valley and the Bavarian plateau were occupied by Palaeolithic man at two different times. Here the Aurignacian with split-base points has a rich and original assemblage of stone tools. This includes numerous pieces with Aurignacian retouch, numerous burins (but no busked ones), and a few carinate scrapers or nosed scrapers. Animal statuettes, the oldest known, have been found in the Vogelherd Cave in Württemberg. The second occupation occurred very late, in the last Magdalenian. There are many open-air sites and caves in the Swabian Jura. The assemblage, which includes needles and harpoons, differs widely from the French assemblage with respect to stone tools. Borers (Fig. 4–6, No. 6) are plentiful and varied; there are no parrot-beak burins; shouldered points, with an oblique truncation (Fig. 4–7, No. 15) show a relationship with the cultures of the open-air sites to the north, near the edges of the glaciers around present-day Hamburg (the Hamburgian culture).

Paradoxically, the greatest demographic expansion occurred during the periods of more severe climate—the Aurignacian and Upper Magdalenian. During the periods of milder climate—for example, the Lower and even the Upper Périgordian, the Solutrean, and the Lower Magdalenian—the cultures were less widespread and the populations were more scattered. Except in Spain, the Lower Périgordian, the Solutrean, and the Lower Magdalenian are totally unknown outside the classical region and the Rhone Valley. During the Upper Magdalenian, the density of population was relatively high in France, as evidenced by the great number of sites occupied for the first time, and by the richness of the sites in animal remains, tools, and works of art. Open-air sites, sometimes with pavements of pebbles, are often found, along the rivers, near the fords where herds of game crossed the stream, or in places where salmon fishing was good (the vertebrae of salmon are very common in the archaeological layers).

During this period there was extensive penetration of mountainous areas, where previously only a few tribes of Aurignacian or Upper Périgordian peoples had made their way. In the Massif Central, small sites are scattered along the high parts of the Loire and Allier valleys. In the Pyrenees, man occupied the valleys he had penetrated during the Aurignaco-Périgordian but had relinquished during the Solutrean. The alpine regions, liberated by the receding glaciers, were progressively occupied. Peoples coming from Jura, Dauphiny, and Savoy settled the south of what is now Switzerland, while others from the Rhine and Danube valleys settled in the north of that country. This late Magdalenian, without parrot-beak burins and with some shouldered points with oblique truncation, seems closely related to the late Magdalenian in Germany. Some of the bone tools, such as the barbed harpoons (Fig. 4–7, No. 6), are very original.

The end of glacial times was fatal to this striking human expansion. The disappearance of the cold fauna and the replacement of the steppe, rich in game, by forests was followed by a demographic regression and breakup of the Upper Palaeolithic cultures, resulting in the traditions that are grouped together under the general names of Mesolithic and Epi-Palaeolithic. The relative cultural unity of western Europe was broken, and man's first great period of creative activity came to an end.

Ways of Life

Man of the Upper Palaeolithic lived by hunting and fishing, as man of the Mousterian did. The coexistence of numerous tribes in neighboring sites suggests a delimitation of hunting territories and a fairly high level of social organization. The fact that huge or dangerous animals were frequently killed suggests that there was a high degree of coordination of activity in the society. The habitation sites are at the entrance of caves and in rock shelters, so very few implements are to be found inside the deep, painted caves such as that of Lascaux. Fire was in general use and had been at least since the Mousterian. Open-air encampments, numerous mainly during the Aurignacian, the Upper Périgordian, and the Magdalenian, were perhaps hunting camps or seasonal habitations.

The custom of burying the dead, known since the Mousterian, was now general, and as a result the men of Upper Palaeolithic times are physically well known. The body, buried in a flexed position, was almost always sprinkled with red ocher and adorned with necklaces and armrings of shells and pierced teeth. It is sometimes a woman holding a very young (perhaps stillborn) child in her arms. A single burial is known from the Lower Périgordian, from Combe-Capelle, Dordogne—a man of small stature, with prominent brows. Numerous complete skeletons of men belonging to the Cro-Magnon race have been found in Aurignacian layers. Human remains from the Upper Périgordian are generally fragmentary, as are those from the Solutrean, and this gives great importance to the skeleton found in the Proto-Magdalenian at Pataud, by Movius, as a representative of man at the end of Würm III. From Magdalenian III on, burials were more common, either because there was an increase in population or because there was increased concern over the fate of the deceased. Magdalenian man of the Chancelade race is different from his predecessors but is still dolichocephalic.

Art, the most impressive innovation of the Upper Palaeolithic, was not constantly and everywhere at the same height. In England and Belgium, examples are few or mediocre. Art objects are well dated by the assemblages which are to be found with them in the archaeological layers. The oldest examples are the animal statuettes of the Aurignacian of Vogelherd, Germany. The statuettes of fat women, called "Venuses," are Aurignaco-Périgordian. Their geographical distribution is very wide, from the Atlantic Ocean to the Ukraine, but there are none in Spain. In Magdalenian times, representations of more slender women are engraved on stone or bone. The last one to be found was discovered in 1962 by F. Bordes in the Magdalenian VI of Couze, Périgord (Fig. 4–10). Beautiful realistic or stylized art objects abound in the sites of France, Spain, and Switzerland, mainly in the Upper Magdalenian levels, but elsewhere they are scarce or totally lacking. Cave art is more limited in distribution. The bas-reliefs of animals and, less often, of human figures, which begin in the Upper Solutrean and continue into the Magdalenian, are found only between the Loire and the Pyrenees. Wall engravings and paintings, invented by the Aurignacians, are found throughout the Upper Palaeolithic, especially during the Magdalenian. They are numerous in France and Cantabrian Spain, but are represented in Italy only by some schematic figures in the southern part of the peninsula. They are unknown in Great Britain, Belgium, Germany, and Switzerland. The cave art of eastern Spain, attributed to the Upper Palaeolithic by Breuil, is probably more recent.

Conclusion

Despite the appearance of continuity, western Europe in the Upper Palaeolithic did not have a constant cultural unity. Except in the extraordinarily active and creative center of southwestern France and, up to a point, Cantabrian Spain, this brilliant civilization knew only two great periods of expansion—the Aurignacian, at the beginning of Würm III, and the Magdalenian, during Würm IV. These two great cultures differed considerably east and west of the Rhine, which seems to have been a cultural border. It was between the Loire and the Pyrenees that the art of Palaeolithic man knew its longest and most complete development.

BIBLIOGRAPHY

D. de Sonneville-Bordes, "Le Paléolithique supérieur en Périgord," thesis, Faculté des Sciences de Paris, Delmas, Bordeaux (1960). D. de Sonneville-Bordes, *L'Age de la Pierre* (Presses Universitaires de France, Paris, 1961).

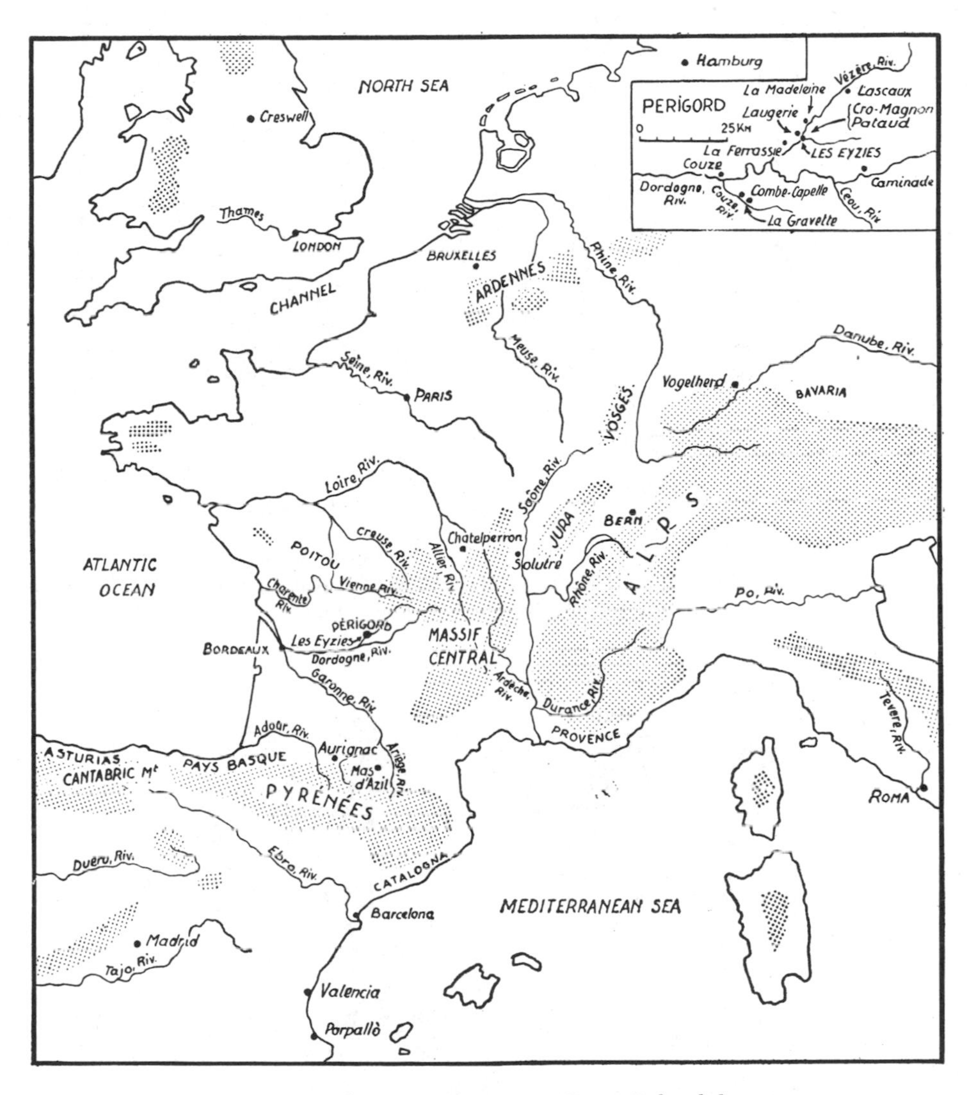

Fig. 4–1. Map of western Europe in Upper Palaeolithic times.

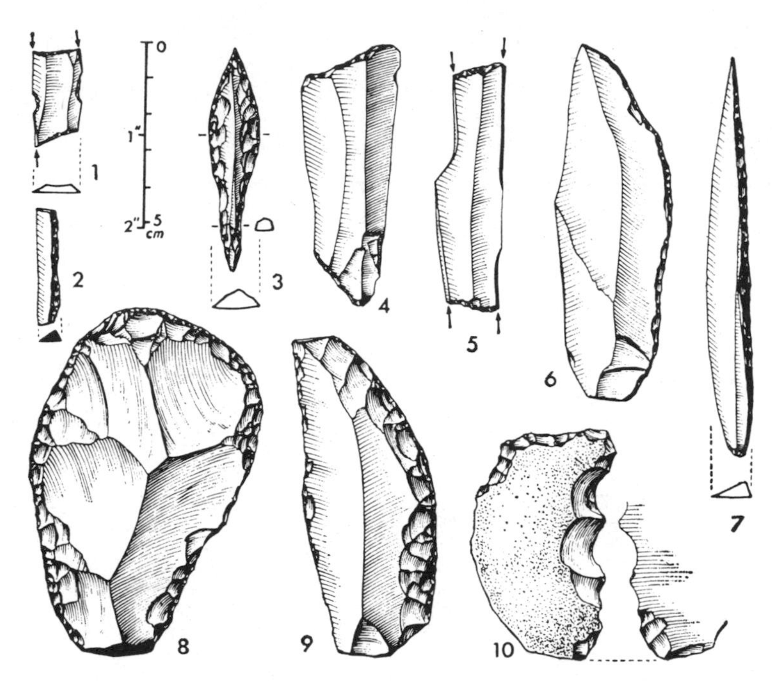

Fig. 4–2. Perigordian tools. 1, Noailles burin; 2, backed bladelet; 3, Font-Robert point; 4, *"elément tronqué"*; 5, quadruple burin; 6, Châtelperron point; 7, Gravette point; 8, end scraper; 9, side scraper; 10, denticulate tool.

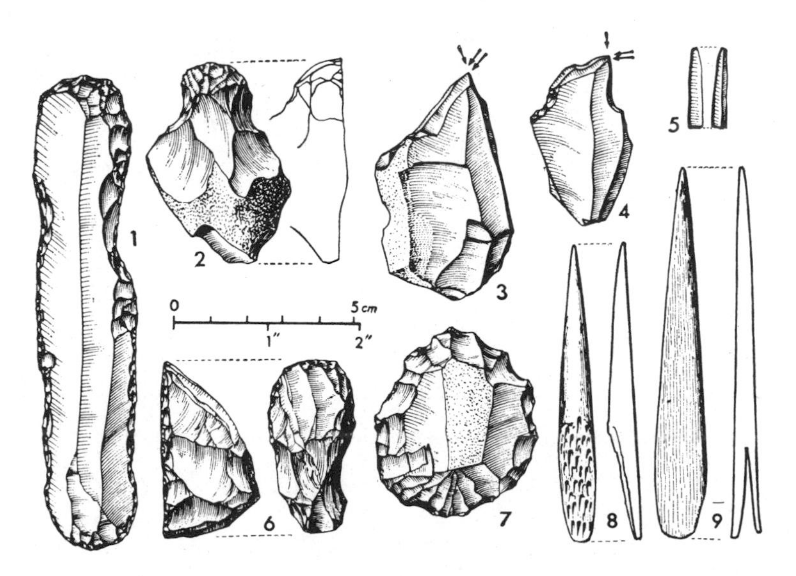

Fig. 4–3. Aurignacian tools. 1, End scraper on an Aurignacian blade; 2, nosed scraper; 3, dihedral burin; 4, busked burin; 5, bladelet with semi-abrupt retouch; 6, carinate scraper; 7, denticulate scraper; 8, bone point with beveled base; 9, bone point with split base.

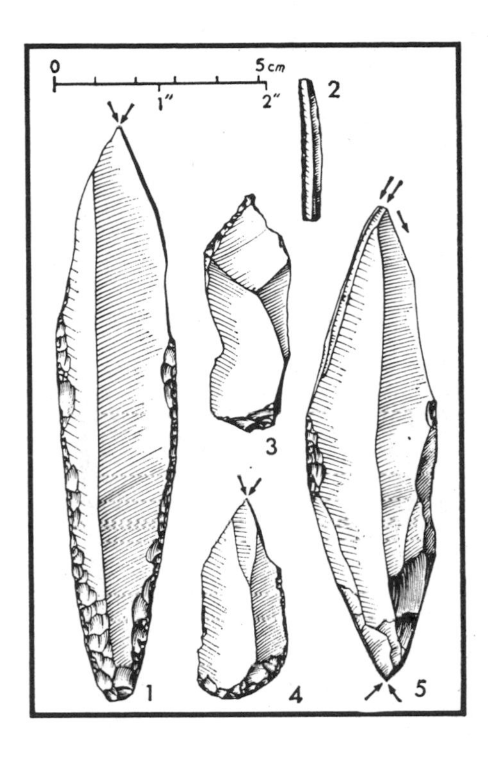

Fig. 4–4. Proto-Magdalenian tools. 1, Dihedral burin; 2, backed bladelet; 3, borer; 4, burin-scraper; and 5, double dihedral burin.

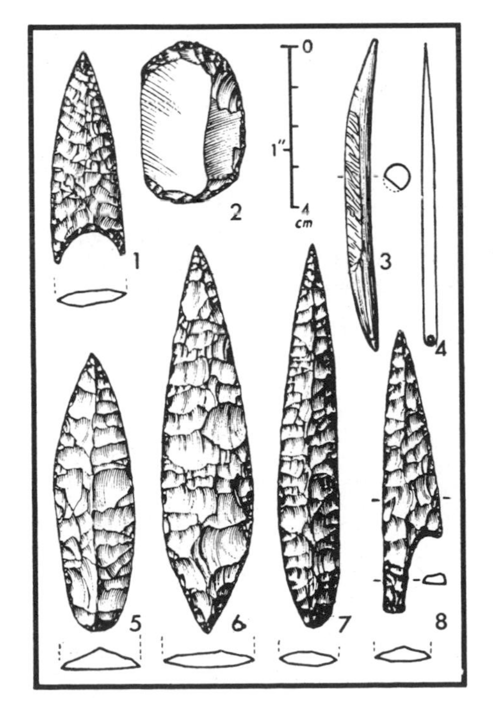

Fig. 4–5. Solutrean tools. 1, Laurel-leaf point with concave base; 2, double-end scraper; 3, curved bone point; 4, needle; 5, unifacial point; 6, laurel-leaf point; 7, willow-leaf point; 8, shouldered point.

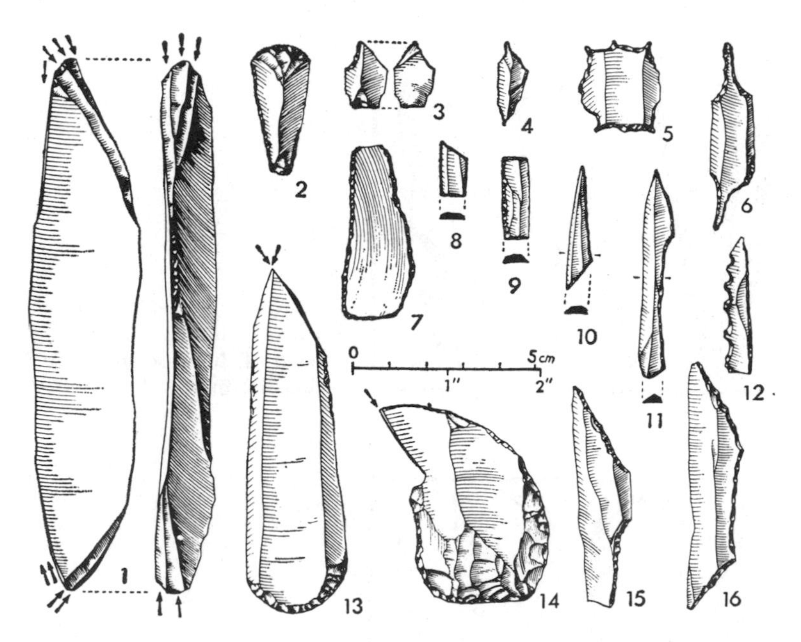

Fig. 4–6. Magdalenian tools. 1, Double dihedral burin; 2, end scraper; 3, "microburin"; 4–6, borers; 7, "raclette"; 8, trapezoid; 9, rectangle; 10, triangle; 11, shouldered point; 12, backed denticulate bladelet; 13, burin-scraper; 14, parrot-beak burin; 15, shouldered point with oblique truncation; 16, trapezoid.

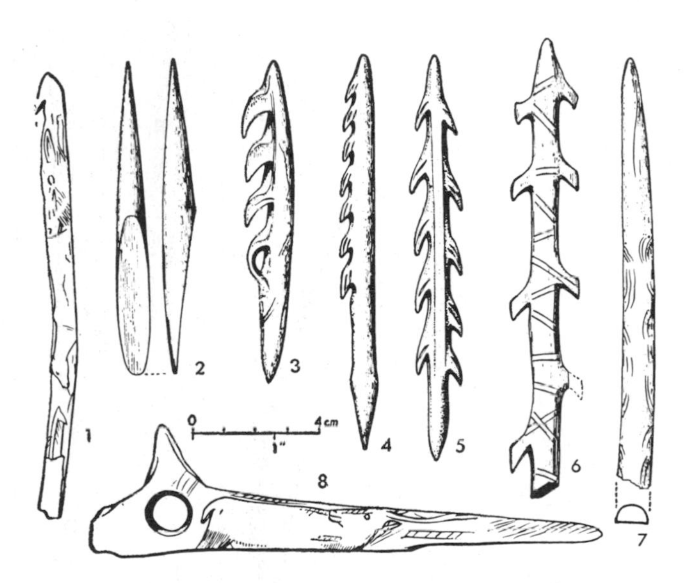

Fig. 4–7. Magdalenian tools. 1, Spear-thrower; 2, bone point with beveled base; 3–6, harpoons; 7, half-cylindrical wand; 8, arrow-shaft straightener.

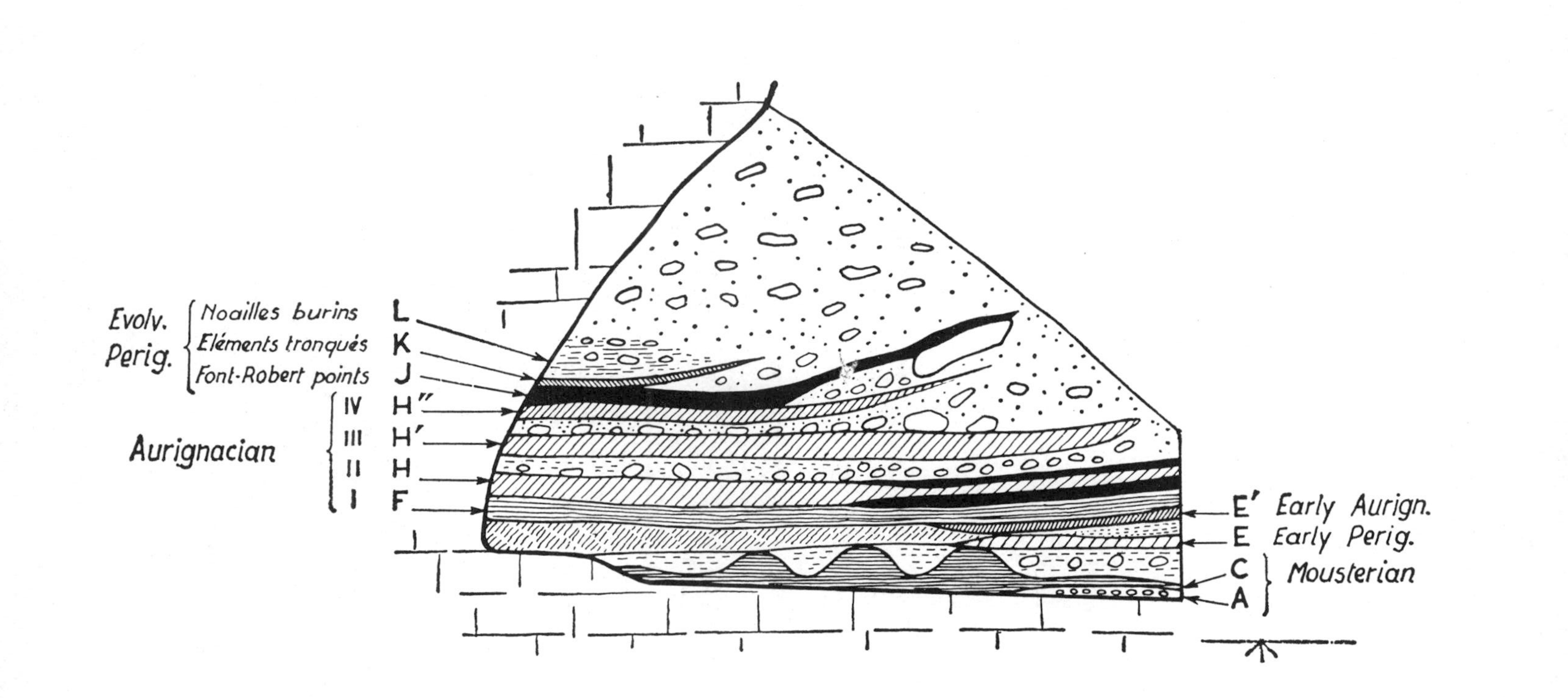
Evolv.
Perig.
Noailles burins
Eléments tronqués
Font-Robert points
Aurignacian
IV
III
II
I
L
K
J
H″
H′
H
F
E′ Early Aurign.
E Early Perig.
C
A
Mousterian

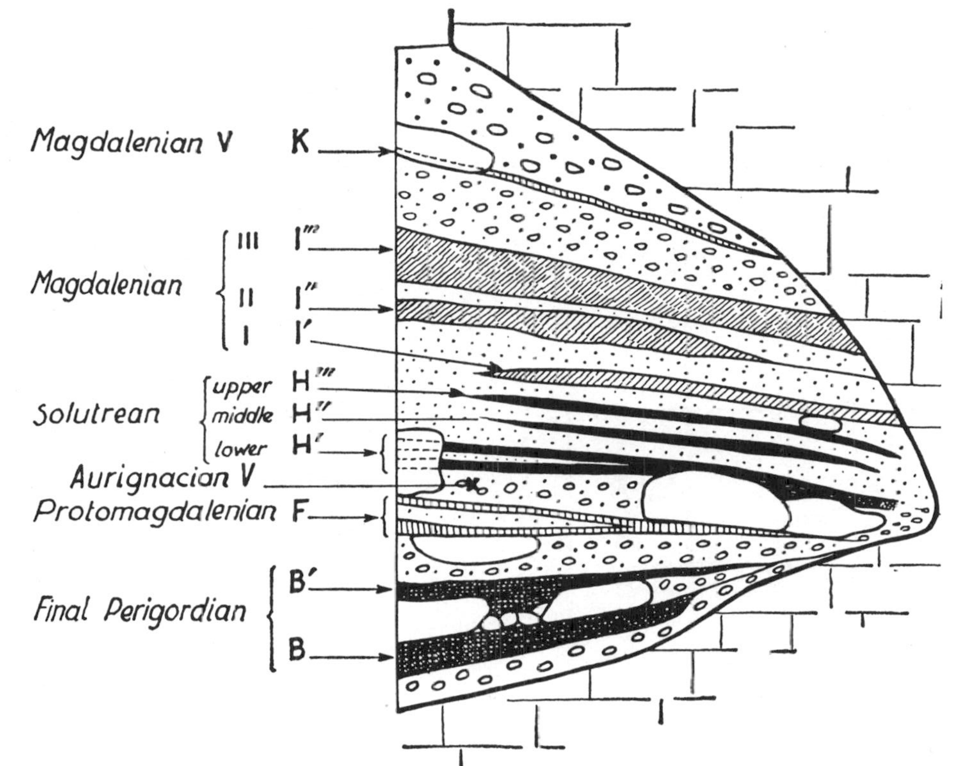

Fig. 4–8. (Top) Schematic section at La Ferrassie (Dordogne), showing the stratigraphic sequence of Aurignacian and Perigordian cultures. (Bottom) Schematic section at Laugerie-Haute East (Dordogne), showing the stratigraphic sequence of Perigordian, Proto-Magdalenian, Solutrean, and Magdalenian cultures. (After D. Peyrony and E. Peyrony.)

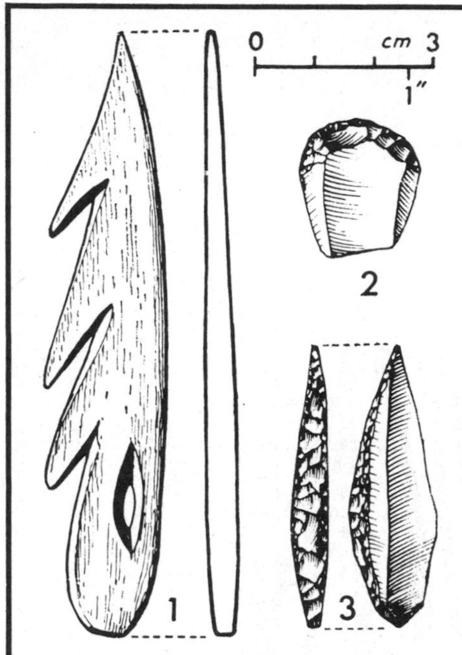

Fig. 4–9. Azilian tools. 1, Flat harpoon; 2, end scraper; 3, Azilian point.

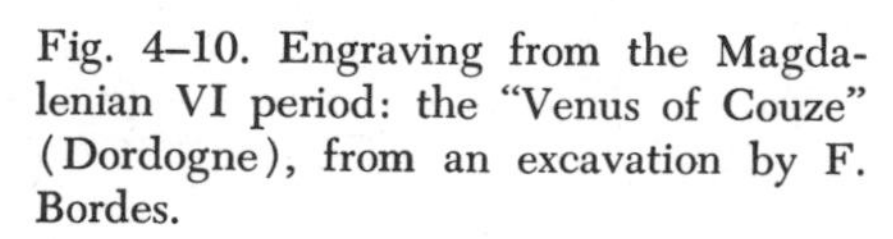

Fig. 4–10. Engraving from the Magdalenian VI period: the "Venus of Couze" (Dordogne), from an excavation by F. Bordes.

Chapter 5

Near Eastern Prehistory

Robert J. Braidwood

The following was written in early January, 1958. My primary interest lay with the background for and the achievement of an effective village-farming community way of life in southwestern Asia. I selected and summarized the evidence, as I then understood its pertinence, and briefly presented my notion of what it indicated in the way of culture-historical reconstruction. However, considerable field research has since been undertaken in the area. Were the essay to be written anew, various details of evidence could be added, and certain of my own emphases and conclusions would be somewhat different. (See the brief appendix at the end.)

The Near East (or Middle East—I have no preference) is traditionally taken to mean the area that stretches from the Libyan flanks of Egypt to include all of southwestern Asia as far as the rim of the Baluchi Hills, which overlook the Indus Valley. The Indus itself, parts of Transcaspian Turkestan and Transcaucasia, and even Greece and the Sudan might also be included, but this is not usual. The geographic core of the area is the drainage basin of the twin rivers, the Euphrates and Tigris, and the highlands and plateaus which immediately flank this drainage basin (Fig. 5–1). In this sense, the Nile and Indus basins lie on the western and eastern boundaries of the core area, as do the Mediterranean, Black, Caspian, and Red seas and the Indian Ocean.[1]

The Area and Its General Problems

Since all human prehistory is restricted to the Quaternary period, it is sufficient to say that the Pleistocene physiographic history of the area has been essentially one of superficial erosion and deposition, sometimes on a large scale.[2] The area shows traces of such world-wide climatically determined features as high marine and river terraces and localized glaciation in the higher mountains, but the over-all structural geography and the positions of the major land masses and seas were essentially set in pre-Pleistocene times. It now appears that extreme climatic change during the late glacial to early postglacial range of time was not *the* important factor in the appearance of plant and animal domestication. C. E. P. Brooks's much quoted "propinquity theory,"[3] which attempted to explain the appearance of food production through the concentration of men, plants, and animals in oases and river valleys as the Atlantic rain winds withdrew northward at the end of the last glaciation, is no longer tenable.[4]

Unfortunately, the geochronological details of Pleistocene events in the Near East may not yet be directly equated with those of western Europe, save in a most general way. This lack of intercontinental geochronological precision allows differences of opinion among prehistorians about how this or that range of Pleistocene artifacts in the Near East may be related to more or less similar types in Europe. Sometimes these disagreements have bearing on the construction of grand syntheses of culture-historical evolution.[5]

For the prehistorian—for any culture-historian for that matter—the area was the scene of three great culture-historical events:

1. The earliest appearance (on present evidence, if we take the more probable geochronological long view) of the blade-tool tradition. This relatively sophisticated set of habits in the preparation of long parallel-sided flint tools seems to have been roughly coincident, in Europe, with the appearance of anatomically modern men, about 40,000 years ago. The Palestinian ("nonclassic" or "sapiensized") Neanderthals may be regarded as ancestral to modern men,[6] and the blade tools make a tentative appearance in the Syrian and Palestinian stratigraphy even earlier than do these unspecialized physical types. It is not impossible, therefore, that the general Near Eastern area was the focus of differentiation and eventual spread of anatomically modern man and of his earliest characteristic habits in the preparation of flint tools.

2. The earliest appearance of the settled village-farming community, based on small-grain agriculture and animal domestication, about 10,000 years ago. The word *agriculture* is here used in a more restricted sense than that given it by Carl O. Sauer.[7] This was V. Gordon Childe's[8] "food-producing revolution" par excellence, and its consequences were momentous. It is probably very difficult for us now to conceptualize fully (or to exaggerate) the consequences of the first appearance of effective food production. The whole range of human existence, from the biological (including diet, demography, disease, and so on) through

the cultural (social organization, politics, religion, aesthetics, and so forth) bands of the spectrum took on completely new dimensions.

3. The earliest appearance of urban *civilization*, first in alluvial Mesopotamia, about 5,500 years ago, and only slightly later in Egypt. This is usually categorized archaeologically by certain reclaimable artifactual criteria[9] such as cities, monumentality in art and architecture, public works, and writing; but the general social and cultural implications of the achievement were even broader.[10] In fact, there is no general agreement with Childe in considering this step a further "revolution" on technological-economic grounds alone.[11, 12] Civilization appeared as a special intensification of cultural activity which effective food production made possible, but it was not necessarily the predetermined consequence of food production.

The subject of this chapter does not include the third event, which needs separate delineation. Nor shall I examine here the other experiments in the achievement of effective food production and of civilization that occurred, at slightly later times, in other parts of the world.

It will quickly become apparent that the reclamation and interpretation of the culture-historical evidence for Near Eastern prehistory is only in its infancy. The broad outline and the major problems are beginning to come into focus, and the research tools are being sharpened, but we still have a very long way to go. In the range of time we deal with here, each of the levels of culture involved required a very intimate balance with its environment. Superficially, it does not seem necessary for our own culture to maintain such a balance, due to vastly more sophisticated means of production, transportation, and distribution. But the expert in prehistoric archaeology faces the duty of reclamation and interpretation in two realms: culture history and natural history. He will have been trained, more or less well, to cope with and be thoughtful about the evidence for culture history, and this in itself is a full-time job and more. But archaeologists' excursions into natural history have usually ended in disaster; evidently competence in the biological and earth sciences also demands a full-time commitment![13]

During our last (1954–1955) field season in Iraqi Kurdistan, for work on the problem of the appearance of the settled village-farming community, we were enabled[14] to take out a skeleton team of natural scientists: a botanist, a geologist, a radiocarbon and ceramic-soils technician, and a zoologist. What we learned together, in daily communication in the area itself, about the reconstruction of an ancient environment doubtless marks a new departure in the study of prehistory. It is probably also worth saying that such teamwork between archaeologists and natural scientists is not without contemporary importance. Both the Israeli and Iraqi governments are utilizing such teams in gaining knowledge about how ancient irrigation and land-usage patterns functioned (and eventually failed to function) in making their plans for modern land-reclamation projects.[15] What is important for our present purposes, however, is that the archaeologist (both in the range of prehistory and of conventional ancient history, for that matter) is faced with problems which have dimensions that go into sciences far beyond his competence. A joint attack on these problems, with at least some field participation and the establishment of easy communication with interested natural scientists, does pay off handsomely.

Pleistocene Prehistory of the Near East

The basis for subdivision of earlier Pleistocene times is somewhat confused, but a working definition might be that the Lower Pleistocene proceeds from the end of the Villafranchian fauna to the end of the Mindel glaciation. The Middle Pleistocene runs thence to the end of the Riss glaciation, and the Upper Pleistocene runs from the Riss/Würm interglacial to about 10,000 years ago. Fleisch[16] assigns a few rolled flint tools to the + 45-meter marine terraces near Beyrouth, Lebanon (which some authorities take to be late Lower Pleistocene), but a general Lower Pleistocene occupation of the Near Eastern area is not yet evidenced. Even Middle Pleistocene flint-tool occurrences, again on marine terraces of the east Mediterranean littoral and on the highest Nile terraces yet examined, have only geological (not archaeological in the sense of "living site") context at best. There is little question but that men, who prepared their flint tools according to the persisting habits of both the core-biface and flake-tool traditions,[17,18] had already arrived in the Near East by Middle Pleistocene times, but we have, so far, little knowledge of their culture history. Really early traces of Pleistocene men, such as have been found in southern and northwestern Africa, have not yet been noted in the Near East.

In the geochronological long view, the archaeological sequences in several caves near the east Mediterranean littoral began to be deposited early in Upper Pleistocene times, if not with the recession of the Riss glaciation itself. An excellent sequence from fossil springs in the Kharga Oasis west of the Nile[19] parallels the littoral sequence in its earlier ranges, and the tools from the Nile terrace fit this same picture in a general way. On both sides of Suez there were fluctuations in utilization of various types of tool: coarse flake-tool industries (Tayacian), developed core-bifaces (Acheulean and Micoquian), and developed flake tools (Levalloiso-Mousterian, and so on). Clark Howell's[20] detailed synthesis of these developments will soon be available.

Three remarkable things appear in our present knowledge of the earlier portion of Upper Pleistocene times in the Near East. The first is the tentative occurrence of the blade-tool tradition, in the Tabūn Cave on Mount Carmel, in contexts which include Acheulean core-biface tools; blades also appear in the Yabrud Cave near Damascus soon thereafter. The second is the appearance, in the just-subsequent Levalloiso-Mousterian levels on Mount Carmel and in nearby caves, of fossil men who show a trend toward anatomically modern morphology.[21] The third is the apparent long persistence, in Egypt and its environs, of the Levalloiso-Mousterian industries, after—at the end of the earlier sub-phase of the Upper Pleistocene—the blade-tool tradition had taken over in southwestern Asia. If the geochronology is as we expect, the early appearance in southwestern Asia of the blade tools and of human beings with anatomical tendencies toward modern man (at a time when "classic" Neanderthal man was flourishing in western Europe) makes this area a focus of some interest. There is not, of course, complete agreement that either the blade tools or anatomically modern men did first appear in the area.[22] The long persistence and diminution in size of Levalloiso-Mousterian tools in Egypt remain inexplicable in the light of our knowledge of southwest Asia, but this trend parallels what happened in the rest of Africa. There is some promise that work in caves in Libya[23] may help elucidate the Egyptian situation, which is still poorly known for later Upper Pleistocene times.

Traces of the earlier aspects of the Upper Pleistocene are now being recovered in Iran and Turkey and in the Tigris-Euphrates Basin. A typologically quite early open site, Barda Balka in Iraqi Kurdistan, worked by the Iraq-Jarmo project staff for the Iraqi Directorate General of Antiquities,[24, 25] yielded tools of the earliest of the standardized traditions—the pebble tools—along with Upper Acheulean core-bifaces and a flake-tool facies. Bruce Howe compares the industry, on typological grounds, with so-called Lower Palaeolithic occurrences in both northwest Africa and in the Punjab, and Herbert Wright suggests that its geochronological position is probably contemporaneous with the onset of the Würm glaciation. Core-bifaces and earlier aspects of the Levalloiso-Mousterian flake tools are reasonably common surface finds in the core of southwestern Asia, as they are on its Mediterranean littoral. The cave sequences in the interior, with the exceptions of the core-bifaces at the bottom of Jerf Ajla, near Palmyra,[26] begin with a developed Levalloiso-Mousterian industry. This industry, first discovered by Dorothy Garrod in Iraqi Kurdistan in 1928, has since been tested in several caves by the Iraq-Jarmo project. B. Howe[27, 28] and Ralph Solecki[29] have recovered the remains of several fossil men in the same horizon at the Shanidar Cave. It appears to at least some human palaeontologists that the physical types involved are of the Mount Carmel rather than the European "classic" Neanderthal type.

As time went on in the Upper Pleistocene, blade tools began to make a more persistent appearance in the higher levels of the caves along the east Mediterranean littoral. The now-developed Levalloiso-Mousterian flake-tool industry began to include blade tools and a peculiar long, thin flint point; and this horizon is the earliest of a six-phase developmental scheme, proposed by Neuville and followed by Dorothy Garrod,[30] although the latter prefaces the scheme with a "Phase 0." The details of this "Upper Palaeolithic" sequence are not critical for our present purposes; it is enough to say that the Levalloiso-Mousterian industry is completely superseded by the developing blade-tool industries and that microbladelets (microliths) presently appeared. Garrod believes that the sixth or Kebaran phase was immediately followed by the Natufian; this point is not completely clear from the evidence, but I do feel justified in considering the Natufian postglacial in time.

The general cultural picture is still not so well known as that for the roughly equivalent range in western Europe. It does, however, suggest the same transition from an earlier, more "natural" food gathering to a more intensified collecting type of activity. Two interesting remarks of Garrod's might summarize this six-phase range: the climate and fauna of the littoral changed very little in latest Upper Pleistocene times, and in the immediately succeeding range—which G. Haas[31] assesses as being essentially modern; and with the speeding up of change and development, detailed similarities between the blade-tool sequences of western Europe and the Near East need not be expected, as cultural evolution now starts to outstrip diffusion.[32]

Strangely, there is not yet a radiocarbon date for this late glacial range in the littoral.

Even within the interior of southwestern Asia, there were blade-tool industries differing from those of the littoral. In Iraqi Kurdistan (figs. 5–2 and 5–3) the Zarzian "extended Gravettian" industry, with microliths,[33, 34, 35] is known now to be prefaced at Shanidar by the earlier Baradostian industry.[36, 37] The Baradostian has two radiocarbon dates: 29,500 ± 1,500 years (sample W-178) and older than 34,000 years (sample W-180).[38] The base of the Zarzian at Shanidar is dated at 12,000 ± 400 years, or about 10000 B.C. (sample W-179). A new date of about 8650 B.C. (sample W-667) for the upper part of the Zarzian at Shanidar is now announced.[39] Howe finds it increasingly impressive, as more caves are tested in Kurdistan, that no post-Zarzian materials have appeared in caves (save the oddments left by occasional transients). Evidently the transition to year-around open-settlement living immediately followed the Zarzian range. Charles A. Reed's[40] preliminary examination of the faunal remains from several Zarzian horizons has convinced him that an essentially modern climate had already been established.

On the eastern flank of the core area, in Afghanistan, the Kara Kamar

Cave has yielded blade tools and steep scrapers with radiocarbon dates comparable to those of the Baradostian,[41] but a developmental sequence in the area is not yet available. To the west, in the Libyan cave Haua Fteah,[42] on the other hand, blade tools appear to have arrived late; this seems to be in keeping with the curious flake-tool conservatism noticed earlier for Egypt.

There are doubtless at least several disconformities (for which industries have yet to be discovered and intercalated) in the archaeological sequences of the Lebano-Palestinian littoral and of Iraqi Kurdistan. These are, so far, the only areas known in any detail. While there is a gratifying increase in the attention now being given to the climatic and environmental history of the late Pleistocene to early postglacial time range in the area (H. E. Wright, Jr.[43, 44]), it appears increasingly certain that much more effort will have to be given to the reconstruction of the natural history of the region. It might be said in this connection that a liberalization—in the interest of prehistory—of the national antiquities laws of some of the countries in the Near East would stimulate more field research. Many of these laws had as their purpose the very justifiable prevention of exploitation, by foreigners, of spectacular sites of the historic range; but the laws have been applied to the detriment of prehistorians and their colleagues in the natural sciences (who need to study materials in their home laboratories).[45] But enough is already known of parts of the area to suggest that, at least in its Upper Pleistocene range, it will yet yield answers to many of the more meaningful questions about how man became what he was 10,000 years ago.

Postglacial Prehistory

There is increasing agreement among some geologists[46, 47] that the late glacial to early postglacial time boundary, in what is now the North Temperate Zone, is to be set at about 10,000 years ago, or 8000 B.C.[48] There is also an increasing number of radioactive carbon dates for sites in the era of the settled village-farming communities in the Near East, which shows that this era must already have been established by about 9,000 years ago, or 7000 B.C. Between the earliest village sites known to us and such terminal Pleistocene industries as the Kebaran and the Zarzian, mentioned above, there are clear hints of a range of materials probably best conceived of as the traces of incipient cultivators.[49] If Solecki's single radiocarbon date for the beginning of the Zarzian is essentially correct (sample W-179, 12,000 ± 400 years before the present) and some time is allowed for the flourishing of this industry (as the newly announced date of sample W-667 suggests[50]) and of its possible Kebaran equivalent, then the sites (Fig. 5–4) of the incipient cultivators probably were in use about 10,000 years ago, or 8000 B.C.—at the onset of early postglacial times. Within a thousand years, this experimental cultivation and—in the Kurdish area, at least—year-around life in the open were succeeded by the settled village-farming community.

The chronology suggested in the above sketch—the correctness of which is not yet guaranteed—could not have been given prior to January of 1958 and depends primarily on a new but modest-sized cluster of radiocarbon dates, from samples in northern Iraq and from two sites on the littoral, counted by Meyer Rubin of the U.S. Geological Survey in Washington.[51] Unfortunately, all the problems of the "geobiochemical" contamination of radiocarbon samples, before they reach the counter, beset the use of this and several other series of radiocarbon dates from the Near East. In figures 5–5 and 5–6 are plotted the available radiocarbon dates for the Near East (save for samples W-667 and W-681[52]), each date being shown as a time-bar to indicate the counter's plus-minus factor. It is clear that at the present moment (and this will be true until many more samples are counted, from many more different sites), the available fabric of radiocarbon dates can give us no more than a *general* indication of the late prehistoric time ranges of the area. This will throw us back primarily upon our old-style typological assessments of the comparative archaeological stratigraphy[53] of the various sites in the area. To these assessments we may then add our own judgments of the dating probabilities based on the general pattern of the radiocarbon dating fabric. For the Near East, at least, the cutting edge of radiocarbon dating as a research tool is still blunt because of our difficulties with the "geobiochemical" contamination factor. Understanding of this contamination factor will demand competences in a middle ground lying between archaeology and nuclear physics, which badly need to be developed.

The chronological sketch at the beginning of this section was made by selecting a cluster of three radiocarbon dates (samples W-607, -651, -652) of about 8,500 years ago, or 6500 B.C., as the probable true general date (out of a series of 11) for the early village site of Jarmo (figs. 5–7 and 5–8) in Iraqi Kurdistan.[54, 55, 56] Jarmo was a single-phase manifestation which cannot have had a time duration of more than a few hundred years.[57] The next phase of the early village-farming community era, in terms of comparative archaeological stratigraphy, as seen at Matarrah (sample W-623) and Hassuna (sample W-660) in the upper Tigris piedmont and at Mersin (sample W-617) on the Cilician coast of Turkey, seems to cluster at between 500 and 1,000 years later, say at about 5750 B.C. Since each of the pertinent phases, on the sites mentioned, will probably have had durations of several hundred years, no essential gap between the Jarmo phase and the Matarrah-Hassuna and Mersin phase need be postulated. The group of five Jarmo dates (samples C-113, -742, -743, and F-44, -45) will not work if the Matarrah,

Hassuna, and Mersin dates are correct, since Jarmo clearly precedes the pertinent basal materials of these sites in terms of comparative archaeological stratigraphy, and has several categories of technological descendants in Matarrah and Hassuna. The two earliest Jarmo dates (samples W-657, -665), of over 11,000 years ago, are simply not conceivable in terms of comparative archaeological stratigraphy as we now understand it. Jarmo must lie near, but not at, the very beginning of the era of village-farming communities; in my judgment this beginning should be put at about 7000 B.C.

It should be made clear that Jarmo is *not* conceived of as *the* spot where the village-farming community level of existence came into being—we do not even believe that there ever was one single such spot—but only that Jarmo represents the earliest example of settled village life which the accident of its prior discovery has allowed us to use as a basis for description.[58] To my mind, however, it is not an accident that Jarmo was found in the hilly-flanks zone of the Fertile Crescent. This zone of upper piedmont and intermontane valleys, stretching at least from Syro-Cilicia into Iran, flanking the Taurus-Zagros arc and still receiving ample rainfall, appears increasingly to have been *the* natural habitat of the potentially domesticable plants and animals.[59, 60, 61]

There is a complication. Early in 1956, the excavators of Tell es-Sultan (usually taken to have been the site of Joshua's Jericho), in the Dead Sea Valley in Jordan, published a pair of radiocarbon dates (samples GL-28, -38) for the second phase above base in that site.[62] More recently, two further dates for the same level (samples GR-942, -963) have become available, as well as a pair of dates for the first phase which parallels the latter ones.[63] From the point of view of comparative archaeological stratigraphy, the two basal phases of Tell es-Sultan are enigmatic, and there is now word[64] that the "first" phase may in fact have been preceded by some simpler materials. There is clear evidence of considerable architectural complexity in the basal layers, which include thick stone-founded fortifications, with a tower, and formed mud-brick house walls, but the remainder of the catalog of materials is relatively primitive and includes neither pottery nor metal objects. Taking her cue from the relatively large area of the site and its architectural complexity, Kathleen Kenyon[65] used the words *urban* and *civilization* in describing its cultural level, and these implications were strongly contested by Childe[66] and Braidwood.[67, 68] Also, in view of the then available radiocarbon dates for other roughly comparable materials in the Near East, which were all considerably later, Kenyon was forced to see the Tell es-Sultan material as something which developed without respect to the chronological and developmental framework of the area in which it lay.[69]

With earlier radiocarbon dates now available elsewhere, this view is

TABLE 5-1

Radiocarbon determinations used in Figure 5-5.

SAMPLE	DESCRIPTION	AGE (YR)
	*Chicago dates**	
C-12	Sneferu tomb	4802 ± 210 (av.)
C-113	Jarmo village (I-7)	6707 ± 320
C-183	Alishar "chalcolithic"	4519 ± 250
C-267	Hemaka tomb	4883 ± 200 (av.)
C-457	Fayum A village	6095 ± 250
C-463	el-Omari village	5256 ± 230
C-550	Fayum A village	6391 ± 180
C-742	Jarmo village (I-7)	6606 ± 330
C-743	Jarmo village (II-5)	6695 ± 360
C-744	Jarmo village (II-2)	5266 ± 450
C-753	Shaheinab village	5060 ± 450
C-754	Shaheinab village	5446 ± 380
C-810	Predynastic tombs, S.D. 34-38	5744 ± 300
C-811	Predynastic tombs, S.D. 36-46	5619 ± 280
C-812	Predynastic tombs ("Nagada II")	5020 ± 290
C-813	Predynastic (?) tombs, S.D. 58-67	4720 ± 310
C-814	Predynastic tombs, S.D. 34-38	5577 ± 300
C-815	Mundigak "Bronze Age"	4580 ± 200 (av.)
C-817	Tepe Gawra 17 +, Ubaid	5400 ± 325 (av.)
C-819	Byblos, "first urban"	5317 ± 300
C-919	Beersheba "Ghassulian" village	7420 ± 520
	Caspian Foreshore dates††	
CC-B.	Belt Cave "ceramic neolithic"	7280 ± 260
CC-B.	Belt Cave "preceramic neolithic"	7790 ± 330
CC-B.	Belt Cave "gazelle mesolithic"	8570 ± 380
CC-B.	Belt Cave "seal mesolithic"	11,480 ± 550
CC-H.	Hotu Cave "software neolithic"	6385 ± 425
CC-H.	Hotu Cave "sub-neolithic"	8070 ± 500
CC-H.	Hotu Cave "vole--eaters" (3 skeletons, 2 samples)	9190 ± 590 9220 ± 570
CC-H.	Hotu Cave "seal-hunters"	11,860 ± 840
	Davy-Faraday dates‡‡	
F-40	Tell es-Sultan lower ("hog-backed brick") phase, two different pretreatments	8725 ± 210 8805 ± 210
F-44	Jarmo village (II-5)	6650 ± 170
F-45	Jarmo village (I-8)	6570 ± 165

SAMPLE	DESCRIPTION	AGE (YR)
	Geochronological Laboratory London dates †	
GL–24	Tell es-Sultan "chalcolithic"	5210 ± 110
GL–28	Tell es-Sultan upper ("plastered floor") phase	8200 ± 200
GL–38	Tell es-Sultan upper ("plastered floor") phase	7800 ± 160
	Groningen dates ‡	
GR–942	Tell es-Sultan upper	
	Tell es-Sultan "chalcolithic"	5210 ± 110
GR–963	Tell es-Sultan upper ("plastered floor") phase	8785 ± 100
	Heidelberg dates §	
H–138/123	Warka, basal "Ubaid" town	6070 ± 160
	Lamont dates ‖	
L–180A	Kili Ghul Mohammed village, I	5300 ± 500
	Pennsylvania dates #	
P–53	Kara Kamar Cave "mesolithic"	10,580 ± 720
	Washington (U.S. Geological Survey) dates **	
W–89	Haua Fteah Cave (evolved blades and microliths)	7300 ± 300
W–97	Haua Fteah Cave (evolved blades and burins)	12,300 ± 350
W–98	Haua Fteah Cave ("primitive 'Neolithic' ")	6800 ± 350
W–104	Haua Fteah Cave (compare W–97)	10,600 ± 400
W–179	Shanidar Cave, B (basal "Zarzian")	12,000 ± 400
W–245	Beersheba "Ghassulian" village	5280 ± 150
W–607	Jarmo village (PQ–14, 2.5 m)	9040 ± 250
W–617	Mersin village (basal layer)	7950 ± 250
W–623	Matarrah village (VI–4)	7570 ± 250
W–627	Byblos A village	6550 ± 200
W–651	Jarmo village (II–4) (1950–51)	8830 ± 200
W–652	Jarmo village (I–7a) (1950–51)	7950 ± 200
W–657	Jarmo village (PQ–14, 2.25 m)	11,240 ± 300
W–660	Hassuna village (5th level)	7040 ± 200
W–665	Jarmo village (N–18, 2.0 m)	11,200 ± 200

* W. F. Libby, *Radiocarbon Dating* (Univ. of Chicago Press, Chicago, ed. 2, 1955).
† See note 62.
‡ H. de Vries and H. T. Waterbolk, *Groningen Radiocarbon Dates, III, Science,* **128**, p. 1555.
§ K. O. Münnich, *ibid.*, **126**, 194 (1957).
‖ W. S. Broecker, J. L. Kulp, C. S. Tucek, *ibid.*, **124**, 154 (1956).
C. S. Coon and E. K. Ralph, *ibid.*, **122**, 921 (1955).
** H. E. Suess, *ibid.*, **120**, 467 (1954); M. Rubin and H. E. Suess, *ibid.*, **121**, 481 (1955); M. Rubin and H. E. Suess, *ibid.*, **123**, 442 (1956); M. Rubin, personal communication.
†† See note 26.
‡‡ From F. E. Zeuner, in correspondence. See also note 26.

no longer necessary. My own tendency, in assessing the dichotomous complexity and primitiveness of the Tell es-Sultan catalog, along with the peculiarity of the ecological niche in which the site lay (some 900 feet below sea level, in an arid valley), is still to suspect that there is some "geobiochemical" contamination in the radiocarbon samples. The site lies in an area of tectonic activity, and faults have been noted, both in the site[70] and near it,[71] as well as upward seepage of radioinactive natural gases.[72] The contamination possibility, however, clearly calls for competences in assessment which the archaeologist does not possess. In addition, on the archaeological side, there would certainly not be general agreement with Kenyon's reading of the comparative archaeological stratigraphy of her site.

The controversy has been of some culture-historical importance for the reason that Kenyon and her colleagues[73] have raised again the old issue of the "propinquity theory" for an oasis origin of agriculture and animal domestication. In spite of its arid, below-sea-level situation, Tell es-Sultan does lie adjacent to an excellent fresh-water spring. But the evidence for the origins of domestication—while still limited and badly in need of further bolstering—has increasingly pointed rather toward the upper piedmont and intermontane valley zone of the "hilly flanks of the Crescent." This not only appears to have been the natural habitat for the potentially domesticable wild plants and animals, but also seems to have had no important climatic variation since later Upper Pleistocene times. While to take this view may be to ignore certain minor depositional features, especially in the first millennium B.C., which may have been climatically determined,[74] one does feel justified in making a general assessment of the environmental situation of some 10,000 years ago in terms of the present situation. In fact, it is not clear that, in the core area of the Near East, the late glacial to early postglacial time boundary was at all an "event" in climatic or environmental terms. Some allowance must naturally be made for the loss of vegetation through overgrazing and charcoal burning, and for the extinction of certain wild animals (hunting from Ford model T's finished off the onager in Iraq about 1928!). Our reconstruction here is founded on the proposition that the available evidence does locate the natural habitat in the hilly-flanks zone, and that domestication took place within this natural habitat.

There are, of course, more excavated sites upon which to base our reconstruction than those which are indicated in Figure 5–5.[75] But there are by no means enough. Our knowledge of the potentially rich (from the point of view of natural habitat) districts of Iran, Turkey, and Syria is almost, if not completely, blank, and little has been done in this range of interest in Egypt in some years. This situation reflects in part the prevailing unfavorable circumstances discussed above for making an up-to-date prehistoric excavation (Fig. 5–9) and in part the great cost of mounting a well-rounded expedition (with a proper staff complement of natural scientists). Hence the reconstruction may be, to a degree, an artifact that reflects the incompleteness of the record. A fairly recent assessment of much of the area is available,[76] and it will therefore not be necessary to name too many of the sites involved.

Incipient Cultivation

Figure 5–6, which was developed as an overlay on Figure 5–5, indicates how the principle of sloping horizons[77] for the successive eras must be involved in thinking about the subregions of the Near East. I have suggested above the general picture given by the archaeological materials for the range of Pleistocene times in the Near East. The means of obtaining food during the whole range was through gathering or collection. The range is usually referred to as the food-gathering stage. If the stage had substages or eras, the last of these was one of more intensive collection—a more intimate "living into" a given environment. We noted that the Kebaran of Palestine and the Zarzian of Iraqi Kurdistan appear to have terminated this era, sometime after 10000 B.C. (The newly announced date of the upper part of the Shanidar Zarzian is about 8650 B.C.[78]) The data in Figure 5–6 suggest, however, that food collection was continued in areas adjacent to the central core of the Near East, and certain materials in caves on the Caspian and Libyan foreshores suggest even further intensifications in collecting activities. In connection with these peripheral areas, the notion of a "Mesolithic" stage has been advanced, as in the case of northwestern Europe, to describe archaeological materials showing cultural readaptations—still on a food-collecting level—to the postglacial environment. The notion will gain validity only if a significant environmental change can be shown to have occurred.

It appears increasingly doubtful, however, whether this or any other meaningful concept of "Mesolithic-ness" can be applied to the core area. In Iraqi Kurdistan, the next materials (following the Zarzian), from Karim Shahir and comparable sites, which are simple open-air establishments, suggest an incipience of cultivation. One of these open sites, Solecki's Zawi Chemi,[79] now has a newly announced radiocarbon date of about 8900 B.C.[80, 81] In Palestine, these are paralleled by the Natufian materials, still primarily in caves but perhaps slightly more convincing as evidence of an era of incipient cultivation. An important Natufian open-air site has just been announced for northern Palestine by Jean Perrot.[82] This era prefaces the swing from the food-gathering to the food-producing stage. Its catalog includes some suggestion of animal domestication, some authorities claiming domestication of the dog. There are flint sickles for reaping, crude milling stones for grinding seeds, and celts; the latter may have been used as either hoes or axes, or as both. Further delineation of this era is very badly needed, and since the era was one of transition and, doubtless, of making-do with some old tool types, it will be an exceedingly difficult one to substantiate fully. The era is still characterized by flint blade tools and microliths. The probability is that the natural scientists will do better here than the conventional archaeologists.

Village-Farming Communities

Next, in the core area, comes the first phase of fully settled village sites, of which Jarmo is simply the earliest example which happens so far to have been found. In the next phase of the village-farming community era, which rather quickly succeeds the Jarmo phase, there are at least five regionally different village assemblages (catalogs of artifactual materials): those of Hassunan type in the upper Tigris piedmont; those of the Amouq A–Mersin type of Syro-Cilicia; those of the third (?) Tell es-Sultan–Abou Ghosh type in inland Palestine; those of the Fayum A type in Egypt; and those of the Sialk I type in northern Iran.[83, 84] Unless the radiocarbon dates on the Fayum A of Egypt (samples C-457, -550/1) are wrong—and more samples should be counted—the principle of the sloping horizon is clearly involved. This, of course, has a bearing on the actual chronological position of the Tell es-Sultan materials.

The earliest of the village-farming communities appear to have clustered still within the natural habitat zone of the upper piedmont and intermontane valleys of the "Crescent," where the wild wheats, barley, sheep, goats, pigs, cattle, and some kind of equid were all at home in nature. It has been suggested that the development was bound to this zone until permissive mutations,[85] or introgressive hybridization,[86] operated, especially on the plants, to allow the domesticates to be removed from their natural area. The curve in Figure 5–6 is inflected to suggest more general spread after this had taken place.[87]

One consequence of this spread was the diffusion of the wheat-barley-sheep-goat-cattle complex, and much of the generalized cultural know-how which had developed with it, to the boundaries of the Near East and far beyond, wherever the environmental situation allowed such spread. We have hints, through radiocarbon dating, that the new way of life had extended well up the Danube Valley by about 4000 B.C.[88] and that by 2500 B.C. it had pretty well covered Europe. It also went eastward; wheat, at least, was being grown in China by at least 1500 B.C., although it does not appear to have been the earliest domesticated plant there. A different consequence of the spread from the hilly-flanks zone of the natural habitat—given the mutations or hybridizations—was the apparent "fingering" movement of early farmers down the mud flats of the Tigris and Euphrates into classic southern Mesopotamia.[89] This probably took place toward the end of the Hassuna phase or early in the succeeding Halaf phase. It is our suggestion that the principles of canalization were learned on these mud flats; canalization made the occupation of classic southern Mesopotamia by farmers feasible. The data in Figure 5–6 suggest that a new era arose on this basis in southern Mesopotamia, and one radiocarbon date (sample H-138/123) indicates that this era was well under way by 4000 B.C. This was an era which is archaeologically manifested by town-sized settlements, temple structures of some degree of monumentality, metallurgy as a specialized craft, and evidently (since they are already present at the beginning of the next era) the use of draft animals and the plow. Even in the first or Ubaidian phase of this era of towns, the strength of the new cultural potential of southern Mesopotamia is suggested by the *oikoumenē* of the spread of its painted pottery style—from the Mediterranean coast to the rim of the Anatolian plateau to the uplands of Iran.[90]

This is the place to end our survey; the next era is that of the appearance of urban civilization in southern Mesopotamia, about 3500 B.C., followed by the beginning of the Egyptian dynasties around 3000 B.C., and by that time prehistory per se is theoretically ended in the Near East.

Conclusions

In summary, it needs to be repeated once more that what is offered here is only one prehistorian's interpretation of very incomplete evidence. For late Upper Pleistocene times especially, much more must be learned of the environments which were available, of the human physical types (only one juvenile example and various fragmentary bits exist), and of the different cultural levels. Only snatches of evidence are now available for the era of incipient cultivation, which prefaced the great swing from the food-gathering to the food-producing stage, and very sophisticated environmental reconstructions will be necessary before the cultural achievements of this era can gain meaning. The same holds particularly for the earlier phases of the era of the settled village–farming community. In reconstructing the general culture history of the Near East, for late glacial to early postglacial times, the concept of sloping horizons appears to be a useful one. It also appears that the zone of the natural habitat may have been a focus of "nuclearity," and that some eras and phases of cultural development may have been manifested there but not elsewhere.[91]

It must be obvious how much the prehistoric archaeologist needs the aid of his interested colleagues in the natural sciences. First and foremost, however, the prehistorian's business is with men—with the anthropology of extinct cultures. He needs to discover all he can about the plants and animals that lived with the men, but the plants and animals did not domesticate themselves. Men domesticated them. The prehistorian is very much aware of the innumerable "how" and "why" questions which still confront him. In the Near East, it is simply a matter of his requiring much more information from the good earth, and some help in interpreting it.

Appendix: April 1963

The following new information, with its hints of shift in emphases, should be taken into account in considering Chapter 5:

1. Artifacts of the earliest or Villafranchian phase of the Pleistocene are now known from Israel and possibly Syria. The Shanidar Neanderthal fossils are apparently more "classic" than those of the Skhul-Qafseh types of Palestine. Another Baradostian occurrence is now known from the cave of Warwasi, near Kermanshah, Iran. In a group of caves on the south central coast of Anatolia near Antalyia, an interesting long sequence at Karain is complemented at its upper end by important finds in Beldibi and Belbasi.

2. The Beldibi Cave hints that a generalized "Natufian" may have extended well beyond Palestine as an early post-Pleistocene littoral adaptation along the eastern Mediterranean. The restricted "Natufian" occurrence at Tell es-Sultan now has a C-14 determination at about 9500 B.C. The domestication of sheep has been reported from the roughly contemporary Zagros flanks level of incipient cultivation and domestication (the Karim Shahirian) at Zawi Chemi Shanidar. There is somewhat less inclination to accept claims (on the basis of dry context pollen) for cereal domestication at this site. Other variants of the Karim Shahirian have appeared at Asiab near Kermanshah and at Ali Khosh near Deh Luran, in Iran, the latter being at a remarkably low elevation on the foothills of the flanks. The sequence of the Ali Khosh site is bound to have a bearing on speculations as to how effective food production first reached alluvial Mesopotamia.

3. New aspects of the settled village-farming community level

have appeared. The ambiguities regarding the two preceramic phases of Tell es-Sultan have not yet been resolved (at least to my satisfaction), for example, as to whether their assemblages actually imply effective food production. The same may be said for the roughly comparable materials from Beidha, in the east Jordanian desert. A preceramic horizon of somewhat different character, with carbon-14 determinations of ca. 7000–650 B.C., has been tapped at Ras Shamra on the northern Syrian coast. Two remarkable new sites on the south central Anatolian plateau, Haçilar and Çatal Hüyük, at least the former with a preceramic horizon, suggest a development which differs markedly from that of both the Zagros flanks and from Palestine, although it may prove to have points in common with that of basal Ras Shamra. Further (and slightly later) aspects of a Jarmo-like phase have been examined at Sarab near Kermanshah, Hadji Firuz near Lake Urmia and Ali Khosh near Deh Luran, all in Iran. Agonizingly comparable material has also appeared on Tepe Djeitun, near Ashkhabad in Soviet Turkmenia; whether this indicates the general spread of a Jarmo-like assemblage across and beyond the Iranian plateau is not yet clear.

4. Manifestly, the conception that "the hilly flanks of the Crescent" alone included the natural habitat zone of the potential plant and animal domestication now appears to be too restrictive.

5. It is also clear that the mere presence or absence of pottery alone, in an early assemblage, is not sufficient to allow immediate classification as to the level of incipient cultivation and domestication or to the subsequent level of settled village-farming. The term "preceramic" may be convenient for archaeologists, but *primary* evidence as to developmental level must depend on the presence and variety of well-attested domesticates.

6. New evidence is in hand to suggest that food production had already spread to Cyprus, Crete, Greece, and Macedonia by about 6000 B.C. Various C-14 determinations also indicate that food production had reached northwestward to the Netherlands as early as 4200 B.C.

7. There is some suggestion, again by C-14 determinations, that the appearance of literate urban civilization in southern Mesopotamia may not have come until slightly later than 3000 B.C.

The most recent general summarizing of most of the points listed above, with a bibliography, is Robert J. Braidwood, "The Earliest Village Communities of Southwestern Asia Reconsidered," M. Pallottino, in *Atti del VI Congresso Internazionale delle Scienze Preistoriche e Protostoriche*, Sansoni, Roma, 1962, pp. 115–126.

NOTES

1. A good general geographical treatment of the area is W. B. Fisher, *The Middle East: A Physical, Social, and Regional Geography* (Methuen, London, ed. 2, 1952).

2. See H. E. Wright, Jr., *Sumer*, **11**, 83 (1955).

3. See V. G. Childe, *New Light on the Most Ancient East* (Routledge and Kegan Paul, London, ed. 4, 1952), p. 15.

4. R. J. Braidwood, *Antiquity*, **31**, 73 (1957).

5. Compare F. Bordes, *Anthropologie (Paris)*, **59**, 486 (1955) with D. Garrod, *Bull. Soc. préhist. franç.*, **54**, 439 (1957) and the general discussion by F. C. Howell in *Trans. Intern. Neanderthal Centen. Congr.*

6. F. C. Howell, *Am. J. Phys. Anthropol.*, **9**, 379 (1951); *Quart. Rev. Biol.*, **32**, 330 (1957); *Trans. Intern. Neanderthal Centen. Congr.* There is not yet, of course, complete agreement on this score, nor—Howell would maintain—complete understanding of the morphological positions of Swanscombe and Fontéchevade and of the full implications of the liquidation of Piltdown. For the older viewpoint, see, for example, M. Boule and H. V. Vallois, *Fossil Men: A Textbook of Human Paleontology* (Thames and Hudson, London, 1957).

7. Compare R. J. Braidwood and C. A. Reed, *Cold Spring Harbor Symposia Quant. Biol.*, **22** (1957) with C. O. Sauer, *Agricultural Origins and Dispersals* (American Geographical Society, New York, 1952).

8. V. G. Childe, *Man Makes Himself* (Mentor, New York, ed. 3, 1952).

9. V. G. Childe, *Town Planning Rev.*, **21**, 3 (1950).

10. H. Frankfort, *The Birth of Civilization in the Near East* (Indiana Univ. Press, Bloomington, 1951); R. Redfield, *The Primitive World and Its Transformations* (Cornell Univ. Press, Ithaca, N.Y., 1953).

11. R. J. Braidwood, *The Near East and the Foundations for Civilization* (Univ. of Oregon Press, Eugene, 1952).

12. G. R. Willey and P. Phillips, *Method and Theory in American Archaeology* (Univ. of Chicago Press, Chicago, 1958), p. 72.

13. R. J. Braidwood, in *The Identification of Non-artifactual Archaeological Materials*, W. W. Taylor, Ed. (Natl. Acad. Sci.–Natl. Research Council Publ. 565, Washington, D.C., 1957), pp. 14–16.

14. The major enabling grant—over and above the Oriental Institute's basic archaeological field budget for its Iraq-Jarmo project—was made to the Department of Anthropology of the University of Chicago by the National Science Foundation. Supplemental grants came from the American Philosophical Society, the Baghdad School of the American Schools for Oriental Research, the Guggenheim Foundation, and the Wenner-Gren Foundation for Anthropological Research, and from friends of the Oriental Institute.

15. J. C. Russel, soil scientist, chairman of the department of soils and agricultural chemistry of the University of Arizona, and recipient of a University of Arizona–Iraq College of Agriculture grant, has thoughtfully considered the matter in "Historical aspects of soil salinity in Iraq," a paper presented at the ICA–FAO conference of 30 Jan. 1956 in Baghdad.

16. H. Fleisch, *Quaternaria*, **3**, 101 (1956).

17. R. J. Braidwood, *Prehistoric Men* (Chicago Natural History Museum, Chicago, ed. 3, 1957).

18. H. L. Movius, Jr., in *Anthropology Today*, A. L. Kroeber, Ed. (Univ. of Chicago Press, Chicago, 1953).

19. G. Caton-Thompson, *The Kharga Oasis in Prehistory* (Univ. of London Press, London, 1952).

20. F. C. Howell, *Proc. Am. Phil. Soc.*, in press.

21. See note 6.

22. See note 6.

23. C. B. M. McBurney et al., *J. Roy. Anthropol.*, **83**, 71 (1953); C. B. M. McBurney and R. W. Hey, *Prehistory and Pleistocene Geology in Cyrenaican Libya* (Cambridge Univ. Press, Cambridge, 1955).

24. H. E. Wright, Jr., and B. Howe, *Sumer*, **7**, 107 (1951).

25. R. J. Braidwood, B. Howe, et al., *Prehistoric investigations in Iraqi Kurdistan* ("Oriental Institute Studies in Ancient Civilizations," No. 31; Chicago, Chicago Univ. Press, 1960).

26. C. S. Coon, *The Seven Caves* (Knopf, New York, 1957). I have used this work as a direct source for radioactive carbon dates (see Table 5–1), since Coon has selected the single average dates (from a much larger number of date clusters) which seem most comprehensible to him.

27. Braidwood, Howe, et al., *loc. cit.*

28. R. J. Braidwood et al., *Sumer*, **10**, 120 (1954).

29. R. S. Solecki, *ibid.*, **13**, 59, 165 (1957).

30. D. Garrod, *Bull. soc. préhist. franç.*, **54**, 439 (1957).

31. M. Stekelis and G. Haas, *Israel Exploration J.*, **2**, 35 (1952).

32. D. Garrod, *J. World History* (*Cahiers hist. mondiale*), **1**, 14 (1953).

33. Braidwood, Howe, et al., *loc. cit.*

34. Garrod, *loc. cit.*

35. R. S. Solecki, *Smithsonian Inst. Publs. Rept. No. 4190* (*1954*) (1955), p. 389.

36. R. S. Solecki, *Sumer*, **13**, 59, 165 (1957).

37. R. S. Solecki, *Smithsonian Inst. Publs. Rept. No. 4190* (*1954*) (1955), p. 389.

38. For the meaning of the notation used with the counter numbers, see Table 5–1.

39. R. S. Solecki and M. Rubin, *Science*, **127**, 1446 (1958).

40. R. S. Solecki, Sumer, **13**, 59–60 (1957).

41. Coon, *loc. cit.*

42. McBurney et al., *loc. cit.*; McBurney and Hey, *loc. cit.*

43. Braidwood, Howe, et al., *loc. cit.*

44. H. Bobek, *Geograph. Jahresbericht Österr.*, **25**, 1 (1953–54); K. W. Butzer, *Erdkunde*, **11**, 21 (1957); M. Pfannenstiel, *Das Quartär der Levante* (Akad. Wiss. Lit. Mainz, Abhandl. math-naturw. Kl. No. 7, Wiesbaden, 1952), vol. 1.

45. It is worth pointing out that the flourishing of activity in the field of prehistory in Iraq is a direct reflection of the intelligent spirit in which the Iraqi Directorate General of Antiquities administers its antiquities law.

46. R. F. Flint, *Glacial and Pleistocene Geology* (Wiley, New York, 1957).

47. G. W. Barendsen, E. S. Deevey, L. J. Gralenski, *Science*, **126**, 908 (1957); H. E. Wright, Jr., *Am. J. Sci.*, **255**, 447 (1957).

48. Some authorities [for example, E. Antevs, *J. Geol.*, **65**, 129 (1957)] interpret the evidence as calling for an earlier late glacial–early postglacial time boundary, and Flint (*op. cit.*, p. 283) considers the difficulties involved if the time boundary is taken to be a unitary chronological "event" of world-wide validity.

49. R. J. Braidwood and L. Braidwood, *J. World History* (*Cashiers hist. mondiale*), **1**, 278 (1953).

50. Solecki and Rubin, *Science, loc. cit.*

51. M. Rubin, personal communication.

52. Solecki and Rubin, *Science, loc. cit.*

53. Such typological assessments are based on the principle of assessing relative age by means of the relative technological complexity of items in the catalogs of several different archaeological sites or phases [see G. Clark, *Archaeology and Society* (Harvard Univ. Press, Cambridge, 1957), p. 151]. The more restricted the region, the better the principle usually works, and it works best of all if "trade objects" link one phase to another. The principle need not imply physical (geological) stratigraphy at all, although sometimes this is also available for parts of the sequence.

54. Braidwood, *The Near East and the Foundations for Civilization, op. cit.*

55. Braidwood, *Prehistoric Men, op. cit.*

56. Braidwood, Howe, et al., *loc. cit.*

57. Because of the internal consistency of its catalog of materials, from top to bottom, Jarmo can only be assessed as a "one-period" site. It seems inconceivable to me that its duration can have been over 500 years at most, and I believe, in fact, that it was probably less. Thus, the scatter of Jarmo dates, from C-744 (5266 ± 450 years before the present) to W-657 (11,240 ± 300 years before the present), is archaeologically quite unrealistic. This seems to have no relation either to the counters [compare Beersheba in Palestine, where the Chicago counter's date (C-919) ran early and the Washington counter's date (W-245) ran late for the same horizon], or to the way the samples were collected from Jarmo (for example, W-651 was of the same batch of samples, collected in the same way, as the C-742, -743 cluster). My own conclusion is that "geobiochemical" contamination *in situ* must have something to do with the 6,000-year scatter of Jarmo dates, and M. Rubin writes that he wholeheartedly concurs.

58. The finding of a new site with Jarmo type material at its base, superimposed by Hassuna-Samarran phase material, has recently been reported from a valley adjacent to that of Jarmo [see H. Ingholt, *Sumer*, **13**, 214 (1957)]. I believe that Solecki's (note 29) typologically earlier open-air site, Zawi Chemi, is of the preceding era of incipient cultivation and is to be considered as contemporary with Karim Shahir and several other Iraq-Jarmo project sites (note 25).

59. Braidwood, *Antiquity, loc. cit.*

60. Braidwood, *The Near East and the Foundations for Civilization, op. cit.*

61. Braidwood, Howe, et al., *loc. cit.*

62. F. E. Zeuner, *Antiquity*, **30**, 195 (1956).

63. K. Kenyon, *Palestine Exploration Quart.*, **89**, 101 (1957).

64. K. Kenyon, personal communication.

65. K. Kenyon, *Antiquity*, **30**, 184 (1956).

66. V. G. Childe, *ibid.*, **31**, 36 (1957).

67. Braidwood, *Antiquity, loc. cit.*

68. Miss Kenyon is due an apology from me, and this is freely given. Basing my opinion on the only cluster of radioactive carbon dates from Jarmo available at that time (C-113, -742, -743), I strongly favored a "short" chronology [*Antiquity*, **31**, 73 (1957)]. The newer dates (see above) for Jarmo and other sites make a general "short" chronology untenable.

69. K. Kenyon, *Antiquity*, **31**, 82 (1957).

70. J. Garstang, *The Story of Jericho* (Hodder and Stoughton, London, 1940).

71. F. E. Zeuner, *Geol. Rundschau*, **45**, 2 (1955).

72. M. Wheeler, *Walls of Jericho* (Chatto and Windus, London, 1956), p. 117.

73. Braidwood, *Antiquity, loc. cit.*

74. H. E. Wright, Jr., *Bull. Am. School Oriental Research*, **128**, 11 (1952).

75. Data in Fig. 5–5 are restricted to the available occurrences which now have radiocarbon dates; compare figs. 5–5 and 5–6 with Table 1 in note 49.

76. R. J. Braidwood and L. Braidwood, *J. World History, loc. cit.*

77. Willey and Phillips, *Method and Theory in American Archaeology, op. cit.*, p. 34.

78. Solecki and Rubin, *Science, loc. cit.*

79. R. S. Solecki, *Sumer, loc. cit.*

80. Solecki and Rubin, *Science, loc. cit.*

81. See note 58.

82. J. Perrot, *Antiquity and Survival*, **2**, 91 (1957).

83. Braidwood, *The Near East and the Foundations for Civilization, op cit.*

84. R. J. Braidwood and L. Braidwood, *J. World History, loc. cit.*

85. H. Helbaek, see note 25 and also *Ann. Rept. Inst. Archaeol. Univ. London*, **9**, 44 (1953).

86. J. R. Harlan, personal communication; see also *Brookhaven Symposia in Biol. No.* 9 (1956), p. 191.

87. Obviously, the character of the curves in Fig. 5–6 depends, among other things, on the way the geographical columns (compare Fig. 5–5) are arranged. Thus, the sharp rise, followed by inflection, of the right side of the village-farming community curve depends on placing the "south Mesopotamia" column next to the "Tigris" column of Fig. 5–5. Early villages have not appeared in the rainless alluvium of southern Mesopotamia.

88. G. R. Willey and P. Phillips, *Am. Anthropologist*, **57**, 723 (1955); in an earlier version of their useful book (see note 12) they used the idea of a "preformative stage" in categorizing certain New World materials. They have now eliminated the "preformative," in part–if I understand them correctly (note 12, p. 105)–because of "the difficulty of finding criteria that would hold for all major areas of New World archaeology." I think they may have been overly timid and that some "stages," eras, or phases may not have been affairs of universal validity but may rather have been manifestations of a given environment alone.

89. Braidwood, *The Near East and the Foundations for Civilization, op. cit.*

90. Braidwood, *Prehistoric Men, op. cit.*

91. See note 88.

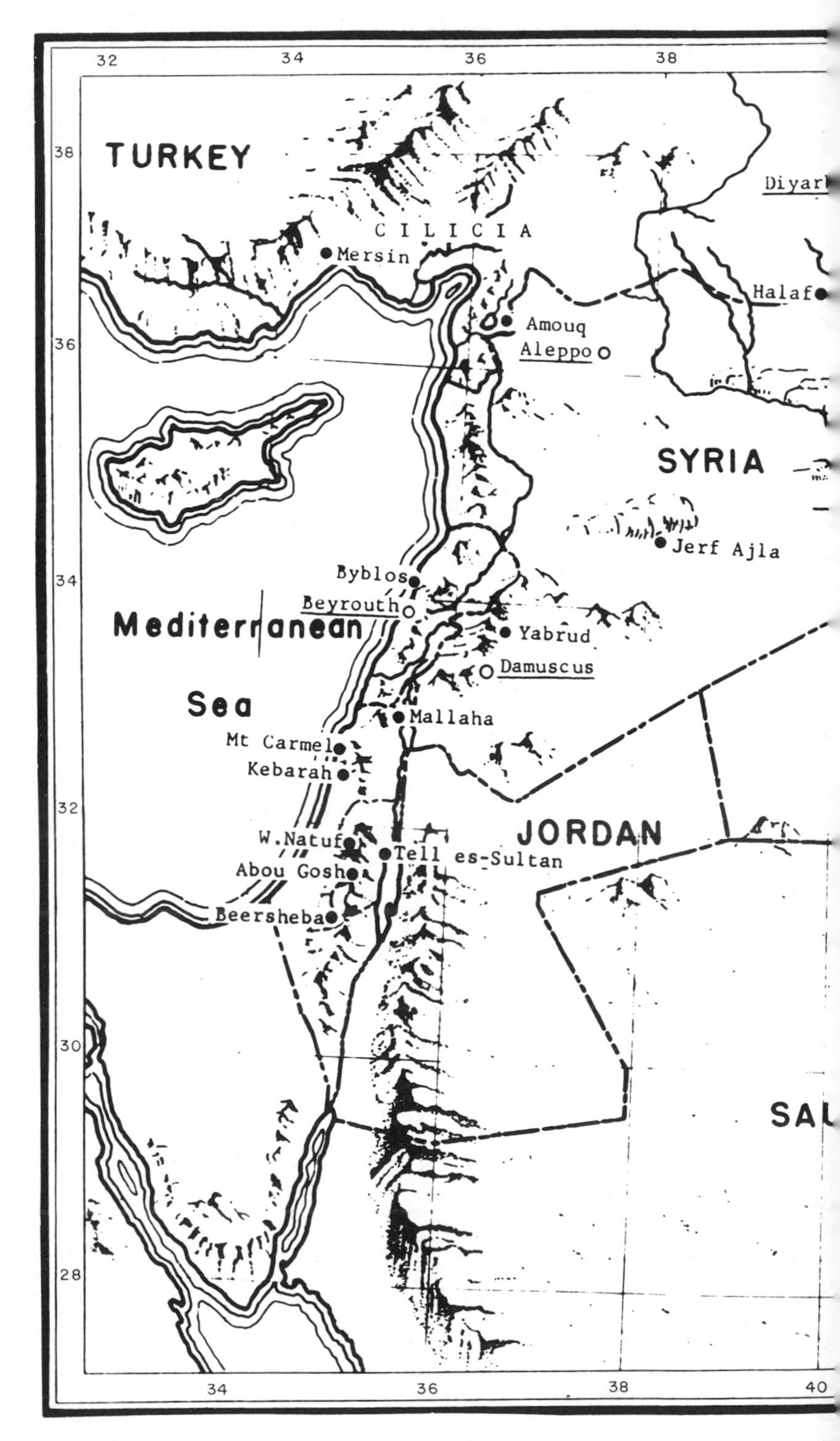

Fig. 5–1. Map of the core area of the Near East in prehistoric times, showing positions of sites for which radiocarbon determinations are available and of cer other key sites. Modern cities are underlined and designated with an open c (for example, ○ *Baghdad*). The "hilly-flanks" natural habitat zone follows an from Kurdistan to north of the city of Diyarbekir, to Cilicia, and thence down Syro-Palestinian littoral.

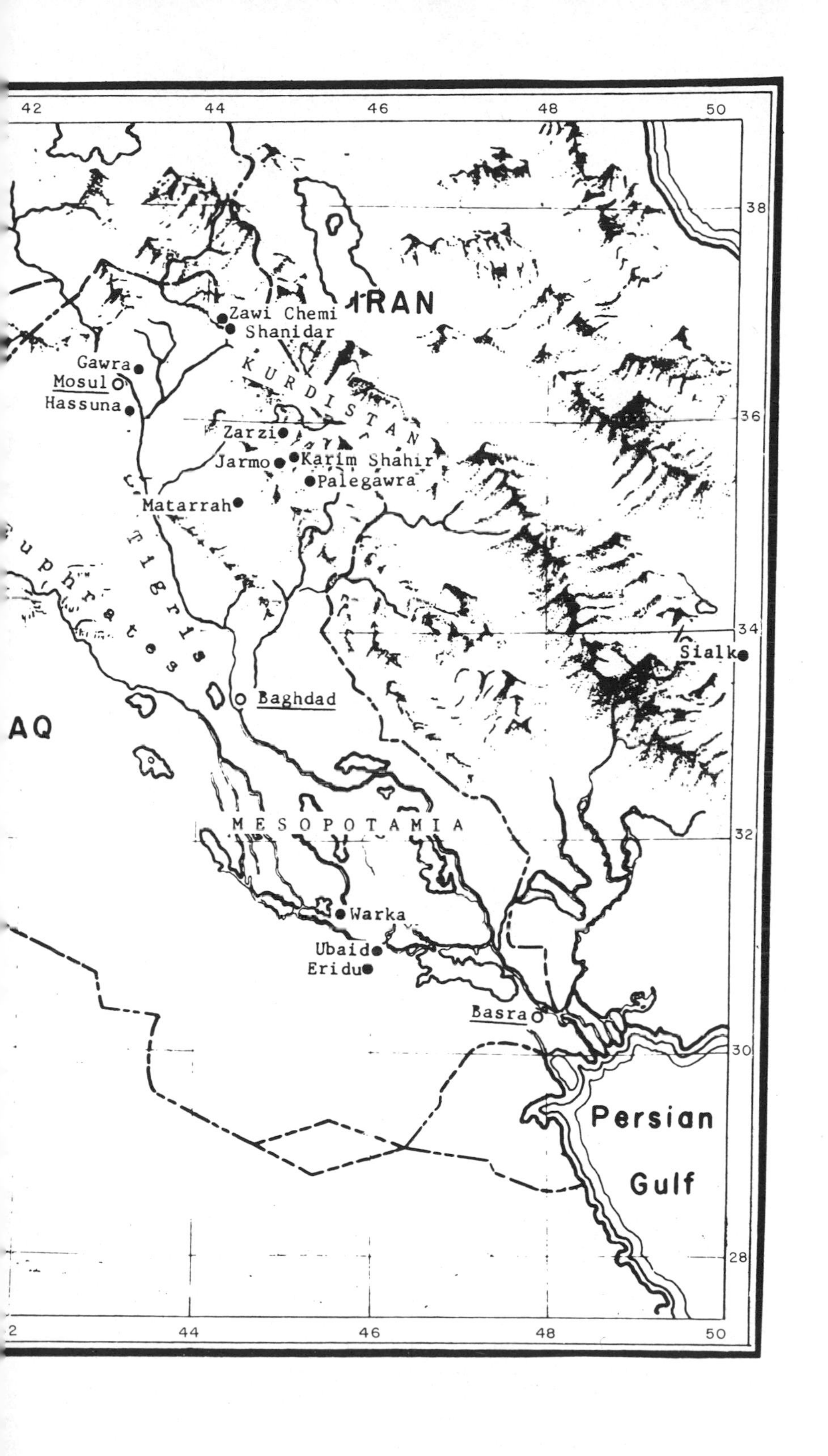

42
44
46
48
50
38
36
34
32
30
28
IRAN
Zawi Chemi
Shanidar
KURDISTAN
Gawra
Mosul
Hassuna
Zarzi
Jarmo
Karim Shahir
Palegawra
Matarrah
Euphrates
Tigris
Sialk
Baghdad
AQ
MESOPOTAMIA
Warka
Ubaid
Eridu
Basra
Persian
Gulf
2
44
46
48
50

Fig. 5–2. Site of the Zarzi Cave (late or food-collecting era of the food-gathering stage) in the intermontane valley south of the Dukan Gap, Iraqi Kurdistan.

Fig. 5–3. The Palegawra Cave (late or food-collecting era of the food-gathering stage) in the Bazian intermontane valley, Iraqi Kurdistan.

Fig. 5–4. The Karim Shahir open site (era of incipient cultivation) in the intermontane valley of Chemchemal, Iraqi Kurdistan, above white bluff, center.

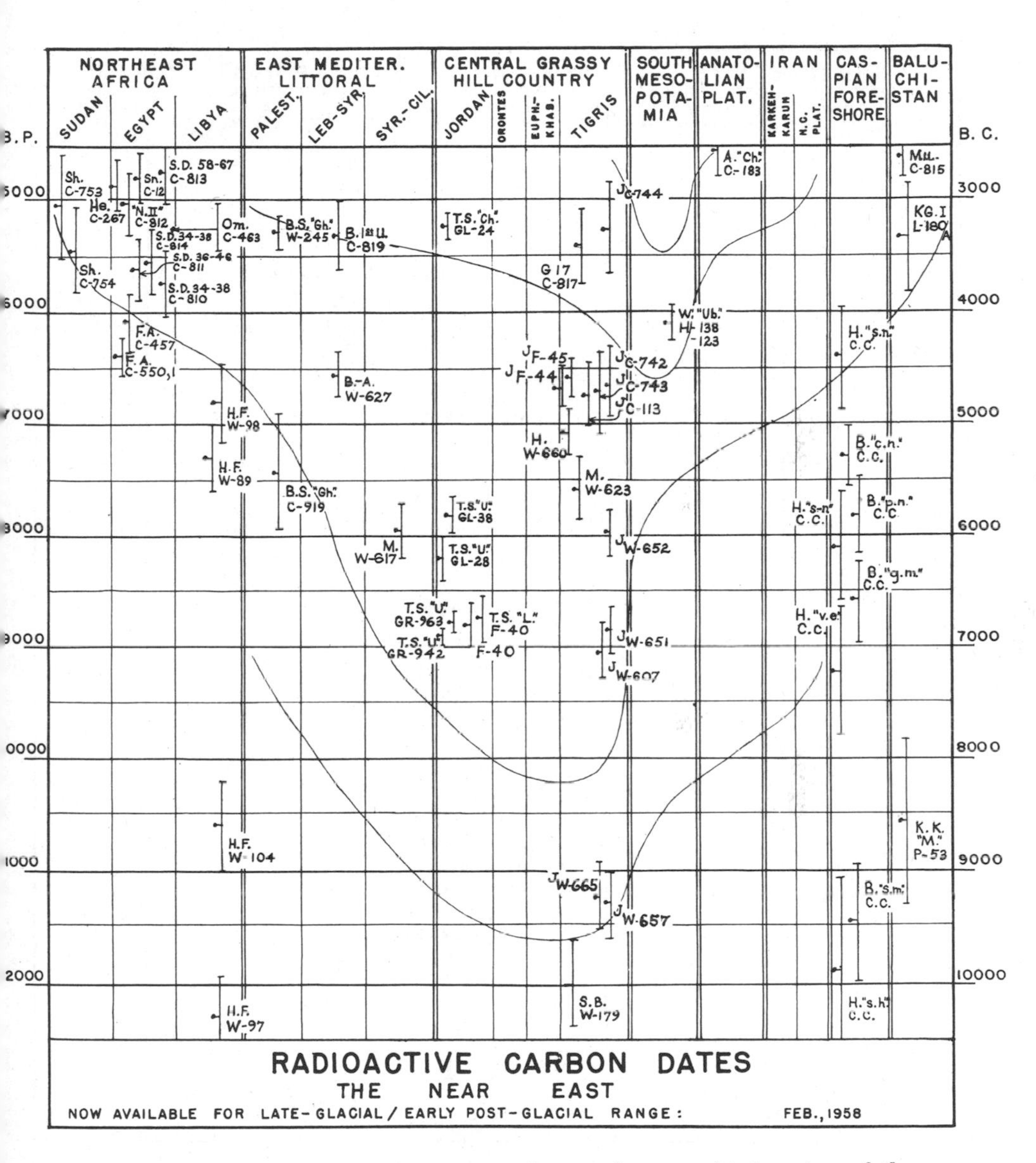

Fig. 5–5. Summary of the positions, in time and general geographical region, of the now available radioactive carbon dates in the Near East, for the range from circa 5,000 to 12,000 years ago. The curves suggest levels or eras of food-getting practices, as shown in Fig. 5–6. Most of the archaeological sites involved are shown in a previously published table. See also Solecki and Rubin (*Science*, **127**, 1446 (1958). A key to these dates appears in Table 5–1.

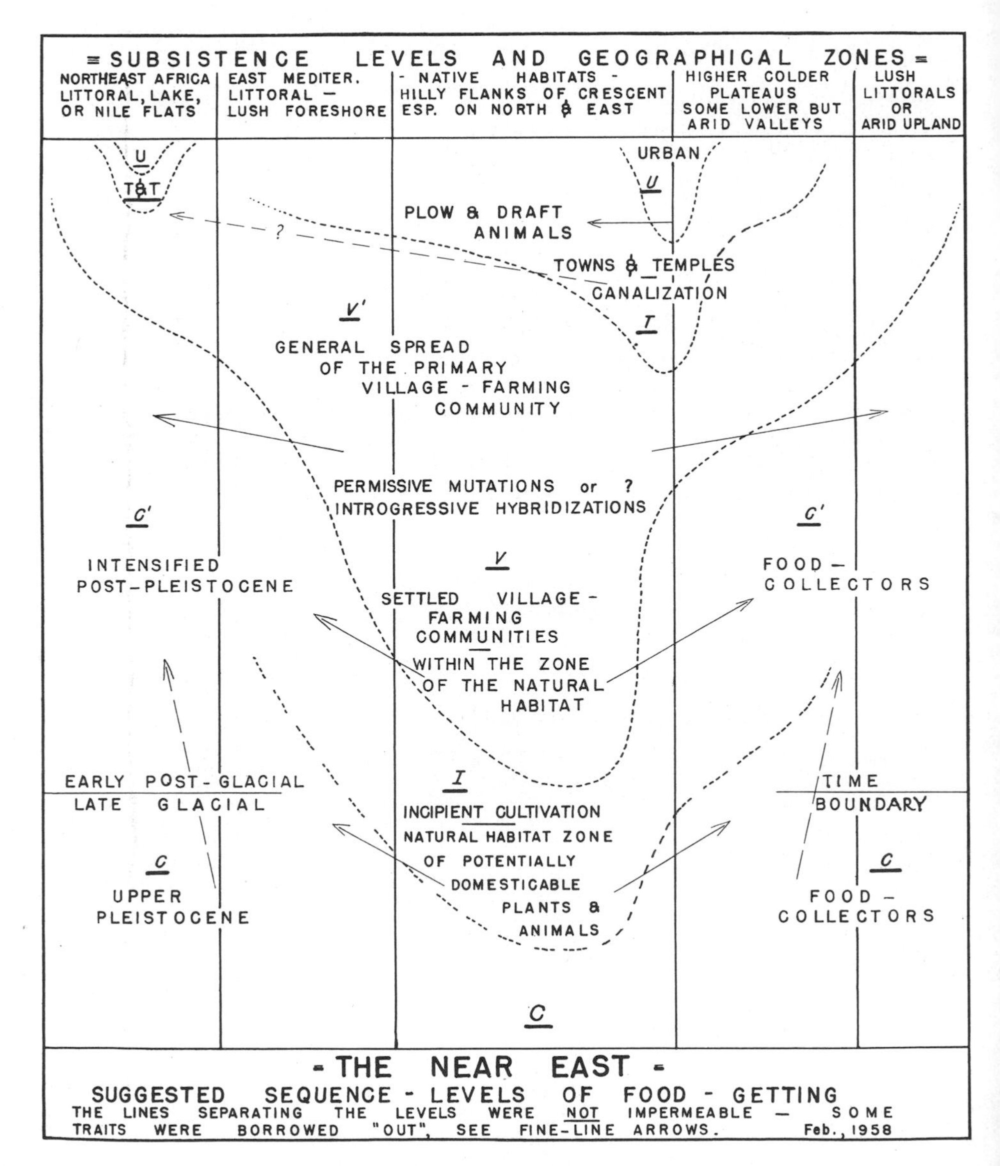

Fig. 5–6. Summary of the levels or eras of food-getting practices in the Near East with respect to the times and geographical regions shown in Fig. 5–5: *C*, food-collecting; *C′*, intensified food-collecting; *I*, incipient cultivation; *V*, village-farming communities; *V′*, intensified village-farming communities; *T* + *T*, towns and temples; *U*, urban communities.

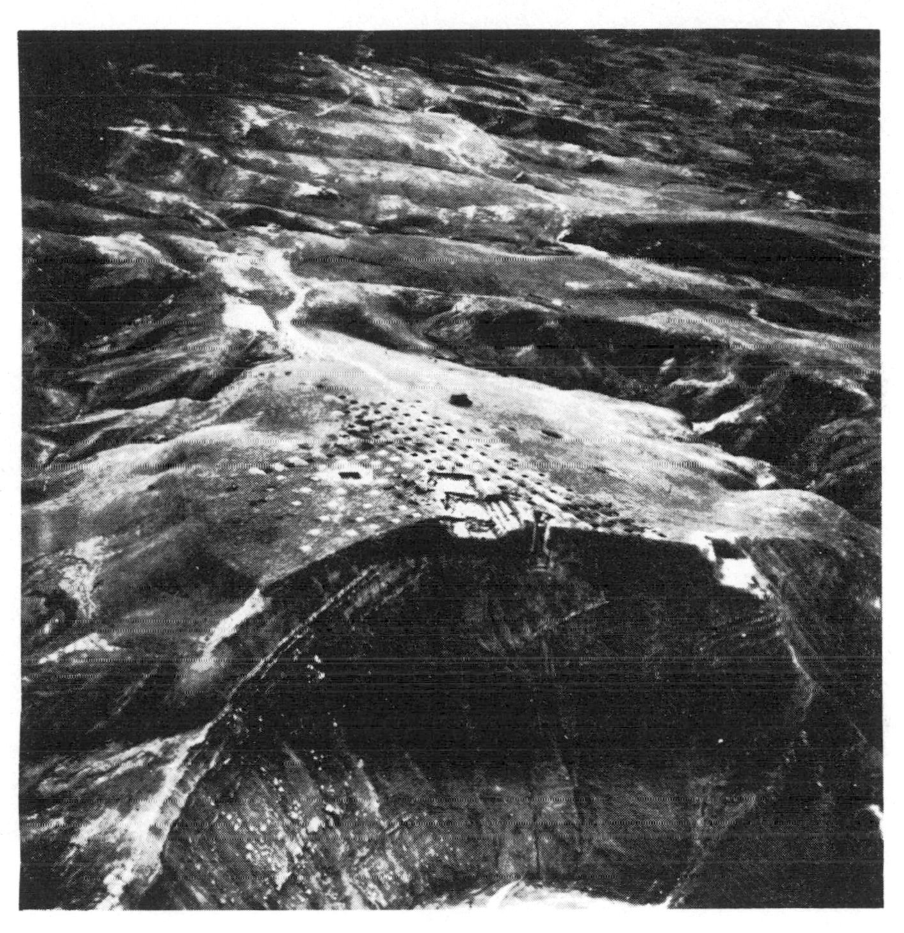

Fig. 5–7. Air view of the village-farming community cite of Jarmo, in the intermontane valley of Chemchemal, Iraqi Kurdistan. The base exposed in the cut at the extreme right is virgin soil; the grid squares in the center were for the purpose of exploring the village plan in the uppermost levels. (Courtesy of Iraq Petroleum Co., Ltd.)

Fig. 5–8. (Top) Partial plan of a mud-walled house in the fifth level in the village site exposed at Jarmo. The white streaks on the room floors are the traces of reeds. (Center) Stone foundations of a house in the second level exposed at Jarmo. (Bottom) Incomplete plan of a mud-walled house in the sixth level exposed. In this level and at deeper levels portable pottery vessels were not in evidence.

Fig. 5–9. Konservator Hans Helbaek of the Danish National Museum, Abdullah Said Osman, field superintendent of the Oriental Institute's Iraq-Jarmo project, and Mrs. Robert Braidwood examine a wheat field in the Chemchemal Valley.

Fig. 5–10. The "division" of antiquities excavated at the village site of Jarmo. H. E. Dr. Naji al-Asil, director general of antiquities for the Iraqi Government, is in charge.

Chapter 6

Animal Domestication in the Prehistoric Near East

Charles A. Reed

The long path in time leads to ourselves from a hominoid group which abandoned forelimb brachiation for hindlimb bipedalism. Once on that path, we can say with the wisdom of hindsight, man was unique as no other animal group ever had been. Combining ever greater skill at abstraction and communication with ever increasing utilization of energy sources, the main pattern of human culture has led through the successive major steps of tool invention, tool improvement, plant cultivation, animal domestication, urbanization and political integration, and so finally to the industrial revolution. Looking forward, this path bids to lead us to other planets and other planetary systems.

Seen thus in the long perspective, the initiation of cultivation and domestication—the Neolithic or "food-producing revolution"[1]—was one of three or four great cultural innovations, and a fundamental and necessary prelude to civilization.[2, 3] (I claim no originality for the above ideas; they are discussed at length in many anthropological writings.) It is true, however, that in spite of our certainty that agriculture and stock-breeding must have had beginnings, changing man from a roving hunter and gatherer to a settled village-farmer, and in spite of our knowledge of the vast ultimate consequences of this technical revolution, we know as yet very little about the details of these origins. Archaeology, the discipline upon which we have traditionally depended for our understanding of these beginnings, has been either uninterested—spending its vigor instead in more glamorous pursuits—or incapable of the fine analyses necessary.[4] For we have here a difficult field of palaeonatural history, where the geomorphologist, climatologist, soil scientist, palaeoethnobotanist, agronomist, ecologist, geneticist, taxonomist, and comparative anatomist must add their skills to those of the field archaeologist. Not all of them have yet done so, and certainly the future will witness greater coordination all along the line.

The literature that shows the attempts of an earlier generation of Near Eastern archaeologists to be their own natural history experts, particularly in the field of zoological identification, is generally a sorry one and should be quietly disregarded. Inevitably and unfortunately, however, the conclusions published in these primary sources are those which have become crystallized into subsequent review papers and textbooks. Too often, also, the intriguing discoveries of the cultural complexity uncovered in the daily digging (particularly in the mounds representing remains of prehistoric towns) argued so strongly for an agriculturally based economy with assured food production that flocks and cultivated fields were assumed. Since the actual proof of the presence of the plants and animals was not thus regarded as necessary, the carbonized grain and the broken animal bones, which should have been considered of primary importance whenever they were uncovered, were too often shoveled onto the dump heap.[5]

Often, simply, the archaeologists of an earlier day—trained as they were in the arts, and in the literature of classical or Biblical history—simply did not know what to look for, and the institutions financing them were not interested in excavating for "natural" (nonartifactual) materials which yielded merely ideas. Instead, the archaeologists sought what their home institutions expected of them: display objects, written records, sculptures, and monuments. The more subtle interpretations that are made possible by the cooperation of teams of archaeologically oriented natural scientists, working at the excavation, have thus only recently become intellectually and financially possible, and even now the budgetary problems involved in including all of the desired personnel in a major archaeological expedition are very great.[6]

With the exception of a recent book (not yet seen by me) by F. E. Zeuner,[7] none of the major works on the origins of animal domestication[8] are in English; this field of study has been pre-eminently a German one. Previous summaries were limited (as is always the case) to the knowledge available at the time; most such information was of the European Neolithic and the *historical* periods (as gleaned from writings and pictures) of the ancient Near East. Except for the peripheral Iranian site of Anau,[9] little was known of the all-important late *prehistoric* cultures of southwestern Asia—the cultures actually representative of the period and the area of incipient domestication.

The more recent summaries on domestication by W. Herre[10] are those of an experienced morphologist with full knowledge of modern taxonomic practice and evolutionary theory; yet these do not attempt to make an evaluation of the kind attempted here—a critical analysis of our present knowledge of particular phases of the origin of animal domestication by one who has collected and worked in southwestern Asia, who has excavated much of the pertinent material at several of the most important sites, and who is studying the collections from these and other important sites.

Difficulties

In spite of a prolific literature, dating well back into the nineteenth century, the central problems concerned with the origins and early history of animal domestication remain unsolved. In large part, as mentioned, this unfortunate situation is due to archaeology's not having asked itself the right questions, or, if it has done so, to its having assumed the answers without having saved the evidence. Thus, too many of the reports on prehistoric Egyptian and southwestern Asian cultures merely assert the presence of domestic animals without offering any anatomical proof.

Even if saved, the evidence may well run to tens of thousands of broken animal bones, which have to be cleaned, sorted, and individually studied in an effort to identify the bone and to determine the age, sex, and species of the animal (when this is possible). Where wild and domestic forms existed together in the same area, attempts must be made to segregate them.

Heretofore, a major deficiency in attempting in America to study the faunal remains from archaeological sites in the ancient Near East has been the almost complete lack of comparative skeletal material with which to make correct identifications. Ideally, one should have complete skeletal series of all the species that existed in the area of the prehistoric culture in order to study age, sexual and individual differences, and differences between wild and domestic forms of the same species. But far from having such series, we had, until recently, practically no study skeletons from the Near East in the Western Hemisphere; indeed, several of the species have become extinct within historical times, and others are perilously near that state. However, a good beginning of such a skeletal collection has now been gathered and is available for study in Chicago.[11]

Piles of dirty broken bones have little appeal to most zoologists, busy with their own researches, nor is the upper Quaternary (particularly the sub-Recent, with its modern-type fauna) of interest to most palaeontologists. Such problems really, then, must be undertaken by zoologists who know the area concerned, and who have collected in it, who have worked cheek-by-trowel with the archaeologists, and who are not only ecologically sensitive to the environmental problems presented but are also anthropologically oriented to the nuances of evolving human cultures. Such zoologists are few (although the field, open and new, will be a promising one once it acquires the respectability of institutional support[12]).

Under these conditions, even when osteological collections from important sites have been made, the bones have sometimes lain around for years while the archaeologist vainly tried to get someone to study them. Perhaps finally succeeding, he has in turn too often been handed a list of generic and specific names, meaningless to him, to be duly published as an appendix to the site report.

Without interpretation, both environmental and cultural, biological studies related to prehistoric sites have practically no meaning.

Another basic difficulty—aside from the fundamental one of the tremendous expense of putting properly staffed expeditions into the field halfway around the world—is the paucity of fundamental evidence to date. We have less than a dozen sites in the time range immediately prior to incipient domestication (and not all of these have been studied in detail or published completely), and we have fewer yet for the suspected crucial period of actual domestication. Furthermore, due to political accidents of modern history, these sites cluster in Palestine and the Zagros region of Iraq and Iran, with a distant one recently discovered in southwestern Turkey. The intervening gaps are archaeologically unexplored, insofar as our problem is concerned.

A difficulty of the late 1940's and the 1950's was the occasional failure of the (carbon-14) technique to yield reliable determinations ("dates"). A classic example of absurdity that would have resulted from a strict adherence to such "dates" is the 6,000-year spread of 11 determinations for Jarmo,[13] a prehistoric village in northeastern Iraq; on the basis of all archaeological evidence, Jarmo was not occupied for more than 500 years. The basic problem seems to have been failure to remove contamination of older or younger carbon from the samples, but newer techniques of chemical pretreatment are making such errors less frequent. One should always be wary, however, of accepting single radiocarbon determinations as being the final authority.

The Problem

What is needed, and what has recently been attempted by several expeditions to southwestern Asia, is a thorough analysis of all the evidence bearing on the origins of agriculture, animal domestication, and the village-farming way of life. The parts of the problem have different degrees of dependence upon each other (for instance, most of the geological and climatic events would have transpired in the absence of man), but all are intertwined. One cannot think of domestication, thus, as happening independently of the geographical factors (terrain, climate, flora) that always determine animal distribution, or independently of the culture—including the primitive agriculture—of the domesticators.

We must then, like good reporters, try to answer the five W's and the lone H: When, Where, Who, What, Why, and How?

When

Although it has been suggested that reindeer were domesticated during the Upper Palaeolithic in western Europe,[14] no real evidence of animal domestication can be shown for any Pleistocene period;[15] we are dealing entirely, so far as is known, with terminal-Pleistocene and post-Pleistocene phenomena.

Stockbreeding provides some assurance of a supply of animal fat and proteins, but evidence from all early Old World villages with domestic animals still shows dependence on wild game. All evidence of early domestication is associated with primitive villages containing grinding stones; the inference is that incipient cultivation[16, 17] and incipient domestication came together, or, perhaps more logically, that domestication followed by some unknown period of time the settling down into primitive village life by early cultivators. The presence of grinding stones, however, does not necessarily prove cultivation; in the North American Southwest, grinding stones (*manos* and *metates*) were used for seeds of wild plants, and cultivation (of maize, beans, squashes, and so forth) developed elsewhere. In southwestern Asia, however, the general pattern seems to be that cultivation followed closely upon, even if not yet proved to accompany, the presence of the earliest grinding stones.

If Dexter Perkins' conclusions[18, 19] are correct that the sheep at Zawi Chemi Shanidar[20] were already domestic, almost 11,000 years ago, then the earliest known grinding stones in southwestern Asia are found at the same incipient village with the earliest known domestic animals. The archaeological record is woefully incomplete for this critical end-Pleistocene period, true, but that which we have is certainly suggestive that grinding of seeds (whether gathered or grown), development of the early village, and first domestication were closely associated events in southwestern Asia. However, there is little evidence as yet that any of the other known "incipient-cultivation" sites—Karim Shahir,[21] Mallaha,[22] and the various Natufian sites in central Palestine—had domestic food animals, nor can a domestic dog now be accepted.[23, 24]

Although we must be properly cautious in accepting as valid any lone carbon-14 determination from an individual locality, the four available dates for sites of the period of incipient cultivation have a comforting closeness in time, being close not only to each other but also to what we had expected on the basis of accumulating evidence of the last few years. The date for the short-time occupation site of Zawi Chemi is 10,870 ± 300 years and that for a typologically contemporaneous level in nearby Shanidar Cave is 10,600 ± 300 years.[25] Two determinations for Early Natufian levels at Jericho are 9,850 ± 240 years and 9,800 ± 240 years.[26] Since milling stones were present at Zawi Chemi and mortars and pestles, plus flint sickles, are known from early Natufian sites in Palestine, we can say, in easily remembered round numbers, that by approximately 10000 B.P. (before the present), reaping and milling of wild cereals was most probably a reality, with purposeful planting a possibility.

If sheep were already domestic 10,500 or possibly almost 11,000 years ago, one would expect that the other food animals (goats, pigs, cattle) as well as the dog might have been soon domesticated in the Near Eastern "nuclear center."[27] However, the evidence is not conclusive; indeed, following Zawi Chemi there is a 2,000-year gap in our history, before we find domestic animals at the village-farming community of Jarmo in the Zagros foothills and at the town of Jericho (Tell es-Sultan) in the Dead Sea Valley. Both contain domestic goats, an identification based on the shape of the male horn cores[28, 29] (Fig. 6–3), and domestic pigs appear suddenly and coincidentally with the introduction of pottery in the upper levels of Jarmo.[30, 31]

Thus by 8,500 years ago, or soon thereafter, three (sheep, goats, pigs) of the four primary food animals had been domesticated, but for the fourth (cattle) and for the dog we have much poorer evidence as to time of domestication.

The answers to our chronological problems still lie in the ground; once excavated, the material must receive detailed study in the laboratory.

Where

All archaeological work to date in the Near East suggests that both agriculture and animal domestication (with the possible exception of that of the dog) had their origins in the hilly, grassy, and open-forested flanks of the Zagros, Lebanese, and Palestinian mountains (see Fig. 6–2). These data have been treated fully elsewhere[32, 33, 34, 35, 36] and need not be repeated here. On the basis of the data assembled by Dyson[37] and of recent archaeological evidence from central Asia, the highly respected ethnologist Christoph von Fürer-Haimendorf[38] has strongly discounted the old notion that animal domestication arose during an early stage of pastoralism. He stressed that, although the dog appeared with pre-agricultural hunters, the basic food animals always appeared in a context of early village-farmers. Further, he said, the domestication of the horse and reindeer, it must now be realized, came relatively late and had no influence on the earliest agricultural communities or their immediate historical derivatives.

From the primary center in the open-forest hills of southwestern Asia, the village-farming way of life diffused in all directions, carrying with it its trade-marks: the village, cereal agriculture (primarily wheat and barley), and the basic domestic food animals. In Egypt, in Thessaly, in Baluchistan and the Indus Valley, probably even in China (at least in northern China), the beginnings of village-farming life came later and seem to have received a cultural stimulus from southwestern Asia.

The case of Egypt is particularly instructive, as wheat, goats, and sheep do not occur wild in Egypt (nowhere in Africa do true sheep and goats occur wild[39]) and so, obviously, were introduced as cultivated and domesticated species. If the radiocarbon dates for the Egyptian Fayum are accurate (possibly they are not, as we do know of some later radiocarbon dates for Egypt which are obviously too recent), the earliest and simultaneous appearance of cereal agriculture and domestic goats (or goats *and* sheep?) in Egypt, at Fayum, was considerably later (about 6200 B.P.) than the probable time of their earliest known associations in Asia (about 8500 B.P.).

Who

The people who first grew grains and domesticated hoofed mammals were, on the basis of skeletal evidence, modern-type men of the Mediterranean race. Doubtless they would pass unnoticed, if suddenly resurrected, among the people of today in the hill country where they lived.

Questions arise, to which we have no answers: Would the "agricultural revolution" have had its start where and when it did if another people, of different color or head shape, had lived there? Or would these important events have occurred if our same Mediterranean peoples had had, by a historical accident, some slightly different cultural pattern?

What

It seems logically probable—although we have as yet no direct evidence—that the cultivation of wheat and barley (or, at first, possibly of wheat alone) induced (or should we be more cautious, and say "allowed"?) the formation of the permanent villages. Probably both agriculture and village development were a necessary prelude to domestication of the basic food animals, although there are contrary views.[40] These food animals, which undoubtedly contributed so much to the evolution of late prehistoric cultures in the Near East, were goats (*Capra hircus*), sheep (*Ovis aries*), cattle (*Bos taurus*), and pigs (*Sus scrofa*).[41]

In addition, the zebu (*Bos indicus*) was certainly present prehistorically in Baluchistan,[42] and subsequently in the Indus Valley. This whole area, however, is peripheral to the central (or "nuclear") Near East with which we are mainly concerned, and the earliest suggested date for domestic cattle (about 5000 B.P.) is late by Near East standards. The donkey (*Equus asinus*), domesticated from the Nubian wild ass, is of a similar antiquity, having been reported from the site of Maadi in Lower Egypt.[43] By this time, too, the Syrian onager (*Equus hemionus hemippus*) probably had been domesticated in Mesopotamia. The other domestic animals, both birds and mammals (ducks, geese, chickens, horses, camels, yaks, water buffaloes, reindeer, rabbits, and so on), which we rightly consider to be and to have been important in various human cultures, were not present as domesticants in the late prehistoric of the Near East and so are not here considered.

Dog (*Canis familiaris*). Since Bate[44, 45] announced that a domestic dog was present in the Natufian period, prehistorians have generally assumed that the dog was the Near East's first domestic animal, ubiquitously present for a period of nearly 10,000 years. However, the independent studies of J. Clutton-Brock[46] and Charles A. Reed[47] have shown that the specimens of supposed dogs fall within the normal range of variation of the Palestinian wolf. An even earlier "dog," announced as such without corroborative evidence, was found in Belt Cave in Iran,[48] and had a carbon-14 determination of 11,480 ± 550 years.[49] If truly a dog, this would at present rank as the world's oldest domestic animal, but unfortunately the specimen never received proper study and seems to have been lost. Instead, the world's oldest known dog[50] is probably that found at Star Carr, England, from the early Mesolithic (about 9,000 years old).

I doubt that dogs were known in the Near East at this time, or for a considerable period thereafter. There are dogs reported at Jericho, true, from the prepottery "plaster-floor level,"[51] but until the material (mostly teeth) has been compared in detail with the small Arabian wolf (*Canis lupus arabs*) no final conclusion is possible.

I have not been able to convince myself that there were dog bones among the midden remains from Jarmo, although wolf and fox have been identified. Since the bones are all extremely fragmented, the remains of a large dog might be mistaken for those of a wolf. Indeed, the dentition of the better preserved fragments of large canids is slightly small for the average of that of the local wolves of the area today, but still within the normal range of such wolves. There is, thus, no zoological evidence for a dog at Jarmo. The best evidence for a Jarmo dog is cultural, not zoological; several clay statuettes or figurines of what certainly appear to be dogs (the tail is curled over the back) have been found among several thousand figurines, many of them identifiable as mammals native to the area. Even when identifiable as goats or sheep, however, these figurines are too crudely modeled to yield any clues about domestication.

In Egypt, the first valid evidence of the dog is also artifactual; four dogs, led on leashes by one man, are represented on a pottery bowl[52, 53] of the Amratian period. They already show characters of the greyhound or seluki type, which by this time is also known (although the build is somewhat sturdy) from the Ubaid period in Mesopotamia by skeletons and, from a somewhat later time, by carvings on cylinder seals.[54, 55] The presence[56] of this specialized breed at this time at both ends of the Fertile Crescent indicates a long, although undocumented, period of artificial selection in the Near East. Not until the late Gerzean period do we find definite skeletal evidence of the dog in Egypt.[56]

The general lack of skeletal evidence of prehistoric dogs in southwestern Asia and in Egypt is probably in part a reflection of the lack of attention given to such skeletal materials when they were found by archaeologists during the last century, but perhaps in part it reflects the fact that dog carcasses were more likely to be available to scavengers than were the bones of the food animals. Perhaps, too, dogs were relatively rare as compared with the hoofed domesticants.

The wolf (*Canis lupus*) has generally been regarded as the ancestor of the dog. This supposition has been based on (1) the great morphological similarities, particularly as to dental details, between the wolf and the earliest dogs of the Mesolithic of western Europe; (2) the complete interfertility of dogs and wolves (with fertile hybrids); and (3) the great similarity of behavior.[57]

In spite of such evidence, however, several authors have suggested the golden jackal, *Canis aureus,* as the dog's ancestor, usually admitting later admixture with wolves for the more boreal breeds.[58] Certainly D. M. A. Bate[59] regarded the supposed "dog" from the Natufian of Palestine as jackal-ancestored (although, as mentioned above, the animal was almost certainly a wolf). However, dog and jackal are dissimilar with respect to certain definite dental characters, and they have a different chromosome number (dog, 78; jackal, 74).[60] While dogs and jackals do interbreed, the fertility of the resulting hybrids seems not to have been established with the certainty usually assumed. Although the problem cannot be said to have been settled, and there may have been some interbreeding of dogs (once established) with jackals, the preponderance of evidence indicates the wolf as the primary ancestor of the first dogs. The third possibility, that a hypothetical "wild dog"[61] or the pariah dog[62] actually represents an ancient stock from which the domestic dog was derived, lacks any historical evidence and fails to find the necessary palaeontological support.

Goat (Capra hircus) and sheep (Ovis aries). Most of the bones—although not the horn cores or metapodials—of these two species are so similar that the species are often included together as "sheep/goat" or "caprovid" in archaeological reports. Even when they are supposedly distinguished, one must always be suspicious of the validity of the identification.

Much careful archaeological and zoological work remains to be done before we can be sure of whether sheep or goats were domesticated first, but present evidence indicates it was the sheep[63] (as based on studies by Dexter Perkins). With sheep, as with other animals, the earliest domesticants would be identical with, or very similar to, the wild form. Only after many generations could mutations accumulate that would so mark the domestic population that their broken bones would be distinguishable from those of wild individuals brought into the village by hunters. (Even now, where it is available, wild game is typically brought into the villages in the Near East for food.) However, population-age analysis based on the bones may show a shift from a stratigraphically lower level with random age distribution to a higher, and thus later, level with a greater proportion of young and near-mature animals. Such a shift would certainly suggest a change from wild-killed animals to domesticated ones, most members of the herd being harvested at optimum times. It is on the basis of this type of evidence that Perkins (unpublished data summarized by Reed[64]) determined that domestic sheep were being kept at the incipient village of Zawi Chemi Shanidar in northern Iraq, almost 11,000 years ago. Using the same type of evidence, although on a rather limited series, Coon[65] concluded that the goats and sheep from Belt Cave in northern Iran (carbon-14 determination of about 8,000 years ago) were domestic.

If sheep were present in northern Iraq almost 11,000 years ago, we would expect to find them generally distributed in subsequent village sites. The evidence, however, is inconclusive, due in part to the difficulty of distinguishing between even intact bones of goat and sheep,[66] and in part to the extremely fragmentary nature of the majority of bones of food animals recovered from Near Eastern prehistoric sites. Even the metapodials (cannon-bones), the one truly diagnostic bone (shorter and broader in goats than in sheep) is difficult to distinguish; rarely is one found intact, and domestication was accompanied by a shortening of metapodials in both sheep and goats. These factors, combined with the naturally smaller size of bones of females in both species, present one with the probability of eight classes in an overlapping series, that is, male and female for both wild and domestic sheep and goats. Additionally, most of the bones of the domestic animals are from immature individuals, so that full growth and adult proportions were not attained.

The characters of the skull whereby neo-mammalogists easily distinguish between *Ovis* and *Capra* are of little help to the student of osteoarchaeology; the skulls are generally too fragmented. Even if identifiable as to genus, a piece of broken skull will rarely yield evidence as to its domestic or wild origin.

The shape of the horn core (the bony inner skeleton of the horn) is almost always as diagnostic of the species as is the horn itself, and it is on the basis of the horn cores that we distinguish domestic from wild goats. However, a most peculiar type of horn core has been appearing in a number of sites (Jarmo, Sarab, Sialk), which has the surface texture and general appearance (outer shape and cross-section) of horn cores of wild goats, but internally has the structure (small air sinus with bony cross struts) of sheep. If these peculiar horn cores, so goatlike superficially, truly belong to sheep, the sheep represented cannot be wild, but may instead represent the puzzling "goat-horned sheep" discussed by J. Ulrich Duerst at Anau.[67] Duerst had the horn cores of such sheep attached to crania, so his diagnosis is not open to doubt. Unfortunately, he did not discuss the morphology in detail, and the photographs are not sufficiently clear to allow satisfactory comparisons. The Anau material has seemingly since been discarded, so direct comparisons cannot be made.

If these particular horn cores do represent "goat-horned sheep" the breed was a highly specialized one, at least with regard to the long scimitar-like horns, which probably curved over the back like the horns of an ibex or wild goat.

If, further, these peculiar horn cores are truly those of sheep (a tentative diagnosis, note, based only on their internal structure) then we have the evidence for the domestic sheep we would expect to find in Near Eastern village sites. Further, this type of sheep (if such it was) persisted at least for some 3,000 years, from the time of Jarmo to Anau II. Correlated with this expectation and possible evidence of Near Eastern sheep, we have the recovery of woolen textiles at James Mellaart's site of Çatal Hüyük in southwestern Anatolia, dated to about 6300 B.C.[68]

J. Wolfgang Amschler[69] has reported both domestic and wild sheep from the Amouq sequence, but the sheep bones were rare in comparison with the other domestic artiodactyls. By 6000 B.P., we find, finally, widespread and convincing evidence of domestic sheep, from the Gerzean period in Upper Egypt,[70] Warka in Sumerian Mesopotamia,[71] and the Anau II level[72] in what is now Turkoman SSR. However, by 6000 B.P. almost half the known history of domestic sheep seemingly had been passed.

The exact ancestry of the earliest domestic sheep must be decided more on geographical than on anatomical data. The population generally called *Ovis orientalis* is the wild sheep of southwestern Asia, where domestication undoubtedly first occurred. Additionally, however, students of domestication have argued endlessly about which kinds of sheep were evolved from which species of wild sheep, without ever really knowing what a species of sheep is, how many valid species occurred (if there was more than one in central and western Eurasia), how much actual interbreeding (and thus gene-flow) occurred between the different populations variously described as species or subspecies, or what genetic factors underlie the characters of horn, head, tail, and fleece that have been so ardently discussed.

The genetics of most of these characters is still largely unknown, and a true classification of Old World *Ovis* is now extremely difficult, due to dwindling numbers of many of the populations.[73] However, if it were sufficiently comprehensive, a gene-frequency study[74] of ovid blood factors (potassium and sodium concentrations, hemoglobin types, blood groups), of both wild and domestic sheep would undoubtedly help clarify the muddled taxonomic situation and would also aid in tracing the ancestries and interbreedings of the different races of domestic sheep. Additionally, detailed study of many bones from many archaeological sites would give valuable collaborative evidence with historical depth. Until such data are forthcoming, I prefer the simplified taxonomic scheme of V. J. Tzalkin,[75] who believed that, aside from *Ovis canadensis* of far eastern Siberia, all the Old World sheep belong to several subspecies of but one species, *O. ammon*. Thus, the detailed anatomical differences that have been so thoroughly studied and discussed by many students of sheep domestication in tracing the phylogeny of different breeds would never have had more than subspecific value.

Perhaps goats were domesticated as soon as sheep, but we have no evidence of domestic goats until about 8500 B.P., when they occur in prepottery levels at Jericho, throughout the site of Jarmo, and somewhat later are also found at Sarab and Sialk. The identification is made on the basis of the shape of the male horn core, which, as in the wild form, was scimitar-shaped and projected up over the back. (The screw horn of the modern goat is a Bronze Age development.) Horn cores of female wild and domestic goats are seemingly not distinguishable, but in males (Fig. 6–3) the definite medial flattening is considered valid evidence of domestication.

Throughout the succeeding 2,000 years, until 6000 B.P., our record for goats is little better than that for sheep, but by the latter part of this period domestic goats were being reared not only in Egypt but near the Danube as well,[76] and only a little later in Baluchistan.[77] The goat was seemingly late in reaching north into the oases of west-central Asia; it is reported only from the upper levels at Anau, whereas domestic sheep are definitely known earlier there. Up the Nile, however, domestic goats—albeit dwarfs—seemingly preceded sheep; goats are known from the Sudan about 5300 B.P., while the contemporaneous evidence for sheep is meager and uncertain.[78]

If we assume, as we must on the basis of present evidence, that the earliest domestication of the goat occurred in southwestern Asia, there is little problem concerning the identity of the wild ancestor, as there is only one population (*Capra hircus aegagrus*) of wild goat in southwestern Asia. The ibex, various species of which occur in Europe, Africa, and Asia, has presumably never been domesticated and so does not complicate the problem, and the only other goat, *C. falconeri*, lives farther east.

Cattle (*Bos taurus*). The large, long-horned, wild *Bos primigenius* illustrated with such magnificent artistry at Lascaux, hunted and portrayed by the Assyrians, described with wonder by the Romans, and extinct in 1627, was distributed throughout the forested regions of Europe, North Africa, and southwestern Asia into historic times. Whether or not a second, short-horned species (*B. brachyceros*) occupied much of the same area has been disputed at length. C. Gaillard,[79] for instance, thought that both short-horned and long-horned species of wild cattle occupied the same area of Upper Egypt during the late Pleistocene, but here as elsewhere perhaps the short-horned individuals were merely the females of *B. primigenius*.[80]

In addition, the European bison (*Bison bonasus*) extended its range into southwestern Asia, and in Iraq, at least, a wild water buffalo (*Bubalus*) undoubtedly existed.[81] In the marshes of the Nile of prehistoric Egypt, in addition to true wild cattle, probably at least one kind of African buffalo (possibly two) existed.[82]

The simultaneous presence of these several Bovini in the Near East, the nuclear area of animal domestication, is important—primarily because of the very fact that it has been generally disregarded. The result has been that any large bovines from prehistoric sites have usually, in the archaeological literature, been labeled "domestic cattle" if, in any particular archaeologist author's opinion, the time range fell within the limits of expected animal domestication. Generally, the real problem—the great difficulty of distinguishing between these various genera, particularly on the basis of a few teeth or broken bones—has simply not been recognized. Additionally, there is the much greater problem, even if the animal is *Bos*, of determining its status—wild or domestic.

The value of the scientific material relative to large bovines that has been thrown away unstudied is fantastic; in some cases the "identification" of the native workmen at the excavation has been accepted on the spot, and the skeletons or skulls have been discarded. The result is a woeful ignorance about the origins of cattle domestication; instead of evidence we have sweeping fictions by archaeologists and culture historians concerning the increasing complexity of human cultures throughout later prehistoric times, as based upon the presumed utilization of cattle and other livestock. A religious motif as a basis for domestication has been suggested more strongly for cattle than for most other animals,[83] but all such attempted cultural reconstructions have lacked associated archaeological and zoological validity for the crucial period during which cattle were being domesticated; instead, evidence is drawn for the most part from practices of people who already had domestic cattle. However, the recent excavations at Çatal Hüyük in southwestern Anatolia,[84] dated at about 8000 B.P., may yield evidence of a people with an emotional (religious?) obsession with wild or possibly incipiently domestic cattle.

The evidence is inconclusive as yet, but I would think logical the idea that cattle were domestic by or soon after 8000 B.P. By 7000 B.P., or slightly earlier, we know that bulls were important in the emotional life of the Halafian people, as shown by their art and deduced for their religion;[85] this emotional attachment of people to their cattle is a very real thing, with multiple manifestations, in all cattle-breeding peoples. Probably the Halafian and other Near Eastern peoples of the period *did* have domestic cattle, but the only evidence I can find has been hitherto overlooked (at least, so far as I can discover; certainly evidence has not been demanded in the archaeological literature!). The particular item is a small but clear reproduction of a cow's head, from a basal Halafian level at Arpachiyah,[86] which has horns that are short and curve forward, quite like those of some cows today.

It is not until very late prehistoric times (about 6000 to 5000 B.P.) that we find actual proof, both zoological and cultural, of domestic cattle. The beautifully clear delineations on the cylinder seals of Warka and other early Sumerian towns testify to the importance of cattle in these communities, as do the careful anatomical studies of Duerst[87] and Amschler[88] on the cattle bones from the roughly contemporaneous Iranian sites of Anau II and Shah Tepé III, respectively.

In Egypt, throughout this same fourth millennium B.C., most prehistorians discuss with confidence the cattle-breeding cultures of the Badarian, Amratian, and Gerzean periods, without realizing that valid evidence of domestic cattle is lacking. As with the Halafian and some other Mesopotamian periods, the conclusions were too often assumed, while the need for evidence was ignored. Only at the Gerzean site of

Toukh[89] was a careful study made of the faunal remains; here Gaillard emphasized the resemblance of the excavated bones of the short-horned cattle to those found in adjacent but earlier Palaeolithic sites, and also stressed their resemblances to bones of known domestic short-horned cattle, both prehistoric and modern. He regarded the bones found in the midden as being those of domestic cattle, even though wild cattle of similar type were, in his opinion, living in Egypt then and later. The relative youth of most the cattle killed can be correlated with their supposed domestic status. In this connection, it is interesting to note that the gazelle bones from the same midden were also mostly from subadult individuals; this latter situation might be due to an unknown type of hunting practice or preference, or perhaps there actually was in Gerzean times in Upper Egypt an early experiment in domestication of gazelles.

The prehistorians are probably correct in thinking that domestic cattle were present and important in the human cultural evolution of Egypt of the fourth millennium B.C., prior to dynastic times and the beginning of written history, but these same prehistorians must become aware of the lack of zoological or cultural evidence for most of their assumptions.

Pig (*Sus scrofa*). During the late prehistoric times here considered, many subspecies of wild pig were native to North Africa and much of Eurasia. In spite of this wide distribution, the ancestor of *all* domestic pigs has been singled out as one southeastern Asiatic subspecies, *S. s. vittatus.*[90, 91] If it is true that this subspecies is the common ancestor, domestic pigs must have been moved westward, presumably slowly, to reach the Near East and most of Europe in prehistoric times. As yet I have not investigated this problem, but the general pattern seems illogical. I suggest instead the probability that domestication of pigs may have occurred several times in different places; A. Pira,[92] for instance, has made a good case for such local domestication of pigs in southern Sweden.

Pigs are not so difficult to tame as one might imagine; an adult wild boar or sow, it is true, is not an animal one approaches casually, but several people have easily reared the young of wild pigs to adulthood, the females having then produced litters to be reared in captivity.[93] Such pigs are surprisingly docile.

Although a domestic pig has been mentioned for the Natufian,[94] the evidence—a single phalanx—is unacceptable. The earliest known record of domestic pigs is from Jarmo in northern Iraq, where animals with relatively small molar teeth and shortened tooth-rows were seemingly introduced quite suddenly, coincident with the first appearance of pottery.[95, 96] Presumably neither the pottery nor domestic pigs were "invented" at Jarmo, but came from elsewhere.

Domestic pigs are not, however, usually recorded from prehistoric Near Eastern villages; probably the criterion of domestication—some shortening of the jaws and especially diminution of molars—was not understood or detected by the investigators. Amschler[97] did list both wild and domestic pigs from the Amouq sequence, but without any explanation for his basis of differentiation.

Other than for Jarmo and the Amouq, there is little osteological evidence for the presence of domestic pigs in the prehistoric Near East except in the north across Iran, near the base of Anau II,[98] where domestic pigs were suddenly introduced with no prior, and little subsequent, evidence of wild pigs having been hunted. By this time (about 3800 B.C.) or before, the pig was quite probably an important food animal in southern Mesopotamia, although this conclusion is based on what I consider to be slight cultural evidence.[99] Certainly, pigs are known to have been important in Sumer in early historic times. However, the only study[100] on the osteological remains from a Sumerian city (Tell Asmar) is from a time so late as not to appear on my chronological chart.

Egypt, it would seem to me, might well have been an independent center of pig domestication, considering its semi-isolated position and late cultural development. It is difficult for me to imagine pigs being driven across the desert of Sinai, but the *idea* of domestication could pass readily, perhaps by way of a Syrian visitor. There are numerous pig bones from the sites of Merimde and Maadi in northern Egypt, but there is no published study of them known to me to vindicate Menghin's often-quoted claim[101] that pig breeding represented an important cultural difference between the late prehistoric cultures of Upper and Lower Egypt. Indeed, present evidence,[102, 103] as based on a re-evaluation of the data of Gaillard,[104] confirms the latter's conclusion that domestic pigs were present in the Gerzean period of Upper Egypt. Domestic pigs may well have been present and important in the economy of prehistoric Lower Egypt, too, but until we have zoological or cultural evidence for such domestication we must assume that the numerous bones of pigs found in the remains of prehistoric villages near the Nile Delta represent wild pigs from the adjacent marshes.

Why

Why did men domesticate animals at all? A religious motif has often been suggested,[105] and may well have been important in the initiation of cattle domestication, but probably at first there was little realization of what was occurring. There was merely a gradual strengthening of an association between two species of social animals (man and dog, man and goat, and so on), preadapted by their respective evolutions to be of mutual benefit. Everything we know about preliterate cultures argues against a sudden realization of the potential values of animal domestication, followed by planned action. Man could have had no concept of the future values of animals' milk, or of wool not yet of useful length on the hairy wild sheep. Later, in literate societies, there *were* purposeful efforts at domestication. Some, such as the Egyptian Old Kingdom domestication of the hyena and of certain antelopes, were seemingly successful but were later abandoned. The era of planned domestication was not limited to peoples of ancient history, however, for we note the successful nineteenth-century domestication of the budgerigar parrot and of the laboratory rat (*Rattus norvegicus*). Today, planned domestication of two large mammals is in the experimental stage—that of the eland (*Taurotragus*) in Rhodesia and of the musk ox (*Ovibos*) in northern North America.[106] Both experiments show promise of great success, but may well fail because of lack of finances on the part of the experimenters and lack of interest by the respective governments, which might most profit from such new domesticants.

How

Man probably entered into a state of beneficial mutualism with certain animal species because, to put it in very general terms, the animals were already socially and psychologically preadapted to being tamed without loss of reproductive abilities. A second factor was the necessary one that the human culture milieu had evolved to a state of organization such that the animals could be controlled and maintained generation after generation in a condition of dependence. At least to some degree, the animals must be protected from predators and provided with food —the latter perhaps only in times of scarcity. The detailed pattern of the process leading to domestication naturally varied with both the particular species and the human culture that were interacting. The domestication of the wolf to the dog by the Maglemosian hunter-collectors of northwestern Europe was different in detail from the domestication of the hoofed food animals by the post-Pleistocene cultivators. Unfortunately, we know nothing of the details of either process, partly because of our inability to reconstruct the behavior and cultural environment of the people involved and partly because of our ignorance of the psychology of the various wild animals involved.

With the exception of one of the most recently domesticated mammals, the laboratory rat, we know little enough about the behavior patterns of our common domestic animals, but we know much less about the behavior of their wild progenitors. Furthermore, detailed comparative observations of wild and domestic *Rattus norvegicus* emphasize the tremendous behavioral changes undergone by a species during domestication.[107] Thus, psychological studies on domesticants probably cannot yield the total behavior pattern of the wild ancestors. It was, however, these wild ancestors that man first tamed and reared.

The social enzyme that activated the union of man and beast was undoubtedly the human proclivity, not only of children but of women also, to keep pets,[108, 109, 110] although purposeful capture of young animals by men, to serve as hunting decoys, may well have been another avenue toward domestication.

The psychological factor of "imprinting," explored particularly by K. Z. Lorenz in a notable series of animal experiments, was undoubtedly a major influence in the domestication of birds with precocial young (chickens, ducks, geese, turkeys, and so on). *Imprinting* refers to the tendency, most pronounced in such precocial birds, to recognize, and psychologically to attach themselves to, the most frequently seen and heard living thing during an early and short "critical period." Typically, this would be the mother, and we have thus an instinctive mechanism for recognition of the parent by an active newborn.

For mammals, we probably cannot speak of "imprinting" in as complete a sense as we do for birds. There are, of course, definite sequences of actions whereby mother and young learn to recognize each other; for the young mammal this is certainly a "critical period." Such recognition of the mother is then enlarged to include other members of the species. A lamb reared in isolation, for instance, rather thoroughly ignores other sheep for the remainder of its life,[111] even though it will mate and produce young. We would seem to have here, in correlation with the above-mentioned tendency to keep pets, a mechanism for the switching of psychological recognition and social dependence from a real mother to a human foster mother.

The "critical period" for hoofed mammals—whose behavior is similar in some respects to that of precocial birds—is within a few hours of birth, but for helpless-born young it comes several days or weeks later (three weeks and later for the dog, for instance—a phenomenon associated with myelinization of cephalic neurons[112]). In such mammals, the critical period is probably not so limited in time or so well defined as to pattern as in the hoofed animals. The essential point, however, is that in the domestic mammals that have been studied, and presumably in the others, there is such a patterned behavior system as is here discussed, a biological mechanism so basic that it remains essentially unchanged in the transition from wild to domestic status.

Since the "critical period" in mammals always comes prior to weaning, we must assume that there was a human wet nurse for whatever small helpless suckler might be brought into the village; there are women of primitive tribes who still act thus and provide the proper model.[113] Once the domestication of sheep and goats had been accomplished and the practice of milking had been established, milk would have been available for orphaned calves and colts, and thus the way for domestication of larger species would have been opened.

It is not, however, only the young of many mammals that can be kept and reared; even the adults of some artiodactyls seem to seek domestication. J. A. Arkell[114] tells of a female wart hog, with young, that made a nuisance of herself about one of his camps during a famine period; and I have had the experience of having my car stopped (*not* during a famine period!) on a major American highway by two large males of that supposedly wild species, the big-horn sheep, who then stuck their heads in the open windows begging for tidbits. These animals may not have known it, but they were *asking* to be domesticated.

Once the nuclei of herds had been established, human selection against the aggressive and unmanageable individuals would have been automatic, resulting in the decrease in production, generation by generation, of the adrenocortical steroids (with multiple attendant physiological changes)—a process that has been studied in detail for the short history of the laboratory rat.[115] Eventually submissiveness becomes genetically ingrained in the population (although some species, such as the sheep, seem more susceptible to such manipulation than others). Furthermore, those animals naturally adapted to breed best in captivity would contribute their characters in larger numbers to the gene pool of each succeeding generation. Such unplanned selection of various sorts must have long preceded the methods of purposeful artificial selection that led eventually to the establishment of different breeds within a domestic species.

However it originated, once domestication had occurred, the idea could be transferred to species other than the original ones—a type of cultural shift which seems to account for the domestication of the reindeer. I find no reason, either, to believe that domestication of the same species could not occur in different places at different times, probably as a result of diffusion of the idea. Thus, pigs and cattle could have been domesticated in both southwestern Asia and in Egypt, the stimulus having been transferred from the former area to the latter in the mind of a human migrant.

A last factor that must be considered in a discussion of the origin of domestication of animals in the Near East is the "propinquity" or "riverine-oasis" theory of domestication.[116] Briefly, the increasing desiccation of the Saharan and Arabian areas during the post-Pleistocene supposedly enforced the juxtaposition of man and the potentially domesticable animals around the disappearing water sources, leading to conditions of beneficial mutualism and thus to domestication.

Aside from the fact that a variety of ecological and distributional data argue against the validity of such a view,[117] accumulating evidence indicates that the known climatic sequence itself makes the idea untenable. I suspect that the adherents of the "riverine-oasis propinquity theory" have been overly impressed by the grand sweep of the very real desiccation of North Africa since the Allerod (about 11,000 years ago) without having given due regard to the fluctuating climatological conditions[118, 119] that existed. There were, beginning in the late Pleistocene, several fluctuations of temperature and rainfall which had profound ecological consequences for the biologically sensitive area of North Africa, where the evidence is best known. However, there is no evidence of domestication during the periods in question (about 15000 to 7000 B.P.) in this or in any other desert area. There then began the "Neolithic wet phase" (Butzer's Subpluvial II[120] and Alimen's "second wet phase"[121]), lasting from about 7000 to 4500 B.P. During much of this time[122] domestic bovids (sheep, goats, and cattle) were present all across the Sahara, as shown by innumerable engravings and paintings,[123] and the subsequent dramatic desiccation to present conditions thus occurred long after the full pattern of domestication had been established.

The "oasis theory," based as it originally was on an idea of continuous desiccation during the post-Pleistocene North African climatic sequence, loses all meaning when transferred to southwestern Asia, the actual site of original bovid domestication. Here, data on Saharan rainfall and temperature fluctuations *may* apply to the central desert areas proper (the evidence is scant and inconclusive) but seemingly have much less meaning elsewhere.

Particularly throughout the hills of the Zagros-Palestinian chain there was relatively little climatic change within the transition period from the Upper Pleistocene to the early Recent;[124] in fact, these terms have relevance in a climatic sense only as we can correlate them with regions of former continental glaciation.

My own unfinished studies on the bones collected from half a dozen sites in northern Iraq and adjacent Iran, which bridge some 90,000 years of the late Quaternary, show that an essentially modern fauna has occupied the area during this period. This does not mean that there has been no climatic change during this time in these hills and mountains (we have indeed the glacial studies of Herbert E. Wright[125] and the pollen studies of Solecki and Leroi-Gourhan[126] to show that profound climatic changes may well have occurred), but it does mean that such variations as have occurred in temperature and precipitation have done little more than simultaneously depress and/or elevate the upper and lower tree lines. The fauna (including prehistoric man, undoubtedly) moved slowly with the flora to the extent necessary to maintain a fairly static ecologic situation.

We must then face the seeming enigma that extremely important cultural evolution (first incipient villages, first grinding stones, first domestication) was occurring about 11,000 years ago, approximately at the end of the Pleistocene, even though the Hilly Crescent of southwestern Asia probably passed at this time through no such dramatic end-Pleistocene environmental crisis as was experienced by Europe, North Africa, and North America. For Europe particularly, with the correlated cultural change from Palaeolithic to Mesolithic (a degree of change perhaps often overemphasized), the idea that there was intensive post-Pleistocene human adaptation to changing environments is generally accepted, usually accompanied by the concept (even though unexpressed) of the development of greater cultural complexity ("progress") in answer to the changing conditions.

In southwestern Asia, however, we have at approximately this time the profound cultural change to incipient cultivation, if not to actual cultivation, within that millennium which includes the Karim Shahir and Zawi Chemi materials of Iraqi Kurdistan and the Natufian of Palestine. But here we cannot point to a dramatic climatic change, furnishing a stimulus for sudden cultural evolution.

Indeed, the available evidence is quite the contrary; true, the Natufian had a more complex set of tools than any of its Upper Palaeolithic predecessors in southwestern Asia, and the culture was marked particularly by large numbers of very small flake tools (microliths) and by the introduction of mortars and pestles for seed grinding, but the whole assemblage is in the blade-tool tradition of some 40,000 years of Levantine history and undoubtedly evolved *in situ*, with a minimum of external influence.[127]

Still eluding us are the factors that led these particular peoples to inaugurate cereal agriculture, however incipiently, and thus, by way of many changes, to furnish the food base of today's technological society. But increasingly the archaeologist is looking for a greater variety of data from his excavations and asking different questions of those data. Increasingly, too, natural scientists are helping him collect and interpret that evidence. It is certain that, under these circumstances, we shall be getting more and better answers to our questions concerning the many unsolved problems in the study of the relationships between climate, man, and the origins of agriculture and domestication.[128]

Conclusions

Concerning the animal aspect of the "food-producing revolution," present evidence indicates that domestication of sheep had occurred in northern Iraq by the early part of the ninth millennium B.C., with goats (the next known domesticant in the Near East) not occurring until 2,000 years later. By this latter time (8500 B.P.) domestic pigs were also present (Jarmo), and cattle were probably domesticated within the next few hundred years.

Domestication of the food-producing animals probably occurred in village-farming communities in the Hilly Flanks area of southwestern Asia; thus, cereal agriculture and the settled village are considered to antedate the domestication of all the original food animals (sheep, goats, pigs, cattle). The domestication of the dog was a different kind of event, associated with Mesolithic hunters and gatherers of northwestern Europe.

Present archaeological data indicate (although many archaeologists have tended to ignore or discard the evidences) that relatively intensive and successful agricultural and stockbreeding (mixed-farming) societies developed in the Zagros hills and their grassy forelands (as well as in the lower Jordan Valley) prior to the appearance of the earliest societies of this type elsewhere; similar Iranian and Egyptian cultures seemingly developed later and peripherally. At least for Egypt, this seeming lateness—a matter of two thousand years or more—is probably not just a reflection of accidental or incomplete sampling.

No sudden and intensive end-Pleistocene environmental change has been detected for southwestern Asia; thus, the all-important "food-producing revolution," while correlated in time with the end of the Pleistocene, a period of profound climatic change elsewhere, was seemingly not stimulated by the challenge of end-Pleistocene climatic change.[129, 130] Yet, one wonders: the correlation between the time, the new cultural innovations, and the known important environmental changes elsewhere in the world make one think that the correlation may be due to more than coincidence. We have much to learn.

NOTES

1. We are indebted to V. G. Childe for stressing the importance of this fundamental concept. The development of Childe's ideas, which have had such a profound influence on the history of thought, are interestingly outlined in an autobiographical note: V. G. Childe, *Antiquity*, **32**, 69 (1958).

2. R. J. Braidwood, *The Near East and the Foundations for Civilization* (Univ. of Oregon Press, Eugene, 1952).

3. R. J. Braidwood, *Science*, **127**, 1419 (1958).

4. R. J. Braidwood, *Natl. Acad. Sci.–Natl. Research Council Publ.*, **565**, 16 (1957).

5. C. A. Reed, *Studies in Ancient Oriental Civilization* (Univ. of Chicago Press, Chicago, 1960), vol. 31.

6. W. W. Taylor, Ed., *The Identification of Non-artifactual Archaeological Materials* (National Academy of Sciences–National Research Council, Washington, D.C., 1957).

7. F. E. Zeuner, *A History of Domesticated Animals* (Hutchinson, London, 1963).

8. *Domestication* to me means simply that the animals are under the control of man to such a degree that, if he wishes, their choice of mates is determined. Artificial selection is thus possible and usually, to some degree at least, inevitable. In a very important sense, then, domestic animals (as well as plants) are a type of human artifact, since they exist in a form changed by man.

9. J. U. Duerst, *Carnegie Inst. Wash. Publ. No. 73* (1908), p. 339.

10. W. Herre, "Domestikation und Stammesgeschichte," in Heberer, *Evolution der Organismen* (Stuttgart, 1955); "Die geschichtliche Entwicklung der Haustierzüchtung," in Zorn, *Tierzüchtungslehre* (Ulmer, Stuttgart, 1958).

11. Skeletons of both wild and domestic animals were collected for the Chicago Natural History Museum by Dr. Henry Field for many years as part of his anthropological researches in the Near East; by the present author in Iraq in 1954–1955, and in Iran in 1960; and by the William S. Street Expedition to Iran, 1962–1963.

12. Braidwood, *Natl. Acad. Sci.–Natl. Research Council Publ., loc. cit.*

13. Braidwood, *Science, loc. cit.*

14. E. Patte, *Compt. rend.*, **246**, 3490 (1958).

15. I follow the common practice of most European geologists and of the U.S. Geological Survey [H. B. Morrison et al., *Am. J. Sci.*, **255**, 385 (1957)] in dividing the Quaternary into Pleistocene and Recent, although at least one eminent authority rejects any such differentiation [R. F. Flint, *Glacial and Pleistocene Geology* (Wiley, New York, 1957)]. The Recent is generally considered to have begun between 10,000 and 11,000 years ago, the difference depending only upon whether the several hundred years of the Younger Dryas is included in the Pleistocene [H. E. Wright, Jr., *Am. J. Sci.*, **55**, 447 (1957)] or in the Recent [W. S. Cooper, *Bull. Geol. Soc. Am.*, **69**, 941 (1958)].

16. Braidwood, *Science, loc. cit.*

17. R. J. Braidwood and C. A. Reed, *Cold Spring Harbor Symposia Quant. Biol.*, **22**, 19 (1957).

18. D. Perkins, "The Post-Cranial Skeleton of the Caprinae: Comparative Anatomy and Changes under Domestication," Ph.D. Thesis, Biology Dept., Harvard Univ. (1959).

19. C. A. Reed, "Osteological Evidence for Prehistoric Domestication in Southwestern Asia," *Zeitschrift f. Tierzüchtung und Züchtungsbiologie*, **76**, 31–38 (1961).

20. R. S. Solecki, *Quaternaria*, **4**, 1 (1957); *Trans. N.Y. Acad. Sci.*, **21**, 712 (1959).

21. Braidwood, *The Near East and the Foundation for Civilization, op. cit.*

22. J. Perrot, *Antiquity and Survival*, **2**, 91 (1957).

23. J. Clutton-Brock, "Near Eastern Canids and the Affinities of the Natufian Dog," *Zeitschrift f. Tierzuchtung und Züchtungsbiologie*, **76**, 326–33 (1962).

24. Reed, "Osteological Evidence for Prehistoric Domestication in Southwestern Asia," *loc. cit.*

25. R. S. Solecki and M. Rubin, *Science*, **127**, 1446 (1958).

26. K. M. Kenyon, *Antiquity*, **33**, 8 (1959).

27. But see C. O. Sauer, *Agricultural Origins and Dispersals* (American Geographical Society, New York, 1952) for the conflicting opinion that the pig was domesticated earlier, in southeastern Asia, and was then brought westward.

28. F. E. Zeuner, *Palestine Explor. Quart.*, **1955**, 70 (1955).

29. F. E. Zeuner, personal communication, 22 Oct. 1958.

30. K. V. Flannery, "Skeletal and Radiocarbon Evidence for the Start and Spread of Pig Domestication," M.A. Thesis, Dept. of Anthropology, Univ. of Chicago (1961).

31. Reed, "Osteological Evidence for Prehistoric Domestication in Southwestern Asia," *loc. cit.*

32. Braidwood, *The Near East and the Foundations for Civilization, op. cit.*

33. Braidwood, *Science, loc. cit.*

34. Reed, *Studies in Ancient Oriental Civilization, op. cit.*

35. R. J. Braidwood and L. Braidwood, *J. World Hist.*, **1**, 278 (1953).

36. R. H. Dyson, Jr., *Am. Anthropol.*, **55**, 661 (1953).

37. *Ibid.*

38. C. von Fürer-Haimendorf, *Yearbook Anthropol.*, **1**, 149 (1955).

39. D. M. A. Bate (47) has accepted a description of native goats in Algeria [A. Pomel, *Service Carte Geol. Algeria, Paleontol. Monogr. No. 13* (1898), p. 1], but the age of the materials from which Pomel described these goats is uncertain [C. Arambourg, *Bull. soc. hist. nat. Afrique du Nord*, **20**, 78 (1929)], and there is a distinct possibility, thus, that Pomel's "*Capra promaza*" is based upon Neolithic domestic goats.

40. F. E. Zeuner, in *A History of Technology*, Charles Singer et al., Eds. (Oxford Univ. Press, London, 1954), vol. 1, p. 327. This paper is an important one on the "why" and "how" of domestication and forms a foundation for much of what I am saying here, even though I do not agree with Zeuner in every detail.

41. It is interesting to note that all of these animals belong to the same order, the Artiodactyla, and that except for the pig (which is an omnivore and a scavenger) they are not only all ruminants but are all Bovidae. Other bovids (water buffalo, zebu, yak) were subsequently domesticated, and many species of gazelles and antelopes, particularly, seem to be potentially domesticable. No other family looms so importantly in the history of domestication. Psychological factors, as yet not evaluated, plus the efficiency of the bovid stomach, undoubtedly play a major role in the bovid's versatile adaptation to domestication. Actually, there may be no bovid which could not be domesticated.

42. W. A. Fairservis, Jr., *Am. Museum Nat. Hist. Anthropol. Paper No. 45* (1956), p. 167.

43. O. Menghin, *Oestrr. Akad. Wiss. Wien, Phil.-Hist. Kl. Anz.*, **70**, 82 (1934).

44. D. M. A. Bate, *The Stone Age of Mt. Carmel* (Clarendon, Oxford, 1937), vol. 1, part 2.

45. D. M. A. Bate, *Proc. Prehist. Soc.*, **8**, 15 (1942).

46. Clutton-Brock, *loc. cit.*

47. Reed, "Osteological Evidence for Prehistoric Domestication in Southwestern Asia," *loc. cit.*

48. C. S. Coon, *Cave Explorations in Iran 1949* (University Museum, Univ. of Pennsylvania, Philadelphia, 1951).

49. E. K. Ralph, *Science*, **121**, 149 (1955).

50. M. Degerbøl, "On a Find of a Preboreal Domestic Dog" (*Canis familiaris* L.) from Star Carr, Yorkshire, with remarks on other Mesolithic dogs: *Proc. Prehistoric Society*, **27**, 35–55 (1961).

51. F. E. Zeuner, *Palestine Explor. Quart.*, **1958**, 52 (1958).

52. M. Hilzheimer, *Antiquity*, **6**, 411 (1932).

53. E. Massoulard, *Trav. et Mem. inst. ethnol. Univ. Paris*, **53**, 1 (1949).

54. A. J. Tobler, *Excavations at Tepe Gawra* (Univ. of Pennsylvania Press, Philadelphia, 1950), vol. 2, plate 37b.

55. H. Frankfort, *Cylinder Seals* (Macmillan, London, 1939), pl. iv*a*.

56. Y. S. Moustafa, *Bull. inst. Egypte*, **36**, 105 (1955).

57. J. P. Scott, *J. Natl. Cancer Inst.*, **15**, 739 (1954).

58. K. Z. Lorenz, *Man Meets Dog* (Houghton Mifflin, Boston, 1955). Lorenz has subsequently withdrawn his advocacy of a jackal-ancestored dog.

59. Bate, *The Stone Age of Mt. Carmel, op. cit.*

60. R. Matthey, *Mammalia*, **18**, 225 (1954).

61. C. O. Sauer, *Agricultural Origins and Dispersals* (American Geographical Society, New York, 1952).

62. B. Vesey-Fitzgerald, *The Domestic Dog: An Introduction to Its History* (Routledge and Kegan Paul, London, 1957); F. S. Bodenheimer, *Bull. Research Council Israel*, **7B**, 165 (1958).

63. Reed, "Osteological Evidence for Prehistoric Domestication in Southwestern Asia," *loc. cit.*

64. *Ibid.*

65. Coon, *op. cit.*

66. Perkins, *op. cit.*

67. Duerst, *loc. cit.*

68. H. Helbaek, personal communication.

69. J. W. Amschler, *Identification of the Animal Bones from the Amouq* (unpublished manuscript in the personal files of R. J. Braidwood). This collection of bones from the excavations of the Oriental Institute's Syrian Expedition is now in Chicago and is to be restudied before final publication of the results.

70. C. Gaillard, *Arch. musée hist. nat. Lyon*, **14**, 1 (1934).

71. Frankfort, *op. cit.*, pl. iii*a*.

72. Duerst, *loc. cit.*

73. F. Harper, *Extinct and Vanishing Mammals of the Old World* (New York Zoological Park, New York, 1945).

74. J. V. Evans, *Advan. Sci.*, **13**, 198 (1956).

75. V. I. Tzalkin, *Materials for the Recognition of the Fauna and Flora of the U.S.S.R.*, published by the Moscow Society of Naturalists, n.s., *Zoological*, **27**, 1 (1951).

76. V. G. Childe, *Antiquity*, **31**, 36 (1957).

77. Fairservice, *loc. cit.*

78. D. M. A. Bate, in A. J. Arkel, *Shaheinab* (Oxford Univ. Press, London, 1953), pp. 11–19.

79. Gaillard, *loc. cit.*

80. F. Koby, *Bull. soc. prehist. franç.*, **51**, 434 (1954); F. E. Zeuner, *Man*, **53**, 68 (1953).

81. R. T. Hatt, "The Mammals of Iraq," *Misc. Publ. Museum Zool. Univ. Michigan*, **106**, 1 (1959).

82. Reed, *Studies in Ancient Oriental Civilization, op. cit.*

83. E. Isaac, "On the Domestication of Cattle": *Science*, **137**, 195–204 (1962).

84. Mellaart, personal communication.

85. M. E. L. Mallowan and J. C. Rose, *Iraqi*, **2**, 1 (1955).

86. M. E. L. Mallowan, *ibid.*, **2**, plate vi(a) (1935).

87. Duerst, *loc. cit.*

88. J. W. Amschler, "Report of Scientific Expedition to the Northwest Province of China under the Leadership of Dr. Sven Hedin (The Sino-Swedish Expedition)," *Archives*, **4**, 35 (1939).

89. Gaillard, *loc. cit.*

90. See note 27.

91. B. Klatt, *Enstehung der Haustiere* (Berlin, 1927).

92. A. Pira, "Studien der Geschichte der Schweinerassen usw.," *Zool. Jahrbuch, Suppl.*, **10**, 233–246 (1909).

93. Reed, *Studies in Ancient Oriental Civilization, op. cit.*

94. R. Vaufrey, *Mem. arch. inst. paleontol. humaine*, **24**, 189 (1951).

95. Flannery, *op. cit.*

96. Reed, "Osteological Evidence for Prehistoric Domestication in Southwestern Asia," *loc. cit.*

97. Amschler, *Identification of the Animal Bones from the Amouq, op. cit.*

99. E. D. Van Buren, *Analecta Orientalia*, **18**, 1 (1939).

100. M. Hilzheimer, *Studies in Ancient Oriental Civilization*, **20**, 1 (1941).

101. Menghin, *loc. cit.*

102. Flannery, *op. cit.*

103. Reed "Osteological Evidence for Prehistoric Domestication in Southwestern Asia," *loc. cit.*

104. Gaillard, *loc. cit.*

105. Sauer, *Agricultural Origins and Dispersals, op. cit.*

106. M. Burton, *Illustrated London News*, **229**, 234 (1956); J. Desmond Clark, personal communication; J. J. Teal, Jr., *Exploration J.*, **36**, 13 (1958).

107. C. P. Richter, *Am. J. Human Genet.*, **4**, 273 (1952).

108. Zeuner, *A History of Technology, loc. cit.*

109. Fairservis, *loc. cit.*

110. R. M. Gilmore, *Bull. Bur. Am Ethnol.*, **143**, 345 (1950).

111. J. P. Scott, in *Interrelations between the Social Environment and Psychiatric Disorders* (Milbank Memorial Fund, New York, 1953), pp. 82–102.

112. J. P. Scott, E. Fredericson, J. L. Fuller, *Personality: Symposia on Topical Issues*, **1**, 163 (1951).

113. Fairservis, *loc. cit.*

114. J. A. Arkell, *Kush*, **5**, 8 (1957).

115. Richter, *loc. cit.*

116. V. G. Childe, *The Most Ancient East* (Kegan Paul, Trench, Trubner, London, 1928); M. Wheeler, *Antiquity,* **30**, 132 (1956); R. J. Braidwood, *ibid.*, **31**, 73 (1957).

117. Reed, *Studies in Ancient Oriental Civilization, op. cit.*

118. K. W. Butzer, *Erdkunde,* **11**, 21 (1957); *Bonner Geograph. Abhandl.*, **24**, 1 (1958).

119. H. Alimen, *The Prehistory of Africa* (Hutchinson, London, 1957).

120. Butzer, *Erdkunde, loc. cit.; Bonner Geograph. Abhandl., loc. cit.*

121. Alimen, *op. cit.*

122. This North African "Neolithic wet period" is to be correlated approximately with the central part of the "Atlantic" period of some authors, the "Hypsithermal" of others. See W. S. Cooper, *Bull. Geol. Soc. Am.*, **69**, 941 (1958), for a summary of a confusing terminology.

123. Alimen, *op. cit.*

124. L. Picard, *Bull. Geol. Dept., Hebrew Univ.*, **4**, 1 (1943); N. Shalem, *Research Council Israel, Spec. Publ. No. 2* (1953), p. 153; H. E. Wright, Jr., *Studies in Ancient Oriental Civilization;* C. A. Reed and R. J. Braidwood, *ibid.*

125. H. E. Wright, Jr., "Pleistocene Glaciation in Kurdistan," *Eisz. und Gegenw.*, **12**, 131–164 (1961).

126. R. S. Solecki and A. Leroi-Gourhan, "Palaeoclimatology and Archeology in the Near East," *Ann. New York Acad. Sci.*, **95**, 729–739 (1961).

127. F. C. Howell, *Proc. Am. Phil. Soc.*, **103**, 1 (1959).

128. Sauer, *Agricultural Origins and Dispersals, op. cit.*

129. See note 124.

130. Howell, *loc. cit.*

131. My acceptance of the presence of five domestic animals for southern Mesopotamia at the beginning of Sumerian culture is based on a number of sources. All five are depicted on cylinder seals of immediately subsequent periods, and skeletons of dogs are known [S. Lloyd, *Illustrated London News*, **123**, 303 (1948)].

132. C. Caton-Thompson and E. W. Gardner, *The Desert Fayum* (Royal Anthropological Institute of Great Britain and Ireland, London, 1934), vol. 1, p. 34.

133. M. E. L. Mallowan, *Iraq*, **8**, 128 (1946).

134. R. Vaufrey, *Musée Louvre, dept. antiq. orient., ser. Arch.* (1939). The horn cores Vaufrey assigned to *Ovis vignei* I had previously (original writing of this article, 1959) assigned to domestic goats. They may well, however, represent the horn cores of "goat-horned sheep" and should be re-studied with this possibility in view.

135. G. Brunton, *Qua and Badari I* (Quaritch, London, 1927); G. Brunton and G. Caton-Thompson, *The Badarian Civilization and Pre-dynastic Remains near Badari* (Quaritch, London, 1928); G. Brunton, *Mostagedda and the Tasian Culture* (Quaritch, London, 1937).

136. T. Josien, *Israel Explor. J.*, **5**, 246 (1955).

137. T. E. Peet, *Mem. Egypt Explor. Fund,* **34**, 6 (1914); J. W. Jackson, *Mem. Egypt Explor. Soc.*, **42**, 254 (1937).

138. F. Debono, *Ann. service antiq. Egypte,* **48**, 561 (1948); E. A. M. Greiss, *Bull. Inst. Egypte*, **36**, 227 (1955).

139. H. Junker, *Oestrr. Akad. Wiss. Wien. Phil.-Hist. Kl. Anz.*, **66**, 156 (1930).

140. S. Lloyd and F. Safer, *J. Near East. Studies*, **4**, 284 (1945). In spite of numerous subsequent statements by different authors concerning the presence of various domestic animals at Hassuna, the original report makes no such claim. However, in view of the geographical location and the chronological position, I believe that domestic dogs and goats, at least, were "probably present."

141. See discussion in text and in *Studies in Ancient Oriental Civilization, Univ. Chicago, op. cit.*

142. C. A. Reed, *Sumer*, **10**, 134 (1954). I can state definitely now that there is no evidence to support the preliminary claims, made in 1954, for domestic animals at M'lefaat.

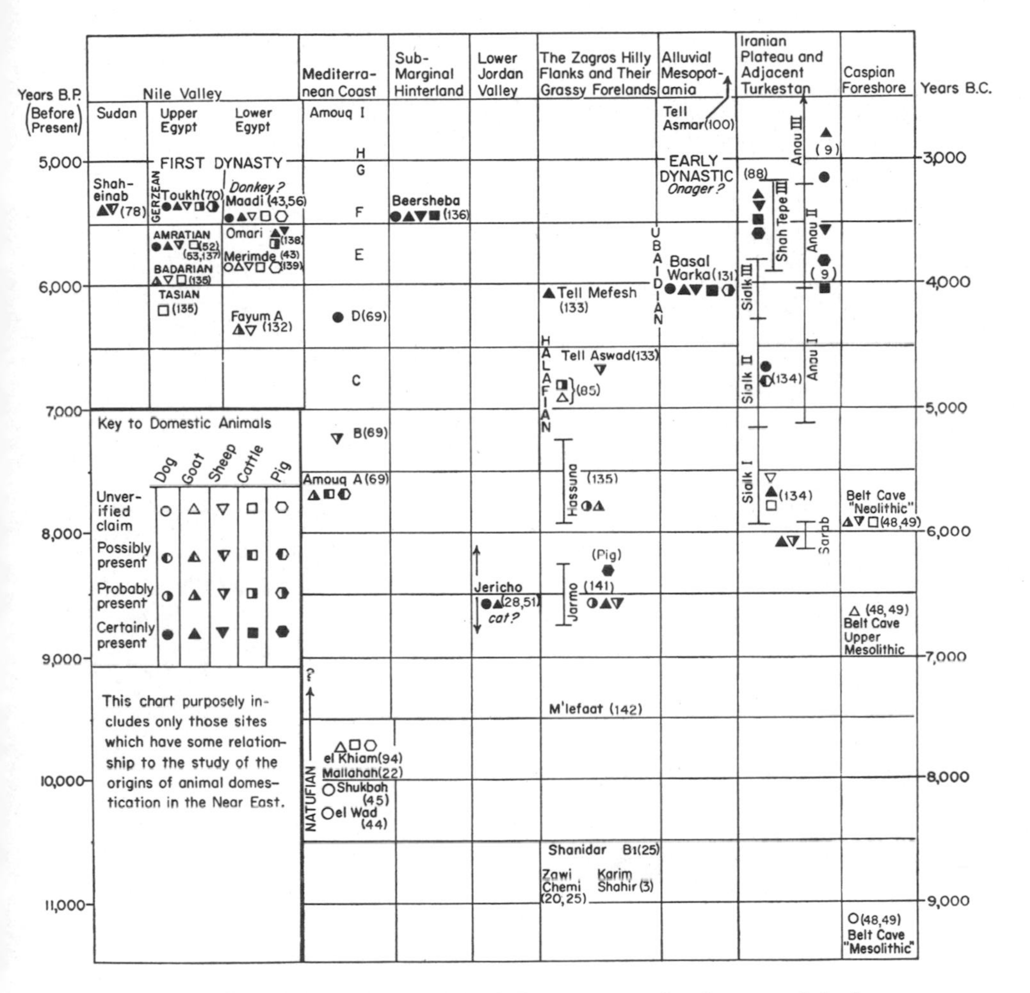

Fig. 6–1. A chronological chart, subdivided into geographical areas, of the known history of animal domestication in the prehistoric Near East. The estimated time is not to be regarded as absolute; the top of each column is fairly well fixed, temporally; but any part of any column may become elongated or shortened as a result of future discoveries. The chronological chart was based primarily on *Relative Chronologies in Old World Archaeology*, R. W. Ehrich, Ed. (Univ. of Chicago Press, Chicago, Ill., 1954.) Radiocarbon dates, to mid-1958, are summarized by R. J. Braidwood (*Osterr. Akad. Wiss. Phil.-Hist. Kl. Anz. Jahrg. 1958* No. 19, 249 [1959]). Additionally, I had advice from staff members of the Oriental Institute, University of Chicago. Other sources consulted were E. Massoulard (*Trav. et Mem. inst. ethnol. Univ. Paris,* **53**, 1 [1949]); I. Rizkana, *Bull. inst. du desert,* **2**, 117 (1952); and R. P. Charles, *J. Near East. Studies,* **16**, 240 (1957).

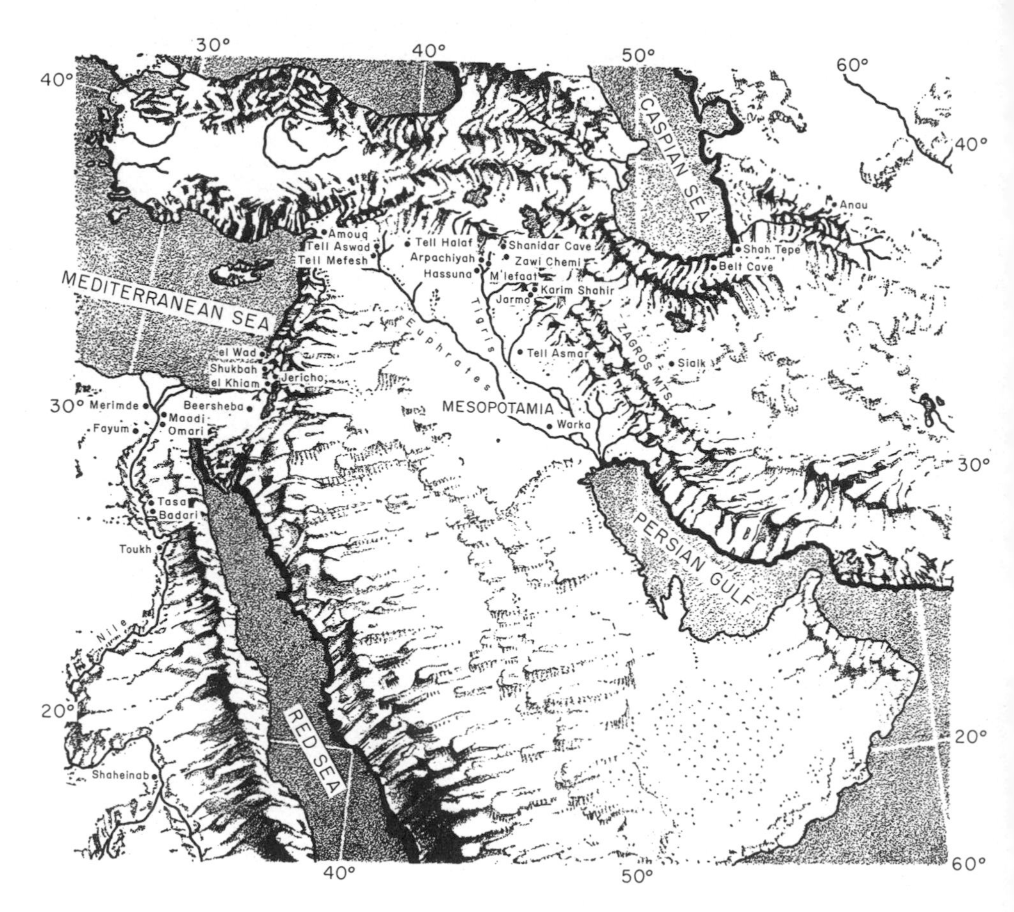

Fig. 6–2. The Near East. Of the numerous archaeological sites that have been excavated in the area shown, only those that have some relation to the study of the origins of animal domestication are included here.

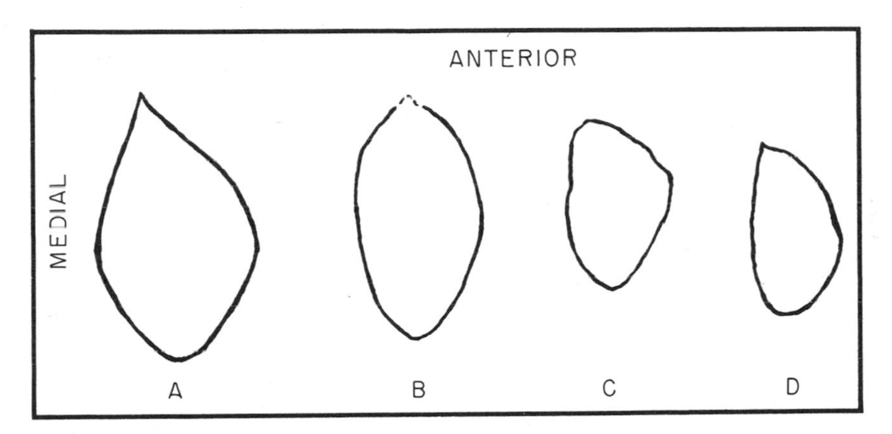

Fig. 6–3. Cross-sections of male horn cores of typical wild and domestic goats, at approximately one-third of the distance from the base of the core toward the tip. All are from northeastern Iraq. (A) Right horn core of a wild goat (*Capra hircus aegagrus*). Note the general quadrilateral shape (Chicago Natural History Museum specimen No. 84493). (B) Left horn core of a domestic goat from prehistoric Jarmo, drawn as if from the right side. This core is similar in its lozenge shape to the core of the first domestic goat from prehistoric Jericho (C. A. Reed, "Öesteological Evidence for Prehistoric Domestication in Southwestern Asia." *Zeitschrift für Tierzüchtung und Züchtungsbiologie*, **76**, 31–38 [1961]). (Iraq-Jarmo field specimen No. J55-194). (C) Right horn core of a domestic goat from prehistoric Jarmo. Note the flattening of the medial surface (Iraq-Jarmo field specimen No. J55-191). (D) Right horn core of a domestic Kurdish goat, killed in 1955. Note the flattening of the medial surface (Chicago Natural History Museum specimen No. 57253). (× 0.82)

Chapter 7

Archaeology and Geology in Ancient Egypt

Karl W. Butzer

Archaeological sites of moderate or great antiquity generally present problems susceptible to investigation by geologists or geomorphologists, and often correct geological interpretation is essential to effective understanding of such sites.[1] The physical environment of the immediate site and of the wider habitat of its occupants may require reconstruction for the period of settlement—at least if the archaeologist aims at full interpretation of all categories of materials.

Interdisciplinary contact between prehistory and geology or geography[2] has a long, although often tenuous, history, dating back well into the nineteenth century. The most frequent occasion for such cooperation was provided by cave excavations, and the digging archaeologist was as often as not a geologist himself. And in recent years earth scientists have frequently participated in archaeological excavations and surveys on various continents.

However, the archaeologists concerned with the younger, postglacial aspects of prehistory have often shown less appreciation of the need for a comprehension of the environment as a functioning whole. From the anthropologist's standpoint, R. J. Braidwood[3] has adequately emphasized the full interpretative potential of the evidence of natural history on the part of the earth scientist; however, there is inadequate awareness of this potential. Most geomorphologists involved in archaeological work have little interest in post-Pleistocene events and often insist that nothing of note has happened during the Recent epoch. The basic difficulty is probably that a great deal of microstratigraphy and a patient search for apparently insignificant pieces of evidence are necessary for this period.

Accordingly, I shall concentrate my present remarks upon the Neolithic and Chalcolithic (Predynastic) settlements of the Nile Valley, sites which date from the earlier part of the Recent, some 7,000 to 5,000 years ago.[4] Although all Pleistocene investigations in Egypt have had Palaeolithic man well within their scope,[5, 6, 7] the only Recent sites subjected to any detailed geological interpretation have been the Fayum Oasis[8] and Maadi, a few miles south of Cairo, examined by S. A. Huzayyin[9] in 1941. This should not be taken to imply that the archaeologists in question have not paid attention to the physical settings. The great pioneer Jacques de Morgan[10] left proof of such attention, and later reports on some sites, particularly those investigated after 1918, are not without comment on or appreciation for the geomorphological background.

Some problems related to the Nile Valley settlements of Neolithic and Predynastic times are widely recognized by prehistorians—chiefly, that many archaeologic sites have been buried by Nile alluvium in the course of flood-plain aggradation. Another problem, ably presented by S. Passarge,[11] has been the physical environment posed by the Nile flood plain in prehistoric time. Passarge indicated that a natural flood plain was no jungle swamp and was in no way comparable to the perennial Sudd marshes of the Upper Nile.

The problems discussed here are more specific:

1. What is the immediate geologic-geographic setting of the late prehistoric sites in the Nile Valley?
2. What are the relations of such settings to the surficial deposits of the valley margins? What situations are likely to have been deliberately selected by man or accidentally preserved from natural obliteration?
3. What regional generalizations can be made about the likelihood of occurrence of sites? Are some of the cultural gaps,[12] in particular between the Lower and Upper Egyptian cultures in pre-Gerzean times (prior to about 3500 B.C.), related to unfavorable preservation over broad areas or to lack of former habitation in frontier marches?
4. What proportion of the late prehistoric sites is actually preserved, or, are the known sites representative of the density of actual settlement?
5. What were the physical conditions dominant during the period of Neolithic-Chalcolithic settlement?

Geologic-Geographic Setting

The Nile Valley consists of three major land-form elements: the fertile alluvium of the seasonally inundated flood plain; the low-lying sand or gravel wastes bordering the flood plain; and, lastly, the mountainous escarpments along the margins of the valley. The latter represent the dissected edge of the horizontal sedimentary strata of the Libyan Plateau. In the Younger Tertiary the predecessor of the modern Nile began to incise its present course[13] and so excavated a great channel averaging some 400 or 500 meters in depth, 10 to 15 kilometers in width. During a marine transgression of mid-Pliocene age this valley was submerged and filled with marine or lacustrine sediments, which were partially re-excavated at the beginning of the Pleistocene period about a million years ago.

The subsequent evolution of the Nile Valley in the course of the Pleistocene is one of alternating gravel aggradation by the Nile and its now-defunct local tributaries and of vertical incision and downcutting. The sum total of semi- or nonconsolidated Pliocene sediments and Pleistocene terrace gravels exposed on the outer margins of the flood plain comprises what the archaeologists designate as the "low desert." The youngest deposits are a relatively thin sheet of clayey silt, averaging some 6 to 11 meters in thickness—the alluvium (Fig. 7–1).

With the exception of the Neolithic settlements along the shores of the ancient Fayum Lake (10 meters above mean sea level),[14] all of the late Stone Age and Copper–Stone Age (Chalcolithic) village- or townsites are located on the low-desert edge, immediately beside the flood plain (Fig. 7–2). The advantages of such locations in terms of water supply, proximity to the agricultural land, and flood-free elevations are obvious.

The oldest such low-desert site studied was that of Merimde, on the western margins of the Nile Delta. One carbon-14 date, possibly some 300 years too young, is 3820 ± 350 B.C.[15] The townsite, thought to have been occupied for a few centuries, covered some 180,000 square meters and is characterized by cultural debris attaining an average depth of two meters.[16] If the whole site was occupied at any one time, it would appear that a population estimate of some 16,000 would not be illogical;[17] Merimde would thus have been the largest prehistoric settlement in Egypt and, at the same time, one of the oldest.

The geological setting of Merimde is relatively simple (Fig. 7–3). The basal sediments are sandy gravels of Lower Pleistocene (pre-Palaeolithic?) age rising as low bluffs some fifty meters above the flood plain. Banked against these are Middle Palaeolithic silts, at least three meters above the alluvium, dating from the late Pleistocene. The townsite is limited to these unconsolidated deposits and may have extended farther northward onto the flood plain; this area is now obscured by sand deposition however. This edge of the townsite may also have been reduced by lateral planation of the Nile in the course of annual flooding and deposition during sixty centuries. The surface aeolian sediments on the alluvium are recent, but bores in the area indicate extensive sand lenses in the lower alluvium. These features are younger than the settlement.

Of interest within the site is the thin but fairly continuous gravel horizon above the lowest settlement stratum.[18] The pebbles suggest a period of sheetflooding after appreciable rainfall. The trenches are unfortunately buried in sand and are no longer accessible. The wild fauna preserved—hippopotamuses, crocodiles, antelopes, tortoises, fishes—and the domestic animals—cattle, sheep, goats, pigs, and dogs[19, 20]—are fully compatible with the setting on the flood-plain margins, probably during a phase of slightly moister local climate.[21] During this wet spell the present semidesert vegetation was probably replaced by moderate seasonal pastures on the Pleistocene gravels.

The next oldest sites of interest are the small Badarian villages on the low desert of the east bank, southeast of Asyut (Fig. 7–4). The "classical" Badarian village, "Hemamieh North Spur," was of very similar size. According to Gertrude Caton-Thompson,[22] the cultural debris covered some 200 square meters to a depth of 150 to 180 centimeters, for which area a population estimate of 20 can be made.[23] In the area considered here, quite analogous to the main site, the low desert is some 200 to 250 meters wide and consists of fluvial gravels and local detritus intercalated with scree, resting upon spurs of limestone bedrock. The edge of these coarse, semiconsolidated deposits to the alluvium is a 3-meter bank; the beds rise to 10 meters a little in the lee. Within the cultural deposits are two horizons of limestone scree; the older of these has been cemented to a tough breccia of up to 30 centimeters in thickness.[24, 25] This indicates a period of greater moisture dating from the Badarian period; the carbon-14 date, probably at least 275 years too young, is 3155 ± 160 B.C.[26] The upper scree within the younger Gerzean horizon (about 3300 B.C.) is unconsolidated. The post-Badarian settlements preserved on the low desert are very few by comparison with the profusion of cemeteries, so it must be supposed that the major settlement location after the Badarian period was on the flood plain.

Of the great number of Gerzean sites (about 3500 to 3000 B.C.) I will discuss only two here—namely, Hierakonpolis, a little north of Edfu, and Nagada, northwest of Luxor in the Thebaid. Both sites are on the western bank in Upper Egypt. The former was a town of religious and political distinction, probably representing the capital of the Predynastic kingdom of Upper Egypt.[27] Kaiser believes the "painted tomb" may represent a royal grave. Settlement remains, probably of one central town and many subsidiary villages, cover a total area of a million square meters. In my opinion this large area may be misleading, as the debris is often very thin. Figure 7–5, showing the final results of the 1958 site survey,[28] permits an assessment of the denser remains, seldom more than a superficial horizon of pottery sherds, as 50,800 square meters. The former population may have been of the order of 4,700, at most 10,000.[29]

Figure 7–6 illustrates the broader geographic and geologic setting of Hierakonpolis: the maturely dissected Nubian sandstone, bordered by flats, one to two kilometers wide, of Late Pleistocene Nile silts. These so-called Sebilian silts average some five to seven meters in depth; they rest uncomformably on the sandstone and pose a steep embankment of several meters to the alluvium. The major Gerzean settlement was located on the semi- or nonconsolidated silts between two shallow wadis dissecting these. Obviously, apart from a perennially "dry" location, the site enabled easy excavation of the pits used as sunken dwellings.

Of particular interest at Hierakonpolis is the evidence of wind deflation and deposition on the southeastern margins of the settlement. A Gerzean cemetery[30] was denuded, and parts of the settlement were eroded or buried, so the aeolian activity responsible must have occurred after 3000 B.C. It very probably was contemporaneous with dune invasions of the Nile Valley farther north, dating from about 2350 to 500 B.C.[31]

Before the First Dynasty and the historical unification of Egypt (about 3000 B.C.), the settlement site had been transferred to the alluvium (Fig. 7–6). The new site was occupied until at least the close of the Sixth Dynasty (about 2180 B.C.). Cemeteries were still laid out on the Late Pleistocene silts, however (Fig. 7–5); these accompany the great structure of sun-dried brick, the so-called fort of Chasechmui (Second Dynasty, about 2700 B.C.). The Kula pyramid, consisting of quarried rock and dating from between the Third and Sixth dynasties, is similarly located some six kilometers to the northwest. Even in later Dynastic times the area remained important, to judge by the temples of Amenophis III (1410–1372 B.C.) and Ramses II (1301–1234 B.C.), situated in a broad wadi incised in the Nubian series to the northeast. Lastly, the site of El Kab (the later town of Nikhab or Eileithyiaspolis) can be seen in the flood plain on the east bank.

Although the flood plain widens to four kilometers at this point, John A. Wilson[32] has pointed out that the general economic potential of the whole area is low. The early importance of the region, therefore, must have been based partly on cultural factors.

The last site to be considered is Nagada-Tukh, *locus typicus* for the Predynastic cultures of Egypt. Figure 7–7 indicates the relation of the Gerzean and Dynastic townsites in relation to the Pleistocene deposits of the area. The low desert, some 3 kilometers wide, consists of Pleistocene wadi gravels on grayish yellow marls, presumably of Pliocene age. Whereas Nubt extends from the flood plain onto a 4-meter terrace, part of it being submerged under alluvium, Nagada-South Town is located on a fan 1.5 to 2 meters above wadi floor at the embouchure of the Wadi Ibeidalla. To judge from samples collected for me by Werner Kaiser, the settlement remains overlie a thin and incomplete veneer of gravels, below which yellowish marls are exposed. The locality is above the wadi floor and the flood plain, and it provided a fine-textured sediment for ready excavation. Although a few contemporary burials were located on the wadi floor, the greater part are concentrated on the low terrace. Only the latter graves were not exposed to subsequent fluviatile activity.

Like Hierakonpolis, the Nagada area was long a focal point of settlement. Remains found here of the Middle Palaeolithic industries of the Pleistocene terraces are the richest of the Luxor area *in situ*; surface finds of Epi-Levallois II flakes are common, and it is apparent that even after the Gerzean settlement the location was quite important, as evidenced by the Old Kingdom townsite nearby, a Sixth Dynasty pyramid, and an Eighteenth Dynasty temple.

The fauna of the Gerzean sites at Nagada is again quite compatible with the situation: Isabella gazelle, a buffalo, tortoise, and various fish,[33] as well as the usual array of domestic species.[34] It is the almost typical combination of steppe and gallery woodland–flood-plain biotopes found at Nile Valley sites from the Upper Pleistocene on.

Geomorphic Situations Typical for Predynastic Settlements

With the exception of the somewhat stony surface at Badari, each Predynastic settlement was located on soft or fine-textured, semi- or nonconsolidated deposits: Merimde, Hierakonpolis, Armant,[35, 36] and Maadi on silts; Nagada on marl; Mahasna[37, 38] and Abydos[39, 40] on sandy Nile gravels. Nowhere was coarse gravel or bedrock used. The reason for this apparently deliberate choice was the type of house construction in use—namely, wattle-and-daub structures set up around shallow pits, the huts being half below ground level.[41]

The second general feature of all sites is a location immediately beside the present flood plain, invariably above embankments or scarps standing several meters over the alluvium. Such location is the result of accident; it was only sites at these heights that escaped the annual inundations of recent centuries and the lateral expansion of cultivation. There has been a rise in the flood-plain level by at least two to three meters since the Predynastic era[42] and a lateral extension of the flood plain, through alluviation and human activity varying from several meters to one or two kilometers during the same period. Consequently, sites at lower elevations, and especially those on gentle slopes, have been buried or destroyed. So, for example, the Predynastic cemetery of Gerzeh, excavated some seventy years ago, has disappeared; and at many localities ancient pottery sherds in cultivated fields give evidence of recent encroachment at the expense of the desert.[43, 44]

It is possible to state generally that all prehistoric settlements and cemeteries today preserved are located on wadi terraces embanked against the alluvium; differentiated Nile deposits immediately in contact with the flood plain; and Upper Pleistocene silts also forming terrace-like steps to the cultivated fields. Virtually no remains were located where small, formerly active wadis have dissected ancient Nile deposits into broad fans; slow horizontal shifts of the Nile have left undifferentiated deposits with very gentle slopes (5 per cent and less); and the low desert is limited to a narrow, talus-strewn belt between the flood plain and the escarpment of the limestone plateau. The principles involved are illustrated in Figure 7–1.

Lastly, there are the complications relating to the alluvium–marginal gravel complex—namely, the third variable, aeolian activity. At Hierakonpolis, deflation and redeposition at one end of the site affected an area of 250,000 square meters. Such features are more local, however, than the evidence of aeolian activity in western Middle Egypt, where dunes border the alluvium over a 175-kilometer stretch from north of Asyut to near the Fayum.[45, 46, 47] Here any possible archaeologic sites of older date would have been located on a gently sloping low desert, now buried by many meters of alternating dunes and beds of Nile alluvium.

Regional Land Forms and Distribution of Archaeologic Sites

Bearing in mind the deliberate choice of location on fine-textured, unconsolidated sediments and the accidental factor of preservation strongly limited to specific geomorphic situations, what can we say about the over-all relations of regional land forms to archaeology in the Nile Valley? To begin with, prehistoric settlements and cemeteries are rather unevenly distributed in Egypt. One complex of sites is located between the Fayum region and the Nile Delta, the Lower Egyptian cultural province *sensu lato*. The other complex occupies the Nile Valley from about Asyut southward—namely, Upper Egypt. With one exception, the intervening section of 175 kilometers has not preserved any prehistoric sites.[48]

On the basis of the principles discussed above, a map was compiled, indicating the distribution of geomorphic features having archaeological significance in the intervening zone—namely, Middle Egypt (Badari-Asyut to the Fayum margins).[49] The salient points can be summarized briefly. On the west bank of the Nile the low desert from Asyut to about Meir consists of Nile gravels sloping very gently to the alluvium. Here the cultivated fields have advanced at least 50 to 100 meters in the course of the present century alone. Northward from Meir to Deshasheh a belt of dunes overlies the alluvium at the edge of the Pleistocene gravels, often merging with dune fields on the open desert. Any existing prehistoric sites on this western margin of the valley would be long buried under several meters of sand or mud, and in fact no settlement or cemetery can be found here antedating the fourth century B.C. A profusion of Ptolemaic and Roman materials stands in contrast to the archaeologic sterility of the preceding periods. This, again, can be explained readily in terms of physical features: between about 500 B.C. and A.D. 300 the Nile arm known as the Bahr Jusef shifted westward, removed the valley dunes by lateral planation, and deposited some two meters of Nile silt to the edge of the Pleistocene gravels, burying older aeolian deposits.[50, 51] During this time, settlement of the area was intense, as manifested by the archaeological remains.

The eastern margins of the valley are more complex in character, but only very locally are the physiographic features conducive to good preservation. The greater part of the area is characterized by a narrow belt of alluvium bordering almost immediately the escarpment of the limestone plateau, often obscured by talus fans. Other stretches are occupied by recently eroded soft bedrock or wadi fans sloping imperceptibly toward the encroaching flood plain. Most of the few favorable locations are occupied by modern villages or cemeteries. The stretch in question is almost sterile in terms of Predynastic remains.[52]

In the light of these considerations, older hypotheses that the cultural gap in Middle Egypt indicated a lack of prehistoric inhabitation seem to lack support. It would be a little surprising if Predynastic remains *could* be found here, at least at the surface, today. The geomorphic conditions are simply inimical to preservation of older archaeological sites. This anthropologically important example demonstrates the applicability of regional land-form analyses in archaeological surveys. After detailed study of a representative number of individual sites, the geomorphologist can assess the significance of various physiographic features, in terms of local conditions, relevant for a larger region. Such generalizations may even permit a direct conversion of surficial geology maps into archaeologically significant units.

Predynastic Sites and Predynastic Population

From the foregoing discussion one can already conclude that only a small proportion of the late prehistoric sites once situated on the desert margins have survived to this day. But the next question is, were all settlements originally located on the desert margins and not in the alluvium, on, for example, the levees?[53, 54]

Two lines of argument can be presented in favor of dense Predynastic settlement right in the flood plain. Firstly, the archaeological evidence indicates many hundreds of Predynastic cemeteries on the low desert but no corresponding settlement sites. The corresponding villages must have been located on the flood plain. Merimde, on the other hand, indicates a similar phenomenon: here, a single, short-lived, but very large town is preserved from an entire cultural epoch of a larger cultural province. It must have had countless predecessors, if not successors, and, above all, there must have been countless complementary village farming communities. Yet not a trace of these is preserved; they must exist under the delta alluvium.[55, 56]

The second line of argument is theoretical: simply that the sites preserved, even if they were all contemporaneous, would indicate a total Egyptian population of no more than 30,000 inhabitants.[57] Actually these sites are spread out over a whole millennium in time. Moreover, one must allow for 16,500 square kilometers of relatively drained, fertile land in the valley at the time in question. So, with an advanced primary village farming economy, the Egyptian population in later Predynastic times must be thought of in terms of 100,000 to 200,000 inhabitants.[58, 59] In other words, most of these people must always have lived in the flood plain, on natural elevations offering ideal location for early settlement.[60, 61, 62] These innumerable villages and towns are no longer readily accessible today. Many probably lie at the base of existing larger townsites, some of which have remained in use for many millennia.

Physical Conditions in Egypt during the Neolithic and Chalcolithic

The macrosetting of Predynastic Egypt in its palaeoenvironment has been discussed in considerable detail already.[63, 64, 65] Briefly, the period 5000 to 2350 B.C. was a time of variable climate, but, in general, there were heavier or more frequent winter rains than there are today. A savanna fauna including the giraffe, elephant, and rhinoceros was not uncommon in large parts of the more elevated Egyptian deserts. From this more humid period, which is geologically verified, there is historical and archaeological evidence of tree growth, of an open park-land character, on large parts of the low desert. This probably indicates that the desert hinterland of the marginal desert sites had economic significance, specifically for a pastoral subsistence of some proportions. A reflection of this favorable palaeoenvironment was the expansion of Neolithic populations into the desert hills and wadis of Egypt after 5000 B.C., areas which have been largely uninhabited since the close of the third millennium B.C.

This brief sketch of the methods and potentialities of geomorphologic analysis of archaeological sites and settings will serve its purpose if it illustrates a means of effective cooperation between the earth scientist and the digging archaeologist. Depending upon physical and human factors, the problems involved will vary from country to country; the ones discussed here are peculiar to the lower valley of the Nile. To recognize these problems the geomorphologist must have some familiarity with archaeology and must actively exchange ideas and notions with the anthropologist. In other words, the "straight" geologist with little direct interest in the cultural aspects cannot fully apply himself to problems which can only be formulated in interdisciplinary discussion. For example, regional studies of Pleistocene tectonics or climate will be of only limited use to an archaeologist excavating a Bronze Age site. Whatever the area, the basic work should be directed to an intensive and comprehensive study of the immediate location, applied to as large a number of representative sites as possible. When the typical geomorphic situation is known, regional land forms can be evaluated as to their possible significance for contemporary settlement. And into this picture should be introduced any detail bearing upon differences in the physical environment—climate, vegetation, and fauna. Only on this foundation can the geography of prehistoric settlement be effectively understood or analyzed.

NOTES

1. H. E. Wright, Jr., *Natl. Acad. Sci–Natl. Research Council, Publ. No. 565* (1957), p. 50.

2. Methods of work would involve surficial deposits, microstratigraphy, general geomorphology and land-form analysis, and sediment and soil examinations. So, depending upon where the division between geography and geology is drawn from country to country, or university to university, the investigator may be "labeled" differently. On account of this difficulty in departmentalizing research in Quaternary geology and geography, the terms *geology* and *geomorphology* as employed here are not intended to specify what category of earth scientist may be involved.

3. R. J. Braidwood, *Natl. Acad. Sci.–Natl. Research Council, Publ. No. 565* (1957), pp. 14, 26.

4. Some of the data may be referred to in greater detail in K. W. Butzer, *Mitt. deut. Archäol. Inst., Abt. Kairo,* **17**, 54 (1961). The material presented here is derived from a geomorphological and pedological field survey carried out under the auspices of the Deutsche Forschungsgemeinschaft (Godesberg) in 1958. The primary purpose was the study of Pleistocene stratigraphy, although I spent six weeks in informal association with an archaeological survey then in progress. The sites described were all examined in collaboration with Dr. W. Kaiser, an Egyptologist, although the evaluation of the evidence and publication of my geological and geographical investigations proceeded independently. I am grateful for the financial

support rendered by the Deutsche Forschungsgemeinschaft, the German Academy (Mainz), and the Canada Council, as well as for certain field facilities provided by the German Archaeological Institute.

5. K. S. Sandford and W. J. Arkell, *Univ. Chicago Oriental Inst. Publs. No. 10* (1929), *17* (1933), and *46* (1939); K. S. Sandford, *Univ. Chicago Oriental Inst. Publ. No. 18* (1934); G. Caton-Thompson and E. W. Gardner, *Geograph. J.*, **80**, 369 (1932).

6. G. Caton-Thompson and E. W. Gardner, *The Desert Fayum* (Royal Anthropological Institute, London, 1934).

7. K. W. Butzer, *Erdkunde*, **13**, 46 (1959).

8. Caton-Thompson and Gardner, *op. cit.*

9. M. Amer and S. A. Huzayyin, *Pan-African Congress on Prehistory, First, Nairobi 1947, Proceedings*, L. S. B. Leaky and S. Cole, Eds. (Blackwell, Oxford, 1952), pp. 222–24.

10. J. de Morgan, *Recherches sur les origines de l'Egypte* (Paris, 1896–97).

11. S. Passarge, *Nova Acta Leopoldina*, **9**, 77 (1940).

12. W. Kaiser, *Mitt. deut. Archäol. Inst., Abt. Kairo*, **17**, 1 (1961).

13. Butzer, *Erdkunde*, loc. cit.

14. Caton-Thompson and Gardner, *op. cit.*

15. I. Olsson, *Am. J. Sci., Radiocarbon Suppl.*, **1**, 87 (1959); H. G. de Vries and G. W. Barendsen, *Nature*, **174**, 1138 (1954); compare E. K. Ralph, *Am. J. Sci., Radiocarbon Suppl.*, **1**, 45 (1959).

16. H. Junker, *Oestrr. Akad. Wiss. Wien, Phil. Hist. Kl. Anz.*, **66**, 156 (1929); **69**, 36, (1932); **70**, 82 (1933); **77**, 3 (1940).

17. K. W. Butzer, *Bull. Soc. Géogr. Egypte*, **32**, 43 (1959).

18. Junker, *op. cit.*

19. *Ibid.*

20. C. A. Reed, *Science*, **130**, 1629 (1959).

21. K. W. Butzer, *Bull. Soc. Géogr. Egypte*, **32**, 43 (1959).

22. G. Brunton and G. Caton-Thompson, *The Badarian Civilization* (British School of Archaeology in Egypt, London, 1928), p. 69.

23. K. W. Butzer, *Bull. Soc. Géogr. Egypte*, **32**, 43 (1959).

24. Brunton and Caton-Thompson, *The Badarian Civilization, op. cit.*, pp. 73–6.

25. K. W. Butzer, *Bull. Soc. Géogr. Egypte*, **32**, 43 (1959).

26. Olsson, *loc. cit.*; de Vries, *loc. cit.*; and E. K. Ralph, *loc. cit.*

27. W. Kaiser, *Mitt. deut. Archäol. Inst., Abt. Kairo*, **16** (1958).

28. Topography was surveyed by K. W. Butzer and W. Kaiser, February 1958; surficial deposits were mapped by Butzer, archaeological remains by Kaiser and Butzer. For earlier topographic-archaeologic maps of the area, see the inaccurate map (1:15,000) by J. E. Quibell and F. W. Green in *Hierakonpolis* (Archaeological Survey of Egypt, London, 1900–1902) and a slightly modified version in W. Kaiser, *Mitt. deut. Archäol. Inst., Abt. Kairo*, **16** (1958). The architectural base of Early Dynastic Hierakonpolis has been transferred from Quibell and Green.

29. K. W. Butzer, *Bull. Soc. Géogr. Egypte*, **32**, 43 (1959).

30. Butzer, *Mitt. deut. Archäol., op. cit.*

31. K. W. Butzer, *Abhandl. Akad. Wiss. Liter. (Mainz), Math.-Naturw. Kl.*, **1959**, No. 2 (1959); *Geograph. J.*, **125**, 75 (1959).

32. J. A. Wilson, *J. Near Eastern Studies*, **14**, 209 (1955).

33. Butzer, *Abhandl.*, op. cit.

34. Reed, *loc. cit.*

35. Butzer, *Mitt. deut. Archäol.*, *op. cit.*

36. K. S. Sandford and W. J. Arkell, *Univ. Chicago Oriental Inst. Publ. No. 17* (1933); R. Mond and O. H. Myers, *Cemeteries of Armant* (Egypt Exploration Fund, London, 1937), vol. 1. See note 9 for Maadi.

37. Butzer, *Mitt. deut. Archäol.*, *op. cit.*

38. J. Garstang, *Mahasna and Bet Khallaf* (Archaeological Survey of Egypt, London, 1903).

39. Butzer, *Mitt. deut. Archäol.*, *op. cit.*

40. T. E. Peet, *Abydos* (Archaeological Survey of Egypt, London, 1914).

41. A. Badawy, *A History of Egyptian Architecture* (Cairo, 1954), vol. 1.

42. Butzer, *Abhandl.*, *op. cit.*

43. Butzer, *Mitt. deut. Archäol.*, *op. cit.*

44. W. Kaiser, *Mitt. deut. Archäeol. Inst.*, *Abt. Kairo*, **17**, 1 (1961).

45. Butzer, *Mitt. deut. Archäeol.*, *op. cit.*

46. Butzer, *Erdkunde*, *loc. cit.*

47. Butzer, *Abhandl.*, *op. cit.*

48. W. Kaiser, *Mitt. deut. Archäeol. Inst.*, *Abt. Kairo*, **17**, 1 (1961).

49. Butzer, *Mitt. deut.*, *Archäol.*, *op. cit.*

50. Butzer, *Abhandl.*, *op. cit.*

51. K. W. Butzer, *Bull. Soc. Géogr. Egypte*, **33**, 5 (1961).

52. W. Kaiser, *Mitt. deut. Archäeol. Inst.*, *Abt. Kairo*, **17**, 1 (1961).

53. Passarge, *loc. cit.*

54. Butzer, *Abhandl.*, *op. cit.*

55. K. W. Butzer, *Bull. Soc. Géogr. Egypte*, **32**, 43 (1959).

56. The significance of the flat sand banks or "turtle-backs" of the Nile Delta for prehistoric settlement is considered in detail in K. W. Butzer, *Abhandl. Akad. Wiss. Liter.* (*Mainz*), *Math.-naturw. Kl.*, **1959**, No. 2, 29–36 (1959).

57. K. W. Butzer, *Bull. Soc. Géogr. Egypte*, **32**, 43 (1959).

58. *Ibid.*

59. R. J. Braidwood and C. A. Reed, *Cold Spring Harbor Symposia Quant. Biol.*, **22**, 19 (1957).

60. Passarge, *loc. cit.*

61. K. W. Butzer, *Bull. Soc. Géogr. Egypte*, **32**, 43 (1959).

62. K. W. Butzer, *Abhandl.*, *op. cit.*

63. K. W. Butzer, *Bull. Soc. Géogr. Egypte*, **32**, 43 (1959).

64. K. W. Butzer, *Abhandl.*, *op. cit.*

65. K. W. Butzer, *Quaternary Stratigraphy and Climate in the Near East* (Dümmler, Bonn, 1958), 110 ff.

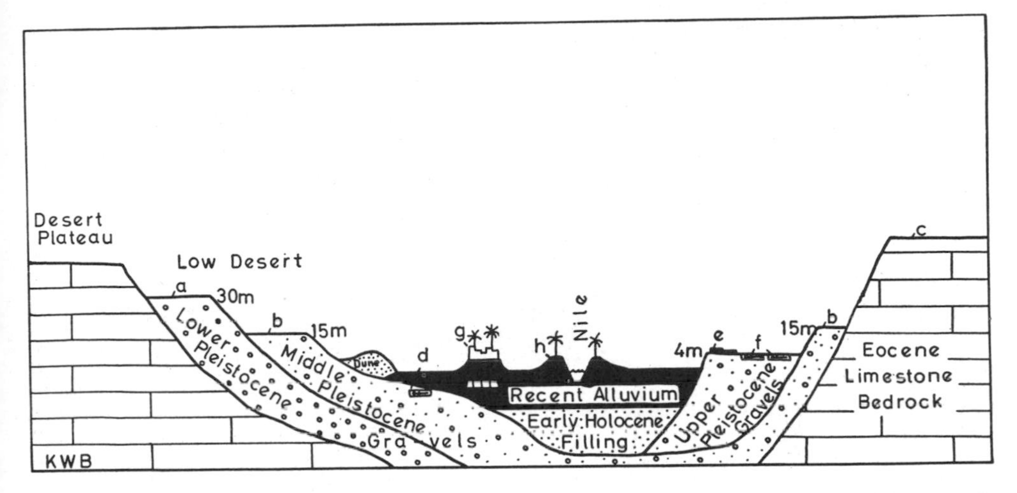

Fig. 7–1. Schematic diagram of the relation of archaeologic sites to geological features of the Nile Valley. a, b, Nile gravels containing Lower Palaeolithic implements with scattered Middle Palaeolithic artifacts on the surface; c, Predynastic flints scattered over desert surface; d, possible "buried" Predynastic cemetery, under subsequent silt deposits; e, remains of Predynastic settlement; f, Predynastic burials; g, modern village on cultural mound (ancient site at base?); h, roads on levee embankments bordering low-water channel of the Nile. (Not to scale.)

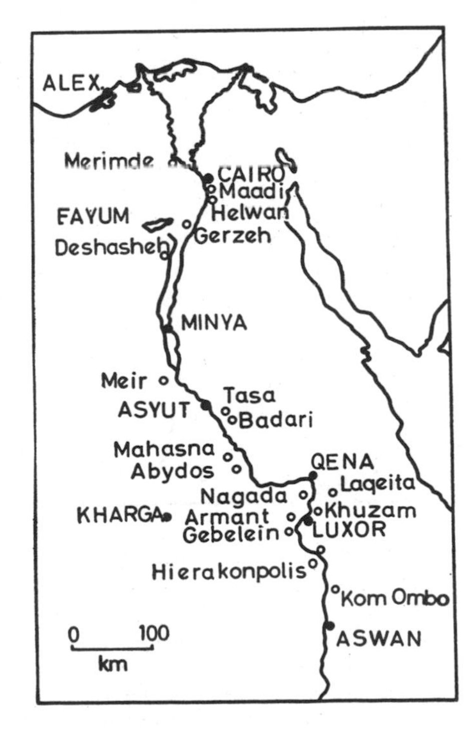

Fig. 7–2. Location of major late prehistoric settlement sites in the Nile Valley.

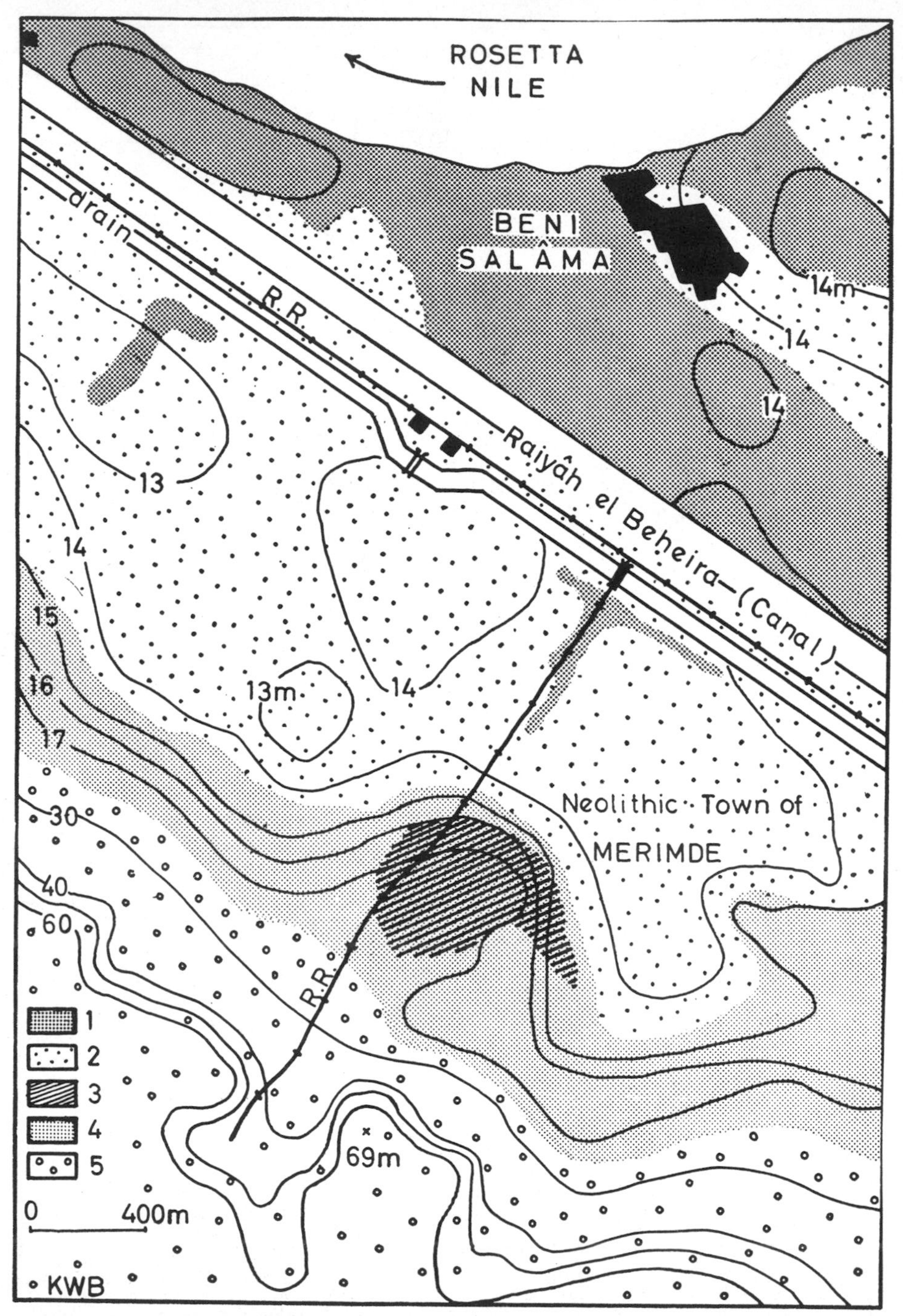

Fig 7–3. Topographic geologic map of Merimde-Beni Salama, Md. el Giza, Mz. Imbaba (Delta). (Md., *mudiriyet*; Mz., *markaz*; administrative units.) The Neolithic townsite is situated on the Upper Pleistocene Nile silts, the Nileward portions being obscured by drifting sand. The paved highway is located near the 30-meter contour. 1, Recent alluvium; 2, aeolian sand and downwash on Pleistocene silts and modern alluvium; 3, approximate extent of townsite; 4, Late Pleistocene Nile silts; 5, Lower Pleistocene gravels. Both 4 and 5 are superficially veneered by fine downwash. Basic map compiled from an excellent map (about 1:8000) by K. Bittel in H. Junker, *Anz. Akad. Wiss. Wien, Phil.-hist. Kl.*, 69, 36 (1932), and from survey of Egypt sheets Nos. 84/54 (1927) and 84/60 (1925, revised 1940) (1:100,000). Geology after geological map (1:150,000) by K. S. Sandford and W. J. Arkell (*Univ. Chicago Oriental Inst. Publ. No. 46* [1939]) and from personal observations.

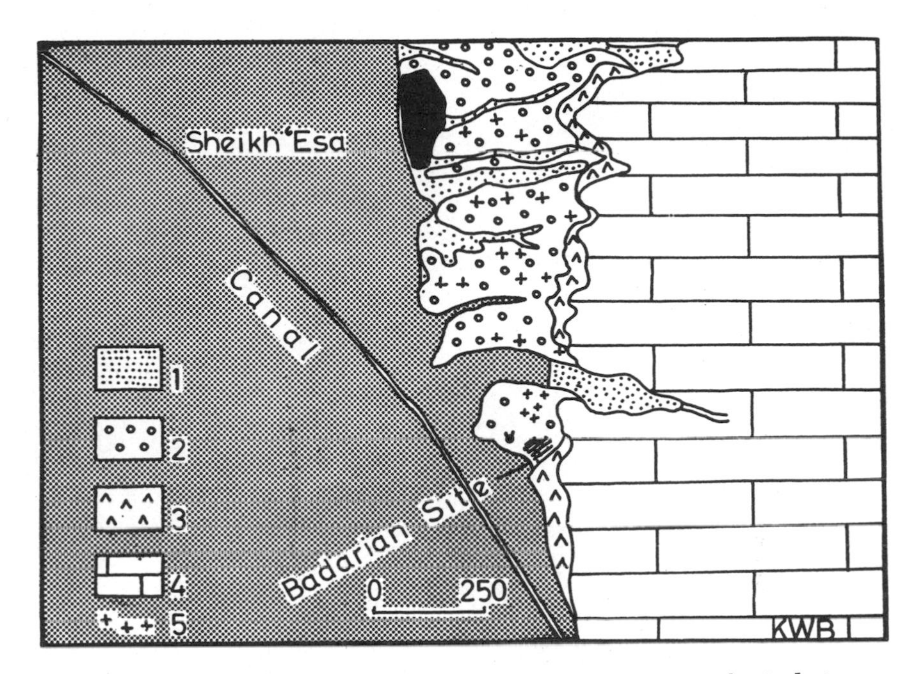

Fig. 7–4. Geologic sketch of some of the Chalcolithic sites in the Badari area, Md. Asyut, Mz. Badari. Indicated are numerous prehistoric cemeteries and one Badarian village near Sheikh Esa. Topography modified after map in G. Brunton and G. Caton-Thompson, *The Badarian Civilization* (British School of Archaeology in Egypt, London, 1928); geology by Butzer. 1, Recent wadi wash; 2, fluvial gravels to 10 meters above flood-plain level on limestone base (Middle Pleistocene?); 3, Recent scree on limestone bedrock; 4, Eocene (Upper Libyan) limestone; 5, prehistoric cemeteries.

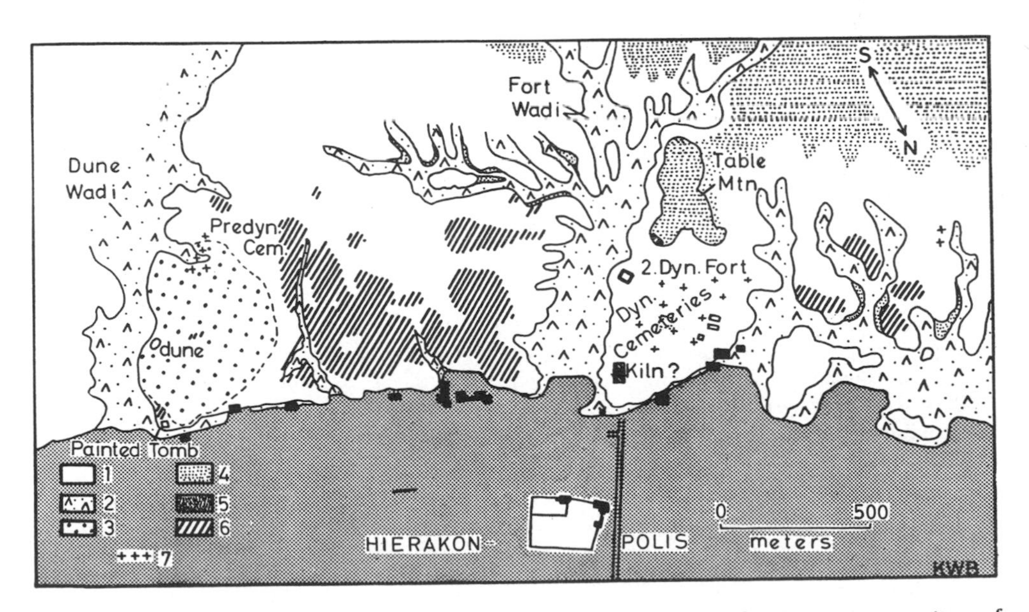

Fig. 7–5. Topographic-geologic map of the Predynastic and Early Dynastic townsites of Hierakonpolis, Md. Aswan, Mz. Edfu. See Note 28. 1, Late Pleistocene Nile silts; 2, wadi wash and detritus; 3, unconsolidated aeolian sand; 4, Nubian sandstone outcrops, locally obscured by wash and detritus; 5, Recent alluvium; 6, approximate extent of major cultural debris of Gerzean settlement on the Late Pleistocene silts; 7, cemeteries and burials.

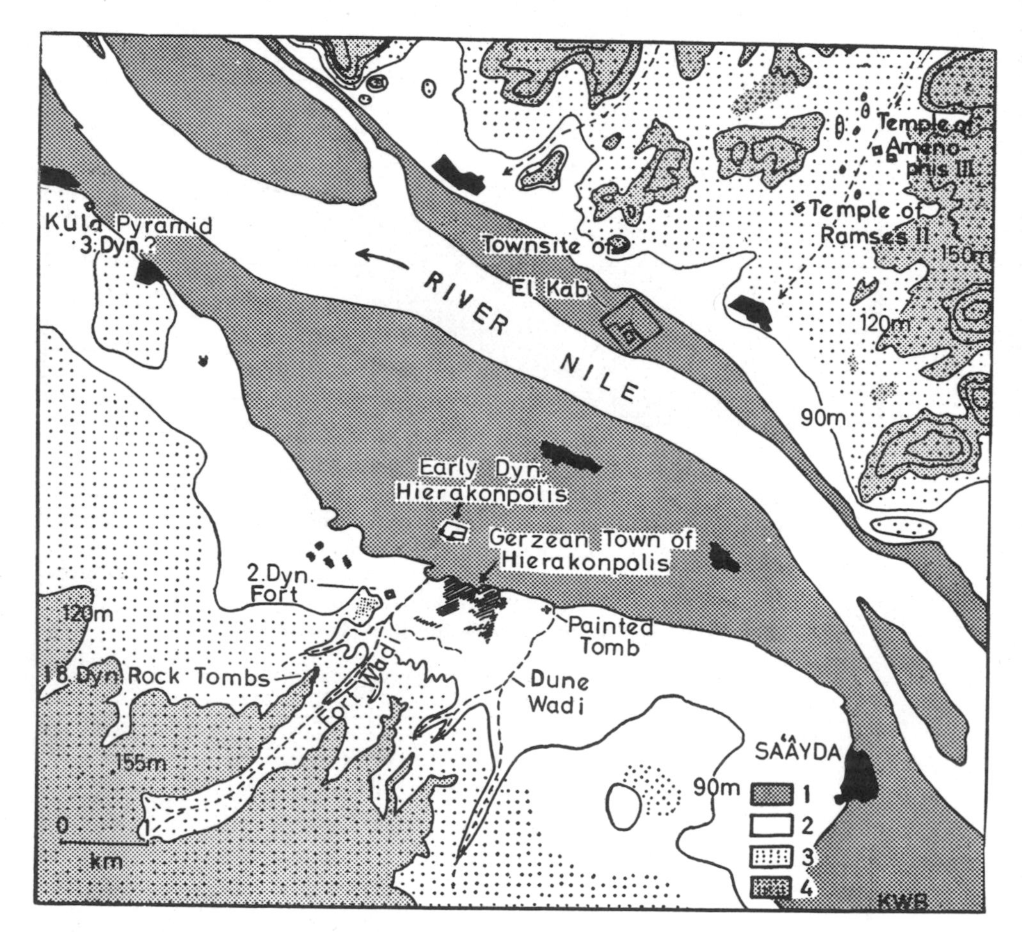

Fig. 7–6. Geographic and geologic setting of the Hierakonpolis-El Kab area. Map based on survey of Egypt map No. 24/72–78 (1928, revised 1940) (1:100,000), on personal observations, and on geological map (1:1,000,000) in K. S. Sandford and W. J. Arkell, *Univ. Chicago Oriental Inst. Publ. No. 17* (1933). 1, Recent alluvium; 2, Late Pleistocene silts; 3, Nubian sandstone with wadi wash and surficial detritus; 4, Nubian sandstone (Mesozoic) outcrops. Flood plain, 82 to 83 meters above mean sea level; contour interval, beginning at 90 meters, is 30 meters.

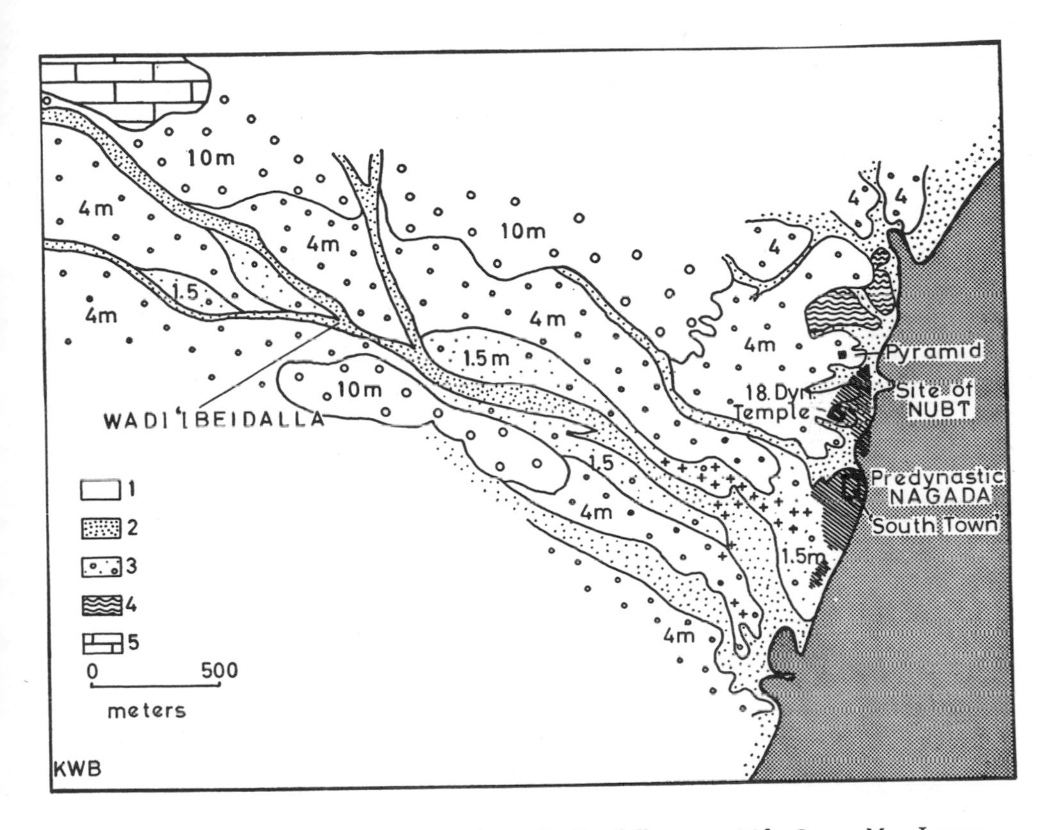

Fig. 7–7. Geologic sketch of the Nagada-Wadi Ibeidalla area, Md. Qena, Mz. Luxor. Part of the topography marginal to the flood plain after the sketch (1:15,000) (greatly modified) by W. F. M. Petrie and J. E. Quibell in *Nagada and Ballas* (British School of Archaeology in Egypt, London, 1896); wadi topography and surficial deposits mapped by Butzer. Compare K. W. Butzer, *Erdkunde,* **13**, 46, Fig. 3 (1959). 1, Recent alluvium of the flood plain; 2, Recent wadi wash; 3, various Pleistocene gravels (on marl): terrace, 10 meters above wadi sole, with scarce Acheulio-Levallois industry, Middle Pleistocene (?); 4- and 1.5-meter terraces with Levallois and Epi-Levallois II, respectively, Late Pleistocene; 4, yellow Pliocene marls; 5, Eocene limestone. (Old Kingdom townsite of Nubt and Gerzean townsite of Nagada after W. Kaiser. *Mitt. deut. Archäol. Inst., Abt. Kairo,* **17**, 1 (1961); "South Town" after Petrie and Quibell. Part of the topography marginal to the flood plain after the sketch [1:15,000] [greatly modified] by W. F. M. Petrie and J. E. Quibell in *Nagada and Ballas* [British School of Archaeology in Egypt, London, 1896]; wadi topography and surficial deposits mapped by Butzer. Compare K. W. Butzer, *Erdkunde,* **13**, 46, Fig. 3 [1959].)

Part II

THE NEW WORLD

Chapter 8

Polynesian Origins

Edwin N. Ferdon, Jr.

When the raft "Kon-Tiki" crunched down on the reef off Raratoria atoll in 1946, the story made the front page. Later, the voyage was the subject of a best-selling book, and the moving picture of the expedition won an Oscar award. To the skipper of the raft, Thor Heyerdahl, these were fringe benefits, for the real goal had been to prove to a world of doubting and experiment-shy anthropologists that the prehistoric Peruvians, at least, had had seaworthy craft capable of taking them to Polynesian islands. The idea that American Indians might have reached, and even populated, some of the Polynesian islands was not new. However, the concept had been laid aside because it was believed, though not shown, that no adequate craft had existed in prehistoric America and that the American Indians lacked the necessary navigational skills.

If the joys of successful experimentation were pleasant, they were equally short, for with the publication of his theoretical book,[1] the skipper of the "Kon-Tiki" opened a Pandora's box of conservative, tradition-bound anthropological argument.

The controversy that ensued over East versus West as *the* source of the people and culture of Polynesia has tended not only to obscure the complexity of the problem, but to conceal the numerous other possibilities. Although this controversy has stimulated research on Polynesia, an unfortunate secondary effect threatens the eventual results of these new and vigorous efforts. There is a tendency to interpret new knowledge in terms of the old East-versus-West argument, as if the twain should never meet, and as if no other interpretations were possible. To continue within this rigid frame will mean losing the healthy effect of revitalization which the initial controversy precipitated.

The origin, or origins, of the people and culture of Polynesia has been the concern of seafarers, missionaries, and scientists since the days of Captain Cook. The range of theories is only slightly exceeded by the variety of evidence mustered to support particular concepts. Mainland Asia, Indonesia, Melanesia, Micronesia, the Philippines, and the west coasts of the Americas have all been proposed as possible points of departure for one or more migration waves into this island world.

By the beginning of World War II, anthropological interest in Polynesia, especially as regards origins, had waned noticeably. Perhaps because the Bernice P. Bishop Museum in Honolulu dominated the field of Polynesian anthropology, the views of its director, Sir Peter H. Buck, were generally accepted as the best that could be arrived at from available data.

According to this widely accepted pre-"Kon-Tiki" concept, the Polynesians originally were a group of people of one composite race located somewhere in Indonesia.[2] They were thought to have broken away from their original homeland at some early date, and, with their then existing culture complex, to have migrated in one or more waves to their present island domain by way of Micronesia. According to other versions, they migrated along the north coast of New Guinea.[3] Except for minor Fijian influence in extreme western Polynesia and for certain food transmissions from Fiji,[4] these people were seen as having evolved biologically and culturally in isolation until their discovery by Europeans.

Although broad, ethnologically based culture areas had been established for Polynesia, their existence had been explained as in no way reflecting occasional and varied outside influences but rather as the result of splinter migrations that broke away from the original migratory band. As a starting point for their own isolated cultural evolution, it was thought, these splinter groups had taken with them the bulk of the cultural content of the original band, plus the various elements which had evolved locally up to the time of departure. Thus, the difference between one culture area and another was considered to reflect the period when each broke from the main (or an ancillary) migratory body, the cultural inventory at the time of the break, and the internal cultural evolution of each splinter group after its isolation on some other island or group of islands.[5] It was recognized that there was cultural exchange, other than of domesticated plants and animals, *within* Polynesia, but apparent parallels in traits beyond this insular area were regarded as the result of independent invention or of parallel development.

So firmly entrenched was the foregoing concept that the lack of supporting evidence in subsurface archaeological excavations made before 1950 was brushed aside, with the simple statement that such deposits were too shallow and too recent to be of use. It was this archaeologically unsupported concept of a single group origin from one location and of isolated, independent development that Heyerdahl challenged, with his "Kon-Tiki" journey, and his theory of an American Indian origin. This challenge, however, was a case of the pot calling the kettle black, for Heyerdahl, also without supporting archaeological evidence, concluded that two specific areas on the continent of South America, Peru and the northwest coast, were dominant sources of Polynesian population and culture.[6]

Although parts of Heyerdahl's book suffer from his having been a novice in anthropology, his exhaustive coverage of the pertinent literature brought together a vast amount of widely dispersed information on the problem. Although some of his arguments could not be regarded as valid, others, if they did not wholly support his specific migration theory, certainly appeared to uphold a thesis that contact between prehistoric Polynesia and America had been made. Also, his trip aboard the raft had shown that the sailors and craft involved in such contacts were not necessarily Polynesian. If many of his arguments were not fully acceptable to anthropologists, his discussions of the weaknesses of other theories were sound enough. These discussions cast a brilliant light on the inadequacy of the existing data to support the thesis that the Polynesians were, racially and culturally, a single homogeneous group.

Of all the types of evidence yet brought to bear upon the question of human migration, probably the strongest has been linguistic evidence. The Polynesian language has been placed by linguists in the Malayo-Polynesian family. Peoples speaking these related tongues are spread in a near-continuous pattern from easternmost Polynesia westward through Melanesia (except for the interior of New Guinea) and Indonesia and across the Indian Ocean to Madagascar. Northward, related languages are spoken by certain peoples in Southeast Asia and the Philippines.[7] Firm as this evidence would appear to be, reflecting, as it must, a strong influence throughout Polynesia of people speaking a common language related to languages spoken to the west, we know that that language is culturally transmitted and that, therefore, one language may be supplanted by another in a variety of ways. Thus, although the linguistic proof of a connection to the west seems sound enough, it is not acceptable per se as proof that no other people existed and that no other languages were ever spoken on any or all of the islands of Polynesia.

The use of genetic information was another approach through which it was originally hoped the question of Polynesian origins would be resolved. With the increasing use of blood typing in genetic studies, it was felt that here, at last, was a biological approach that might more clearly indicate the probable source, or sources, of the Polynesian people. Contrary to the linguistic evidence, the results of a study of blood groups and gene frequencies of the Cook Islanders, made by R. J. Simmons, J. J. Graydon, N. M. Semple, and Ernest I. Fry, prompted these workers to state,[8] "The blood groups and gene frequencies presented here for the Cook Islanders do not invalidate the conclusions reached, that there is a close blood genetic relationship between American Indians and Polynesians, and that no similar relationship is evident when Polynesians are compared with Melanesians, Micronesians, and Indonesians, except mainly in adjacent areas of direct contact." Later, after working with blood samples from eastern Polynesia collected by the Norwegian Archaeological Expedition to Easter Island and the east Pacific, Simmons and Graydon concluded,[9] "The results obtained are comparable with those previously reported for Maoris of New Zealand and Cook Islanders, and in a number of characters are comparable with some South American Indian tribes. No such similarity is evident when comparisons are made with Melanesians, Micronesians and Indonesians."

The results of this latter study have been challenged on the grounds that the blood samples taken by the Norwegian expedition were not from "pure" Polynesians,[10] but such criticism is basically fatuous. In view of the two hundred years of contact with foreign sailors and travelers and the sexual license common in Polynesia, who could honestly expect, let alone prove, purity for any Polynesian sample? Of far more significance, granted the impurity of the samples, is the question of why the results showed apparent affinity with the American Indians. Obviously, as Robert C. Suggs[11] and E. Goldschmidt,[12] and more recently Simmons,[13] have pointed out, it is not enough to determine the present blood groups and gene frequencies for Polynesians, and for other surrounding racial groups, and from these comparative data draw conclusions about racial relationships. Processes of microevolution, such as genetic drift, mutation, and selection—especially those selective factors that have operated in historic times through the decimation of native populations by European diseases—must be determined and taken into account. Even if the difficulties originally encountered in determining blood types from prehistoric bone material should be resolved, the microevolutionary forces must be taken into account before much more can be said about the biological relationships of the Polynesian peoples.

While claims and counterclaims were being made, and the data were being said to support one theory or its opposite, one of the primary sources of basic evidence, subsurface archaeology, had been virtually neglected. Prior to World War II, only two excavations, one on Tonga[14] and the other in New Zealand,[15] had been made. This lack of professional interest stemmed in part from the difficulty of access to the numerous islands. To a larger extent, however, it stemmed from a misconception, current as late as 1953: "sites are shallow, refuse is sparse, and there seems to have been relatively little change in culture through time."[16] However, with the revitalization of research in the Pacific, it was realized that, shallow or not, Polynesian archaeological deposits must be excavated, and that, for purposes of comparison, such activity must be extended into neighboring areas.

In 1950, as part of a University of Hawaii course in archaeological techniques, Kenneth P. Emory and several of his students undertook the excavation of a series of shelters on Oahu.[17] In 1953 Heyerdahl, with two archaeologists, conducted a brief expedition to the Galapagos Islands, where, to the surprise of everyone, prehistoric Peruvian pottery was found.[18] Two years later Heyerdahl moved his archaeological activities directly into Polynesia by organizing and financing a major expedition to conduct excavations on Easter Island and several other islands of eastern Polynesia.[19] No sooner had this expedition gotten under way than the American Museum of Natural History sent Suggs to the Marquesas;[20] then, somewhat later, Roger Green undertook excavations in the Gambier Islands and the Society Islands. During this same period, New Zealanders became increasingly interested in what their country could reveal of Polynesian prehistory. They now are excavating on other Polynesian islands as well. Other excavation has been undertaken by archaeologists of various nationalities, in neighboring Melanesia and Micronesia. Today, more archaeological expeditions than have ever before been concerned with Oceania are either in the planning stage or already in the field.

As of today, it would be foolhardy to attempt to summarize our knowledge of Polynesian prehistory on the basis of the excavated record. The record is excessively spotty, and the sites are separated by hundreds of miles of ocean; moreover, the results of numerous archaeological excavations now being made are certain to alter the present view, which, at best, is a delicate and highly mobile frame of reference.

Thanks to dating by the radiocarbon technique, we now know that Polynesian prehistory goes back farther than had previously been estimated, and that it does evince change through time. A date of 122 B.C. has been established for human occupation in the Marquesas at the eastern edge of Polynesia, and a date of A.D. 9 has been obtained for Samoa, at the western extremity. An early date of occupation of 46 B.C. has been obtained for neighboring Fiji, and it seems reasonable to expect at least temporally comparable evidence of human occupation on Samoa. Far to the north, in Hawaii, a possibly valid date of A.D. 124 may indicate that this outpost of Polynesia was settled at about the beginning of the Christian Era. To the south, in New Zealand, where thirty-eight radiocarbon samples have been obtained, the earliest date of occupation so far obtained is around A.D. 1000.[21]

Because much of Polynesian material culture was of a perishable nature, the number of artifacts that have remained in the soil is quite limited. Pottery, which readily lends itself to change in form and decoration and is, therefore, especially useful in the finer cross-dating of one archaeological deposit with another, so far appears to be largely restricted to western Polynesia. However, the pioneering efforts of Roger S. Duff of New Zealand[22] in classifying the numerous stone adzes of Polynesia will, as their stratigraphic relationship is gradually determined through excavation, aid in determining cultural relationships. Careful study of the stratigraphic sequence of fishhooks from excavations in Hawaii[23] has already shown that the fishhook is another artifact whose change through time may aid in determining island relationships at different periods. No doubt the stone *poi* (fermented taro root paste) pounders of central and marginal Polynesia will eventually prove of equal value. On Easter Island, where no deep culture-bearing deposits were found, archaeologists turned to excavation of the great ceremonial platforms called *ahu*. Here they found that, as with Mesoamerican ceremonial structures, various platforms had been covered or modified through time, so that the shape and variety of the architectural features provided a means of interpreting changing religious functions and of estimating relative dates.[24] Thus, although the available archaeological record is not so rich in Polynesia as in other parts of the world, nevertheless, the record in the islands can provide ample evidence on which to reconstruct the culture of Oceania.

Problems of Interpretation

Undoubtedly a mass of archaeological information will be revealed in the next several years, and thus the greatest problem facing the culture historian is proper interpretation of the data.

Anthropologists, being essentially landlubbers, have long interpreted the concept that an ocean is a formidable barrier to mean that it is an absolute barrier to all but a highly specialized few. Over the years this restricted interpretation has limited the search for potential sources of cultural inspiration for any given primitive group to other societies on connected, or immediately accessible, land masses. Thus, the anthropological search for possible prehistoric contacts between the Old World and the New, for prehistoric American-Polynesian contacts, and for prehistoric Asiatic-American over-water contacts has been restricted by insistence on the kind of proof that could be found only where a more or less continuous contact, or a series of chain-linked contacts, had occurred between two cultures over an extended period. I do not question the general validity of a requirement for proof of long-continued contact, but the requirement is unrealistic where the route of dispersal involves the crossing of great oceans, especially where one-way voyages may have occurred.

That one-way accidental voyages did occur within Polynesia has been amply documented by Andrew Sharp and G. M. Dening.[25, 26] Also, historic records of derelict junks encountered in the north Pacific as far east as Mexico, many of which carried survivors after months of drifting,[27] testify to the fact that accidental dispersal of mankind in the Pacific has occurred. This is not to say that there were no planned voyages into, and within, Polynesia, but, as Sharp has demonstrated from historic documentation, the Polynesians were largely limited, geographically, to island groups that they had the navigational ability to reach. Once the Polynesian mariner had passed these limits, as a result of storms, wind shifts, or other natural hazards, he was essentially lost, and his final destination was a matter of happenstance.

In addition to unplanned voyages into distant seas, there were probably planned voyages into the unknown, made for a variety of reasons, including the basic one of overpopulation. Unless there was cultural control of population, overpopulation could have become a cyclic phenomenon in this island world and thus have induced a series of migratory resettlements. That the Polynesians did make such planned voyages into the unknown, with the full knowledge that they did not have the navigational skill to return, and that only by chance could they do so, is indicated by the inventory of human and cultural cargo they carried on some of their voyages. This inventory appears to have included as much of the homeland cultural complex as possible, so that a new settlement could be made with the fewest possible adjustments.[28]

In terms of cultural diffusion into, and throughout, Polynesia, the two types of voyager—those who migrated intentionally and those who migrated unintentionally—are distinguished by the fact that the former carried with them as much of their human and cultural heritage as they could, whereas the latter brought to an island refuge only their personal knowledge and concepts of their original cultural world, and any objects which happened to be aboard the vessel when it was carried into unknown seas. Thus, the intentional landing of a group of immigrants on an uninhabited island assured the transplanting of a fairly complete cultural inventory of the parent complex. If the island were already settled, such a landing offered a choice of culture traits. This was a far greater cultural contribution than could be made by a solitary voyager who made an unplanned landing.

It appears likely that at least one population group which entered Polynesia and spoke a Malayo-Polynesian language developed, or maintained, a cultural tradition of intentional migratory voyaging. Because this tradition was probably maintained over several centuries, a basic language and culture were successfully implanted throughout the Polynesian islands. However, as new cultural information is revealed through excavation, it must be kept in mind that other population groups may have been equally successful at an earlier date and may later have been wiped out, subjugated, or amalgamated.

Because intentional migrants would tend to transplant a variety of trait complexes, the similarity of complexes between distant Polynesian islands has been accepted as proof of prehistoric contact and attendant diffusion. Where a variety of specialized developments on various islands do not fit into the total assemblage of interfunctioning traits, they have too often been interpreted as the result of independent invention. However, it is precisely this type of specialized characteristic, which seemingly does not fit into basic Polynesian culture, that could reflect the influence of single voyagers who made unplanned landings.

Although it is seldom realized, there is a parallel between the chance dispersal of animals and the chance dispersal of man onto Pacific islands. Just as faunal associations that spread to an island by the accidental transferal of occasional individuals exhibit an imbalance, so cultural influences that spread in a similar manner can result in the incorporation of one or more culture traits into an already existing, but wholly different, functional complex. Thus, individual components of such a complex may resemble components of a complex in another area of the Pacific and indicate contact and diffusion, even though the complexes are quite different. With cultural as with faunal dispersals of this accidental type, there is no reason to believe that such trait diffusion is not "relatively random or indeterminate." With man as with faunal associations, "groups that might cross do not necessarily do so; crossings may be long delayed and are scattered through time; and the sequence seems to depend in part on chance."[29]

Of course, the insular dispersal of prehistoric man does not wholly fit the pattern of faunal accidental dispersal. Unlike other animals, man has the power to set out intentionally upon the sea, to extend his period of survival by living off the sea, and, weather and currents permitting, to direct his course to the extent of his navigational ability. Because of these factors, the chances of man's reaching and occupying island after island, even in the vast Pacific, are considerably greater than the chances of faunal dispersal from island to island by natural means, and the time required would of course have been very much less.

Environment and Human Dispersals

The essential environmental requirement for the inception of either a migratory or an accidental voyage is a littoral where a maritime culture can arise. However, the factors that precipitate the two kinds of voyage are quite different. The planned voyage presupposes navigational knowledge, equipment for sailing on deep waters, and an urge to seek new lands. Ascertaining the possible points of origin of such planned voyages in the Pacific is basically a problem of determining the locations of prehistoric advanced maritime cultures bordering on the Pacific. In this respect the problem is one of anthropological interpretation and appraisal of circum-Pacific cultures. Such an appraisal has hardly been begun.

As for accidental voyages, these could originate within any maritime culture where there was any kind of seagoing craft. Thus, the possible points of origin of accidental voyages into Polynesia are much more numerous and cover a much greater geographic area than the possible points of origin of intentional voyages. The craft of many primitive maritime societies were suitable only for inshore cruising, but the fact remains that such craft could have been blown into open seas. Although the mortality rate would have been high, there would always have been a chance of survival. Animals have been able to survive on floating trees and other objects,[30] and certainly primitive fishermen, with their knowledge of the sea and their ability to live off it, would stand a considerably greater chance of survival.

Although maritime cultures have developed around the greater part of the perimeter of the Pacific, not all of these culture areas are equally likely points of origin of accidental voyages, because winds of the strength to blow a vessel into unknown seas occur less frequently in some of these areas than in others. Principal among such winds are hurricanes, or typhoons, and gales.

The dominant area of hurricane occurrence in the Pacific is in the western part, immediately to the north and to the south of the equator (Fig. 8–1). Most of the hurricanes that occur to the north of the equator develop over an area between 120 and 160 degrees west longitude. Passing to the north of New Guinea, the Celebes, and Borneo, some of the hurricanes move west and north, striking the mainland of Southeast Asia. Others move first to the west and then veer north and northeast, striking the coast of China, the Philippines, and Japan before advancing into the north Pacific.

To the south of the equator, hurricanes develop over an area from 160 degrees west longitude to 160 degrees east longitude. These pass over the southern islands of Melanesia, east of New Guinea; over the southwestern area of Polynesia; and, to the west, over Australia. Far less frequent but of equal importance are the hurricanes that occur off the Pacific coast of Guatemala and Mexico, as well as those that occur occasionally well off the coast of South America.[31]

Less violent than hurricanes, but capable of driving a ship into unknown seas, are gales with winds of 43½ kilometers (27 miles) per hour or more. Figure 8–2 shows the extreme equatorward distribution of those areas of the Pacific in which winds of this force, or greater, have made up 5 per cent or more of the wind observations (at 12 noon, Greenwich time) for any single month of the year.

The area from New Guinea to Borneo is largely free of such winds, as it is of hurricanes. However, the Asiatic mainland from Southeast Asia northward, as well as the islands of Micronesia and the northern Philippines to Japan and northward, is subject to such winds during at least one month of the year. To the south, the equatorward limits of winds of fresh-gale strength extend from, roughly, Cape Flattery, Australia, northeastward around New Guinea to the Solomon Islands, and eastward in a sinuous line to include most of the southern Polynesian islands south of 12 degrees south latitude. Along the coast of South America, the equatorward limits of such winds in the Southern Hemisphere is in the approximate latitude of Santiago, Chile. In the northern hemisphere the winds have a continuous distribution along the coast from Alaska down to the vicinity of Santa Barbara, California, and there is an isolated area of occurrence at, and to both sides of, the Isthmus of Tehuantepec.

From the premise that weather phenomena of this type are the principal cause of accidental voyages, we may logically reason that the likelihood of occurrence of such voyages is greatest in those areas of the Pacific where both hurricanes and gales occur. Thus we conclude that Southeast Asia, the Philippines, the lands washed by the East China Sea between Formosa and Japan, the lands touched by the Sea of Japan, the east coast of Australia, and Melanesia south and east of the Solomons are the most likely areas of origin of accidental voyages, as both typhoons and gales occur frequently in these regions. Almost as likely an area of origin, in the east Pacific, is the south coast of Oaxaca and Chiapas, Mexico. Hurricanes occasionally occur north of this region to Baja California, and gales occur along the West Coast of the United States and Canada, roughly from Santa Barbara northward, and along the south-central coast of Chile from Santiago to the Chonos Archipelago. Thus, these coasts might have been areas of origin of accidental voyages.

At this point one is tempted to postulate that the Asiatic areas where hurricanes and gales are most frequent are *the* sources of the people and culture of Polynesia. They may indeed be the dominant sources. However, the eastern coast of the Pacific has its area of storms, and since it is accidental dispersal that we are considering, these areas also must be considered.

Disregarding planned voyages, we can see that any Pacific-coast area where hurricanes or gales are frequent, and where man has had a maritime orientation, might have been a source of Polynesian culture. Whether or not it was, when (if at all) its influence was felt, and where the island that received this influence was located are matters that can be determined only through comparison of the culture history and cultural remains of these many coastal and island localities, and through study of the ocean currents and winds between these coasts and islands.

The course of a vessel on a planned voyage, as well as that of one accidentally sailing unknown seas, would have been largely dependent upon the natural elements. Ocean currents and major wind systems probably would have governed the course of the vessel that was lost and would have influenced the choice of direction in the case of a fully operational primitive craft. Heyerdahl has justifiably laid great stress on the effects of these systems in determining the principal over-water routes of prehistoric migrants.[32] However, remarkably little attention has been paid to the opportunities presented by the monsoonal windshifts of the western Pacific and the spiraling winds that accompany the eastward passage of cyclonic storms in both hemispheres. In the Northern Hemisphere these winds blow in a counterclockwise path around the center of a low-pressure storm, so that winds on the equatorial quarter of a cyclonic disturbance tend to blow from the west or northwest—that is, in a direction opposite to that of the easterlies, which they frequently displace in the subtropics and higher latitudes of the tropics. In the Southern Hemisphere the circular path of cyclonic winds is clockwise, resulting in a similar displacement of marginal easterlies by westerly winds. Thus, in the subtropics of both hemispheres, in spite of opposing ocean currents and the normally dominant easterlies, a craft could be borne to the east by a cyclonic disturbance.[33] Thus, the assumption that simple craft could have made a west-to-east crossing only in the cold, higher latitudes of westerly winds and currents is not valid. Because the dominance of the easterlies in the higher tropical latitudes is a seasonal matter, I have indicated in Figure 8–3, by directional arrows, those regions where easterly winds account for 60 per cent or more of the observations for every month of the year. These areas, then, may be said to be virtually dominated by the easterlies the year round; undirected craft entering these regions would normally follow the path of the arrows, while craft under sail could also move easily to the north or south. However, on the poleward edges of these easterlies, movement to either the east or the west would have been possible, depending upon the season. That primitive sailing craft could have made headway against the easterlies is not denied, but to do so would have been time-consuming, and it is obvious that, in sailing an unknown sea, the more ocean the vessel covers each day, the greater is the chance of discovering land and, therefore, the greater is the chance of survival.

Although one might reasonably expect those Polynesian islands that are closest (geographically, and from the standpoint of wind and current) to a continental mass to exhibit the greatest number of cultural parallels, islands are, of course, but dots in the Pacific. Migrants as well as involuntary voyagers might unknowingly pass them by and transplant themselves a thousand miles deeper into the Pacific. In this respect Polynesia differs from most other culture areas of the world, for the possible sources of human cultural influences, on any Polynesian island, are not necessarily influences from adjacent areas. For this reason, culture traits or complexes may not exhibit a continuous island-to-island distribution from their point of origin.

Because there are many possible sources of Polynesian origins and many possible routes of travel, we can never expect to gain a complete picture of this maritime activity. We can learn a good deal, however, through greater understanding of the nature of such voyages and through appreciation of the fact that cultural changes attributable to an unplanned voyage were matters of chance, and that numerous, sometimes evanescent, factors governed the acceptance or rejection of ideas and objects transported in this manner.

Since voyagers, especially involuntary voyagers, may have implanted only fragments of their culture in Polynesia, the picture is a mosaic, and it will indeed be difficult for the culture historian to determine the sources of the pieces. Incorporation of a cultural component into a particular island culture would have resulted in new uses; the component would seem aberrant in its new association, and the culture historian would be likely to overlook it as evidence of cultural diffusion.

The problem of Polynesian origins and cultural diffusions is far too complex to be solved from immediately available evidence. The variety of possible sources and of possible routes is infinite. What routes were chosen, and what routes were forced upon what number of undirected vesesls, can never be completely known. Only through a broad view, and an awareness of these facts, can we eventually arrive at a better interpretation of the meaning of the data derived from archaeological evidence, ethnographic collections, and historic and ethnological observations.

Although the results of many of the more recent excavations are still being compiled, a few have been published. Happily, many of these reflect the authors' understanding of the complex and nascent nature of Polynesian archaeology. Conclusions are limited to the problem at hand, and comparisons, if any, are made only to point up the need for excavations in other, possibly related, areas. Although the field still suffers slightly from pronunciamentos concerning Polynesian origins, the more thoughtful Pacific archaeologists are awaiting the day when enough excavated objects have been accumulated to provide a sound basis for the reconstruction of Polynesian prehistory.[34]

NOTES

1. T. Heyerdahl, *American Indians in the Pacific* (Allen and Unwin, London, 1952).
2. P. H. Buck, *Vikings of the Pacific* (Univ. of Chicago Press, Chicago, 1960), p. 26.
3. *Ibid.*, p. 43.
4. *Ibid.*, pp. 310, 316.
5. *Ibid.*, p. 309.
6. Heyerdahl, *American Indians in the Pacific, op. cit.*, pp. 705–7.
7. A. L. Kroeber, *Anthropology* (Harcourt, Brace, New York, 1948), pp. 215–16.
8. R. T. Simmons, J. J. Graydon, N. M. Semple, E. I. Fry, *Am. J. Phys. Anthropol.*, **13**, 687 (1955).
9. R. T. Simmons and J. J. Graydon, *ibid.*, **15**, 365 (1957).
10. R. C. Suggs, *The Island Civilization of Polynesia* (Mentor, New York, 1960), p. 214.
11. *Ibid.*, pp. 35, 216.
12. E. Goldschmidt, in *Abstracts of Symposium Papers, 10th Pacific Science Congress* (Honolulu, 1961), p. 99.
13. R. T. Simmons, *Oceania*, **32**, 209 (1962).
14. W. C. McKern, *Bernice P. Bishop Museum Bull. No. 60* (1929).
15. R. Duff, *Canterbury Museum Bull. No. 1* (1950).
16. I. Rouse, in *Anthropology Today* (Univ. of Chicago Press, Chicago, 1953), p. 58.
17. K. P. Emory and Y. H. Sinoto, *Bernice P. Bishop Museum Spec. Publ. No. 49* (1961).
18. T. Heyerdahl and A. Skjölsvold, *Soc. Am. Archaeol. Mem.*, **12**, No. 2 (1956).
19. T. Heyerdahl, in "Archaeology of Easter Island," T. Heyerdahl and E. N. Ferdon, Jr., Eds., *Monographs of the School of American Research and Museum of New Mexico, No. 24, part 1* (1961), pp. 15–19.
20. R. C. Suggs, *Anthropol. Papers, Am. Museum Nat. Hist.*, **49**, 1 (1961).
21. R. Shutler, *Asian Perspectives*, **5**, 193 (1961).
22. R. Duff, in *Anthropology in the South Seas*, J. D. Freeman and W. R. Geddes, Eds. (New Zealand, 1959), pp. 121–47.
23. K. P. Emory, W. J. Bonk, Y. H. Sinoto, *Bernice P. Bishop Museum Spec. Publ. No. 47* (1959).
24. R. Mulloy, in "Archaeology of Easter Island," T. Heyerdahl and E. N. Ferdon, Jr., Eds., *Monographs of the School of American Research and Museum of*

New Mexico, No. 24, part 1 (1961), pp. 93–180; C. S. Smith, *ibid.*, pp. 181–219; E. N. Ferdon, Jr., *ibid.*, pp. 223–29.

25. A. Sharp, *Ancient Voyagers in the Pacific* (Penguin, Harmondsworth, England, 1957), pp. 57–78; 21.

26. G. M. Dening, *J. Polynesian Soc.*, **71**, suppl. 137 (1962).

27. J. F. G. Stokes, *Proc. Pacific Sci. Congr. Pacific Sci. Assoc., 5th* (1934), vol. 4, pp. 2791–2803.

28. Buck, *op. cit.*, pp. 39–41, 68–69, 99–100.

29. G. G. Simpson, *Evolution and Geography* (Univ. of Oregon, Eugene, 1962), p. 24.

30. E. C. Zimmerman, *Insects of Hawaii* (Univ. of Hawaii Press, Honolulu, 1948), vol. 1, p. 57.

31. S. S. Visher, *Bernice P. Bishop Museum Bull. No. 20* (1925), pp. 56–60.

32. T. Heyerdahl, *Am. Antiquity*, **28**, 482 (1963).

33. Sharp, *op. cit.* pp. 88, 24.

34. I thank Drs. Emil W. Haury and Raymond H. Thompson of the department of anthropology, University of Arizona, for their critical reading of the manuscript.

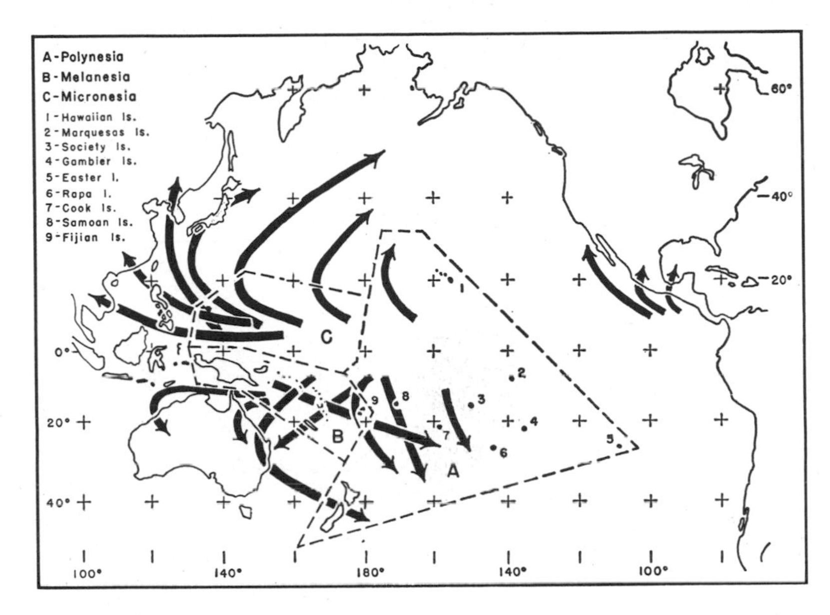

Fig. 8–1. A greatly simplified representation of paths of hurricanes in the Pacific (The generalized routes of the hurricanes were plotted from information in S. S. Visher, *Bernice P. Bishop Museum Bull. No. 20* [1924], and O. W. Freeman, in *Geography of the Pacific* [Wiley, New York, 1951].)

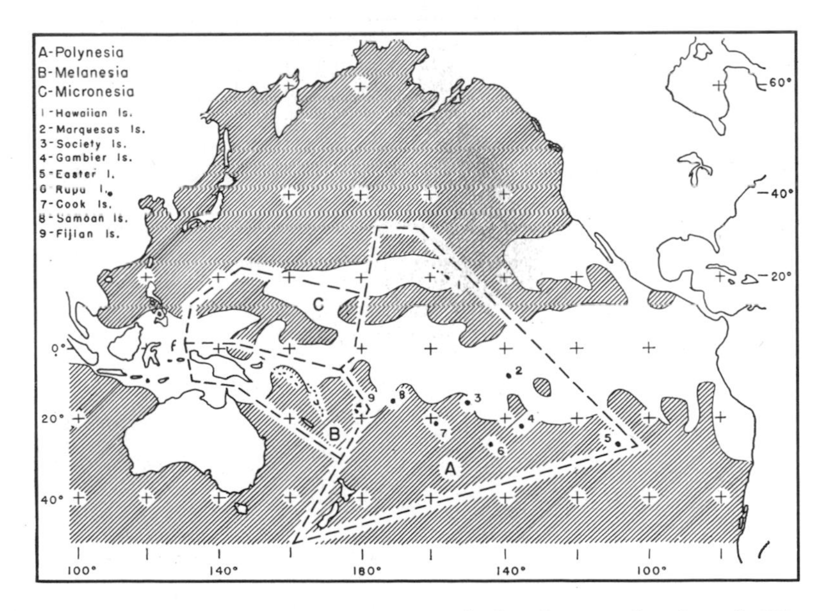

Fig. 8–2. Average maximum equatorward distribution of gales of 43½ kilometers (27 miles) per hour, or more, in the Pacific. (The data were obtained from W. F. McDonald, *Atlas of Climatic Charts of the Oceans* [U.S. Weather Bureau, Washington, D.C., 1938].)

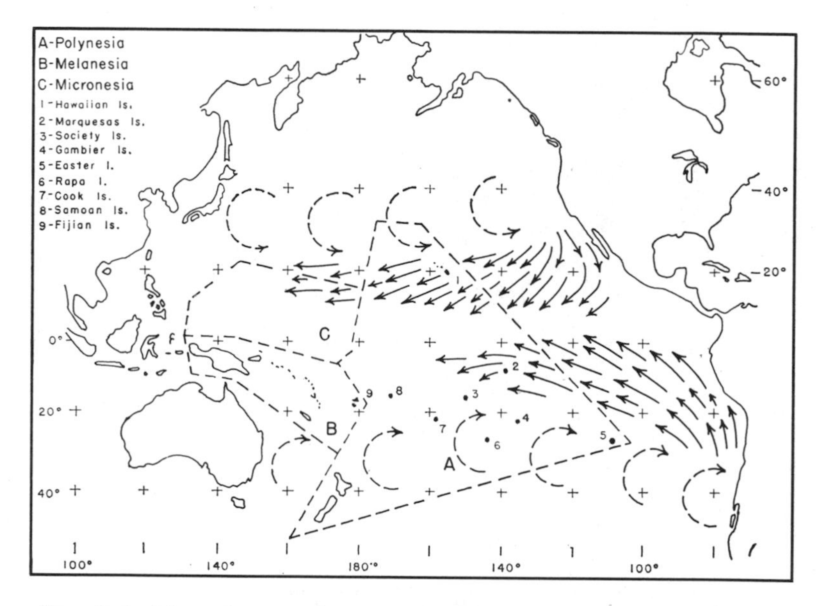

Fig. 8–3. Map showing the approximate equatorward limits (curved, dashed arrows) of the westerly-wind effect from cyclonic storms in the Pacific during the period of low sun in each hemisphere. The solid arrows indicate the extent and the average direction of the easterlies during at least 60 per cent of every month of the year. (Data for the compilation of dominant easterly winds were obtained from W. F. McDonald, *Atlas of Climatic Charts of the Oceans* [U.S. Weather Bureau, Washington, D.C., 1938].)

Chapter 9

Eskimos and Aleuts: Their Origins and Evolution

William S. Laughlin

Eskimos have been known to the European world since A.D. 1000, when Leif Erikson found them on the coast of Labrador. The Greenlandic Eskimos and their relatives the Aleuts and Eskimos to the west and north were by no means newcomers to North America and Greenland. At the time of their discovery by Erikson they had been living in the New World for over 4,000 years.

In the more than five millennia that have passed since these Mongoloid peoples migrated from Siberia to Alaska, they have worked out a remarkable system of adaptation to a series of diverse environments, ranging from the harsh climate and poorly lighted terrain of the polar regions to the more moderate marine environment of the Aleutian Islands. Their adaptations, physiological and cultural, have enabled them to occupy the entire coast of northern North America, from Alaska to the northeastern coast of Canada, and the entire coast of Greenland. Because of their antiquity, the evolutionary changes in successive groups, and the variations throughout their distribution over a long coastal area, the Eskimos and Aleuts provide a unique opportunity for studying microevolution, population history from the standpoint of genetics, and biological and cultural adaptation. Such studies have been facilitated by the excavation of stratified village sites rich in durable artifacts and in faunal remains and human skeletons, many of them showing direct continuity with living groups. Moreover, studies of the blood groups of living individuals show a basic similarity between Aleuts and Greenlandic Eskimos, as distinguished from American Indians, who are much less Mongoloid. In addition, the linguistic diversity within a single stock provides an invaluable means of tracing the impressive florescence of these energetic and practical peoples.

Linguistic Characterization

The three distinctive languages spoken by members of this stock differentiate them from American Indians and indicate a connection with Siberian Mongoloids that is confirmed by the serological and morphological evidence. The two Eskimo languages and the Aleut language differ to the extent that an individual who speaks Eskimo cannot understand Aleut, but phonologically and grammatically they are quite similar. Rasmus Rask first noted the basic similarity between Greenlandic Eskimo and Aleut in 1819. Inyupik Eskimo is spoken in northern Alaska and Canada as well as in Greenland. The second Eskimo language, Yupik, is spoken in western Alaska south of Unalakleet on Norton Sound—the area where most of the Siberian Eskimos live—and by the Eskimos of St. Lawrence Island in the Bering Sea. There are at least four dialects of Yupik, and communication between individuals who speak different dialects is difficult. Edward Sapir noted this diversity and formulated a useful working principle: "The greater the degree of linguistic differentiation within a stock, the greater . . . the period of time that must be assumed for the development of such differentiation."[1] He concluded that the divergence between Aleut and Eskimo and the diversity of dialects within them pointed to southern Alaska as the earliest center of dispersion. Later, a method of dating on the basis of linguistic characteristics (called glottochronology or lexicostatistic dating) was applied by Gordon H. Marsh and Morris Swadesh,[2] who estimated that the Aleut and Eskimo languages had separated some 3,000 years ago. They based this estimate on an assumed retention rate and on the number of words Aleut and the Eskimo languages now have in common. This date has more recently been revised to about 4,500 years ago, the subsequent separation of the two major divisions of Eskimo being placed close to 1,400 years ago.[3] The possibility that there was a connection with the Siberian languages Chukchi-Koryak and Kamchadal some 5,000 years ago has recently been suggested.[4] Other linguists, though not necessarily proponents of the glottochronological method, generally agree that the differentiation took place in southern Alaska and that the speakers of Yupik moved from south to north.[5] These estimates fit well with findings which indicate that, at the time of the exploration of Vitus Bering in 1741, southwestern Alaska was the area of highest population density.[6] At that time some 7,000 Koniag Eskimos lived on Kodiak Island, and 16,000

Aleuts inhabited the Aleutian Islands and the western part of the Alaska peninsula. A few hundred Eskimos lived in the interior of the Alaska peninsula, in northern Alaska, and in the Barren Grounds of Canada.[7]

Anangula

The archaeological evidence also points to southern Alaska as the homeland of the Eskimo-Aleut stock. The oldest known site in the Eskimo-Aleut world, according to dates obtained by the radiocarbon technique, is the lamellar-flake site on Anangula Island, five miles off the shore of Umnak Island, opposite the present village of Nikolski on the Bering Sea side. Three dates have been obtained from a cultural level underlying two layers of ash and humus; these are 8,425 ± 275 years ago (Isotopes Inc. specimen I-715), 7,990 ± 230 years ago (Isotopes Inc. specimen I-1046), and 7,660 ± 300 years ago (U.S. Geological Survey specimen W-1180). Prismatic blades from two to twelve centimeters long, polyhedral cores from which small blades have been struck, core tablets, retouched blades with chipping on one surface to form scrapers, knives, gravers, and burins, and many refuse flakes make up the bulk of this industry[8] (see Fig. 9–1). In this unifacial industry, imported obsidian, greenstone, cherts, and other siliceous, fine-grained stone were used. The obsidian apparently came from the Cape Chagak region ("chagak" is the Aleut word for obsidian) on the north end of Umnak Island, probably the only source of obsidian in the Aleutians. Who made these blades and what they were used for can only be inferred in the absence of human skeletons and faunal remains. However, the location suggests a marine economy. Richard Foster Black, who is currently investigating the geology of this area, has noted that, whereas a 10-meter depth of ocean now separates Anangula from Umnak, this area would have been dry land 8,000 years ago, and that 12,000 years ago Umnak was an extension of the Alaska mainland, having formed, in fact, the end of the Alaska peninsula and the southern corner of the old Bering platform. Early migrants from Asia could have walked along the southern edge of the platform and reached the world's richest hunting ground without losing contact with the sea upon which they depended for their principal food supply. St. Lawrence Island, Nunivak, and the Pribilof Islands, like Umnak and Anangula, are remnants of the higher hills on this now submerged platform. The pass (now the strait) that separates Umnak from the Islands of the Four Mountains was never closed during the Pleistocene. People living on the Umnak-Anangula corner of this platform could hunt the annually migrating whales and fur seals, as well as the resident sea otters, hair seals, and sea lions. The founders of

Anangula may have been a migrant Bering-platform population who reached the site on foot (see figs. 9–2 and 9–3).

Continuity with the village site of Chaluka is suggested by the presence of unifacial tools fashioned on lamellas or prismatic blades. A similar manufacturing technique and similar materials are involved, though the polyhedral cores have not been found at the Chaluka site. The frequency of occurrence of lamellar tools declines rapidly from the bottom level to the top of the Chaluka site, and the frequency of occurrence of obsidian and greenstone declines as well;[9] these matters were defined more fully by the 1962 excavations of C. Turner, G. Boyd, A. McCartney, L. Lippold, and J. Aigner.

This true core and blade industry is more like the lamellar industries in Japan and Siberia between 9,000 and 13,000 years ago than like the somewhat later Denbigh Flint complex of Norton Sound.[10] M. Yoshizaki, who has excavated at Anangula with McCartney and R. Nelson, suggests that the Anangula materials are most like the tools of the Sakkotsu microblade industry of Hokkaido, the Araya site on Honshu, and the Budun site in Siberia. They seem more clearly Asiatic than tools of the Arctic small-tool tradition, which extended over much of Alaska and provided the base for the Dorset culture of the eastern Canadian Arctic and related cultures of Greenland.[11, 12] Both W. Irving and Yoshizaki place them, with confidence, in a separate province from tools of other Alaskan industries, though there are some similarities to materials of the Campus Site at College, Alaska. I infer that the bone artifacts occurred in approximately the same proportion and were of approximately the same kind as those associated with the lamellar tools in the lowest levels of Chaluka, and that the Anangula industry was that of a marine-based people similar to the Palaeo-Aleuts of 4,000 years ago, a long-headed Mongoloid group. American Indians of comparable antiquity, proto-Mongoloid in appearance, have already been found thousands of miles to the south.

Chaluka

Four thousand years of continuous history are lavishly illustrated in the deep, stratified village site of Chaluka, which now forms the southern margin of Nikolski, a village of about fifty-five Aleuts. This is one of the few sites in the Arctic or subarctic which contains the requisite materials for interdisciplinary study of evolution, prehistory, and ecology: human skeletons, abundant artifacts, and remains of buildings and of faunas, all superimposed in such a way as to provide a record of several thousands of years of events in one place and evidence of a connection with the living inhabitants. The reasons for the long-term occupation of this site are apparent. There were fresh-water lakes, vital for spawning salmon (these lakes may not have existed at the time of initial occupation); an enclosed bay with front reef, which provided protected waters for fishing during storms at sea; large reefs, exposed at low tide, which were rich in invertebrates such as sea urchins, mussels, whelks, limpets, and chiton, and in edible seaweeds, kelpfish, and octopus; offshore rocks and islands that provided cliffs for nesting cormorants and puffins, where they were protected from foxes (the Aleuts ate the eggs of these birds); a complex coastline that trapped driftwood, dead whales, and dead sea lions and provided diverse ecological niches attractive to sea otters, which like to live in extensive kelp beds. Cod and halibut could be caught from the shore, as well as from boats in the channels and pockets. This site is on a 10-meter beach of the postglacial thermal maximum and thus cannot be older than 5,000 years; the water level has probably changed very little during this period. The oldest date obtained by the radiocarbon method, 3,750 ± 180 years, is for a sample found above the sterile floor,[13] and an age of 4,000 years has been proposed for the site, on the basis of the lower limits of seven dates obtained by this method.

The earliest Palaeo-Aleuts of the Chaluka site used stone lamps, various kinds of harpoons and spearheads, fishhooks, unifacial (lamellar) tools, and ivory and bone labrets for lip decoration. Adze bits and whalebone wedges are evidence of a woodworking industry. Distinctive harpoon heads are fluted or channeled (Fig. 9–4) and slotted to receive straight-based bifacial chipped stone points. Stone points have been found embedded in sea lion bones in these strata. The elaborate carving of harpoon heads with circle-dot-and-line designs shows great artistry. The incompletely excavated foundations of houses indicate that these were oval, with coursed masonry; that one entered from the side; that there was a slab-lined hearth; and that ribs and mandibles of whales were used for rafters.

The artifacts in the upper levels include many tools and objects needed for life in a subarctic, marine environment, but none of these represent basically new categories, with the possible exception of barbs for fish spears. Many new types do appear. Two-piece bone sockets to receive long harpoon heads with stone points inserted in a basin rather than a slot are found, still in association with Palaeo-Aleut skeletal remains. The forms of fishhook shanks change. The older, elbow shank carved from whalebone is replaced by a curved shank carved from sea otter rib. In general, the artifacts at various levels are of the same categories: lamps, root diggers, splitting wedges, bird-spear side prongs, labrets, weights for fish lines, chipped stone knives, and flaking tools. Recently (only a few hundred years ago at most) four variations appeared at Chaluka: ground-slate ulus, shallow soapstone lamps, single-piece sockets to receive small barbed ivory harpoon heads used in hunting small mammals (see figs. 9–5 and 9–6), and a group of artifacts, including hats, that have been found in mummy caves. The Neo-Aleut skeletal type is associated with these recent artifacts, and this association poses the problem of rapid internal change versus migration.

There is no single change in kind or category of artifact over 4,000 years that appears to have made a detectable change in the system of adaptation or in the way of life. While it is entirely possible that such fundamental changes did in fact take place, it is significant that, despite sensitive excavation techniques and the recovery of large numbers of artifacts and faunal remains and of human skeletons in sufficient numbers to show how the people lived, what they lived on, and how they buried their dead, no major change is detectable. The later, bifacial tools have no obvious advantage over unifacial tools. Line holes in a few harpoon heads of the lowest levels are gouged out rather than circularly drilled. Such traits may be important time markers, but they have little significance from the standpoint of adaptation.

Faunal Remains

The faunal picture at the Chaluka site explains the rise of a large population in this area and the use of many tools. William G. Reeder[14] has sketched the basic relations between mammals, birds, fish, and invertebrate remains and is relating these to modern communities. As Victor B. Scheffer has noted, "There are in fact no small marine mammals of any kind."[15] Large marine mammals were available, and thus the Eskimos and Aleuts had an unusually rich source of food. In addition, their social habits and their numbers assured them a good food supply. Eskimo culture is primarily adapted to marine hunting but is sufficiently flexible to include means of hunting land animals as well.

The principal faunal remains excavated in 1962 (except for invertebrate remains) are shown in Table 9–1. The pinnipeds include harbor seals, fur seals, and sea lions. Though Aleuts on the Alaska Peninsula secured important numbers of walrus, sea lion remains are found more often in the Aleutians. More of the sea lion than of the seal skeletal remains are adults. This may reflect the selective use of hides of mature sea lion for making umiaks and kayaks, and the use of sea lion flippers for boot soles. The fur seal now migrate through the Aleutian straits to the Pribilof Islands, some 400 kilometers north of Umnak. Since they do not ordinarily haul up on beaches while in transit, it is probable that they were killed at sea from boats.

Cod and halibut are the principal fish represented in the 1962 ex-

TABLE 9–1

Preliminary summary of data for excavated bones, Chaluka, 1962. (L. Lippold, G. Streveler, and R. Wallen).

LAYER	BONES (NO.)	BONES (%)				
		BIRD	FISH	PIN-NIPED	SEA OTTER	CETA-CEAN
Upper I	3,626	27.8	44.2	20.4	6.8	2.0
Upper II	5,906	15.2	50.6	25.8	5.8	2.2
Upper III	10,858	19.6	7.3	47.0	25.3	0.1
Lower IV	1,127	29.8	13.3	39.8	16.6	.4

cavations, the cod constituting some 80 per cent of all the fish remains. Salmon are poorly represented—a finding which suggests that the great salmon runs of the last several hundred years had not started 4,000 years ago. Among the bird remains in these deposits, cormorant remains are the most common, with puffin and duck remains less frequent. Albatross remains occur at all levels, perhaps in somewhat greater number at the lower level. Either the habits of the albatross have changed and they formerly nested on Umnak Island, as claimed by Adelgertus de Chamisso,[16] or they were hunted down at sea. Sea urchin remains make up a large part of the deposits; apparently the sea urchin was the basic invertebrate in the diet, as the sea otter was a basic vertebrate. The importance of the sea urchin cannot be overestimated. Women, disabled men, and children could gather sea urchins and other invertebrates during the spring, when other food sources were depleted. Thus, starvation was avoided in many communities. The contrast between this situation and that in the central Arctic, where the inhabitants were unable to get food from the sea because of the impenetrable ice barrier, is reflected in the population profiles. The economic productivity of children, disadvantaged women, and elderly or disabled men is a major factor in community adaptation, with both genetic and cultural consequences.

The human skeletons (figs. 9–7 to 9–11) indicate an evolutionary change in all of southern Alaska—a change in head form from dolichocranial (breadth less than 75 per cent of length) to brachycranial (breadth more than 80 per cent of length). There are also changes in the occurrence of dental cusps and mandibular tori. One Palaeo-Aleut found at the bottom of the Chaluka site displayed vault thickening (hyperostosis cranii) characteristic of some anemias. All the skeletons found in the eastern end of the Chaluka site, with the exception of recent burials, belonged to the longheaded (dolichocranial) Palaeo-Aleut type. The most recent inhabitants—those whose remains are associated with knives of ground slate, with shallow stone lamps, and with single-piece sockets and short harpoon heads—are extremely broad- and low-headed. Aleš Hrdlička, who was the first to note that there were two physical types, one of which succeeded the other stratigraphically, termed the earlier type "pre-Aleut" and the later type "Aleut."[17] Marsh and I suggested the terms "Palaeo-" and "Neo-Aleut" for the physical populations, to indicate the continuity and similarity, and the possible evolution of the later from the earlier population.[18] Dental studies[19] and anthropometric studies of living individuals[20] establish a continuity between the most recent skeletons in Chaluka and the present-day inhabitants. Some characteristics of the earlier population appear in the western Aleuts. They are more narrowheaded than the eastern Aleuts, and they differ from them in frequency of occurrence of discontinuous traits. The pressing problem is that of securing enough skeletons from stratified and dated sites to confirm this apparent evolutionary change and to provide a basis for estimating the rate of change, as well as its extent. Parallel changes on Kodiak Island and along the Kuskokwim River indicate that there was both internal evolution and migration.[21, 22]

The record for 4,000 years of prehistory in the Chaluka site demonstrates that the styles of artifacts change more rapidly than ecological circumstances. The artifacts (harpoons and spears in particular) were necessary for hunting marine mammals, but the particular forms varied considerably. It apparently made no difference, either to the sea lion or to the hunter wielding the harpoon, whether the harpoon had a fluted, stone-tipped head or a four-barb whalebone head. Cod and halibut were caught equally well with an elbow-shank fishhook and a curved-rib-shank fishhook. The 16,000 Aleuts discovered by Bering in 1741 were adequate evidence of the faunal wealth of the Aleutians and of the efficiency of their system of exploitation.

Bering Strait Sequence

The time depth for Eskimo sites decreases as one proceeds from southwestern Alaska to the Bering Strait region. Prior to 2000 B.C. the Bering Strait cultures are represented primarily by lithic industries and yield little information about the way of life or the racial characteristics of the people. James L. Giddings has drawn attention to the persistence of tradition in the constituent areas: the Asiatic, Chukchi Sea, and Bering Sea areas. One site in particular is comparable to Chaluka; this is the great Kukulik mound of St. Lawrence Island, which shows continuous occupation from Old Bering Sea II times (about A.D. 300) to 1884. Walrus hunting, sealing, and whaling have been of continuing importance and, interestingly, the artifacts used in these activities show only a slow change in style and a gradual loss of the art of engraving over 2,000 years. As Giddings comments,[23] "No basic change appears abruptly in the pattern of subsistence, and only a few exotic elements were introduced before the coming of Europeans." The significance of this continuity and stability in these three Bering Strait areas for studies of evolution lies in the indication that there has been little real migration of groups. It may be that many of the later physical variations are due, rather, to internal changes, with gene flow resulting from exchange of mates across the boundaries of isolate groups making a relatively small contribution.

Important to an understanding of the early interconnections between Mongoloids in the Eskimo-Aleut area as a whole is recognition that early cultural traditions were more widespread than the relatively localized cultures of the last several hundred years. Of critical importance is the Arctic small-tool tradition, defined by Irving, which provided roots for the development of the pre-Dorset and Dorset cultures of the eastern Canadian Arctic and Greenland.[24, 25, 26] This tradition, which includes the Denbigh Flint complex, is characterized by large numbers of microblades struck from conical polyhedral cores and by burins, retouched burin spalls (also used as engraving tools), small bifacially retouched blades for insertion in the sides of harpoon heads, and medium-sized (4 to 10 cm.) biface points and knife blades.[27] This tradition appeared in Alaska and Siberia as early as 6,000 years ago, and reflections of it reached the Canadian Arctic and Greenland very early. Henry B. Collins, who initiated the early work on St. Lawrence Island and continued his investigations across the Arctic, points out that the Arctic small-tool tradition is "pre-Eskimo," but that the prefix *pre* in this case connotes "predisposed" or "leading up to."[28] Thus, the evidence, though meager, indicates that the people of the Dorset culture were Eskimo in their morphology.[29]

In general, the more northern cultures used the toggle-head harpoon more often than the harpoon with a simple detachable head that is used in southern Alaska. Toggle-heads presumably turn inside the animal with tension on the harpoon line (Fig. 9–12). (Much the same effect is produced with multi-barbed, detachable harpoon heads, especially those that are asymmetrical.) These harpoon heads are indispensable time markers, just as pottery is in the southwestern United States. They are more closely related to the way of life than pottery is, but there is still by no means a one-to-one correspondence. Other general characteristics of the northern cultures are their greater use of the umiak and their custom of hunting on the ice, with the sled drawn by dogs. Kayaks were inevitably more important in the south, where there is more open water. Hunting from kayaks has many advantages: kayaks provide speed and a means of rapidly scanning complex coastlines. Moreover—

and this is extremely important—the kayak does not have to be fed! The Aleuts and Koniags developed the kayak to a higher degree than any other members of this stock, and much of the material culture reflects the elaboration of open-sea hunting of mammals.

Adaptation to Cold and Glare

A common misconception about Eskimos is that they are fat or chubby. Measurement of the thickness of skin folds confirms the observation that they are in fact lean. Though muscular, with heavy bones, they have little fat even at advanced ages. They are medium-to-short in stature, with long trunks and short legs. The lower leg is particularly short. Their heads are large, and their hands are small-to-average in size. When they are fully clothed for protection against dry cold, only a portion of the face is exposed. Possibly the large face with broad jaws and the bulky clothing have contributed to the notion that they are fat.

Thermoregulation in the Eskimo is characterized by basal metabolism that is higher than clinical standards for normal metabolism and high in view of the lean body mass. When the Eskimos are fully clothed they are living in a tropical microclimate in which sweating accounts for most dissipation of the excess metabolic heat. Frederick Milan has found that Eskimos maintain their warmth even while lying on the winter ice, hunting seal. They probably have more sweat glands than members of other races, but too few counts have been made to confirm this. Blood flow to the hands and legs is greater when the limbs are cooled in water than it is in normal white controls. The perception of pain resulting from cold appears to be less acute in both children and adults than it is in other races, and the Eskimos have higher finger temperatures during cooling in cold air.[30]

Equally important in the adaptation to cold have been the material culture of the Eskimos and the Aleuts and the child-training practices. The ordinary clothing includes undergarments, pants, boots, mittens, and parka with ruff. Where hunting is done from kayaks, the clothing is made of waterproof materials, such as the esophagus of hair seal or sea lion, and the intestines of various mammals, such as whales, walrus, or seals. Parkas made of the skins of birds—cormorants, puffins, and auklets in particular—are used in all areas. Insulative materials such as dried grass and caribou skin are worn inside boots.

Glare from water, ice, or snow is minimized by the used of slit goggles or visors. In the Aleutian Islands and in southern Greenland, kayakers habitually wore visors or shades for protection against spray and glare. Eskimos who use sleds wear slit goggles instead of visors.

Selection for cold-adapted individuals has probably been extensive in the southern area, where there is greatest use of the sea. Heat loss is so greatly accelerated in cold water that a victim who has been immersed for a few minutes often cannot be saved even by rapid rewarming techniques. Ability to withstand wet cold for even a few additional minutes has often meant rescue by another hunter. On land, by contrast, a fully clothed person can survive many days of extreme cold. No studies on the heritability of resistance to cold have been made. There is no experimental evidence to suggest that the Mongoloid face, the long trunk, and the short legs are characteristics that have developed as a result of the climate.

Longevity

A critical variable, about which reliable information is slowly being accumulated, is that of age at death. In this respect there are large contrasts between isolates, with ascertainable genetic and cultural consequences. In general, it appears that the people who lived in the more harsh arctic environments died earlier. Those who inhabited ecologically richer areas—subarctic and more ice-free areas—lived much longer. A sharp contrast, for which the cultural context is known, is that of the extinct Sadlermiut Eskimos of Southampton Island (Northwest Territories), in the northern part of Hudson Bay, who died at a much earlier age than their Aleut counterparts in the eastern Aleutians. The Sadlermiut numbered 57 at the time they became extinct, in 1903. The Unalaska Aleuts (from the Fox Island district), already decimated by disease and massacre, numbered some 1,500 between the years 1825 and 1835. Their chronicler, I. Veniaminoff, reported age at death for 491 Aleuts.[31] These ages are therefore directly comparable with ages at death estimated from the skeletal remains of the Sadlermiut. As Figure 9–13 shows, the maximum age at death among the Sadlermiut was between 50 and 55 years. In marked contrast, the maximum age at death among the Aleuts was between 90 and 100.

Age at death affects not only such things as disease patterns but the genetic composition and cultural complexity of the group as well. There is more wastage in the group in which a higher percentage of the offspring die before the age of reproduction, and each generation is a less adequate sample of the preceding generation. Age at death is closely correlated with population size, and this in turn is correlated with the ecological base and the technological system.

The cultural consequences and the biological consequences may be considered separately for purposes of analysis. The greater overlap between generations associated with greater longevity provides more time for transfer of information. The experience of older people is stored in accessible form for a longer period in more "storage cells." The florescence of medical and anatomical knowledge among the Aleutian Islanders indicates a specific form of feedback. Difficult deliveries could be successfully managed because of the presence of old and skilled individuals,[32] and treatment of serious injuries enabled the injured individual to participate in group life, if not in active hunting, on his recovery. Artistic expression and the larger number of public ceremonials and myths are among the correlative benefits of greater longevity and greater population.

Various patterns of pathology are characteristic of the Eskimos and Aleuts, and in some of these conditions there is probably a high element of heritability. Spondylolysis of the lumbar vertebrae is especially common.[33] Developmental anomalies such as premature or irregular closure of cranial sutures are also common. Arthritis occurs frequently, and there is an interesting sex difference: in arthritis of the elbow, arthritic lesions occur much more frequently on the capitellar surface of the humerus in males and on the trochlear surface in females.

Blood-Group Evidence of Origins

The blood-group data, in addition to their value in studies of population genetics of small isolates, throw much light on Eskimo-Aleut affinities and origins. They show that the members of this stock are clearly distinguished from American Indians and more similar to Asiatic Mongoloids. The percentage of blood type B in Eskimo-Aleut isolates ranges from two to twenty-six, whereas in American Indians it is zero. The percentages in Asiatic Mongoloids are the highest in the world; those for the Chukchi (the lowest in the Asiatic Mongoloid group) are a little higher than the average for the Eskimos.[34] Indians who live in contiguous areas—the Tlingit, Athabascan, and Algonkin—have no gene for blood group B. The percentage of group B is especially low among the Eskimos of the central Canadian Arctic. Bruce Chown and M. Lewis[35] have suggested that this indicates a Dorset-culture residuum. Traces of earlier peoples should show through more frequently in areas of low population density, or in areas where the earlier population made a large contribution to the newer population, as in the western Aleutians. North American Indians have genes for blood groups A and O, and as in all Eskimos and Asiatic Mongoloids, the A is of subdivision A_1. Though Alaskan Athabascan Indians are low in group A, the highest frequencies of group A in the world are found among the Blood and Blackfoot Indians of Alberta and Montana.

An interesting cline shows the distribution of the frequencies of blood type MNSs. Chown and Lewis found the occurrence of type MS higher in the Copper Eskimos of the central Canadian Arctic and lower as one proceeds to the southeast. Significantly, as the Eskimos approached the Indians geographically they became genetically more dissimilar.[36]

Eskimos lack the Diego factor, which is found in highest frequency in Venezuela and reappears in Asia. All Eskimos secrete blood-group substance in their saliva, but many are nontasters of phenylthiocarbamide (PTC), in contrast to Indians, among whom tasters of PTC are fairly numerous. The Rh chromosomes R_1 and R_2 are common, and similarities between Aleuts and Greenlandic Eskimos are marked.[37] Both differ from Indians in having low rates of excretion of β-aminoisobutyric acid and a low incidence of haptoglobin Type 1-1.[38]

Discontinuous Variation

The component isolates and populations of the Eskimo-Aleut stock are self-defining in that they are breeding isolates. They generally choose their mates within their own groups. Eastern Aleuts mate with eastern Aleuts and Polar Eskimos mate with Polar Eskimos, through preference and because of proximity. These groupings exist, regardless of whether we choose to recognize or classify them. The Polar Eskimos are a classic example of a geographically isolated breeding isolate. When they were discovered, in 1818, they thought they were the only people in the world. In addition to cultural and geographic barriers to mating —that is, to gene flow across isolate boundaries—the factor of relative population size has played an important role in minimizing the effects of mixture between Eskimos and Indians. The size and density of the Eskimo populations provided genetic insulation against the smaller groups of contiguous Indians. The physical traits which characterize isolates may be divided into those which vary continuously (stature, intermembral proportions, size of the ascending portion of the mandible) and those which are discontinuous or not present in all the people (particular blood groups, fissural patterns on the teeth, various foramina and sutures). (See Fig. 9–14.)

Among the continuous traits in Eskimo and Aleuts are the very large cranium, the large flat face, the broad mandible with unusually broad ascending ramus (the mandible is broader in Eskimos and Aleuts than it was in Neanderthal man), and the medium-to-narrow nose. Earlier Aleuts and Eskimos had heads that were narrow in proportion to their length (head breadth is usually less than 75 per cent of the length, in cranial series). The greatest head breadth occurs on Kodiak Island, where, among the living, the head breadths are some 86 per cent of the head lengths. This value, the cephalic index, decreases both to the west and to the north. Trunks are long and legs are short, though the total height varies. Eskimos in the interior of northern Alaska and Canada are taller than those along the coast.[39] Interestingly, Eskimos and Aleuts grow over a longer period of time than people of other stocks.[40]

Although all Greenlandic Eskimos can be characterized as a single group, it is more profitable, in research at the microevolutionary level, to recognize and compare the constituent isolates within Greenland. Findings on variation between breeding groups can then be used for the study of traits as such, for inferring the direction of migration, and for estimating rates of change. Blood type B is rare among the Polar Eskimos, more frequent along the west coast, and most frequent among the Angmagssalik Eskimos of the southeast coast. No blood-group comparison with Eskimos of northeast Greenland is possible, for those Eskimos are extinct. However, in studies based on discrete traits of the skull they can be included, and thus it is possible to draw comparisons among peoples whose migrations were limited to coastal areas because of inland ice. The Greenland Eskimos could migrate only clockwise and counterclockwise, or in both these directions, from a single area of entry, and no mating between isolates on opposite sides of Greenland was possible (see figs. 9–15 and 9–16). If they moved in both directions, the terminal isolates should display the greatest differences (Table 9–2). Studying eight discontinuous traits observed in some six hundred skulls representing four isolates, Jørgen B. Jørgensen and I found that the northeast series and the southeast series did in fact show the greatest differences. We inferred, therefore, that the Eskimos migrated in two

TABLE 9–2

Frequency (in percentages) of discontinuous cranial traits in Greenlandic Eskimo isolates. The percentages, based on a total series of 293 male skulls that J. B. Jørgensen and I studied in the Laboratory of Anthropology, University of Copenhagen, provide the basis for estimating similarity, as illustrated in Figure 9–16.

TRAIT	NORTH-WEST	SOUTH-WEST	SOUTH-EAST	NORTH-EAST
Dehiscences	26	32	19	26
Parietal notch bone	22	21	17	14
Supraorbital foramina	60	58	59	48
Mandibular torus	69	65	44	90
Palatine torus	36	32	24	9

directions around the coasts of Greenland.[41] This conclusion is supported by ethnological and archaeological evidence. As early as 1909, Franz Boas,[42] on the basis of similarities between artifacts found in Greenland and artifacts in Canada and Alaska, suggested that a migration movement north around Greenland had taken place. Through measurement of continuous traits, the northeast and southeast series have been identified as the terminal isolates. Measurements for the northeast series are the largest in Greenland, and those for the southeast series are the smallest. The cranial samples represent migrations after A.D. 1000 and roughly indicate the extent of differences which may occur between isolates in some eight hundred years. The southern Norse colony on the southwest coast of Greenland was raided by Eskimos in A.D. 1379, and it disappeared about A.D. 1500. Eskimos who had come from the west coast had been living in southeast Greenland no more than four hundred years when they were discovered in 1884.

Differences between more distantly related peoples are larger, as would be expected (Table 9–3). The mandibular torus (Fig. 9–17) occurs most often in Mongoloids. It also occurs in American Indians and in Europeans. Among the latter groups, however, the proportion of palatine tori is greater. This suggests a different mode of inheritance.[43]

Sinanthropus and Modern Mongoloids

The time depth for contemporary Mongoloid types is short, perhaps on the order of 10,000 to 15,000 years. The record of changes within the last several hundred years is considerable. Therefore, the finding that Middle Pleistocene *Sinanthropus pekinensis* displays traits that recur in Mongoloid and related populations, such as American Indians and Polynesians, is of major importance.

It has not been definitely established that any of the fossil men of China are Mongoloids. The most frequently mentioned as being so are three skulls from the Upper Cave of Chou Kou Tien, in north China. They are thought to be late Pleistocene, but they are probably no older than early American Indian remains such as the "Midland Woman," to which a date earlier than 8000 B.C., and possibly as early as 18500 B.C., has been assigned. These three skulls are quite different from each other and have been individually compared to skulls of Melanesians, Europeans, and Eskimos. The best appraisal that can be made is that they resemble "unmigrated American Indians."[44] Other fossil men of China do not look like Mongoloids of the last 5,000 years. The evidence from China indicates that modern Mongoloids are a relatively recent development.[45]

Japan offers no early materials that can be categorized as Mongoloid.

TABLE 9–3

Frequencies (in percentages) of discontinuous traits in distantly related races. The relatively large differences between (1) Norse and (2) Eskimos and Indians and the smaller differences between Eskimos and Indians parallel the anthropometric and serological differences. Caucasoids and American Indians both have an excess of palatine tori over mandibular tori, in contrast to the Mongoloid Eskimos. (No submedium or ambiguous tori are included in these series.)

	MEDIEVAL NORSE IN GREENLAND		ARIKARA INDIANS OF SOUTH DAKOTA		ESKIMOS OF GREENLAND	
	♂ ($N=38$)	♀ ($N=43$)	♂ ($N=60$)	♀ ($N=40$)	♂ ($N=293$)	♀ ($N=291$)
Dehiscences	6	1	29	43	27	36
Parietal notch bone	15	17	10	12	21	15
Supraorbital foramina	16	38	50	59	59	62
Mandibular torus	37	41	0	0	67	47
Palatine torus	59	58	29	44	32	36

The Pleistocene remains are fragmentary. Ushikawa Man has been assigned a date in the Upper Middle Pleistocene, but the fossil consists only of a portion of the left humerus. The Mikkabi skull fragments are Upper Pleistocene, but no racial assignment is possible.[46] The earliest definitely Mongoloid remains in Japan are from the last few thousand years.

Franz Weidenreich observed that *Sinanthropus pekinensis* displayed twelve traits found in Mongoloids, in which category he included Polynesians and American Indians. Three of the twelve fit into the category of traits that may or may not occur in Mongoloids—the mandibular torus, shovel-shaped incisors, and the Inca bone (figs. 9–18 and 9–19). Two other traits of the twelve are also relevant. Weidenreich noted that in all the temporal bones there was a well-marked notch (parietal incisure). This is common in modern Mongoloids, though it is not limited to them. In many of the fossils a separate bone is found at this site (Fig. 9–20); these separate bones can simply be considered examples of extremely well-marked notches. There are slits in both tympanic plates of Skull III. These slits occur in the same area as the tympanic dehiscence of modern Mongoloids and are probably related to it (Fig. 9–21). The auditory exostoses described by Weidenreich occur most often in American Indians; they do not occur in Eskimos or Aleuts. When we review these traits in *Sinanthropus* and in modern Mongoloids, keeping their general morphology in mind, we cannot consider them grounds for regarding *Sinanthropus* as a Mongoloid. On the other hand, these traits are additional evidence of similarity between *Sinanthropus* and modern man, especially Mongoloids and American Indians. Further evaluation must wait for comparable data from the other representatives of the erectus stage in Java, Africa, and Europe. In discussing the nomenclature and classification of *Sinanthropus*, Weidenreich remarked: "It would be best to call it '*Homo sapiens erectus pekinensis.*' Otherwise it would appear as a proper 'species,' different from '*Homo sapiens*' which remains doubtful, to say the least."[47]

Recent Changes

There is good evidence from Japan, Alaska, and Greenland that appreciable changes in morphology have taken place within a relatively short period. Makato Suzuki has documented differences between protohistoric times (the fourth to eighth centuries A.D.) and modern times.[48] The ancient Japanese had long heads, broad faces, and wide, flat nasal roots, and they were prognathic. The later Japanese had rounder heads, narrower faces, and narrower and higher nasal roots, and they were less prognathic. The possibility that there was admixture with other races is slight. By the seventh century the Japanese people numbered six million; this density of population would, of itself, have been an effective barrier against the effects of mixture with immigrants.

The mean cranial index for Palaeo-Aleut males is 74, and the mean cephalic index for Aleuts now living in the same area is 85. Subtracting 2 index units from the latter value (a step that is necessary in comparing cephalic indexes of the living with cranial indexes), we find a difference of 9 index units, representing a large change in the ratio of length to breadth. The cranial index for early male Koniags of Kodiak Island is 77, in marked contrast to the index of 86 for later Koniags. A similar but smaller change has taken place in east Greenland, where there can be no question of admixture.[49] Similar changes, and also an increase in stature, are reported for many series of American Indians.[50] Local migrations may explain the change in particular places but cannot explain the change for all the areas involved. The question of local evolutionary changes must be given more attention.

Summary

The emerging picture for the immediate origin of New World Mongoloids, the Eskimos and Aleuts, is that of a Bering platform inhabited by contiguous isolates stretching from Hokkaido around to what is now Umnak Island, probably some 10,000 to 15,000 years ago. The lithic similarities between Anangula and Hokkaido and the similarities in human morphology suggest this. The probable linguistic relationship between Eskimo and Aleut on the one hand and Chukchi, Koryak, and Kamchadal on the other is in general agreement with this picture and raises the possibility that the Yupik-speaking Eskimos of Siberia and St. Lawrence Island may be derived from populations that formerly lived on the Bering platform and withdrew toward their present locations as the platform was inundated. Diffusion of traits, both genetic and cultural, from the center in southwestern Alaska became of increasing importance as the population differential between south and north became greater. Once Bering Strait became a channel, major migrations ended; successional continuity is indicated wherever deep stratified sites are found. Palaeo-Indians (proto-Mongoloids or semi-Mongoloids) were clearly established in South and Central America before 10000 B.C. Their separation from ancestral Eskimo-Aleut-Chukchi Mongoloids was probably insured by differences in economic adaptation and therefore by differences in their routes of migration into the New World. The land bridge that connected Siberia and Alaska during early Wisconsin time, as early as 35,000 years ago and as late as 11,000 years ago, was more than 1,000 miles wide. The ancestral Indians, with their land-based economy, could have crossed often, following big game, without coming in contact with the Mongoloids, who worked their way along the coastal edge of the reduced Bering Sea.

Upon reaching the end of the Bering platform, the Umnak Island of today, the Mongoloids flourished, owing to the richness of the marine fauna. As deglaciation proceeded from west to east, they spread in two directions, following the retreating ice and setting out in boats toward the western Aleutian Islands. The earliest known Aleut skeleton is some 4,000 years old. Early Kodiak Eskimo skeletons slightly less old are easily distinguishable from the Aleut skeleton. The populations have not become demonstrably more similar, but they have undergone some parallel changes.

As a distinctive group in their present form, Mongoloids represent a recent evolutionary development that has occurred within the past 15,000 years. They do share more discontinuous traits with Middle Pleistocene *Sinanthropus* than members of any other living racial divisions, though *Sinanthropus* is clearly different from a modern Mongoloid. Inferences concerning long-term connections must remain tentative in view of the small number of fossil remains, the great time spans, and the deficiencies in our knowledge of the modes of inheritance of many traits. However, when we find that significant differences have developed, over a short time span, between closely related and contiguous peoples, as in Alaska and Greenland, and when we consider the vast differences that exist between remote groups such as Eskimos and Bushmen, who are known to belong within the single species of *Homo sapiens*, it seems justifiable to conclude that *Sinanthropus* belongs within this same diverse species.

NOTES

1. E. Sapir, *Can. Dept. Mines Geol. Surv. Mem. 90, Anthropol. Ser.*, **13**(**1916**), 76 (1916).

2. G. Marsh and M. Swadesh, *Intern. J. Am. Linguistics*, **17**, 209 (1951).

3. M. Swadesh, in *Proc. Intern. Congr. Americanists, 32nd, Copenhagen* (1958), pp. 671–4.

4. M. Swadesh, *Am. Anthropologist*, **64**, 1262 (1962).

5. L. L. Hammerich, in *Proc. Intern. Congr. Americanists, 32nd, Copenhagen* (1958), pp. 640–44; C. F. Voegelin, *Univ. Ariz. Bull.*, **29**, 47 (1958).

6. A. L. Kroeber, *Univ. Calif. (Berkeley) Publ. Am. Archaeol. Ethnol. No.* 38 (1939).

7. W. S. Laughlin, *Anthropol. Papers Univ. Alaska*, **1**, 25 (1952); *ibid.*, **6**, 5 (1957).

8. W. S. Laughlin and G. H. Marsh, *Am. Antiquity*, **20**, 27 (1954).

9. W. S. Laughlin, *Anthropol. Papers Univ. Alaska*, **5**, 5 (1956).

10. J. L. Giddings, *Current Anthropol.*, **1**, 121 (1960).

11. W. N. Irving, *Arctic Institute of North America Technical Paper 11* (1962), pp. 55–68.

12. H. B. Collins, *ibid.*, pp. 126–139.

13. W. S. Laughlin and W. G. Reeder, *Science*, **137**, 856 (1962).

14. W. G. Reeder, personal communication.

15. V. B. Scheffer, *Seals, Sea Lions, and Walruses* (Stanford Univ. Press, Stanford, 1958).

16. A. de Chamisso, "Cetaceorum maris Kamtschatici imagines, ab Aleutis e ligno fictas, adumbravit recsuitque," *Verhand. der kaiserlichen Leopoldinisch-Carolinischen Akad. der Naturforscher,* **12** Pt. 1, 249–63. Bonn.

17. A. Hrdlička, *The Aleutian and Commander Islands* (Wistar Institute, Philadelphia, 1945).

18. W. S. Laughlin and G. H. Marsh, *Arctic,* **4**, 74 (1951).

19. C. F. A. Moorrees, *The Aleut Dentition* (Harvard Univ. Press, Cambridge, 1957); A. A. Dahlberg, *Arctic Anthropol.*, **1**, 115 (1962).

20. W. S. Laughlin, in *The Physical Anthropology of the American Indian,* W. S. Laughlin, Ed. (Viking Fund, New York, 1951).

21. W. S. Laughlin, in *Proc. Intern. Congr. Americanists, 32nd, Copenhagen* (1958), pp. 516–30.

22. A. Hrdlička, *The Anthropology of Kodiak Island* (Wistar Institute, Philadelphia, 1944).

23. Giddings, *op. cit.*

24. Irving, *loc. cit.*

25. Collins, *loc. cit.*

26. W. E. Taylor, *Anthropologica*, **1**, 24 (1959).

27. Irving, *loc. cit.*

28. Collins, *loc. cit.*

29. W. S. Laughlin and W. E. Taylor, *National Museum of Canada Bull. 167* (1960), pp. 1–28.

30. F. A. Milan, thesis, University of Wisconsin (1963); L. K. Miller and L. Irving, *J. Appl. Physiol.*, **17**, 449 (1962); B. G. Covino, *Federation Proc.*, **20**, 209 (1961); K. Rodahl and D. Rennie, *Arctic Aeromedical Laboratory Technical Rept. 8-7951* (1957); G. M. Brown and J. Page, *J. Appl. Physiol.*, **5**, 221 (1952); G. M. Brown, J. D. Hatcher, J. Page, *ibid.*, **5**, 410 (1953); J. P. Meehan, A. Stoll, J. D. Hardy, *ibid.*, **6**, 397 (1954).

31. I. Veniaminov, *Notes on the Islands of the Unalaska Division* (St. Petersburg, 1840).

32. G. H. Marsh and W. S. Laughlin, *Southwestern J. Anthropol.*, **12**, 38 (1956); L. S. Laughlin, in "Man's Image in Medicine and Anthropology," I. Galdstone, Ed. (in press).

33. C. F. Merbs and W. H. Wilson, *National Museum of Canada Bull. 180* (1962), pp. 154–80; T. D. Stewart, *J. Bone Joint Surg.*, **35A**, 937 (1953); T. D. Stewart, *Clin. Orthopaed.*, **8**, 44 (1956).

34. M. G. Levin, *Sov. Ethnografiya*, **1958**, No. 5, 8 (1958) (in Russian); M. G. Levin, *ibid.*, **1959**, No. 3, 98 (1959) (in Russian); W. C. Boyd, *Science*, **140**, 1057 (1963).

35. B. Chown and M. Lewis, *National Museum of Canada Bull. 167* (1960), pp. 66–79.

36. B. Chown and M. Lewis, *Am. J. Phys. Anthropol.*, **17**, 13 (1959).

37. W. S. Laughlin, *Cold Spring Harbor Symp. Quant. Biol.*, **15(1950)**, 165 (1950); A. E. Mourant, *The Distribution of the Human Blood Groups* (Blackwell, Oxford, 1954).

38. A. C. Allison, B. S. Blumberg, S. M. Gartler, *Nature*, **183**, 118 (1959); B. S. Blumberg, A. C. Allison, B. Garry, *Ann. Human Genet.*, **23**, 349 (1959).

39. R. Gessain, *Medd. Grønland*, **161**, No. 4 (1960).

40. J. B. Jørgensen and W. S. Laughlin, *Folk*, **5**, 199 (1963).

41. W. S. Laughlin and J. B. Jørgensen, *Acta Genet. Statist. Med.*, **6**, 3 (1956).

42. F. Boas, *Science*, **30**, 535 (1909).

43. C. F. A. Moorrees, R. H. Osborne, E. Wilde, *Am. J. Phys. Anthropol.*, **10**, 319 (1952).

44. W. Howells, *Mankind in the Making* (Doubleday, New York, 1959), p. 300.

45. K. Chang, *Science*, **126**, 749 (1962).

46. H. Suzuki, in *Actes du 6e Congrès International des Sciences Anthropologiques et Ethnologiques* (Paris, 1960), vol. 1; *Zinruigaku Zassi*, **70**, 1 (1962).

47. F. Weidenreich, *Palaeontol. Sinica*, **1943**, No. 10, 127, 246, 256 (1943).

48. H. Suzuki, in *Selected Papers of the Fifth International Congress of Anthropology and Ethnological Sciences* (Philadelphia, 1956), pp. 717–24.

49. Laughlin, in *Proc. Intern. Congr. Americanists, 32nd, Copenhagen, loc. cit.*

50. M. T. Newman, *Am. Anthropologist*, **64**, 237 (1963).

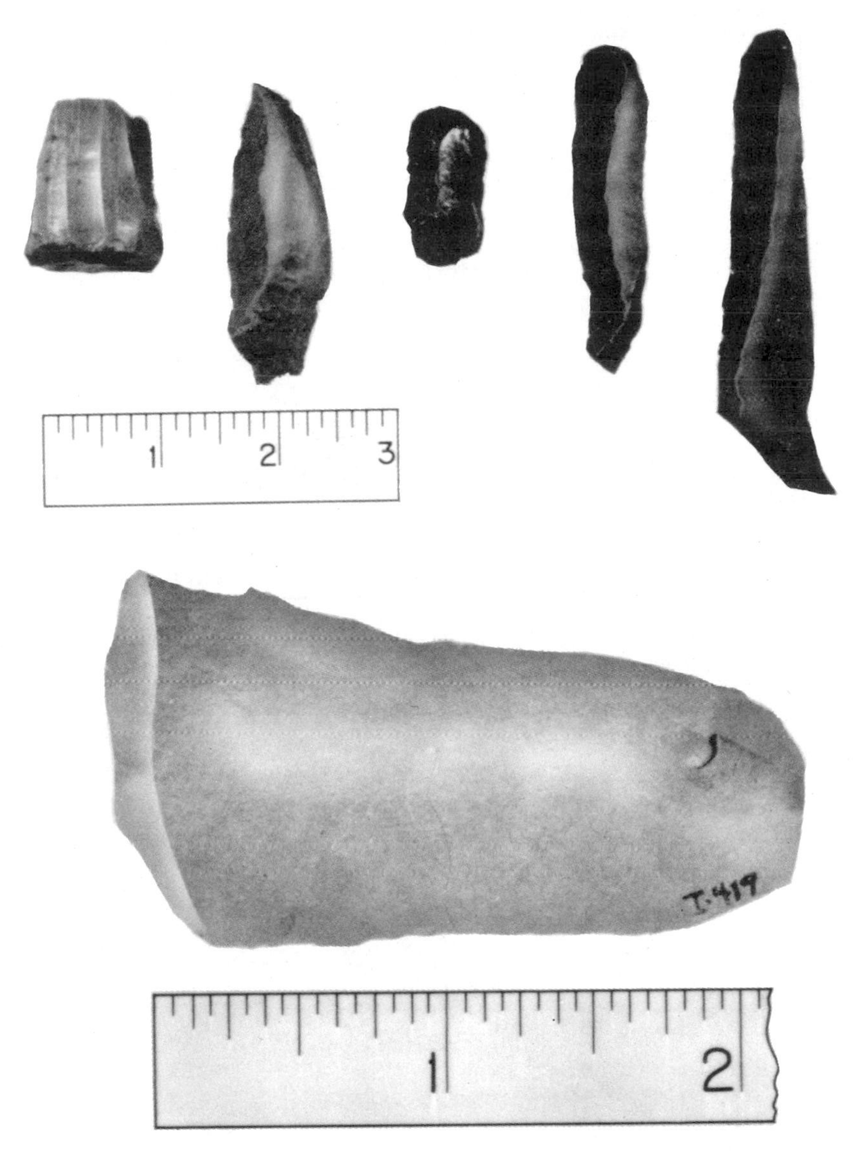

Fig. 9–1. Anangula core and blades. Eight thousand years ago, when Anangula was a portion of Umnak Island, its inhabitants struck microblades from polyhedral cores and also made larger prismatic blades. The pointed blade (top, second from left) is an "Aleutian graver," presumably used for incision. The blade (bottom) with the transverse flake scar is a burin. This industry is similar to industries in northern Japan 9000 to 13,000 years ago.

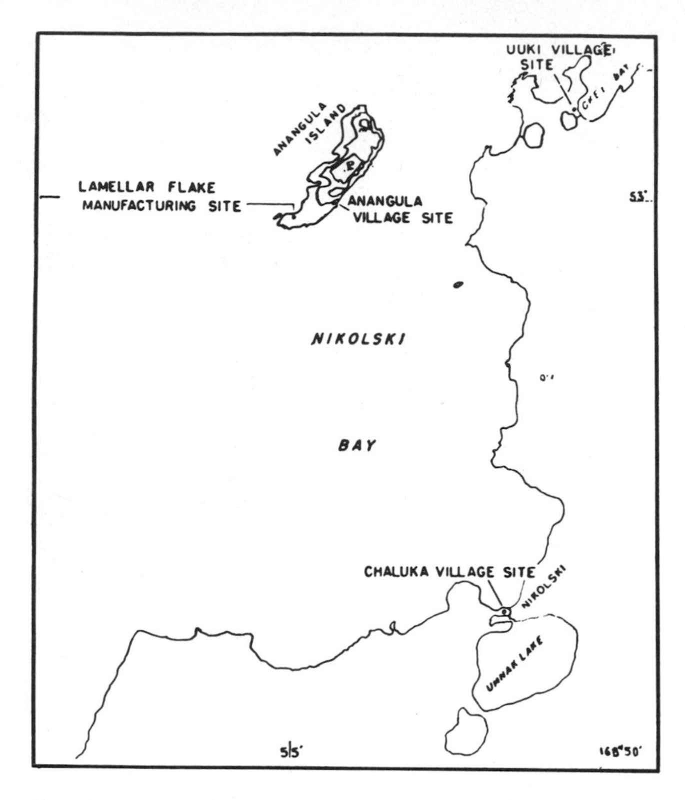

Fig. 9–2. Map of Anangula Island and Nikolski Bay (scale: 1/40,000).

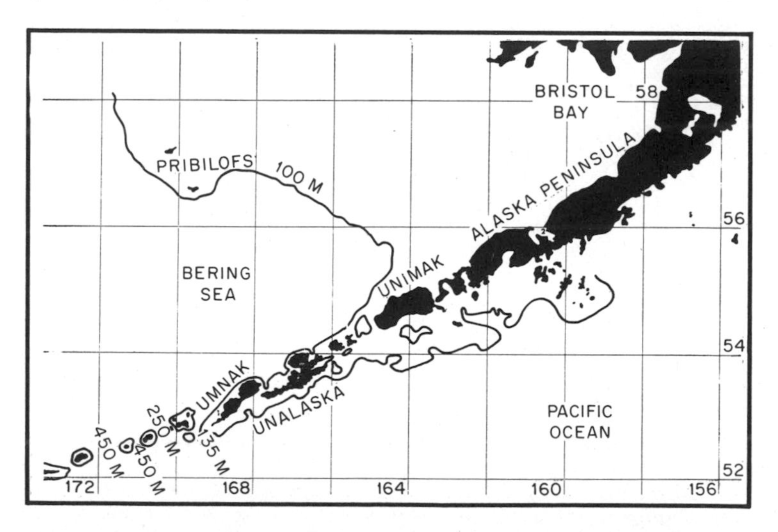

Fig. 9–3. Map of the Eastern Aleutians, showing the outlines of the former Bering platform of 11,000 years ago at approximately the 100-meter contour. Umnak Island was then the terminus of the Alaska Peninsula. The passes west of Umnak are too deep for a land bridge to have formed at any time during the last glaciation as a result of lowered water levels. Presumably, early populations lived on the platform and withdrew as the water level rose. 1, The Chaluka-Anangula area; 2, Port Moller, the point of division between Aleuts and Eskimos. (Courtesy of R. F. Black.)

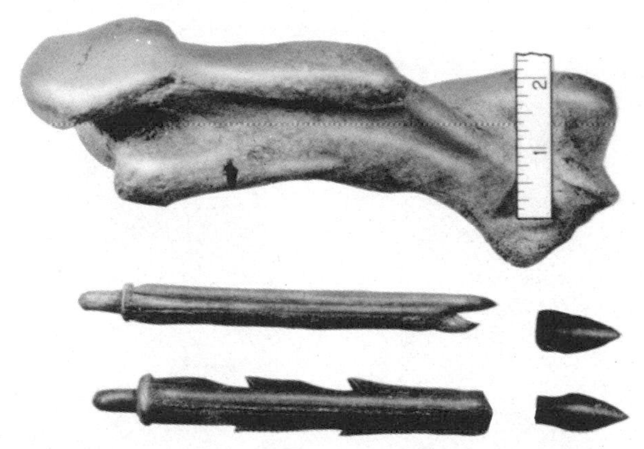

Fig. 9–4. (Top) Sea lion humerus, from the lower levels of Chaluka, with a stone point embedded in it. Detachable whalebone harpoon heads of this type were used for some 2,000 years or more. (Middle and bottom) Early Aleutian harpoon heads with slots in the end for inserting chipped stone points.

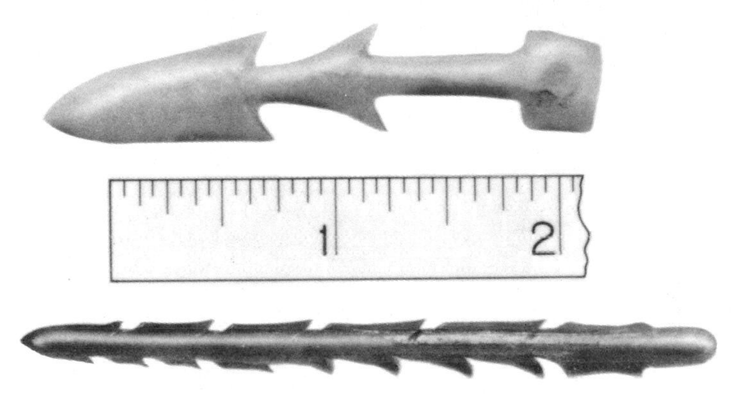

Fig. 9–5. (Top) Harpoon head of a very late type, made of ivory and used by the Umnak Aleuts for hunting sea otter. (Bottom) Chaluka harpoon head of an early, fluted type (about 3,000 years ago), made of ivory and suitable for hunting small sea mammals, such as sea otters, female fur seals, and harbor seals.

Fig. 9–6. (Top) Harpoon head of recent type, in use at the time of the discovery of Alaska by the Russians, in 1741. In contrast to the head of an earlier type (bottom), the chipped-stone end point has a round base and is inserted into a basin on the side of the whalebone head. A harpoon head of this type was inserted into a two-piece socket and lashed to a wood shaft for use as a spear or lance in hunting humans. It was supposedly long enough to penetrate a man's chest and to his spine.

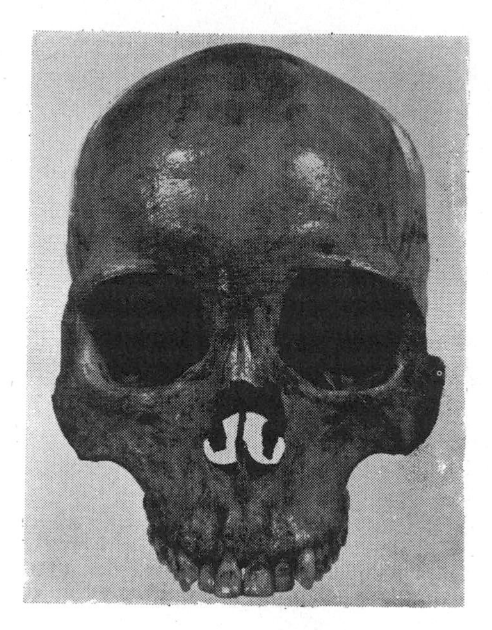

Fig. 9–7. Palaeo-Aleut cranium.

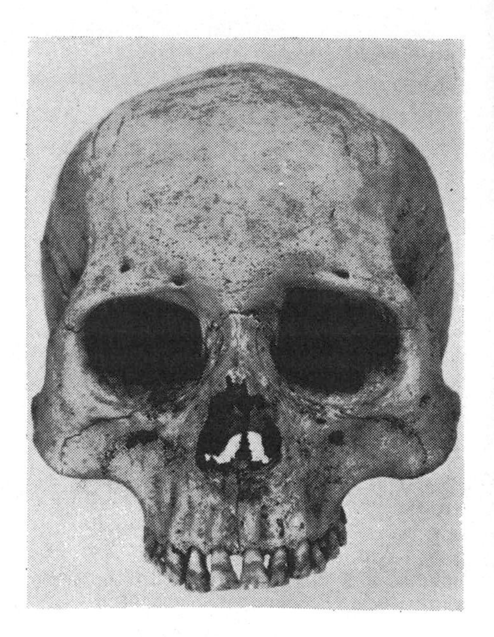

Fig. 9–8. Neo-Aleut cranium, with supraorbital foramina instead of notches, an accessory supraorbital foramen, large infraorbital foramina and an accessory infraorbital foramen, and an accessory zygomaticofacial foramen. Mongoloids characteristically have large or accessory foramina.

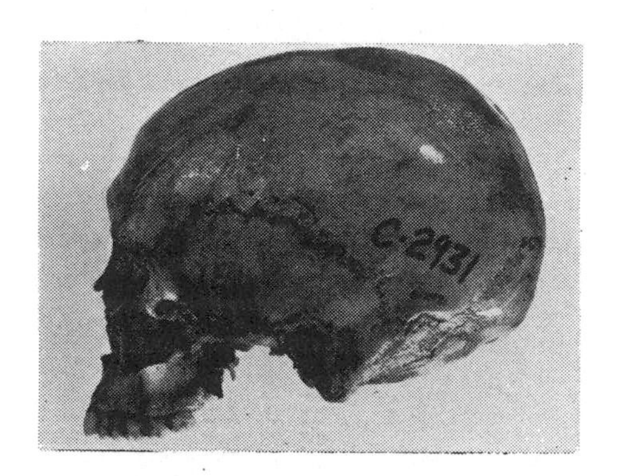

Fig. 9–9. Palaeo-Aleut cranium (side view).

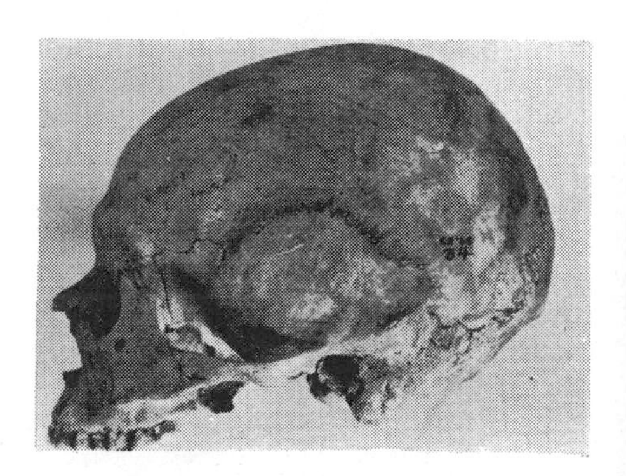

Fig. 9–10. Neo-Aleut cranium (side view). Crania of Neo-Aleuts are among the lowest and most capacious in the world.

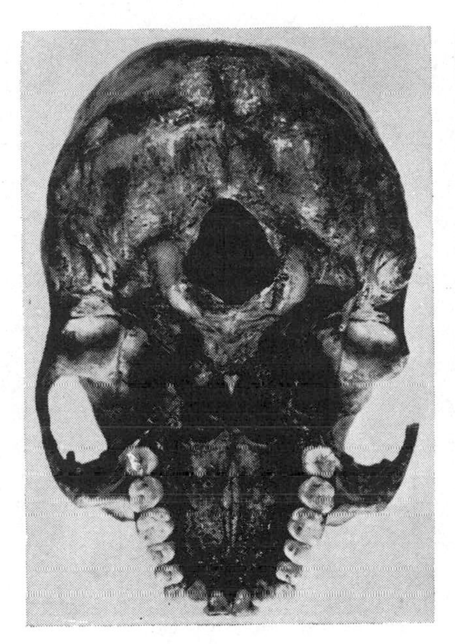

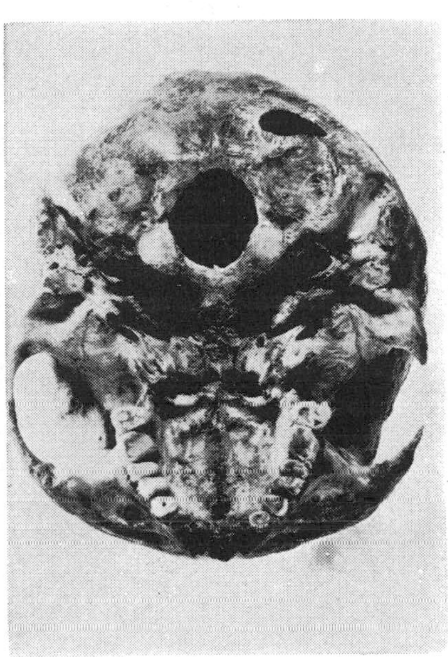

Fig. 9–11. (Left) Base of the cranium of a Palaeo-Aleut. The relative narrowness of this cranium and the long occipital area are characteristic of the earlier Aleutian population. (Right) Base of the cranium of a Neo-Aleut. The great breadth of this cranium and the short occipital area are characteristic of individuals of this more recent population.

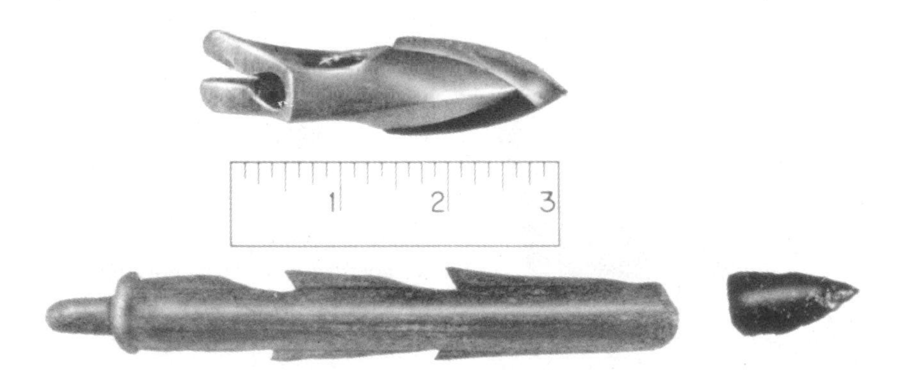

Fig. 9–12. (Top) Modern toggle-head harpoon (for hunting walrus), made of metal by the Aivilik Eskimo, Southampton Island, Canada. Since the sea mammal is retrieved by means of a line attached to the harpoon head, the continuing use of harpoons, along with rifles, is assured, for rifles kill but cannot retrieve. (Bottom) Early Aleutian fluted harpoon head with inset stone point, used for hunting large marine mammals, such as sea lions. The detachable head remains in the body of the mammal, but does not turn at right angles to the line as the toggling harpoon head does.

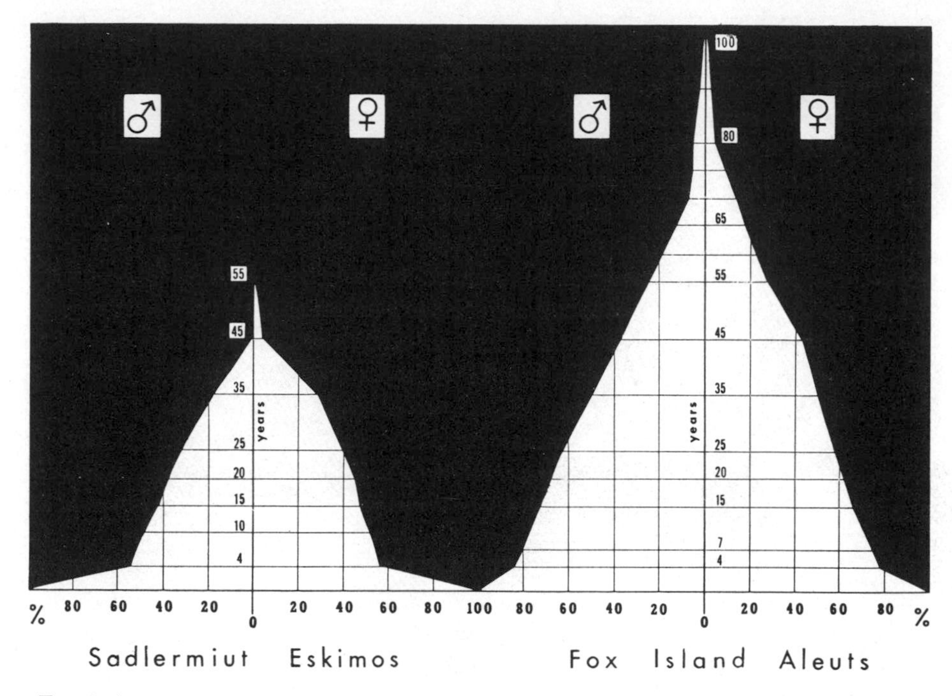

Fig. 9–13. Data on age at death of the Sadlermiut Eskimos of Southampton Island, Canadian Arctic, and Aleuts of the Fox Island district in the eastern Aleutians. The Sadlermiut Eskimos died relatively early in life as compared with these Aleuts. Infant mortality was also greater under the more stringent conditions of the arctic environment.

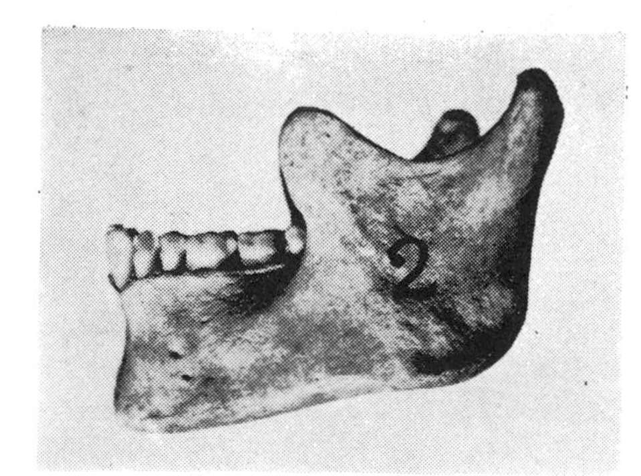

Fig. 9–14. Mandible of a man of the Okhotsk culture, Hokkaido, Japan, about A.D. 1000. The enormously broad ascending ramus is characteristic of many Mongoloid groups. The breadth of this feature in Eskimos and Aleuts exceeds the breadth in Neanderthal man. There are multiple mental foramina in the region of the chin. (Specimen courtesy of Kohei Mitshuhashi, Sapporo Medical College.)

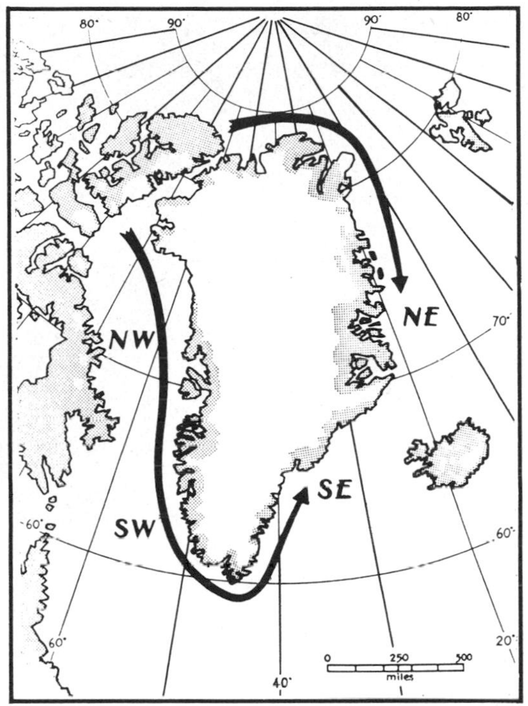

Fig. 9–15. Migration of the Eskimos about Greenland. The migration was confined to the coasts because of the inland ice. It moved in two directions, with the result that the terminal isolates (the Northeast and the Southeast), separated for the longest period, show the greatest morphological differences.

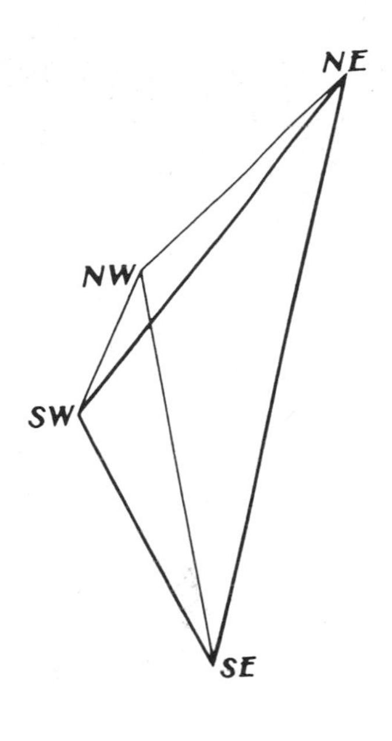

Fig. 9–16. Geometric representation of the relative degrees of similarity between the four Greenlandic Eskimo isolates. The difference between the Northeast and the Southeast isolates is greater than the difference between any other two contiguous isolates. Though geographically as far apart as the Northeast and Southeast isolates, the Northwest and Southwest isolates exchanged mates more frequently and are much more similar to each other. (Courtesy of L. S. Penrose [Professor Penrose adjusted his coefficient of divergence to accommodate discrete traits, made the original three-dimensional model, and provided a photograph from which the figure was drawn].)

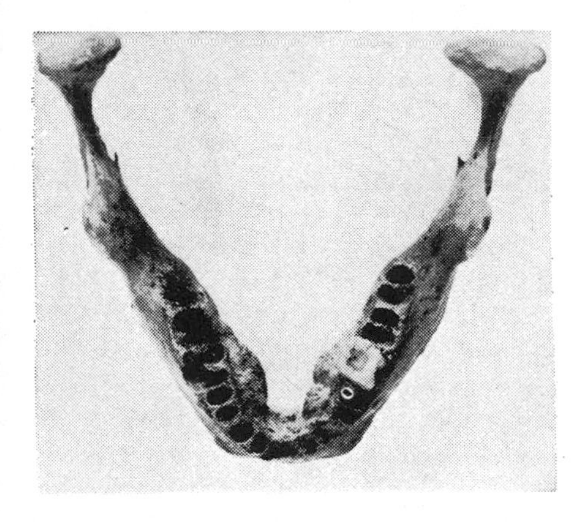

Fig. 9–17. Eskimo mandible with mandibular torus—the bony mound on the lingual surface. A form of this torus is found in *Sinanthropus pekinensis*. In Eskimos and Aleuts the number of mandibular tori exceeds the number of palatine tori; the converse obtains in American Indians and Europeans.

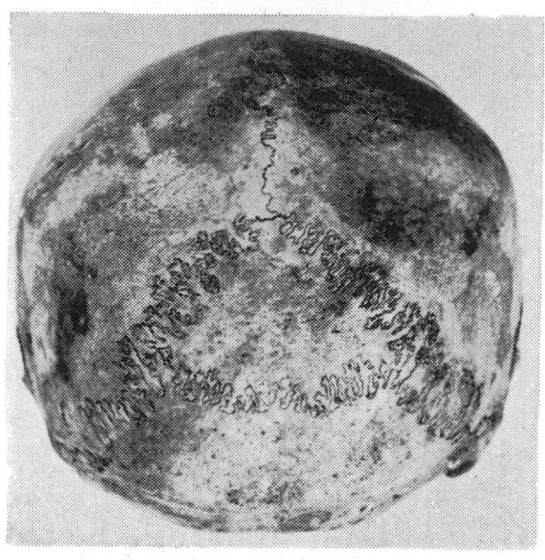

Fig. 9–18. Inca bone in the occipital bone of the cranium of a Neo-Aleut. A horizontal suture separates the upper portion of the occipital into a triangular portion. This is found in *Sinanthropus pekinensis*, Mongoloids, and American Indians in varying but often high frequencies.

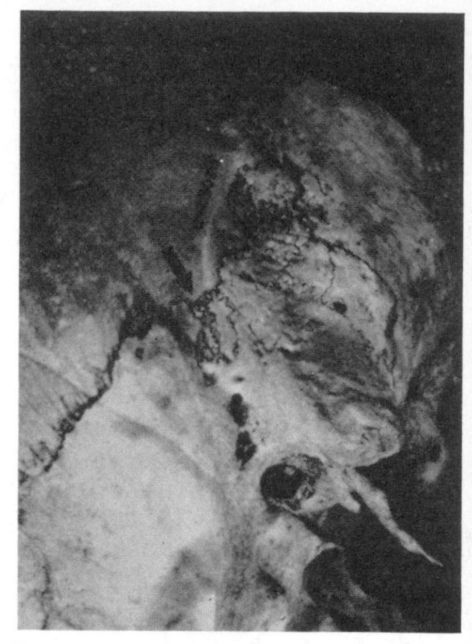

Fig. 9–20. Cranium of a Neo-Aleut male from Chaluka, showing multiple parietal notch bones.

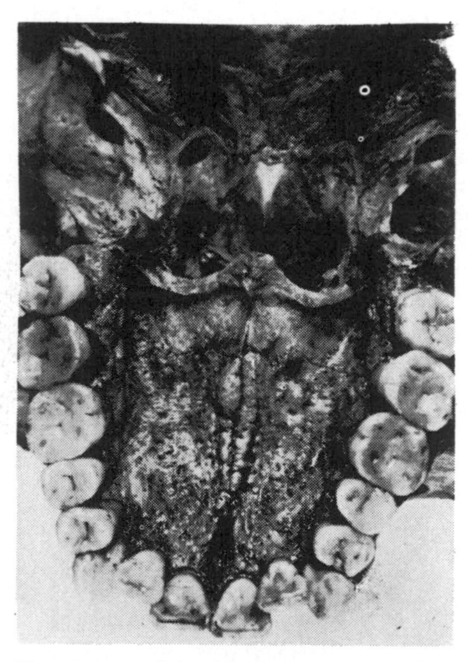

Fig. 9–19. Palate of the cranium of a Palaeo-Aleut, showing the palatine torus, the mound of bone running along the center of the palate, and the shovel-shaped incisors. Marginal ridges on the lingual surface of the incisors create a scooped-out area. These ridges are found in Mongoloids, American Indians, Polynesians, and *Sinanthropus*, and in some members of other races.

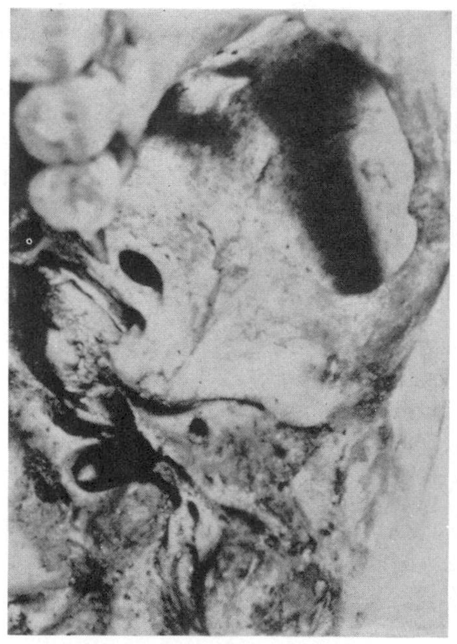

Fig. 9–21. Dehiscences and marginal foramen on the tympanic plate of a Palaeo-Aleut. Both these features are related to the "infantile gap" in *Sinanthropus pekinensis*. The dehiscence, or perforation, is a feature that is often found in the tympanic bone of humans, most frequently in Mongoloids and related races.

Chapter 10

Some Prehistoric Connections between Siberia and America

James B. Griffin

When western Europeans discovered the Americas they not only took on the task of exploring and colonizing the Western Hemisphere, but they also faced the question of the origin of the peoples who inhabited the New World. The name "Indians" was first applied under the misapprehension that the discovery had been made of the outer regions of India. Among the explanations for the peopling of the New World, one of the most common was a connection to the lost tribes of Israel.

One of the major difficulties in obtaining a satisfactory answer to the question of the New World population was the limitation of knowledge available during the sixteenth and seventeenth centuries about the age of the earth and about the age of human beings on the earth. With the exploration of the north Pacific area and the discovery of Bering Strait, a great many people concerned with the question of the origin of the Indians, recognizing the resemblance of the American Indian to the Mongoloid people, suggested that Bering Strait was the nearest or the easiest way by which people could have passed from Asia into North America. During the eighteenth and the early part of the nineteenth centuries, not only the Bering Strait crossing was suggested as a means by which America was peopled after 2000 B.C., but there were also theories of movements across the Atlantic from the Mediterranean area and of movements of people across the Pacific to populate the southern parts of the New World. Some of these outmoded ideas still persist.

During the past one hundred years, with the extraordinary growth of knowledge in various scientific fields, it has been possible to provide much more satisfactory explanations of Indian origins. The study of the geological features of the earth, particularly in Europe during the early 1800's proved the great antiquity of the earth. The development of Pleistocene or Ice Age studies in Europe and America indicated very clearly that implements of human manufacture were associated with this last major stage in the shaping of the earth's surface features. The discoveries and study of fossil men in the Old World from *Pithecanthropus* in Java to Neanderthal in Europe, and the recognition of Cro-Magnon in Europe associated with the late phase of the Old World stone industries gave man an antiquity in Eurasia that was not dreamed of in the centuries preceding. The studies of European prehistory certified the existence of a long period of development of human culture and also certified that by the time of the late Pleistocene in Europe, man had reached a fairly advanced stage of a hunting and gathering economy with a distinctive and spectacular art style. Mural painting begins with the Upper Palaeolithic.

Along with these developments in prehistory, the study of the culture of various non-European groups around the world, including groups in the Americas, indicated that there were a number of major groupings or stages of human culture. There were simple hunting and gathering peoples, more advanced food collectors, small-scale food producers, and finally, the more advanced non-European cultures in the New and Old World that had developed complex and well-integrated civilizations. In the Americas particularly, it could be seen that the cultures of the primitive groups on the marginal areas of South America and North America contrasted sharply with the agricultural civilization of Middle and South America, or even with the agricultural groups in the American Southeast and Southwest. Physical anthropologists have indicated that there is considerable variation among the various American Indian tribes and that the Indians did not belong to a single human physical type. There were also indications that the physical type of some of the marginal peoples of the New World corresponded to the physical type of the older American prehistoric skeletal material. Studies of American Indian languages have shown that there are a great many linguistic groups in the New World. The linguistic complexity was one of the main cultural features that indicated a considerable antiquity for man in the New World. For example, it has been said that there was more linguistic diversity within the present area of the state of California than there is in western Europe. All of the relevant data gathered by scientists on the American Indians implied that the American Indian was indeed derived from northeast Asia, and that he had probably come at a period when he was in a hunting-gathering stage of cultural development. The prevailing opinion is that the major complex agricultural civilizations, which were the wonder of the European adventurers, were developed in the New World.

The discovery of fossil man in the Old World and of ancient Old Stone Age implements was an incentive for scholars and other people of curiosity in the Americas to search for similar materials. During the late 1800's and early 1900's there were claims of fossil men comparable to European forms, and also claims of very ancient stone implement finds. There were even claims that the Indians had developed in the New World without connections to Eurasia. Critical studies of almost all of these purported fossil men and ancient implements proved that they were not of the same order of great antiquity as those of the Old World.

For some time archaeologists in the New World did not think that man had been there for more than a few thousand years before Christ. In the last thirty years, however, archaeological work has been able to produce sound evidence of a considerable antiquity for human cultures not only in the Americas (Fig. 10–1) but also in eastern and northeastern Asia (Fig. 10–2). Excavations in China have indicated that modern man was there in an Upper Palaeolithic stage of development during the closing phases of the last glaciation. A similar situation is found in south central Siberia, where man was present during the closing stages of the Upper Palaeolithic and in association with extinct animal forms of the closing phases of the Ice Age. In North and Middle America a considerable number of finds have human implements in association with extinct American fauna and in association with geological formations attributed to the closing phases of the Wisconsin glaciation in North America.

The Siberian Advanced Palaeolithic

Much of Siberia was not glaciated during the Pleistocene. Northwestern and extreme northern Siberia, as far as the Taimyr Peninsula, were glaciated as an eastward extension of the major European ice mass. The mountain areas of southern and eastern Siberia were glaciated, but there were extensive areas of lowland and plateau where there was not sufficient moisture to allow for the accumulation of a continental glacier. In south central Siberia in the Upper Yenisei and Upper Lena river valleys and in the area around Lake Baikal, there are a number of sites where an Upper Palaeolithic culture has been found that is best called advanced Palaeolithic, for the cultural complexes do not correspond precisely to the sequence of culture types which have been recognized for western Europe. These southern Siberian early cultures are directly connected with the late Palaeolithic complexes to the west, and from the evidence of faunal remains, it is known that people lived here during the closing stages of the last glaciation. The beasts that were hunted, and whose bones are found in the dwelling sites, are the mammoth, the woolly rhinoceros, the arctic fox, the reindeer, the cave lion, and the bison. The people lived in semisubterranean dwellings that afforded permanent shelter and a permanent camp for their hunting forays. Most of their flint tools were manufactured by striking long narrow blades of flint from carefully prepared cores. This technique is one of the diagnostic features of the late Palaeolithic. Their projectile points were then narrowed toward the top with lateral retouching along one side. Their dependence upon hunting is reflected not only by the animal bones found in their dwelling sites, but also by the types of implements they made, which include a large number of flint scrapers for working hides. They also had piercers or perforators made from flint flakes, blades with very sharp points, and graving tools, or burins, for working bone and perhaps wood. The smaller bones from one of their major food sources were also used to provide bone awls, long bone projectile points, bone handles for the flint scrapers and also bone needles. Effigies of birds and, particularly, of the pregnant human female figure, were made from bone and other materials. The artifact styles and the general way of life of these early peoples, identified particularly at the site of Malta (figs. 10–3 and 10–4) near Irkutsk and Bureti in the Angara Valley, are equated with the cultural stage in western Europe of late Solutrean to early Magdalenian. From the type of soil formation as well as the animals existing there, it is reasoned that the people lived in a tundra environment in south central Siberia. The period should be somewhere between 15000 and 10000 B.C.

Some of the Siberian sites have successive levels of human occupation. In deposits above the earliest known remains in south central Siberia, a second stage of the advanced Palaeolithic is recognized at Malta, at the Afontovo Site near Krasnoyarsk, and at a site near Irkutsk. Sites of this period are also found in the Upper Lena Valley between Lake Baikal and Yakutsk. This second stage of the advanced Palaeolithic is connected with loess deposits, the wind-blown soil which accumulated during the warming phase or climatic amelioration that followed the final glacial retreat. These sites are also associated with indications of the initial return of forest conditions, for the charcoal in hearths is from willow and larch. The rhinoceros disappears, but the mammoth and the other cold-weather fauna are still present. From the artifacts found in these sites, it is thought that there is a cultural deterioration, inasmuch as cruder implement types are more frequently found. There is a pronounced drop in the frequency of the long knifelike blades made from prepared cores, and there are very few pendants, beads, or sculptures. There are large stone tools, such as scrapers or choppers, that are made from river pebbles which are flaked only along one crescentic edge. These are probably indicative of influences coming in from the chopper–chopping tool industry of eastern and southeastern Asia. Some of the projectile points are ovoid in shape and resemble bifacially flaked tools of the much earlier Mousterian stage of Europe. Some of these specific implement types are indicative of a connection with the late Upper Palaeolithic and early Mesolithic or Middle Stone Age finds of western Europe. The time period is likely to be between 10000 and 7000 B.C. The estimates of Siberian dates are my own, based on the reported climatic changes.

The third and last stage of the advanced Palaeolithic in south central Siberia is known from sites on the Upper Lena such as Shishkino (Fig. 10–5), as well as in the Yenisei Valley. These sites are located in a clay soil zone immediately below the humus horizon of the coniferous forest. The climate is clearly drier and more continental in type, and the rivers are shallow. The animal remains do not include the mammoth or polar fox, because they have moved north. Most of the animal remains are of the reindeer, with a large number of bones from horses and wild cattle. The cultural deterioration noticed in the preceding stage continues, and it is believed that this is, at least in part, the result of the climatic change from tundra to boreal forest conditions. The most common implements are the crude pebble tools, which in superficial appearance look as though they belong to a much earlier cultural stage than is actually the case. The climatic conditions suggest that this stage of culture was in existence from approximately 7000 to 5000 B.C.

In Europe and in the Near East following the Upper Palaeolithic cultural development, the Mesolithic industries have as a primary flint working tradition the development of a small flint industry with microblades, cores, and burins. This Mesolithic stage small-tool industry spreads east across the steppe and desert country of southern Siberia and Mongolia to Manchuria and to Japan. It does not seem to have established a firm foothold in south central Siberia until shortly before the introduction of more advanced implement types, such as adzes of flint, and the introduction of pottery. There are a number of sites, however, of microlithic implements from the Baikal area which are a part of the closing phases of the advanced Palaeolithic peoples. The present indications are that this cultural stage was rather localized in Siberia and of short duration. It is, of course, possible that future explorations will produce a considerably larger number of sites, and it would be particularly important if these were found in the Lena Valley. Without anything very substantial to go on in the form of evidence, it has been said that the physical type of the advanced Palaeolithic population in Siberia and eastern Asia is closely related to the Upper Palaeolithic Europoid types to the west.

The Palaeo-Indian in North America

On the American side of the Bering Strait the Wisconsin glaciation covered most of the northern United States and practically all of Canada. Glacial ice moved down the eastern slopes of the American and Canadian Rockies and a short distance out onto the plains. In western Canada this ice advance was in contact with the moraines and other glacial debris from the large central Canadian Laurentian ice sheet, which moved west. It is not known for certain whether the maximum advance of the continental sheet and the mountain ice occurred simultaneously, or whether there was an ice-free corridor from north to south along the east side of the Canadian Rockies all through the Wisconsin glaciation. The Wisconsin ice advance is thought by some Pleistocene students to have begun about 70000 B.C., followed by warmer periods corresponding to the Würm interstadial in Europe. This may have provided an ice-free corridor east of the Rockies some 50,000 years ago. The furthest south ice advance, which produced the glacial stages in the Great Lakes area during the Wisconsin glaciation, reached as far as central Illinois about 18000 B.C. and may have closed the postulated gap in western Canada. It is highly probable that by at least 12000 to 14000 B.C. there was a corridor between the mountain and continental glacier in western Canada that would have allowed access from northern Alaska and northwestern Canada to the plains area to the south.

There are a number of finds which support the argument that man may have been in the New World, particularly in the western part of North America, some 30,000 to 20,000 years ago. The evidence for man at this time is, however, not completely convincing, and the majority of American archaeologists are hesitant about accepting it. The reason for hesitation varies from find to find. If a clearly defined cultural complex is radiocarbon dated and is in agreement with the correct geological formations and faunal associations, then certainly the majority of American archaeologists will be willing to accept such evidence. If early man did come in and occupy the North American continent 20,000 or 30,000 years ago, it would mean that he must have come in during an interstadial between the early and later Wisconsin glaciations or in pre-Wisconsin times, or both. This would imply that the physical type of man associated with such early finds could be very close to Neanderthal in appearance. It would further imply that the cultural type associated with man at this stage would be one resembling the Mousterian of Europe, or the chopper-chopping tool industry of eastern Asia. Neither man nor cultural material has been found in sites of the required age in southern and eastern Siberia.

We are certain that the earliest American Indians were in the United

States area by between 10000 to 12000 B.C. This is based on radiocarbon dates of some 11,000 years ago from the Rocky Mountain areas in the west, and the implied evidence that man was in the southeastern and eastern part of the United States at approximately this same time. Furthermore, the degree of cultural diversity between finds in the west and in the east are an indication that no small time must be allowed for such a differentiation of culture types in the New World. While there are no skeletal remains clearly associated with the most ancient definite cultures in the New World, there are crania with an antiquity of around 8000 B.C. that are of the same general physical type as that thought to be associated with the advanced Palaeolithic cultures in Siberia.

Our best evidence of man in the period from 10,000 to 12,000 years ago is in the general high plains area where the economy was that of a hunting-gathering people with heavy emphasis upon the use of game animals for food and for clothing. The general way of life in America was very much the same as in the early advanced Palaeolithic culture of Siberia. Many of the beasts that were hunted are similar, for the mastodon and the mammoth, the bison, horse, camel, and other such herbivores are found associated with remains of early man. The early American hunters of the east and west had about the same inventory of flint artifacts, such as the scraper, the perforator, and the graver, but not the burin, and probably very similar bone implements. These Palaeo-Indian hunters did not make their projectile points from carefully prepared blades, struck off from carefully shaped flint cores, but instead produced their projectile points from large flakes struck off from flint blocks. They also did not make human figurines. A significant new development is the production of bifacially flaked projectiles from which long flakes are removed from both faces of the point, by means of very careful and excellent chipping techniques, from the bases of the projectile (Fig. 10–6). This carefully controlled technique of bifacial flaking is relatively rare in the Siberian area until close to the Neolithic and is one of the reasons why some of the prehistorians were loath to accept a high antiquity for these finds in North America. It must be assumed, on the basis of present evidence, that the culture trait is primarily an American development, along with the distinctive fluting that removed flakes from both faces of the projectile.

Mary R. Haas[1] has presented the view that the Muskhogean and Algonkian languages, which almost blanketed the area east of the Mississippi, are descendants of an ancient common language of some 8,000 or more years ago. This could mean that the Palaeo-Indian hunters of the east spoke related dialects of a common language and that the marked linguistic diversity of modern times was initiated during the long Archaic period.

We have seen that at the close of the Pleistocene in southern Siberia an advanced Palaeolithic culture was established with a Europoid physical type, and with the amelioration of climatic conditions there was a tendency for the cultural type to deteriorate. It may be that as the climatic conditions were modified, people hunting the big game animals followed the animals northward into the tundra ecological zone and on to a much wider Siberian coastal plain, which would have been in existence at that time, because of the lowering of the sea level. From there they could have moved eastward across the Bering Strait, where a land bridge would have been in existence some 10,000 to 15,000 years ago, and then moved to the valley of the MacKenzie.[2] The northern section of Alaska north of the Brooks Range and a significant area of the Alaskan coastal shelf were tundra vegetation at this time, and unglaciated. This is also true of the Seward Peninsula area and the Chukchi Peninsula. They would still have been in a tundra ecological zone in their spread up the MacKenzie until they came into the prairie grassland areas of the eastern slopes of the American Rockies. It is also possible for man to have moved from the Seward Peninsula area into the boreal forest of the Yukon, up the Yukon Valley to the east and southeast, and finally to have come into the MacKenzie Valley and northern plains in northern Alberta and northeastern British Columbia.

From about 10000 to 8000 B.C. the Palaeo-Indian fluted-blade hunters occupied most of the United States east of the Rocky Mountains and south of the retreating Wisconsin ice. These Palaeo-Indians seem to be the earliest people east of the Mississippi. There are distinctive local complexes and a number of significant variants in the shapes of the fluted projectile in the east. The date 8000 B.C. may be said to be a convenient dividing line between the Palaeo-Indian and the early Archaic culture of the east, because after this date the fluted projectile style tends to disappear, and most of the Wisconsin ice has moved north of the Great Lakes.

The American Archaic

The early Archaic of 8000 to 4000 B.C. is a continuation of the hunting-gathering way of life. There are increasing indications of regional differences reflecting local ecological and climatic zones. A wider variety of projectile forms are known and heavy chipped stone choppers or diggers are found. During this period the seasonally migrating small bands of people were becoming familiar with the native mineral, animal, and plant resources. Many of the major flint quarries were discovered, and certain spots were selected as seasonal camping grounds, where deep refuse deposits bear witness to their intermittent occupation.

In the western plains the Plano cultures are known during the early Archaic period. They had a pronounced emphasis upon hunting both the large extinct bison form, during the early Plano, and the modern bison at a later time. The prairie hunting adaptation spread north into Alberta and Saskatchewan with the return of the grassland to those areas. The long, slender, beautifully chipped projectiles of the several Plano styles are known as far east and north as southern Manitoba, along the shores of glacial Lake Agassiz, on the north shore of Lake Superior on a high ancient beach ridge, and as far as northern Lake Huron.

Up to about 4000 B.C. the cultural developments in the east are primarily of American origin and are part of natural cultural changes. Between 4000 B.C. and about 1500 B.C., however, there are new artifact types with both formal and functional resemblances to northern Eurasian forms. Prominent among these are the gouge and the adze in the Great Lakes to New England region and the grooved ax in the area from the Ohio Valley to northern Alabama. These heavy woodworking tools are of ground stone and may have had their prototype in chipped flint and stone choppers known in early Archaic sites. Other artifact similarities to northern Eurasia are various ground slate, knife, spear, and projectile forms in sites from New York to New England, and almost identical forms made of native copper from the Wisconsin-Michigan area. It is possible that some of these artifacts of slate and copper are copies of bone and flint implements.

The artifact similarities of the gouges, adzes, and slate forms are closest between the New York-New England area and northern Scandinavia and Karelia, where on sites indicative of some antiquity the gouge, adze, and slate forms are well represented. The tendency is, however, to date these forms in extreme northern Europe at about the same time period that they are known in the northeastern United States. One of the adze forms from Karelia is very much like the beveled Lamoka adze in New York of 3000 B.C. to 2500 B.C., but it is not found east of the Ob River in western Siberia. These artifacts have often been referred to as a part of an eastern circumboreal spread, but they do not have sufficient continuity across Siberia, Alaska, or northwest Canada at a sufficiently early period to support either a suggested movement of people or diffusion from the Old World to the New. Of the two possible explanations of cultural movement, slow diffusion would appear to be the best hypothesis, which future excavation may document.

There are many polished stone forms which also serve to characterize the late Archaic in the east. Among these are banner stones, boat stones, and birdstones. These are believed to have been attached to the throwing stick that acted as an added lever to propel the spears during the hunt. These forms have no known Old World or American Arctic counterparts. The wide variety of projectile forms and other flint and bone implements of the late Archaic is best regarded as a local development. The projectile forms particularly cannot be derived from Siberian or American Arctic prototypes.

The Arctic Denbigh Complex

If a satisfactory demonstration of cultural movement from Siberia to eastern North America, which would strongly affect the late Archaic cultures, cannot yet be made, there is abundant documentation (in the archaeological sense) for a significant cultural spread from Siberia to the American Arctic from about 3000 B.C. to 1500 B.C. As already indicated, following the advanced Palaeolithic stage in Siberia, there is a movement into southern Siberia from the south of the small-tool complex of blade, core, and burin and small, bifacially flaked arrowpoints, which make their appearance before the introduction of pottery. By the time this complex is *known* to have reached the Middle and Lower Lena, however, pottery is always associated with it.

In Alaska, however, this small tool complex is known from a series of sites from the Brooks Range to the Aleutians where pottery is not in association. The complex is best known from the stratified Iyatayet site at Cape Denbigh, on the south side of the Seward Peninsula (Fig. 10–7). In addition to the small flint tool complex, there are a large number of beautifully flaked projectile-point forms which have been compared with Plano points, and it has been suggested that these have diffused to the Seward Peninsula from the northern plains. It is also possible that they are a local development based on the bifacially flaked points and side blades of the early eastern Siberian Neolithic. This suggestion is favored in this chapter because the Denbigh complex also has small bifacially flaked triangular arrow or harpoon points that are very much like the Siberian Neolithic points.

Former assessments of the age of the Denbigh complex have usually been to a period from 8000 B.C. to 4000 B.C., based on the interpretation of geological evidence and a natural inclination to connect the small core, blade, and burin with the Eurasian Mesolithic culture, and as close to that time period as possible. Such an antiquity was never completely acceptable. The view presented in this article is that the lower occupation level at Iyatayet was in existence at 2000 B.C., and this is supported by a series of radiocarbon dates from the site. The core blade and burin may have appeared in Alaska somewhat earlier. The Denbigh small tool complex, with variations, does spread to the east, where it has been found on the Firth River in extreme northwest Canada; from Knife River in northeastern Manitoba; from the Melville Peninsula; and as far east as Disko Bay in central-western Greenland. The locations of some of these sites in the area around Hudson Bay are such that they could not have been occupied because of ice, or later marine submergence, at an age compatible with a high antiquity for the Denbigh complex. On the other hand, the spread of this complex into the central and eastern Canadian tundra between 2000 B.C. and 500 B.C. would be geologically completely feasible.

Although the full Denbigh complex does not penetrate into the Great Lakes area and the northeast, the core and blade technique, but apparently not the burin, does appear in the Lower Mississippi Valley in the form of the microflint industry of the Poverty Point culture. The radiocarbon dates for this industry would be somewhere around 1000 to 500 B.C. Larger blades struck from a variety of core types are a diagnostic feature of the Hopewell culture of the Upper Mississippi and Ohio Valley and of Point Peninsula II in New York from 300 B.C. to A.D. 300.

The Siberian Neolithic[3]

The most striking and widespread prehistoric material culture trait of eastern North America that can be attributed to an origin in the Siberian Neolithic is the Woodland pottery tradition, which covered a wide area from the Rocky Mountains to the Atlantic, and from the Gulf of Mexico to southern Canada. In Canada, Woodland pottery is known from southeastern Alberta to southern Quebec and Nova Scotia. For many years, the majority of American anthropologists believed that all of the prehistoric pottery of the North American Indians was derived from Mexico and was associated with the northward spread of American agriculture. This view is gradually being abandoned in favor of an Asiatic origin.

The earliest known eastern Siberian Neolithic is in the Baikal area, according to Russian archaeologists, and their use of the word "Neolithic" indicates the presence of pottery, ground and polished stone tools and ornaments, arrowheads, and other bifacially chipped flint forms, but does not mean that agriculture was practiced. The most experienced Russian excavator in central and eastern Siberia, and the most prolific producer of papers on the area, is A. P. Okladnikov.[4] Many of his papers have now been abstracted or interpreted by American scholars. Parenthetically, it may be observed that Okladnikov is conversant with the American archaeological literature.

There are three sequential premetal archaeological phases in the Baikal, named Isakovo, Serovo, and Kitoi. Some traits of the postulated earlier nonceramic Khinskaya culture persist, such as small cores and blades, and the bifacially chipped arrowpoints which become much more common. Some of the characteristic traits of Isakovo are given below. The arrowpoints are trianguloid with a concave base and asymmetrical barbs. One of the distinctive traits derived from Mesolithic cultures to the west is a long bone point which has slots along the sides for the insertion of flint side blades. They are identified as spears, and similar but shorter forms are called daggers. There are composite knives of large side blades inset into bone handles, large flint ovoid and elliptical scrapers, chipped and ground slate and nephrite knives, and large chipped and partly ground adzes and gouges. The pottery has a shape like the lower half of an egg. It is low fired, relatively thin (about 5 mm.), was tempered with crushed rock, and was, of course, handmade. The vessel exterior is covered with net impressions made while the clay of the shaped vessel was still plastic. Impressions made from these pottery fragments clearly show the net knots and connecting threads. Okladnikov thinks that the Isakovo vessels may have been made in net-lined pits in the ground. The pottery of the succeeding period, however, is clearly coiled or ring-built and then paddled to apply a variety of surface finishes. The use of small paddles is well documented ethnologically and archaeologically throughout the distribution of the Siberian Neolithic and in North America. It is one of the important concepts in pottery-making on this primitive level. Okladnikov[5] has proposed a time period of 4000 to 3000 B.C. for the Isakovo phase of the Lake Baikal sequence, but in the chronology of this essay, it is placed at 3000 to 2500 B.C.

The succeeding Serovo phase continues many of the earlier cultural traits, but with some significant additions and changes. The basic pottery form takes on a more rounded base and small, thick lugs, which may have been used for suspension. The most common surface treatment is still with a net, but the surface was usually smoothed over before the vessel was fired. A new decorative treatment is presumably derived from the comb ceramic areas to the west. These dentate stamp impressions were apparently made with small, narrow, slate fragments with grooves cut across the edges. Around the outer rim of the vessel there are one or more horizontal rows of small bosses or protuberances made by pushing a small rod into the inner wall and forcing the boss to appear on the outside. Polished stone knives increase in number and variety of form. Polished stone fish effigies were used as lures. The chipped flint adzes and gouges are well polished. Unilateral and bilateral barbed bone harpoons are found for the first time, as are barbed bone fishhooks. A distinctive braced or composite bow is unique to the Serovo phase, and comes in with small stemmed arrowheads. Most of the arrowheads are small triangular forms. The archaeological evidence, then, suggests a strong emphasis upon fishing in the Serovo culture, for besides the spears, harpoons, and fishhooks, nets were used. There are bone and antler flakers with which the Serovo people produced long side blades with ripple flaking for bone daggers and spears, as well as long flint knives and daggers, some of them also with parallel oblique ripple flaking.

Okladnikov has suggested that a reasonable estimate for the age of the Serovo phase would be 3000 to 2000 B.C. In the chronology adapted here the dates are 2500–2000 B.C. This may be somewhat too early, particularly if published statements on the presence of cord-marked and linear stamped pottery in Serovo are true,[6] but I have not been able to verify this from the translations available or from illustrations. The age of the early Baikal Neolithic is not firmly established. On the hypothesis that the ceramic tradition of this area should be allied to, and in part derived from, the early Neolithic pottery in the belt from China to Japan, the presence of a number of types of surface finish, such as cord-marked, linear stamped, and check stamp, should not be earlier in the Baikal and Lena valleys than it is in China. The Isakovo pottery does not have a direct ancestor in northern China, but would not precede the introduction of pottery into China on an early, simple "Neolithic" level. China may be regarded as a primary center for the hypothetical early Neolithic, as it was for the subsequent late Neolithic of the Yangshao and Lungshan complexes, and for succeeding cultural developments.

The Kitoi phase of the Baikal sequence certainly has cord-marked pottery, but the dominant decorative devices are dentate stamp, punctates, and linear punctates. These are placed on the outer rim in vertical, horizontal, or zigzag patterns. The use of linear punctate with raised rim patterns is reminiscent of certain Japanese styles such as the Shiboguchi type.[7] Distinctive new traits are three-stop bone flutes and bone "panpipes." While bone flutes are a part of the late Archaic in the eastern United States of at least 2500 B.C., panpipes are not known until the Hopewell culture of around A.D. 1. Another new Kitoi trait is the free use of red ochre in graves, either on flexed burials or cremations. An increased emphasis upon fishing is indicated by large numbers of composite fishhooks. Harpoons are unilaterally and bilaterally barbed, and many of them have perforations for a line attached through a flange near the base of the harpoons. Side-bladed knives, spears, and daggers continue, as do the adze and gouge. The true celt makes its first appearance in the Kitoi phase. A distinctive, narrow, hammerheaded bone point has been likened to those from pit and catacomb burials of southern Russia and from a Bronze Age site in Scandinavia of about 1800 B.C. There is an increased use of ground and polished nephrite for knives, points, adzes, and axes. Bone bracelets, pointed instruments, and needle cases are decorated by incised circle and dot decoration, which is also present in eastern European Bronze Age sites. While Okladnikov suggests a time period from about 2500 to 1750 B.C.,[8] in this essay Kitoi is placed between 2000 and 1500 B.C.

While actual metalworking is not known in the next phase, Glaskovo, it is clear that it is temporally on the same level as the Bronze Age cultures to the south and west, and by the following Shivera phase local metallurgy is known in the Baikal area.

The Middle Lena Neolithic complex is significantly different from the Baikal area for a number of reasons. The majority of the ceramic-bearing sites in this area are associated with a strong element of small cores, blades, and burins, as well as the small bifacial arrows, and side blades. Either this small, flint tool complex has survived from an earlier local unidentified nonceramic phase, or it may have moved into the Yakutsk area from Manchuria. The latter area seems to have furnished many ceramic elements, such as the check stamp, linear stamp, and cord-marked surfaces, which tend to supersede the Baikal-derived, net-impressed surfaces. Another possible ceramic trait from Manchuria is seen at the site of Kullaty, where some of the vessel rims are thickened and incised with horizontal lines. There may also be some vessels which had a woven fabric impressed against the walls of the vessel. This type of surface treatment may resemble that from the eastern United States, or it may be more like fabric or cloth impressions from an early ceramic site of about 500 B.C. on the north side of Seward Peninsula.[9] The dentate stamp is also significant as a decorative device. The developed Neolithic of the Middle Lena is regarded here as being about 1500 B.C.[10] This area and the Lower Lena would seem to be the major known concentration of Neolithic pottery and is directly ancestral to the Norton pottery of the Seward Peninsula and the Firth River pottery of northwestern Canada.

In the Lower Lena Valley some of the earliest known sites have long trihedral points with short stems, or stemless forms, which are very similar to those of the Lake Onega area of northwestern Russia, where they are said to date between 2000 and 1500 B.C. In general, the flint industry resembles that of the Middle Lena, with an emphasis on burins, small cores and blades, triangular bifacial arrowheads, side blades, semilunar knives, and a few chipped and polished adze fragments. The pottery includes cord-marked, check and linear stamp, and dentate stamping. Because of the large size of the squares on much of the check stamp pottery, and the organic temper, most of this pottery would seem to be quite late. This is based on the late time position of these traits in western Alaska. There are other sites between the Lena and the eastern part of the Chukchi Peninsula. They have not been fully excavated or reported, and additional work will be required to adequately document the cultural movements which must have taken place into Alaska.

The earliest described pottery in Alaska is the Norton complex at Iyatayet and related sites in western Alaska. This complex is directly derived from the Lena linear stamped and check stamped types. In turn, Norton is the western progenitor of the Firth River pottery in northwest Canada, where it has been reported that in addition to the Norton types of surface finish there are also dentate stamp and cord marking. The Norton complex is dated a few hundred years either side of A.D. 1, and the Firth River pottery should be about the same age. This early pottery in the American Arctic is associated with a flint industry of Denbigh complex origin. The pottery is not known to have moved south through the Boreal forest into the Great Lakes area, nor has any other early ceramic complex appeared in that area.

The Transitional and Early Woodland Complex

During the period of 1500 to 500 B.C., there are a number of new cultural developments in the northern part of the eastern United States. This time period can be called transitional between the Archaic and the Woodland cultures or regarded as early Woodland, particularly when pottery is associated with the rest of the culture. One of these developments is the increased attention to mortuary observances in the Red Ochre, Glacial Kame, Point Peninsula I, and Red Paint phases from Illinois and Wisconsin to New England.

There is a marked emphasis on the burial of cremated human remains in excavated pits. These burials are accompanied by a variety of the polished stone forms previously mentioned, and caches of projectile forms showing a marked preference for triangular points. Another feature was the custom of burying fire-making sets of iron pyrites and a flint striker. The pyrites are usually found as yellow limonite. This is the period of the first certain recognition of percussion fire making, and should be connected not only with the historic distribution of this method in the northeast and American Boreal forest and Arctic, but should also be derived from Eurasiatic percussion techniques, which are certainly known during the Neolithic and Bronze ages, and probably earlier as well. The grave area, the artifacts, the skeletal material, or the cremated bone were covered with a considerable amount of red ochre.

To each of these local complexes is added the earliest known Woodland pottery, which is thick and very coarsely tempered with large particles of crushed rock. The vessels are either cord-marked on both the inner and outer surfaces or, more rarely, may be smoothed on both surfaces. The vessel shapes are conoidal, rounded on the base, or flattened. This early variety of Woodland pottery belongs to the same tradition as the early arctic pottery, but a close counterpart is not known either in America or Asia. Perhaps this is another example of stimulus diffusion, where knowledge of a particular technology moves across a geographical area—in this case from the extreme northern American Arctic to the Great Lakes—and it appears in a variant form without a specific prototype in the Arctic. In any event, knowledge of and the manufacture of pottery spreads south. From the Ohio Valley to the Tennessee Valley the earliest pottery has a fabric-impressed surface and a conical or flat base. The textile that was employed had a wide warp and closely woven weft threads of twisted bast fibers. It is not coiled basketry, nor was the pottery made in a basket. In Georgia the early Woodland pottery is fabric-impressed, dentate stamped, or simple stamped, with linear impressions somewhat like the Siberian and early American Arctic linear stamp. Once the concept of using a carved paddle instead of a cord or fabric-wrapped paddle was adopted in the southeast, there developed a center of check stamp and complicated stamp designs with a great many striking similarities to the pottery of the eastern Chinese area during the Chou and Han dynasties. These similarities can be regarded as parallel developments based on the common possession of a general technique or tradition as to the correct manner of pottery manufacture.

A number of examples will help to demonstrate that formal resemblances between Asiatic and American pottery are not always to be regarded as evidence for a direct connection between these two widely separated areas. Following the first appearance of early Woodland pottery around Chesapeake Bay, a net-impressed pottery becomes the most common type. It is almost identical in surface appearance to the early Baikal pottery, but this style of surface finish is not found on early pottery between the Lena and the Potomac. The linear stamped pottery of the late Neolithic of northern China has its closest parallel in the post-1200 A.D. Plains area grooved paddle pottery. The zone-decorated, fine Hopewell pottery of the northern Mississippi Valley between 200 B.C. to A.D. 250 has its closest stylistic resemblance in middle to late Jomon pottery of Japan. This style is not represented in the pottery of northeastern Siberia or northwestern North America. A very distinctive stamped design is composed of groups of adjoining small diamonds with a raised dot in the center. This has been recognized from sites near Hong Kong and from Manchuria and probably belongs to the Chou to Han periods in eastern Asia. An almost identical design is found in America only from early Woodland sites near Savannah, Georgia.

Another significant cultural addition of the 1500 to 500 B.C. period is the trait of burial mounds in the Illinois and Ohio valleys, which begin as small, low, dome-shaped earth constructions over a burial complex analogous to the specialized interments of the transitional period mentioned above. The mound burial ceremonialism develops over a thousand-year period to about A.D. 500, and very large complex burial mounds containing hundreds of burials were erected by the Adena and Hopewell cultures. Burial mounds are distributed over a wide area in Eurasia, but are not a part of Neolithic or Bronze Age sites from the area east of the Upper Yenisei to the very late mound constructions of southern Manitoba.

A number of perishable products made their first appearance in eastern America about 1500 to 500 B.C. Among these are skin bags, thread, simple fabrics, nets, and fish weirs. They were probably also in use during the Archaic but were not preserved.

During the Archaic the skull form of the burials is predominantly longheaded. A number of regional variants of this archaic population are recognized but not clearly defined. During the late Archaic and early Woodland a roundheaded cranial type appears in the northeast and in the Ohio Valley. Whether this is the result of a gradual change in the resident population or the result of the introduction of a new population from the south or from the northwest is not definitely known. One possibility is that it represents a population movement from northeastern Siberia.

The most significant cultural event which gradually transformed the Archaic societies to the developed Woodland cultures was the addition of agriculture from Mexico. From 500 B.C. to A.D. 500, gourds and perhaps corn were added to the food supply. There are, in addition, a few traits of the developed Woodland culture clearly derived from Mexico. The last major prehistoric stage in the Mississippi Valley from about A.D. 800 to the historic period is known as the Mississippi pattern or culture. It was strongly influenced by concepts of Mexican origin which were in turn integrated and developed in the eastern United States for eight hundred years into the distinctive cultural forms of the early historic period.

Summary

The prehistory of the American Indian in the eastern United States, in spite of an impressive amount of excavation and study, is still in an unsatisfactory state, even though the major outline of cultural change and development is known. The earliest food-collectors of the area, the Palaeo-Indians, possessed a culture type of more or less close connection to the Siberian advanced Palaeolithic groups. During the long Archaic period from 8000 to 1500 B.C. a variety of minor culture changes and adaptations take place which are primarily developments of the native American populations. There may well be significant increments from Asia during the late Archaic, as has been postulated for the heavy woodworking tools such as the gouge and adze, but to definitely establish such influences, their manner of spread needs to be adequately documented.

There are a number of cultural traits that appear in the eastern United States between 1500 to 500 B.C. which are best explained as the result either of diffusion from Asia or to some degree also by population movement. These are added to the resident culture and this merger, along with agriculture and influences from Mexico, produced the developed Woodland cultures of eastern America from 500 B.C. to A.D. 500. Following this period, Mexican influences, but not a migration, shaped the dominantly agricultural societies of the Mississippi Valley in the early historic period.

NOTES

1. Mary R. Haas, "A new linguistic relationship in North America: Algonkian and the Gulf languages," *Southw. J. Anthrop.*, **14**, 231.

2. D. M. Hopkins, "Cenozoic history of the Bering land bridge," *Science*, **129**, 1519 (1959).

3. My interpretation of the Siberian Neolithic was slightly altered in 1963 as the result of study in the USSR in 1962–63.

4. A. P. Okladnikov, "Traces of the Paleolithic in the Valley of the Lena," in *Materials and Researches in the Archaeology of the U.S.S.R.* (in Russian) Moscow (1953), vol. 39.

5. Okladnikov, *loc. cit.*

6. P. Tolstoy, "The archaeology of the Lena Basin and its New World relationships, part II," *Am. Antiquity*, **24**, pt. 1, 63 (1958).

7. G. J. Groot, *The Prehistory of Japan* (Columbia Univ. Press, New York, 1951), plate XI.

8. Okladnikov, *loc. cit.*

9. J. L. Giddings, "Round houses in the western arctic," *Am. Antiquity*, **28**, pt. 1, 121 (1957).

10. G. Bonch-Osmolovsky and V. Gromov, "The Paleolithic in the Union of Soviet Socialist Republics," *Intern. Geol. Congr. Rept. of XVI Session, U.S.A., 1933* (1936), vol. 2, pp. 1291–1311.

11. F. H. H. Roberts, "Additional information on the Folsom complex," *Smithsonian Misc. Collections*, **65**, No. 10 (1936).

12. F. J. Soday, "The Quad Site, a Paleo-Indian village in northern Alabama," *Tenn. Archaeologist*, **10**, No. 1, 1 (1954).

13. S. Byers, "Bull Brook—a fluted point site in Ipswich, Massachusetts," *Am. Antiquity*, **19**, 343 (1954).

14. E. H. Sellards, *Early Man in America: A Study in Prehistory* (Univ. of Texas Press, Austin, 1952).

15. J. L. Giddings, "The Denbigh flint complex," *Am. Antiquity*, **16**, 192 (1951).

	NORTHWEST CANADA	HUDSON BAY	SASKATCHEWAN	MANITOBA MINNESOTA	WISCONSIN	ILLINOIS	MICHIGAN	OHIO	NEW YORK
500 A.D.	Birnirk								
				Anderson Laurel		Late Hopewell		Late Hopewell	Point Peninsula III
1 A.D.	Firth River	Dorset II			Hopewell		Hopewell		
						Early		Hopewell Adena	Point Peninsula II
		Ti-Site	Sandy Creek						
500 B.C.					Early Woodland		Early Woodland		
	Late New Mountain	Thyaszi (Knife River)							
						Early Woodland	Andrews Complex	Adena	
1000	Fisherman's Lake		Pelican Lake	Larter					Orient Point Peninsula I
				Old Copper		Red Ochre			
1500	N.T. Docks		Thunder Creek	White Shell	Old Copper		Glacial Kame	Glacial Kame	
	Early New Mountain	Igloolik		Minnesota Man					
2000			Agate Basin				Old Copper		
	Flint Creek				Old Copper			Raisch Smith	Laurentian
						Ferry Site			
2500	Great Bear								
							Old Copper		Lamoka
3000									
						Modoc Zone III			
3500									
4000	Sandy Lake		Cody Complex	Browns Valley					
									Fluted Points?
	Franklin Tanks				Plano Points				
5000									
			Fluted Points		Fluted Points	Dalton Points	Early Archaic	Early Archaic?	
7000						Fluted Points			
							Fluted Points	Fluted Points	Fluted Points
9000									
11,000 B.C.									

Fig. 10–1. Chronology chart of prehistoric complexes in northern North America.

	WESTERN EUROPE	KANSU	SHANSI-HONAN	JAPAN	BAIKAL	MIDDLE & UPPER LENA	LOWER LENA	KOLYMA CHUCKCHI	NORTHERN & CENTRAL ALASKA
500 A.D.									Birnirk
				Late Yayoi (Iron Age)				Birnirk	
						Iron Age	Iron Age		Old Bering Sea
1 A.D.			Han						
								Welen-Okvik	
						Kullaty I (Bronze)	Bronze		
500 B.C.	Iron Age			Early Yayoi					
						Ymyiakhtakh			
								Chirovoye?	
			Chou						
1000					Shivera			Yakitikiveem?	
		Ssu-Wa	Anyang (Bronze)	Late Jomon		Kullaty II	Kylarsa		
1500		Hsin-Tien			Glaskovo				
		Ma-Chang	Pu-Chao-Chai			Turukta Kullaty III	Kyrdal Uolba Lake	Pomaskino	
					Kitoi				Denbigh Flint Complex
2000		Pan-Shan	Hou-Kang I Yang Shao			Bestyakh	Chokurovka?		
				Middle Jomon	Serovo				
	Bronze Age								Anuktuvak Pass Campus Site
2500	Late Neolithic								
3000	Early Neolithic			Early Jomon					
		Early Neolithic	Early Neolithic		Isakovo				
3500									
4000	Late Mesolithic				Khin				
5000									
6000					Verkholenskaia Gora				
7000						Makarovo			
8000	Early Mesolithic				Afontova II				
						Chastinka			
9000									
10,000					Malta-Bureti				
11,000 B.C.	Late Magdalenian								

Fig. 10–2. Chronology chart of prehistoric complexes in eastern Asia and north and central Alaska.

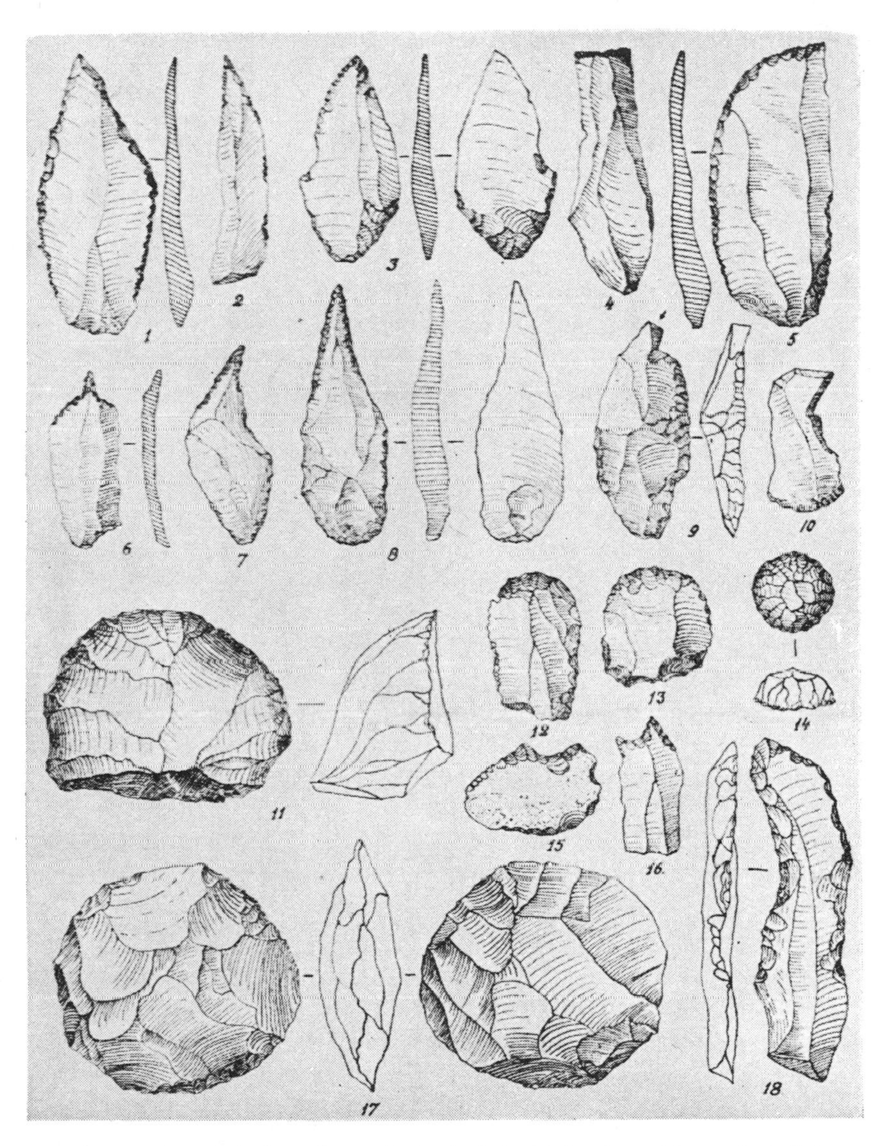

Fig. 10–3. Flint tools from the early level of the Malta Site, Siberia. 1–3, Points worked on one face only; 4, blade with transverse retouch; 5, blade with curved edge; 6–8, perforators; 9, 10, burins; 11–14, scrapers; 15, 16, concave scrapers; 17, disk-shaped implement; 18, notched blade. (From Bonch-Osmolovsky and Gromov, see note 10.)

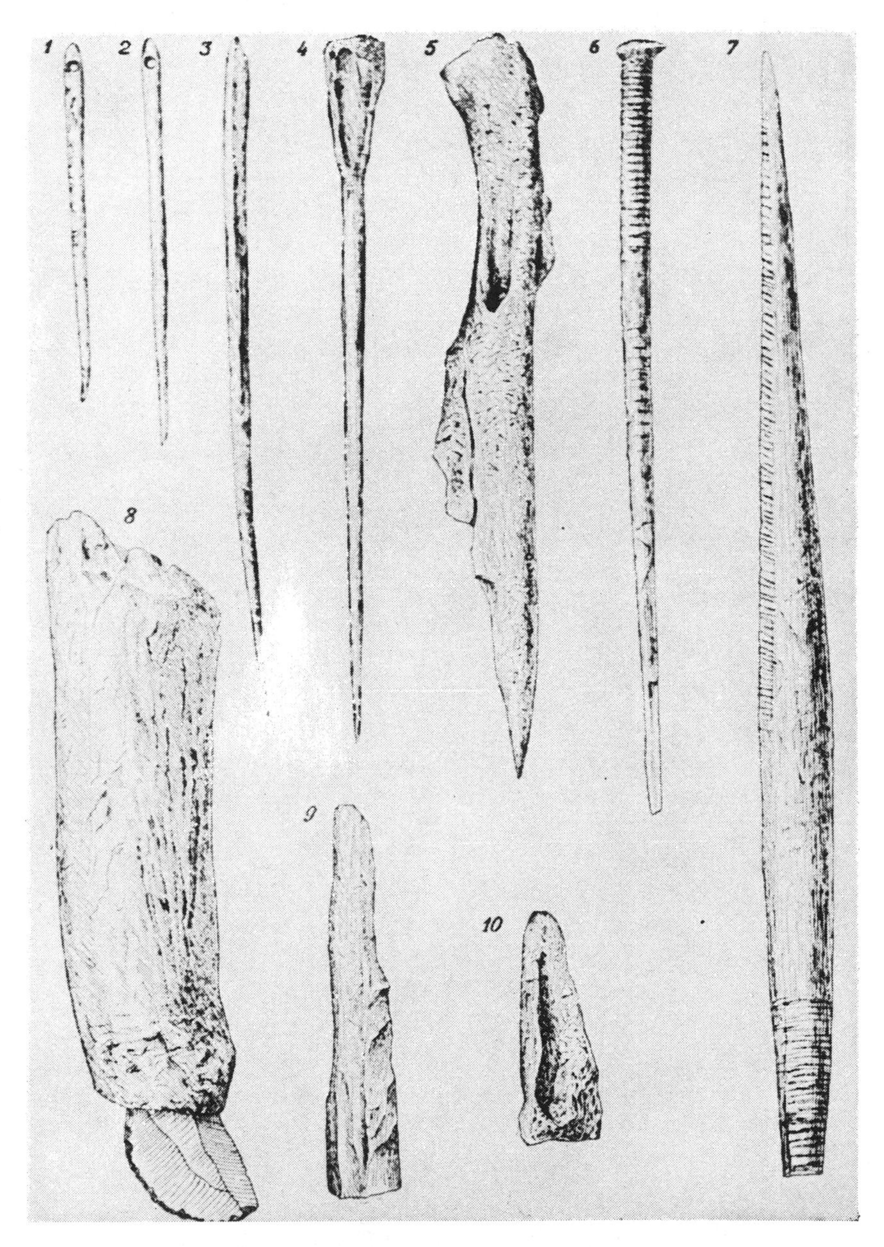

Fig. 10–4. Bone and antler objects from the early level of the Malta Site, Siberia. 1, 2, Needles; 3–6, points or awls; 7, spearpoint with beveled proximal end; 8, reindeer antler haft and scraper; 9, 10, worked bone fragments. (From Bonch-Osmolovsky and Gromov, see note 10.)

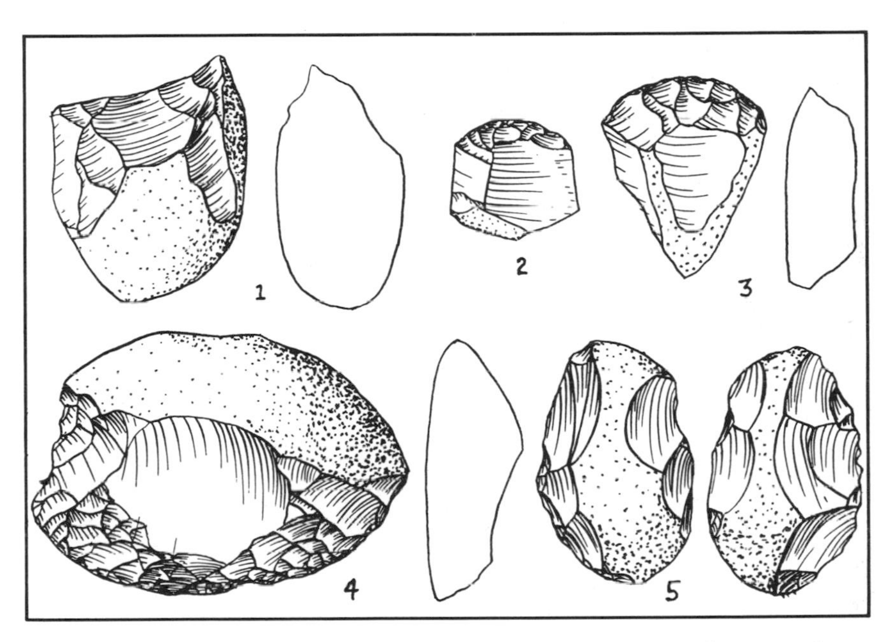

Fig. 10–5. Flint tools from Shishkino, a late Palaeolithic site in the Upper Lena Valley. 1, 4, 5, Pebble core scrapers or choppers; 2, 3, end scrapers. (After Okladnikov, see note 4.)

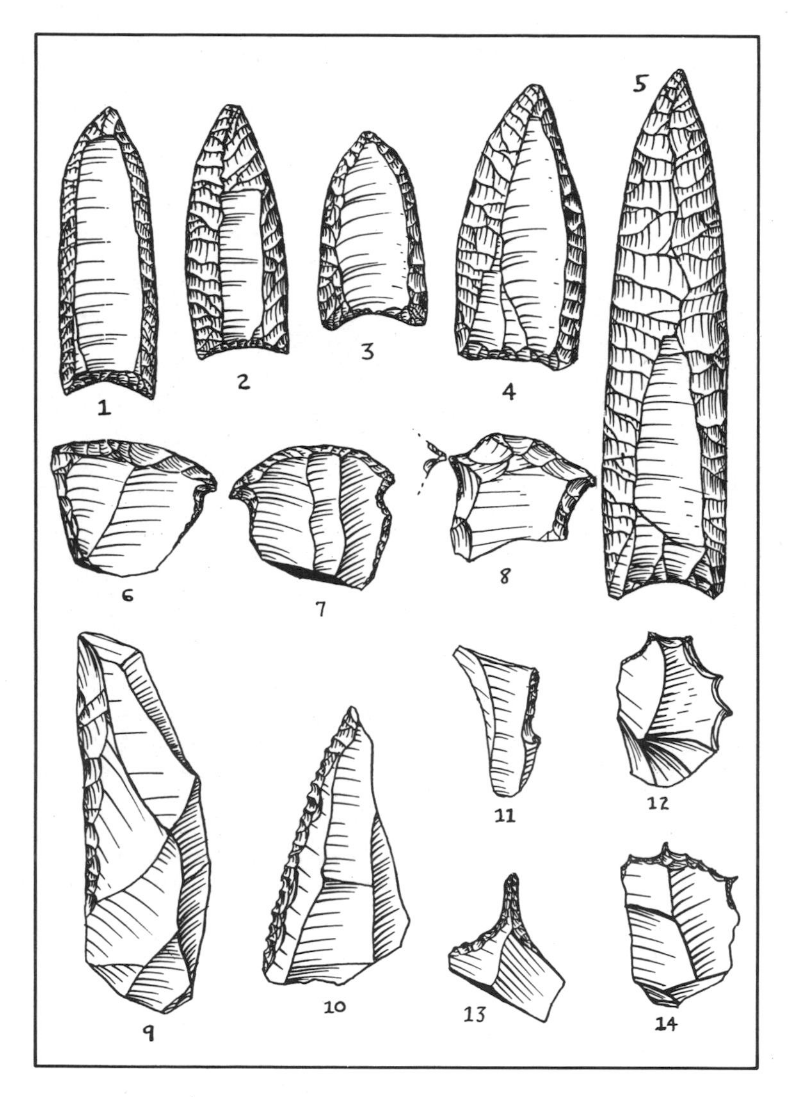

Fig. 10–6. Selected flint implements of the fluted blade complex in the United States. 1–5, Fluted projectile points from western and eastern United States; 6–8, end scrapers combined with gravers; 9, 10, side scrapers; 11, used flake; 12–14, gravers and drill. Implements 1, 6, 9, 11, and 13 are from the Lindenmeier Site, Colorado (see note 11). Implements 2, 7, 10, and 14 are from the Quad Site, Alabama (see note 12). Implement 4 is from Bull Brook, Massachusetts (see note 13). Implement 5 is from Black Water No. 1 locality, New Mexico (see note 14). Implements 3 and 8 are from Michigan. (Museum of Anthropology, University of Michigan.)

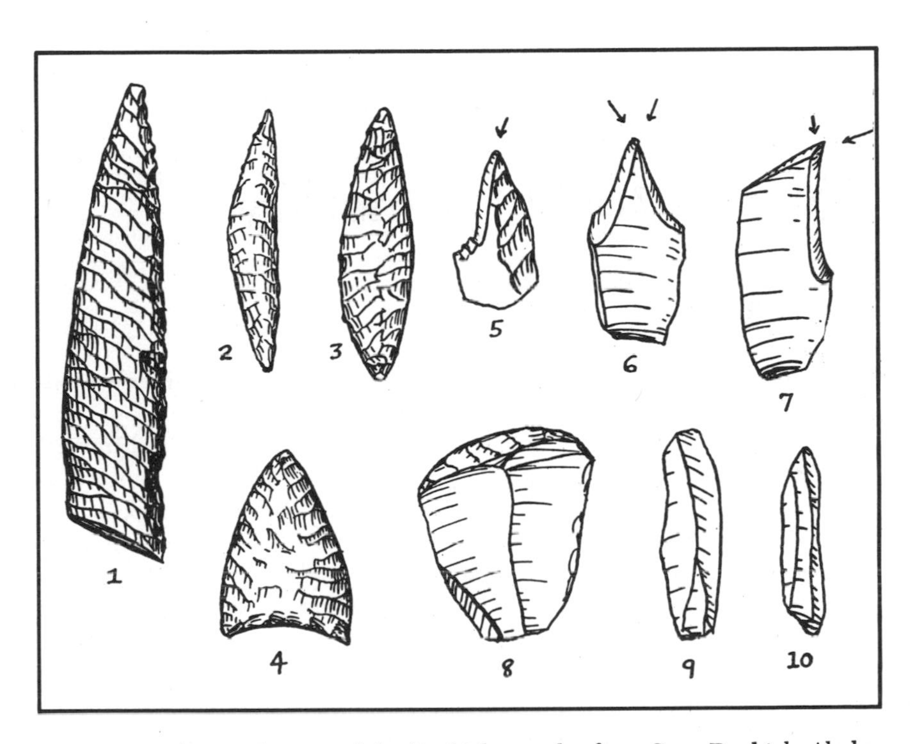

Fig. 10–7. Flint implements of the Denbigh complex from Cape Denbigh, Alaska. 1, Side blade about 3.3 cm long; 2, crescentic side blade; 3, plano-convex point; 4, small triangular point (harpoon or arrow?); 5, small angle burin; 6, chisel burin; 7, angle burin; 8, small end scraper; 9, 10, retouched lamellar blades. (Drawn from illustrations in Giddings, see note 15.)

Chapter 11

New World Prehistory

Gordon R. Willey

The prehistory of the New World is so multifaceted and complex that synthesis demands not only compression but rigorous selection. What strands of human activity can be followed most easily through the maze of the past? Which elements are the significant ones? These are always troublesome questions for the archaeologist, and in the present case they are made more so by the tremendous range of space and time and by the quantity and quality of the data with which we are dealing. It is difficult to fix consistently upon criteria of comparison. The best we can do is to adhere to those universal themes of man's existence that leave their mark in or upon the earth: technology, environmental adaptation, subsistence, and settlement. These were not necessarily determinative of the form and elaboration of other aspects of man's life, but they provide a background and a base that are necessary to the understanding of societies and cultures in pre-Columbian America.

Major Problems in New World Archaeology

Before beginning this account of New World prehistory it will be well to review some of the major problems confronting the American archaeologist, for it will be evident that the tentative conclusions that I have reached about these problems give the outline and structure to the present chapter. They are problems not unlike those of Old World prehistory[1, 2] in that they are concerned with the great changes in man's adaptations to his natural and social environments.

Most briefly, and in approximate chronological order, these problems are as follows:

1. Who were the earliest inhabitants of the New World? Were they food gatherers comparable in their simple subsistence technology to the peoples of the Old World Lower and Middle Palaeolithic?

2. Where and at what time did the American big-game-hunting specialization of the Pleistocene arise? What were its relationships to the possible earlier food gatherers mentioned above? What were its relationships to the big-game-hunting tradition of the Old World? What happened to the pattern?

3. What were the origins and relationships of the specialized food-collecting subsistence patterns of the post-Pleistocene? Did Asiatic diffusions and migrations play a part in these developments, especially in the Arctic and Boreal zones?

4. Where and when were food plants first domesticated in the New World, and what was the effect of this on society and culture?

5. What is the history of pottery in the New World?

6. At what period and in what regions did sedentary village life based upon farming arise in the New World, and what was the history of the spread of this pattern in native America?

7. What was the nature of sedentary village life in the New World in those areas or regions where plant cultivation was poorly developed or lacking, and when did it occur? To what extent were such cultures and societies dependent upon the diffusion of ideas and elements from the village-farming pattern?

8. When and how did the native civilizations of Nuclear America come into being? What were their relationships within the Nuclear sphere? What were their relationships to non-Nuclear America?

In the statement of these problems and in the discussion that follows, certain terminology is used that needs explanation. This terminology also relates to the three diagrammatic charts (figs. 11–1 to 11–3) that summarize New World prehistory in broad eras or stages of subsistence technology (earlier chronological ranges) or settlement types (later chronological ranges). The term *food gathering* is applied to subsistence patterns where the gathering of wild plant foods or the hunting of animal life lacked regional specialization or technological diversification. This usage follows that of Braidwood in Old World archaeology.[3] *Food collecting*, in contradistinction, implies both specialization and diversification in the taking and utilization of wild plant and animal foods. The other terms descriptive of types of subsistence and settlement—*incipient cultivation, village farming, towns and temples, cities*, and a few other special terms of this nature—are defined below.

The geographical arrangements and the designations of the charts deserve a word. Figure 11–1 is a cross section for an area that runs north and south through the western axis of the hemisphere. The name *Nu-*

clear America refers to the southern two-thirds of Mexico, all of Central America, and Andean and coastal Colombia, Ecuador, and Peru, with adjacent portions of Bolivia. This was the heartland of native American agriculture and the seat of the two pre-Columbian centers of civilization, one in Middle America (Mexico-Guatemala) and the other in Peru-Bolivia.[4] There is a column for each of these two centers on the chart, and the column between, headed "Intermediate," refers to what I am calling the "Intermediate area" of southern Central America, Colombia, and Ecuador.[5] To the north of Nuclear America is western North America, divided into the Southwest culture area and the adjacent Great Basin area. Under "Southern South America" are columns headed "South Andes" and "Pampas-Patagonia." Fig. 11–2 is a cross section for an area extending from the Intermediate area of Nuclear America eastward across Venezuela, then southeastward to the Amazon drainage basin and eastern Brazil, and finally south to the Pampas-Patagonia region. In Fig. 11–3 the "Middle America" column is repeated under "Nuclear America," and the cross section is extended to include the North American eastern woodlands and plains areas. The charts are highly schematic, and only a small number of archaeological cultures, or phrase names, has been entered in the columns for various areas. (These names appear in small letters.)

The point should be made that the diagonal and curving lines which mark off the major subsistence and settlement types on the charts are not impermeable ones.[6] (See Fig. 11–6.) Influences and traits crossed these lines, frequently moving outward from areas of cultural complexity and intensity into areas of simpler cultures. Such traits were often assimilated by the receiving groups without effecting basic changes in subsistence or settlement. In some instances suspected diffusions of this kind are indicated on the charts by means of arrows.

Pleistocene Food Gathering (?)

There are scattered finds in the Americas which suggest by their typology and chronological position that they may be the remains of early food-gathering societies.[7,8] These artifacts include rough, percussion-chipped flint choppers, scrapers, and possibly knives or points, and occasional worked bone splinters. In some places, such as Tule Springs, Nevada, or Friesenhahn Cave, Texas, these crude weapons and tools have been found associated with the bones of extinct Pleistocene mammals, so it is likely that some hunting, even of large game, was practiced.[9] In general, however, the technological aspects of the implements show a lack of specialization toward hunting or toward any other particular means of obtaining food. In this the artifacts, and the inferences made from them, are analogous to those for the food-gathering cultures of the Old World lower and middle Palaeolithic.[10]

In age and geological placement, such putative early food gatherers in the Americas are not, however, comparable to those of Asia or any part of the Old World. In other localities such as the lowest levels of Danger Cave, Utah,[11,12] or Fishbone Cave, Nevada,[13,14] the assemblage can be no older than the final Wisconsin advance. Still other artifact assemblages that suggest an unspecialized food-gathering economy are not satisfactorily dated.[15]

Pleistocene Big-Game Hunting

Sometime during the last Wisconsin Interglacial era, or possibly even earlier, inhabitants of the North American continent entered upon a way of life that was based upon the pursuit and killing of the great Ice Age mammals, such as the mammoth, the mastodon, the camel, and later the buffalo. The origins of this life pattern are unknown. There are no visible antecedents in the possible earlier food-gathering cultures of the Americas. There is, it is true, a general correspondence between this New World specialized hunting of Pleistocene fauna and what was going on in the Old World in the approximately coeval Upper Palaeolithic stage; yet even this possibility of a connection with the Old World does not provide a reasonable source for the big-game-hunting complexes of the New World, with their distinctive and highly specialized equipment. Apparently the forms which are most indicative of the American big-game-hunting technology are New World inventions.

The technical equipment associated with big-game hunters in the Americas includes lanceolate projectile points shaped by pressure-flaking. These are frequently distinguished by a channel fluting on both faces of the blade. A variety of skin-scraping tools accompanies the points as they are found in camp sites, "kills," and butchering stations.[16] The best documented of these discoveries come from the North American high plains in eastern New Mexico, Colorado, and Texas, and there are others from southern Arizona southward into Mexico. Some finds, such as those of the lower layer of Sandia Cave, New Mexico, may date back to before 15000 B.C.[17,18] The Sandia complex is characterized by a lanceolate single-shouldered projectile point. Other discoveries, such as Clovis and Folsom, appear to be later, ranging perhaps from 15000 to 7000 B.C. The projectile points of both the Clovis (Fig. 11–4) and Folsom complexes are of the fluted form.[19] There are also a variety of lanceolate, unfluted points that appear to mark a horizon subsequent to the Folsom. These include the Angostura, Scottsbluff, Plainview, and Eden types (see Fig. 11–2).[20]

The spread of big-game hunting in the Americas took place during and in the first or second millennium after the final Wisconsin substage, the Mankato-Valders. The total span of time of this dissemination appears to have been from about 9000 to 5000 B.C. Finds of fluted projectile points throughout the eastern woodlands of North America indicate the former prevalence of the pattern there.[21] The Iztapan and Lerma remains in central and northeastern Mexico,[22] the El Jobo points of Venezuela,[23] the Aympitín industry of the Andes and southern South America,[24] and the Magellan I culture of the Straits[25] give the geographical range of the early big-game-hunting societies.

The fate of the big-game-hunting pattern is better known than its beginnings. After 7000 B.C. and the glacial retreats, there was a shrinkage of the total territory in which the big herbivores could be hunted. The intermontane basins and the range country of western North America became more arid, and a similar climatic shift took place in southern South America. After 5000 B.C., with a still greater increase in warmth and dryness, big-game hunting persisted in the central zones of the old continental grasslands, such as the North American plains and the Argentine pampas. In these areas a modified hunting pattern, based, respectively, on the buffalo and the guanaco, continued into later times. Elsewhere, populations of hunters probably were forced into new environmental situations and new subsistence habits.

Later Food Collecting and Hunting

These new subsistence patterns can best be described as food collecting. They are differentiated from the possible earlier food-gathering pattern in that they show specialization in the exploitation of regional environments and much more effective technological equipment. Although the taking of game is a means of subsistence in some of these patterns, it is not the old big-game hunting of the Pleistocene. The food collectors, for the most part, developed cultures of greater material wealth, larger communities, and more stable settlements than their predecessors. There were exceptions to this, particularly in areas or regions of severe natural limitations and in the earlier periods of the food-collecting patterns; but on the average, and certainly at the optimum, these generalizations hold true.[26]

Chronologically, most of the food-collecting patterns had their beginnings in the span of time between about 6000 and 2000 B.C. There were, however, exceptions to this, as in the North American Great Basin, where the specialized collecting of wild seeds was well established as early as 7000 or even 8000 B.C.[27] As this is the same general area where clues to the most ancient food gatherers are found, it may be that there is a continuity in the Great Basin from the unspecialized gathering of the early Pleistocene to the later food collecting. According to this interpretation big-game hunting would be only partially represented or would be absent in an intervening sequence position.[28] This relationship is expressed in Fig. 11–1.

This possibility of continuities between the North American desert food collectors and earlier resident cultures and populations brings attention to the larger question of the origins of the New World food-collecting patterns and peoples in general. There are three logical possibilities: (i) food-collecting societies and cultures were derivative, arising from the earlier food gatherers; (ii) members of such societies were the descendants of big-game hunters who were forced by the changing climatic conditions that followed the end of the Wisconsin glaciation to make readjustments; or (iii) they were more recent arrivals from the Old World by way of the Bering Strait. It seems quite likely that all three explanations may be useful, according to the particular geographical areas involved, and I have already mentioned the first two. The third explanation, that new arrivals from Asia played a part, is very probably correct insofar as the development of food-collecting cultures in northern North America is concerned. I have in mind particularly the northeastern woodlands, the northwest Pacific coast, and the subarctic and arctic. Elsewhere Asiatic influences were almost certainly of less direct account.

There are several major food-collecting patterns in the New World, and we can only skim over these very briefly. I have referred to what has been called a Desert pattern.[29] The long depositional histories at Danger Cave, Utah,[30] Leonard Rock Shelter, Nevada,[31, 32] and Fort Rock Cave, western Oregon[33, 34] are representative, and the basketry and crude milling stones found at these sites testify to a seed-collecting and seed-grinding subsistence. A similar story is recorded in the Cochise culture of southern Arizona-New Mexico,[35] and there are evidences of this Desert pattern in Mexico as well.[36]

In the woodlands of eastern North America there is another collecting pattern that shows an adaptation to forest and riverine conditions in hunting, utilization of wild plants, fishing, and catching shellfish. Such sites as the Graham Cave, in Missouri,[37] suggest that there was a transition in the eastern woodlands area, at about 7000 B.C., from big-game hunting to food collecting. In the ensuing millennia these eastern Woodland collecting cultures, subsumed under the name *Archaic* in much of the literature,[38] underwent progressive adaptations to regional conditions. By 3000 B.C. they were characterized not only by rough grinding stones and specialized projectile points but by numerous items of polished stone, such as vessels, celts, weights for throwing sticks, and various ornamental or ceremonial objects. The Indian Knoll, Kentucky,[39, 40] and Lamoka, New York,[41, 42] phases are typical of their particular regions. Many of the Archaic sites are huge heaps of shells situated along rivers or on the Atlantic Coast. Such locations were undoubtedly suitable for a semisedentary, or even sedentary, existence.

Along the Pacific Coast of North America there was another food-collecting pattern which paralleled in many ways that of the eastern Woodlands. Here, by 2000 B.C. if not earlier, semisedentary societies based upon fishing and acorn gathering were established all along the coast from southern Alaska to southern California.[43] In South America there were also ancient fishing societies along the coasts. The Quiani phase[44] of northern Chile displays this adjustment. On the Brazilian coast are the huge *sambaqúis*, piles of shell refuse containing the skeletons and artifactual remains of food-collecting peoples who lived along these shores probably as much as two millennia before the beginning of the Christian Era.[45] Coastal shell-mound dwellers are also known from Venezuela at about this same period.[46, 47]

I have mentioned that in both the North American and the South American plains there were retentions of big-game-hunting patterns into later times; even these cultures, however, show the result of contact with the neighboring food collectors in their possession of an increasing number of food-grinding implements. This is exemplified in the later North American Plains phases, such as the Signal Butte I,[48] and by the later phases in the Strait of Magellan sequence and on the Argentine pampas.[49]

Incipient Cultivation

The change from food collecting to a subsistence based upon plant cultivation was one of the great turning points in human prehistory. This is true of the New World as well as the Old, and there are indications in both hemispheres that this switch-over was not a rapid one, but that it was effected only over a period of experimentation. It is this era of experimental or incipient cultivation in the New World that I now wish to examine.[50]

In the Americas it would appear that there may be at least four distinct and semi-independent traditions of incipient farming. Two of these are Nuclear American. The northern one, the probable propagator of maize, was located in Middle America and in the adjacent deserts of northern Mexico and the southwestern United States; the southern one had its focus on the Peruvian coast. A third incipient-cultivation tradition centered somewhere in the tropical forests of the Amazon or Orinoco. Its existence is difficult to demonstrate archaeologically, but such a tradition is needed to explain the domestication of manioc and other root crops. A fourth, and distinctly lesser, tradition rose in eastern North America in the Mississippi Valley system.

The earliest evidence for incipient cultivation in any of these traditions comes from northern Nuclear America. The region is the northeastern periphery of Middle America, in the semiarid hill country of Tamaulipas. Here, preserved plant remains were taken from the refuse deposits of dry caves. In the Infiernillo phase, dating from 7000 to 5000 B.C., there are traces of domesticated squash (*Cucurbita pepo*) and of possible domesticates of peppers, gourds, and small beans. The cultural context is that of North American desert food collectors. There are, in addition to flint implements, net bags of yucca and maguey cords and woven baskets of a rod-foundation type. In the succeeding Ocampo phase, from about 5000 to 3000 B.C., beans were definitely domesticates. After this, between 3000 and 2000 B.C., a primitive small-eared maize came into the sequence in the La Perra and Flacco phases. R. S. MacNeish, who excavated and studied the Tamaulipas caves, has estimated the composition of food refuse of the La Perra phase to be as follows: 76 per cent wild plants, 15 per cent animals, and 9 per cent cultigens. The La Perra and Flacco artifact inventories are not strikingly different from inventories of the earlier phases, although they demonstrate a somewhat greater variety of manufactures and an increased concern for seed foods. A few centuries later, at about 1500 B.C., an archaeological complex that is representative of fully settled village farming appears in the region. Thus, the Tamaulipas sequence offers a more or less unbroken story of the very slow transition from food collecting supplemented with incipient cultivation to the patterns of established cultivation.[51]

Early and primitive maize is also found to the north of Tamaulipas, actually outside of Nuclear America, in New Mexico. At Bat Cave, corncobs from refuse of a Cochise-affiliated culture date between 3500 and 2500 B.C.[52] This is as early as the La Perra maize, or even earlier.

As yet, neither archaeologists nor botanists have been able to determine the exact center of origin for domestication of maize in the New World, and it may be that this important event first took place in northern Middle America and in southwestern North America, where the intensive use of wild seeds in a food-collecting economy in a desert area provided a favorable setting. There remains, nevertheless, the very good possibility that a territory nearer the heart of Nuclear America and more centrally situated for the spread of maize in the hemisphere—an area such as southern Middle America—played this primary role in the cultivation of maize. [Chapter 14 describes studies bearing on this point that were done after Willey's chapter was written—Ed.]

Coastal Peru, at the southern end of Nuclear America, provides a rainless climate and splendid conditions for preservation of organic materials in open archaeological sites, and it is in Peru that we have glimpsed what appears to be a second tradition of incipient plant cultivation in Nuclear America. At Huaca Prieta, in a great hill of marine shells, sea urchin spines, ash, and other debris, cultivated squash, peppers, gourds, cotton, and a local bean (*Canavalia*) were found, along with an abundance of wild root plants and fruits. The people who raised and gathered these crops and seafoods lived at Huaca Prieta at least 2,000 years before the Christian Era. Whether there was, however indirectly, an exchange of domesticated plants between these early Peruvians and their contemporaries in Middle America is not certain. Such connections could have existed; or the beginnings of cultivation may have been truly independent of each other in these two areas of Nuclear America. Definite connections between early farmers of Middle America and of Peru appear, however, by 700 B.C. with the sudden presence of maize in Peru.[53] This maize was not, like that at Bat Cave or in the La Perra culture of Tamaulipas, of an extremely primitive kind. It was brought, or it spread, to Peru as a relatively well-developed plant, and it serves as a link to Middle America. We may conclude that Nuclear America possessed, from this time forward, a single major horticultural tradition, but by this time we have also passed beyond the chronological limits of cultivation incipience.

An ancient tradition of plant cultivation in the South American tropical forest[54] is based upon the presumption that a long period of experimentation was necessary for the domestication of such tropical root crops as bitter and sweet manioc (*Manihot utilissima, M. Api*) and the yam (*Ipomoea batatas*). It seems reasonably certain that these domesticates date back to before 1000 B.C. in lowland Venezuela. This is inferred from the presence of pottery griddles, of the sort used for cooking manioc cakes in later times, in the Saladero phase at the Orinoco Delta by this date.[55] Also, the early archaeological phase of Momíl I, in Caribbean Colombia, has the pottery manioc griddle.[56] The dating of Momíl I is debatable, but some of the ceramic traits suggest a date as early as 2000 B.C. Saladero and Momíl I are, however, outside the chronological and developmental range of incipient cultivation patterns. They appear to be village sites based upon the cultivation of root crops, and as such they are comparable to, although historically separate from, village farming based on maize. I shall return to this point farther along. For the present I bring these sites into the discussion because their existence implies centuries, or even millennia, of prior incipient root-crop cultivation in tropical northern South America.

A fourth tradition of incipient cultivation for the New World derives from the cultivation of local plants in the Mississippi Valley by as early as 1000 B.C. These plants include the sunflower, the goosefoot (*Chenopodium*), and the pumpkin (*Cucurbita pepo*).[57] This domestication may have been in response to stimuli from Middle America, or it may have been an entirely independent development. This Eastern Woodland incipient-cultivation tradition was undoubtedly but a minor part of the food-collecting economy for a long time. Just how important it ever became, or how important the early diffusion of maize was to eastern United States cultures of the first millennium B.C., are crucial problems in the understanding of the area. I shall return to them later.

Appearance of Pottery

Before taking up the rise of village farming in Nuclear America and its subsequent spread to other parts of the hemisphere, let us review the first appearances of pottery in the New World. Obviously, the line indicating the presence of pottery on the charts is not comparable to the lines indicating type of subsistence or settlement (figs. 9–1 to 9–3). American archaeologists no longer consider pottery to be the inevitable concomitant of agricultural village life, as was the fashion some years ago. Still, ceramics, because of their very ubiquity and durability, are an important datum in many prehistoric sequences. Their presence, while not a necessary functional correlate of farming, at least implies a certain degree of cultural development and sedentary living.

There seem to be two pottery traditions for native America. Curiously, the ages of these two pottery traditions—in the broadest sense of that term—may be about the same, 2500 B.C.

One of these pottery traditions, which we shall call the Nuclear American, is believed to be indigenous, but we can be no more specific about its geographic point of origin than to state that this is somewhere in the central latitudes of the New World. Actually, the earliest radiocarbon dates on the Nuclear American pottery tradition come from coastal Ecuador, in the Valdivia phase (Fig. 11–5), and are from about 2500 to 2400 B.C.[58] There are also early dates on pottery generally similar to that of Valdivia from Panama (about 2100 B.C.[59, 60]). Thus, these earliest ceramic datings for Nuclear America are not from Middle America or Peru but from the Intermediate area, and this may be significant in following up origins, although the record is still too incomplete to say for sure. Both the Ecuadorean and the Panamanian early potteries are found in coastal shell-mound sites, and in connection with cultures about whose means of subsistence it is not easy to draw inferences, except to say that full village farming was unlikely. Possibly marine subsistence was supplemented with incipient cultivation, although we have no proof of this. The Valdivia and the Panamanian (Monagrillo) pottery is reasonably well made and fired, the forms are rather simple, and the vessels are decorated with incisions, excisions, punctations, and very simple band painting. These early Ecuadorean and Panamanian styles may be part of a stratum of ancient Nuclear American pottery that underlies both Middle America and Peru. There are some indications that this may be the case, although the oldest pottery so far known in the Middle American and Peruvian areas dates from several centuries later.[61] In Fig. 11–1 the interpretation is offered that Nuclear American pottery is oldest in southern Middle America (for this there is as yet no evidence) and in the Intermediate area (for this there is evidence). Whatever the point of origin for pottery in Nuclear America, there is fairly general agreement that the ceramic ideas generated there carried to much of outlying North and South America.

The second major pottery tradition of the Americas is widely recognized by the term *Woodland.* Apparently not indigenous, but derived from northern Asia, it is best known from the eastern woodlands of New York and the Great Lakes region. So far, its presumed long trek from the arctic down through Canada has not been traced.[62] Woodland pottery is generally of simpler design than the early Nuclear American wares. Of an elongated form, it is frequently finished only with cord-marked surfaces (Fig. 11–6). As already noted, the oldest of this cord-marked pottery in the Americas may go back to 2500 B.C.[63] Even if this early dating is not accepted, there is little doubt that Woodland pottery was well established in eastern North America before 1000 B.C.

In spite of the fact that the Nuclear American and Woodland pottery traditions are so radically different, there are, interestingly, a few similarities. The most notable of these is the technique of rocker-stamping combined with incised zoning of plain surface areas, known in Nuclear America and in the eastern United States (Fig. 11–7). The distinctive rocker-stamped treatment of pottery was accomplished by impressing the soft, unfired surface of a vessel with either a small straight-edged implement manipulated rocker-fashion or, possibly, with a fine-edged disk used like a roulette. The impressions left on the pottery may be either plain or dentate, and they always have a characteristic "zigzag" appearance. Rocker-stamping is found in the Valdivia phase in Ecuador, and it also occurs at about 1000 B.C. in parts of Middle America and in Peru.[64] In eastern North America it is not found on the earliest Woodland pottery but is found on vessels which date from just a few centuries before the beginning of the Christian Era. Thus, the Nuclear American rather than the Woodland tradition has chronological priority in this trait in the New World.[65] Again, as with so many other problems that perplex Americanists, we can only refer to this without coming to any conclusions as to the timing and direction of the flows of possible diffusions. Nuclear American and Woodland ceramics may in some way be related, but at the present state of knowledge they appear to have different origins and substantially separate histories (Fig. 11–8).

Village Farming in Nuclear America

Braidwood and others have stressed the importance in the Old World of the threshold of the village-farming settled community.[66, 67] Although in its beginnings the agricultural village had a subsistence base that was no more adequate, if as ample, as that of some of the food-collecting communities, this base offered the potential in certain Old World localities that led, eventually, to civilization. In the New World a similar development was repeated in Nuclear America.

In the New World the line between incipient cultivation and village farming has been drawn at that theoretical point where village life is, in effect, sustained primarily by cultivated food plants.[68] In archaeology this distinction must be made by an appraisal of the size and stability of a settlement as well as by direct or indirect clues as to the existence of agriculture. In Nuclear America the earliest time for which we can postulate the conditions of village farming is the second millennium B.C. For example, in Middle America in the Tamaulipas sequence, the change-over from incipient cultivation to established cultivation takes place at about 1500 B.C.[69] Elsewhere in Middle America the known sequences begin with the village-farming stage, as at Early Zacatenco[70] (Valley of Mexico), Las Charcas[71] (Guatemalan Highlands), Ocós[72] (Pacific coast of Guatemala), and Mamom[73] (Maya lowlands).[74] In Peru the village-farming level is reasonably well defined with the appearance of maize in the Cupisnique phase and the shift of settlements back from the coast to the valley interiors. The date for this event is shortly after 1000 B.C.;[75] this suggests that the horizon for village farming may have sloped upward in time from Middle America to Peru (Fig. 11–1). For the Intermediate area, where I have noted the earliest occurrence of pottery in Nuclear America, the threshold of village farming is difficult to spot. In Ecuador, the phases succeeding Valdivia have a different ecological setting, being inland in the river valleys rather than on the immediate shores.[76] Perhaps, as in Peru, this correlates with the primary economic importance of plant cultivation. In Colombia, the Momíl II phase, which is represented by a stable village site area, is believed to have possessed maize.[77]

The foregoing discussion carries the implication that village farming was a pattern diffused through Nuclear America from a single area or region. Essentially, this is the point of view expressed in this essay. This is not to overlook the possibility that village agricultural stability may have arisen independently in more than one place in the New World. In fact, it apparently did just that in the tropical forests of South America. I am of the opinion, however, that in the Nuclear American zone the maize plant, genetically developed and economically successful, became the vital element in a village-farming way of life that subsequently spread as a complex. For the present, I would hazard the guess that this complex developed in southern Middle America and from there spread northward to Mexico and southward as far as Peru. This was, in a sense, its primary diffusion or spread. Afterward, there were secondary diffusions to other parts of the Americas.

The Village in Non-Nuclear America

These secondary disseminations of the Nuclear American pattern of village farming were responsible for the establishment of similar communities in areas such as southwestern North America, the southern Andes, lowland tropical South America, and the eastern woodlands of North America (see figs. 11–1 to 11–3). This process was relatively simple in southwestern North America and the southern Andes. The agricultural patterns were diffused to, or carried and superimposed upon, peoples with food-collecting economies of limited efficiency. In the Southwest, village farming and ceramics first appear at about the same time in such cultures as the Vahki, the Mogollon I, and the Basketmaker.[78] This was between 200 B.C. and A.D. 300. Moving from the south, the village-farming pattern pushed as far as the Fremont culture[79] of the northern periphery of the Southwest. In the southern Andes there is, as yet, no good hint of an early incipient-cultivation tradition, and, apparently, pottery and agriculture arrive at about the same time, integrated as a village-farming complex. This flow of migration or diffusion was from Peru-Bolivia southward. Pichalo I[80] of northern Chile marks such an introduction, as do the earliest of the Barreales phases[81] in northwest Argentina. The time is about the beginning of the Christian Era. Beyond the southern Andes the village-farming pattern did not diffuse onto the plains of the pampas or Patagonia.

The relationship of Nuclear American village farming to the tropical lowlands of South America was much more complex. There the maize-farming pattern was projected into an area in which village life already existed. This is indicated in Fig. 11–2 by the entry "Village Farming–Manioc" in the columns headed "Venezuela" and "Amazon." Sedentary village life based upon root-crop farming is estimated to be as old as 2500 B.C. This is a guess, and, if it is correct, these villages are older than the Nuclear American village sustained by maize. Perhaps the estimated date is too early; however, at 2000 and 1000 B.C., respectively (see Fig. 11–2), we have the villages of Momíl I and Saladero, which, apparently, were supported by root-crop cultivation. It is of interest to note that Momíl I, near the mouth of the Sinú River in Colombia, lies within the axis of Nuclear America; yet it differs from the succeeding Momíl II phase at the same site in being oriented toward manioc rather than maize. This suggests that, in the Intermediate area at least, tropical-forest farming patterns may have preceded farming patterns for maize in Nuclear America.

Relationships between village farming in Nuclear America and in eastern North America are also complicated. It is unlikely that the local incipient-cultivation tradition in eastern North America ever matured into a subsistence pattern that could have supported fully sedentary village life. Joseph R. Caldwell[82] has argued that, in its place, a steadily increasing efficiency in forest collecting and hunting climaxed at about 2000 B.C. in a level of "Primary Forest Efficiency" (see Fig. 11–3). Such a level, he concludes, offered the same opportunities for population stability and cultural creativity in the eastern woodlands as were offered by village farming. While agreeing with Caldwell that the efflorescence of Adena-Hopewell (about 800 B.C. to A.D. 200)[83] (Fig. 11–9) is the brilliant end product of a mounting cultural intensity in eastern North America that originated in the food-collecting or Archaic societies, I am not yet convinced that plant cultivation did not play an important role in this terminal development. And by plant cultivation I am referring to maize, brought or diffused from Nuclear America. There is, as yet, no good direct evidence of maize associated with either the Adena[84] or the contemporary Poverty Point[85] culture. Maize is, however, found with Hopewellian cultures,[86] although it has been assumed that it was of relatively little importance as subsistence at this time. I would argue that the riverine locations of Adena and Hopewell sites, together with the great size and plan of the ceremonial earthworks that mark many of them, make it difficult to infer an adequate subsistence if maize agriculture is ruled out.

To sum up briefly, the amazing cultural florescence of the eastern Woodlands in the first millennium B.C. has not yet been satisfactorily explained. This florescence rests upon a chronologically deep series of Archaic food-collecting cultures that were at least semisedentary, and it contains elements, such as pottery, which are probably of Asiatic derivation and which added to the richness of the Archaic continuum. But the sudden burst of social and cultural energy that marks the Adena culture cannot be interpreted easily without adding other factors to the equation, and perhaps these missing factors are maize agriculture and other stimuli from Middle America (see Fig. 11–3).

Village life is, of course, present in native America in the non-Nuclear areas under conditions where plant cultivation may be ruled out entirely. Settled villages developed on the northwest coast of North America, with population supported by the intensive food-collecting economy of the coast and rivers. The same is also true for the coast and interior valleys of California. It is significant, however, that in neither of these areas did aboriginal cultivation ever make much headway, while in eastern North America it became a staple of life in the later pre-Columbian centuries.

Temples, Towns, and Cities

In Nuclear America the town and eventually the city had beginnings in the settled farming village. A centralizing factor in this development was undoubtedly the temple. This earliest form of permanent structure usually had a flat-topped pyramidal mound of earth or rock as a base, and these mound bases of temples are found associated with some, but not all, of the village-farming cultures in Middle America.[87] At first, the importance of such a mound, and of the temple that stood on it, was probably limited to the immediate village. Sometimes these villages were small, concentrated clusters of dwellings; in other instances the settlement pattern was a dispersed one, with a number of small, hamlet-like units scattered at varying distances from the temple center. Later on, the temple, or temple and palace structures, became the focal point of what might be called a town[88] (figs. 11–10 to 11–12).

In Nuclear America the towns, like their antecedent villages, were either concentrated or dispersed. The former pattern developed in parts of Middle America, such as the valley of Mexico or the Guatemalan highlands, and in Peru; the latter was characteristic of the Veracruz-Tabasco lowlands or the Peten-Yucatan jungles of Middle America. In the towns the temple or ceremonial precinct was devoted to religious and governmental matters and to the housing of priests and of rulers and their retainers. The surrounding settlement zone, either scattered or concentrated, grew with increase in the numbers of farmers, artisans, or both. Trade was an important function of these towns.

In Nuclear America the town-and-temple community dates back to 800 B.C., a date that is applicable both to Middle America and to Peru. In the Intermediate area, between these two, town life was certainly pre-Columbian, but its date of origin is difficult to determine because there is a lack of adequate archaeological chronologies.[89]

In lowland South America, town-and-temple communities also antedate the Conquest, and it seems likely that these communities were, in part, the result of contact with and stimulus from the Nuclear American axis.[90] In the southern Andes the tightly planned clusters of rock and adobe buildings of the late archaeological periods of northwestern Argentina reflect town and city life in Peru (Fig. 11–13) and Bolivia.[91] Similarly, towns of the prehistoric southwestern United States relate to the Nuclear American zone. Development of these towns dates from sometime after A.D. 500, with an apogee in the Pueblo III and IV periods and in the Classic Hohokam phases.[92]

On the other great periphery of Nuclear America, eastern North America, Middle American town life, with its temple mound-and-plaza complex, entered the Mississippi Valley sometime between A.D. 500 and 1000 and climaxed the Mississippian or Temple Mound cultures shortly afterward.[93] Maize cultivation was an established part of this complex. Thus, in a sense, the thresholds of village farming and of the town-and-temple complex in the eastern woodlands, when these beginnings can be identified indisputably as of Nuclear American inspiration, are synchronous (Fig. 11–3).

There remains, however, as in our consideration of the village-farming level, the puzzle of the Adena-Hopewell cultures. As we have already noted, the Adena-Hopewell ceremonial mounds and earthworks, built between 800 B.C. and A.D. 200, are of impressive size. Some of them are comparable in dimensions, and in the amount of coordinated manpower necessary to build them, with the contemporary mounds of Middle America. Although the mounds of Middle America were usually temple platforms while the Adena-Hopewell tumuli were mounds heaped up to cover tombs and sacred buildings, this dichotomy should not be overstressed. Some mounds of Middle America also were tombs, or combined tombs and temples.[94] In any event, it is safe to conclude that the Adena-Hopewell mounds were structures which memorialized social and religious traditions and served as community nuclei, as the ceremonial building did in Middle America. Was there a historical connection between Middle America and the eastern woodlands at this time, and was Adena-Hopewell ceremonial construction influenced by the emergence of the town-and-temple concept of Middle America? There is no satisfactory answer at present, but the possibilities cannot be dismissed (see Fig. 11–3).

In Nuclear America the city developed from the town and temple, and there is no sharp division between the two. Size is, assuredly, one criterion but not the only one. These cities were the nerve centers of civilizations. They were distinguished by great public buildings and the arts. Formal pantheons of deities were worshipped in the temples under the tutelage of organized priesthoods. Populations were divided into social classes. Trade, in both raw materials and luxury items, was carried on in these cities, and science and writing were under the patronage of the leaders.[95] Not all of these criteria are known or can be inferred for any one city in the New World, but many of them do properly pertain to Middle America and Peruvian sites from as early as the first centuries of the Christian Era.

Cities in the New World seem to have been of two types, and these types may have their antecedents in the earlier dispersed and concentrated towns. The dispersed city, with its ceremonial center and outlying hamlets, appears to have been orthogenetic in its traditions and to have drawn upon, and commanded, a relatively limited geographical territory. The great lowland Mayan centers of the Classic period, such as Tikal or Palenque, are representative.[96] The concentrated city adheres more to the concept of the city in the western European definition of the term. It was a truly urban agglomeration. Its traditions were heterogenetic, and its power extended over a relatively large territorial domain. The city was, in effect, the capital of an empire. Peruvian Chanchan, Aztec Tenochtitlan, and, probably, the more ancient Mexican city of Teotihuacan represent the type.[97]

Although the cities and civilizations which developed in Middle America and Peru in the first millennium A.D. were unique and distinct entities in their own right, it is obvious that they also drew upon a common heritage of culture that had begun to be shared by all of Nuclear America at the level of village-farming life. This heritage was apparently built up over the centuries, through bonds of interchange and contact, direct and indirect. There are substantial archaeological evidences in support of this supposition.[98] During the era of city life these relationships continued, so that a kind of cosmopolitanism, resulting from trade, was just beginning to appear in Nuclear America in the last few centuries before Columbus.

In the outlands beyond Nuclear America, trade and influences from the cities followed old routes of contact and penetrated and were assimilated in varying degrees. In the south Andes there was the very direct impact of the Inca state in the final hundred years before the Spanish conquest,[99] and northward from Mexico, Toltec-derived influences reached the North American Southwest in relatively unadulterated form.[100] But, for the most part, the potentialities of the New World city for influencing and acculturating the "barbarian outlanders" were still unrealized when the Europeans entered the American continents.

Comments

Conclusions are inappropriate to a synthesis which, by its nature, is an outline of opinion, however tentative. Retrospective comment seems more in order.

A few things stand out. The early inhabitants of the New World were not remarkably different in their mode of life from the food gatherers and hunters of the Old World; yet even on these early horizons, and despite the relatively limited cultural inventories available, dissimilarities of form are striking. The interrelationships of the two hemispheres during the Pleistocene are still very vague.

Plant cultivation in the New World—its incipient rise and its culmination as the most effective subsistence base of the Americas—is, of course, analogous to happenings in the Old World. The important American plants, however, are of local origin. In the Western Hemisphere the incipience of cultivation followed the end of the Pleistocene, and was not a great deal later, perhaps, than in the Old World Middle East. Yet the period of incipience was longer here; over 5,000 years elapsed before village life was sustained by crop cultivation. Is this because the first New World cultigens were inadequate as foodstuffs, and it was necessary to develop, first, the cereal maize before agriculture was made profitable?

Although there is a high correlation between village life and agricultural subsistence in the New World, there were New World societies and cultures which maintained villages without plant cultivation. In at least one instance, that of the ancient Adena-Hopewell development of eastern North America, community centers comparable to those of the contemporary farmers of Middle America may have been built and supported without a full-fledged farming subsistence.

I have slighted in this presentation the relationships between Asia and the Americas that were probably maintained from Pleistocene times down to the European conquest. This is particularly true of the cultures of the northern half of North America, where it is certain that there were contacts between the Old World and the arctic, subarctic, and northwest Pacific coasts. For Nuclear America nothing at all has been said of the possibility of trans-Pacific contacts between the Old World civilizations of China and Southeast Asia and those of Middle America and Peru. This undoubtedly reflects my own bias, but I remain willing to be convinced of such events and their importance to the history of culture in the New World.

NOTES

1. R. J. Braidwood, *Science*, **127**, 1419 (1958).

2. G. R. Willey and P. Phillips, *Method and Theory in American Archaeology* (Univ. of Chicago Press, Chicago, 1958).

3. R. J. Braidwood, "Prelude to civilization," paper presented at a symposium on the "Expansion of Society," held at the Oriental Institute, University of Chicago, December 1958.

4. See A. L. Kroeber, *Anthropology* (Harcourt, Brace, 1948), pp. 779–81, and G. R. Willey, *Am. Anthropologist*, **57**, 571 (1955), for a discussion of Nuclear America.

5. The Intermediate area is defined by G. R. Willey in a paper presented at the 33rd International Congress of Americanists, San Jose, Costa Rica (1958).

6. R. J. Braidwood, *Science*, **127**, 1419 (1958).

7. G. R. Willey and P. Phillips, *Method and Theory in American Archaeology* (Univ. of Chicago Press, Chicago, 1958), pp. 82–86.

8. G. R. Willey, paper presented at the Darwin Centennial Celebration, University of Chicago, November 1959.

9. H. M. Wormington, "Ancient Man in North America," *Denver Museum Nat. Hist. Popular Ser. No. 4* (1957), pp. 197–98, 218.

10. R. Linton, *Trans. N.Y. Acad. of Sci.*, **2**, 171 (1949).

11. H. M. Wormington, *op. cit.*, pp. 193–95.

12. J. D. Jennings, "Danger Cave," *Soc. Am. Archaeology Mem. No. 14* (1957).

13. H. M. Wormington, *op. cit.*, pp. 192–93.

14. P. C. Orr, *Nevada State Museum Bull. No. 2* (1956).

15. There are many of these. In figs. 11–1 and 11–2 the Alto Paraná complex of southern South America is representative. See O. F. A. Menghin, *Ampurias*, **17**, 171; *ibid.*, **18**, 200.

16. H. M. Wormington, *op. cit.*, pp. 23–90.

17. *Ibid.*, pp. 85–91.

18. F. C. Hibben, *Smithsonian Inst. Publ., Misc. Collections*, **99** (1941).

19. H. M. Wormington, *op. cit.*, pp. 23–84.

20. *Ibid.*, pp. 107, 118, 138.

21. J. Witthoft, *Proc. Am. Phil. Soc.*, **96**, 464 (1952); D. S. Byers, *Am. Antiquity*, **19**, 343 (1954).

22. L. Aveleyra Arroyo de Anda, *Am. Antiquity*, **22**, 12 (1956); R. S. MacNeish, *Rev. mex. estud. antropol.*, **11**, 79 (1950).

23. J. M. Cruxent and I. Rouse, *Am. Antiquity*, **22**, 172 (1956).

24. A. R. Gonzales, *Runa*, **5**, 110 (1952); D. E. Ibarra Grasso, *Proc. Intern. Congr. Americanists 31st Congr., São Paulo*, **2**, 561 (1955).

25. J. B. Bird, *Geograph. Rev.*, **28**, 250 (1938).

26. These later food-collecting and hunting cultures are discussed by G. R. Willey and P. Phillips (see note 2, pp. 104–43) as the New World "Archaic" stage.

27. See references to level "D-II" in J. D. Jennings, "Danger Cave" (note 12), as an example.

28. C. W. Meighan, in *The Masterkey* [(Los Angeles, 1959), vol. 33, pp. 46–59], discusses the evidences for a big-game-hunting pattern in the Great Basin.

29. J. D. Jennings and E. Norbeck, *Am. Antiquity*, **21**, 1 (1955).

30. J. D. Jennings, *op. cit.*

31. H. M. Wormington, *op. cit.*, pp. 190–2.

32. R. F. Heizer, *ibid.*, **17**, 23 (1951).

33. H. M. Wormington, *op. cit.*, p. 184.

34. L. S. Cressman, *Southwestern J. Anthropol.*, **7**, 289 (1951).

35. E. B. Sayles and E. Antevs, "The Cochise Culture," *Medallion Papers, No. 24* (Gila Pueblo, Glove, Arizona, 1941).

36. The caves in Coahuila, excavated by W. W. Taylor, are representative. See W. W. Taylor, *Bull. Texas Archaeol. Soc.*, **27**, 215 (1956).

37. W. D. Logan, "An Archaic Site in Montgomery County, Missouri," *Missouri Archaeol. Soc. Mem. No. 2* (1952).

38. See *Am. Antiquity*, **24**, No. 3 (1959) (an issue devoted entirely to Archaic cultures of North America).

39. G. R. Willey and P. Phillips, *Method and Theory in American Archaeology* (Univ. of Chicago Press, Chicago, 1958), p. 116.

40. W. S. Webb, "Indian Knoll, Site Oh-2, Ohio County, Kentucky," *Univ. Kentucky Dept. Anthropol. and Archaeol. Publs.* (1946), vol. 4, No. 3, pt. 1.

41. G. R. Willey and P. Phillips, *Method and Theory in American Archaeology* (Univ. of Chicago Press, 1958), pp. 116–17.

42. W. A. Ritchie, "The Lamoka Lake Site," *N.Y. State Archaeol. Assoc. Researches and Trans.* (1932), vol. 7, pp. 79–134.

43. G. R. Willey and P. Phillips, *Method and Theory in American Archaeology* (Univ. of Chicago Press, 1958), pp. 133–37.

44. J. B. Bird, "Excavations in Northern Chile," *Am. Museum Nat. Hist. Anthropol. Paper No. 38* (1943), pt. 4.

45. See G. R. Willey, *Am. Antiquity*, **23**, 365 (1958).

46. J. M. Cruxent and I. Rouse, "An Archaeological Chronology of Venezuela," *Pan American Union Social Sciences Monographs*, **1** (6), 223–233 (1958–59).

47. I. Rouse, J. M. Cruxent, J. M. Goggin, *Proc. Intern. Congr. Americanists, 32nd Congr., Copenhagen* (1958), pp. 508–15.

48. W. D. Strong, *Smithsonian Inst. Publs. Misc. Collections*, **93**, No. 10 (1958).

49. J. B. Bird, *Geograph. Rev.*, **28**, 250 (1938).

50. See R. J. Braidwood [*Science*, **127**, 1419 (1958)] and G. R. Willey and P. Phillips [*Am. Anthropologist*, **57**, 723 (1955)] for discussion of the "Preformative" stage.

51. R. S. MacNeish, "Preliminary archaeological investigations in the Sierra de Tamaulipas, Mexico" [*Trans. Am. Phil. Soc.*, **48**, pt. 6 (1958)] is the basis for this summary of the Tamaulipas sequences.

52. P. C. Mangelsdorf, *Science*, **128**, 1313 (1958).

53. J. B. Bird, in "A Reappraisal of Peruvian Archaeology," *Soc. Am. Archaeol. Mem. No. 4* (1948), pp. 21–28.

54. C. O. Sauer discusses this possibility in *Agricultural Origins and Dispersals* (American Geographical Society, New York, 1952).

55. Cruxent and Rouse, *loc. cit.*

56. G. Reichel-Dolmatoff and A. Reichel-Dolmatoff, *Rev. colombiana antropol.*, **5**, 109 (1956).

57. R. M. Goslin in *The Adena People, No. 2*, W. S. Webb and R. S. Baby, Eds. (Ohio State Univ. Press, Columbus, 1957), pp. 41–46.

58. C. Evans and B. J. Meggers, *Archaeology*, 11, 175 (1958).

59. For discussion of the Monagrillo pottery, see G. R. Willey and C. R. McGimsey, "The Monagrillo Culture of Panama" [*Peabody Museum, Harvard, Papers*, **49**, No. 2 (1954)].

60. For the radiocarbon dating, see E. S. Deevey, L. J. Gralenski, V. Hoffren, *Am. J. Science Radiocarbon Suppl.*, **1**, Y-585 (1959).

61. The problem of the age of pottery in Middle America is complicated and by no means settled. Such relatively well-developed village-farming phases as early Zacatenco Valley of Mexico and Las Charcas (Guatemalan Highlands) have radiocarbon dates that indicate an age of about 1500 B.C. There are also contradictory radiocarbon dates that suggest these phases occurred several hundred years later. For a review of some of these dates for Middle America, see G. R. Willey, *Am. Antiquity*, **23**, 359 (1958) and E. S. Deevey, L. J. Gralenski, and V. Hoffren (see note 60). It may be that other Middle American ceramic complexes, such as the Chiapa I (Chiapas), Ocós (Pacific Guatemala), Yarumela I (Honduras), Yohoa Monochrome (Honduras), and Pavon (northern Veracruz), are older than either early Zacatenco and Las Charcas, although there is no clear proof of this. In Fig. 11–1, the dotted line indicating the inception of pottery has been put as early as 2500 B.C. in Middle America. A 1958 discovery in conflict with this comes from Oaxaca, where a preceramic site, possibly representative of incipient cultivation, has radiocarbon dates of only about 2000 B.C. This has been presented by J. L. Lorenzo, "Un sitio preceramico en Yanhuitlan, Oaxaca," *Inst. nac. antropol. e hist. Publ. No. 6* (1958). For Peru, the earliest pottery appears on the north coast, at an average date of about 1200 to 1000 B.C. See radiocarbon dates for early Peruvian pottery as itemized by G. R. Willey [*Am. Antiquity*, **23**, 356 (1958)].

62. R. S. MacNeish, in "The Engigstciak Site on the Yukon Arctic Coast" [*Univ. Alaska Anthropol. Papers*, **4**, No. 2 (1956)], has contributed to this problem by the discovery of early Woodland-like pottery in the far north.

63. W. A. Ritchie, *N.Y. Museum Sci., Circ. No. 40* (Albany, N.Y., 1955); see J. B. Griffin, "The Chronological Position of the Hopewellian Culture in the Eastern United States," *Univ. of Michigan Museum of Anthropol., Anthropol. Paper No. 12* (1958), p. 10, for a different view.

64. See G. R. Willey, *Am. Anthropologist*, **57**, 571 (1955).

65. However, zoned rocker-stamped pottery decoration appears earlier in Japan than in any part of the Americas. A distributional study of this technique for decorating pottery is included in the Ph.D. thesis of R. M. Greengo (Harvard University, 1956).

66. R. J. Braidwood, **127**, 1419 (1958).

67. See V. G. Childe, in *Anthropology Today*, A. L. Kroeber, Ed. (Univ. of Chicago Press, Chicago, 1953), pp. 193–210.

68. G. R. Willey and P. Phillips (note 2, pp. 144–47) define this as the "Formative" stage.

69. See R. S. MacNeish (note 51) for such culture phases as the Laguna and the Mesa de Guaje.

70. G. C. Vaillant, *Excavations at Zacatenco* [*Am. Museum Nat. Hist. Anthropol. Paper No. 32* (1930)], pt. 1.

71. E. M. Shook, in *The Civilizations of Ancient America* (vol. 1 of selected papers of the 29th International Congress of Americanists), S. Tax., Ed. (Univ. of Chicago Press, Chicago, 1951), pp. 93–100.

72. M. D. Coe, thesis, Harvard University, 1959.

73. For Mamom phase, see A. L. Smith, "Uaxactun, Guatemala: Excavations of 1931–1937," *Carnegie Inst. Wash. Publ. No. 588* (1950).

74. The early ceramic phases, Yarumela I, Yohoa Monochrome, and Pavon, from Honduras and northern Veracruz, may represent village-farming cultures, or they may be coincident with incipient cultivation. For these phases see J. S. Canby, in *The Civilizations of Ancient America*, S. Tax, Ed. (Univ. of Chicago Press, Chicago, 1951), pp. 79–85; W. D. Strong, A. Kidder II, A. J. D. Paul, *Smithsonian Inst. Publs. Misc. Collections*, **97**, 111 (1938); R. S. MacNeish, *Trans. Am. Phil. Soc.*, **44**, No. 5 (1954).

75. J. B. Bird, in "Radiocarbon Dating," *Soc. Am. Archaeology Mem. No. 8* (1951), pp. 37–49, sample 75.

76. C. Evans and B. J. Meggers, *Am. Antiquity*, **22**, 235 (1957); Evans and Meggers, personal communication (1958).

77. G. Reichel-Dolmatoff and A. Reichel-Dolmatoff, *Rev. colombiana antropol.*, **5**, 109 (1956).

78. G. R. Willey and P. Phillips, *Method and Theory in American Archaeology* (Univ. of Chicago Press, Chicago, 1958), pp. 151–55.

79. H. M. Wormington, "A Reappraisal of the Fremont Culture," *Proceedings, Denver Museum of Natural History* (1955), No. 1.

80. J. B. Bird, "Excavations in Northern Chile," *Am. Museum Nat. Hist. Anthropol., Paper No. 38* (1943), pt. 4.

81. A. R. Gonzalez, "Contextos culturales y cronologia relativa en el Area Central del Noroeste Argentino," *Anales arqueol. y etnol.*, **11** (1955).

82. J. R. Caldwell, "Trend and Tradition in the Prehistory of the Eastern United States," *Am. Anthropol. Assoc. Mem. No. 88* (1958).

83. See J. B. Griffin, "The Chronological Position of the Hopewellian Culture in the Eastern United States," *Univ. of Michigan Museum of Anthropol., Anthropol. Paper No. 12* (1958), for a résumé and analysis of Adena and Hopewell radiocarbon dates.

84. R. M. Goslin, *op. cit.*

85. J. A. Ford and C. H. Webb, "Poverty Point: A Late Archaic site in Louisiana," *Am. Museum Nat. Hist., Anthropol. Paper No. 46* (1956), pt. 1.

86. J. R. Caldwell, *op. cit.*

87. R. Wauchope [*Middle American Research Records* (Tulane University, New Orleans, La., 1950), vol. 1, No. 14] states the case for an early village-farming level without ceremonial mounds or constructions. While it is true that in some regions of Middle America the temple mound is absent in the earlier part of the "Formative" or "Preclassic" period, it is not clear that such a horizon prevails throughout all of Middle America. In fact, recent data [see M. D. Coe (note 72)] suggest that temple mounds were present in southern Middle America at the very beginnings of village farming.

88. See R. K. Beardsley, B. J. Meggers et al., in "Seminars in Archaeology: 1955," *Soc. Am. Archaeol. Mem. No. 11* (1956), pp. 143–45, for discussion of an "advanced nuclear centered community."

89. It is possible that such a ceremonial center as San Agustín, in southern Colombia, was, in effect, a town with concentrated ceremonial components and, probably, scattered hamlet-sustaining populations. San Agustín has not been satisfactorily dated, but estimates have been made that would place it as comparable in age to town-temple centers in Middle America and Peru. See W. C. Bennett, "Archaeological Regions of Colombia: A Ceramic Survey," *Yale Univ. Publs. in Anthropol.*, **30**, 109 (1944).

90. The town life of the Caribbean regions of Colombia and Venezuela at the period of the Spanish conquest is described by J. H. Steward in "Handbook of South American Indians," *Bur. Am. Ethnol., Smithsonian Inst. Publ.* (1949), vol. 5, pp. 718 ff.

91. See W. C. Bennett, E. F. Bleiler, F. H. Sommer, "Northwest Argentine Archaeology," *Yale Univ. Publs. in Anthropol.*, **38**, 31 (1948).

92. See H. M. Wormington, "Prehistoric Indians of the Southwest," *Denver Museum Nat. Hist., Popular Ser. No. 7* (1947), pp. 76–102, 107–47.

93. J. B. Griffin [*Archaeology of Eastern United States* (Univ. of Chicago Press, Chicago, 1952), Fig. 205] estimates these events at about A.D. 900 to 1000. There are indications from some parts of the southeastern United States that temple mounds are much older. For example, see H. P. Newell and A. D. Krieger, "The George C. Davis Site Cherokee County, Texas," *Soc. Am. Archaeol. Mem. No. 5* (1949), and R. P. Bullen, *Florida Anthropologist*, **9**, 931 (1956), for a radiocarbon date (about A.D. 350) on the Kolomoki culture.

94. See W. R. Wedel, in P. Drucker, "La Venta, Tabasco, A Study of Olmec Ceramics and Art," *Bur. Am. Ethnol. Smithsonian Inst. Bull. No. 153* (1952), pp. 61–65, for a description of a stone-columned tomb within an earth mound at La Venta. In this connection, the stone tombs covered by earth mounds at San Agustín, Colombia, as described by K. T. Preuss, *Arte monumental prehistoric* (Escuelas Salesianas de Tipografia y Fotograbado, Bogotá, 1931), may be pertinent.

95. See V. G. Childe's criteria of city life in *Town Planning Rev.*, **21**, 3 (1950).

96. Such centers, although serving as foci for the achievements of civilization, continue more in the form and in the homogeneous traditions of the Beardsley, Meggers et al., "advanced nuclear centered community" (note 88).

97. This kind of city, a "true" city in a modern western European sense, corresponds more closely to what Beardsley, Meggers et al. call "supra-nuclear integrated" communities (note 88, pp. 145–46).

98. See G. R. Willey, *Am. Anthropologist*, **57**, 571 (1955), and in *New Interpretations of Aboriginal American Culture History* (Anthropological Society of Washington, Washington, D.C., 1955), pp. 28–45; see also S. F. de Borhegyi, *Middle American Research Records* (Tulane University, New Orleans, 1959), vol. 2, No. 6.

99. W. C. Bennett, E. F. Bleiler, F. H. Sommer, "Northwest Argentine Archaeology," *Yale Univ. Publs. in Anthropol.*, **38**, 31 (1948).

100. Such features as Middle America-derived ballcourts and the casting of copper ornaments are well known in Hohokam archaeology [see Wormington (note 92)].

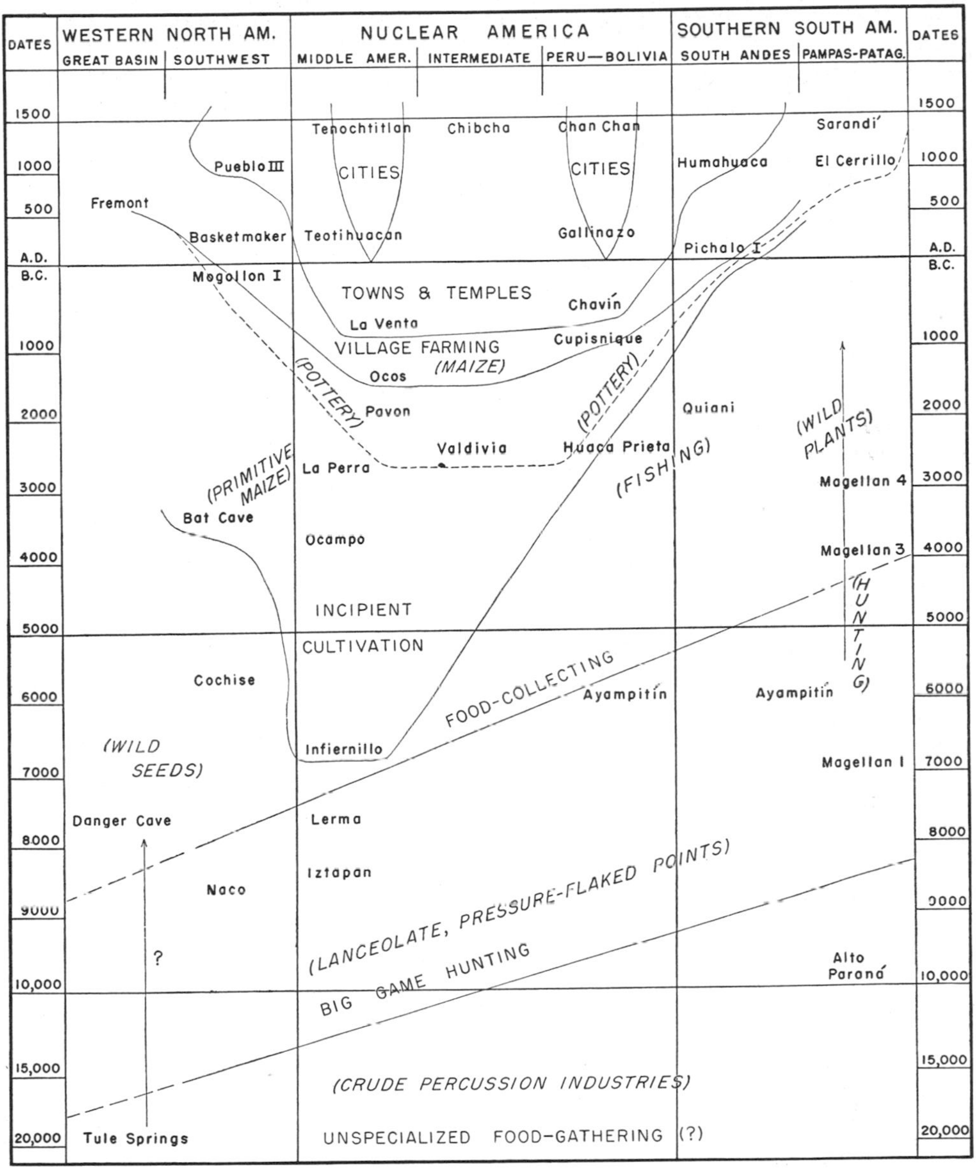

DATES | WESTERN NORTH AM. | NUCLEAR AMERICA | SOUTHERN SOUTH AM. | DATES

GREAT BASIN | SOUTHWEST | MIDDLE AMER. | INTERMEDIATE | PERU—BOLIVIA | SOUTH ANDES | PAMPAS-PATAG.

1500
Tenochtitlan
Chibcha
Chan Chan
Sarandí
1000
Pueblo III
CITIES
CITIES
Humahuaca
El Cerrillo
500
Fremont
Basketmaker
Teotihuacan
Gallinazo
Pichalo I
A.D.
B.C.
Mogollon I
TOWNS & TEMPLES
Chavín
La Venta
1000
VILLAGE FARMING
(MAIZE)
Cupisnique
Ocos
(POTTERY)
(POTTERY)
2000
Pavon
Quiani
(WILD PLANTS)
Valdivia
Huaca Prieta
(PRIMITIVE MAIZE)
La Perra
(FISHING)
3000
Magellan 4
Bat Cave
Ocampo
4000
Magellan 3
INCIPIENT
(HUNTING)
5000
CULTIVATION
Cochise
FOOD-COLLECTING
Ayampitín
Ayampitín
6000
(WILD SEEDS)
Infiernillo
7000
Magellan I
Danger Cave
Lerma
8000
Iztapan
Naco
9000
(LANCEOLATE, PRESSURE-FLAKED POINTS)
?
Alto Paraná
10,000
BIG GAME HUNTING
15,000
(CRUDE PERCUSSION INDUSTRIES)
20,000
Tule Springs
UNSPECIALIZED FOOD-GATHERING (?)

Fig. 11–1. Subsistence and settlement type levels in native America: cross section for western North America, Nuclear America, and southern South America. The first appearance of pottery is indicated by the dotted line.

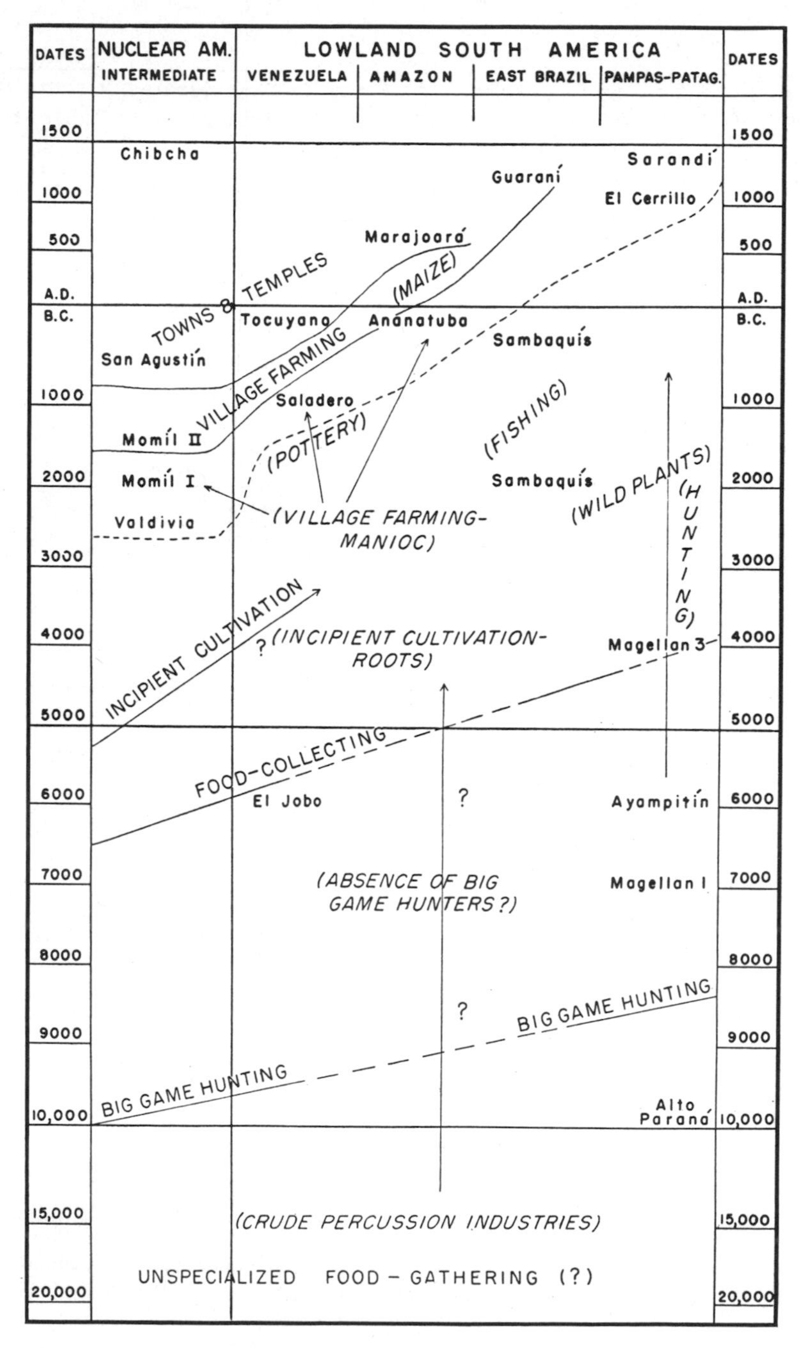

Fig. 11–2. Subsistence and settlement type levels in native America: cross section for Nuclear America and lowland South America. The first appearance of pottery is indicated by the dotted line.

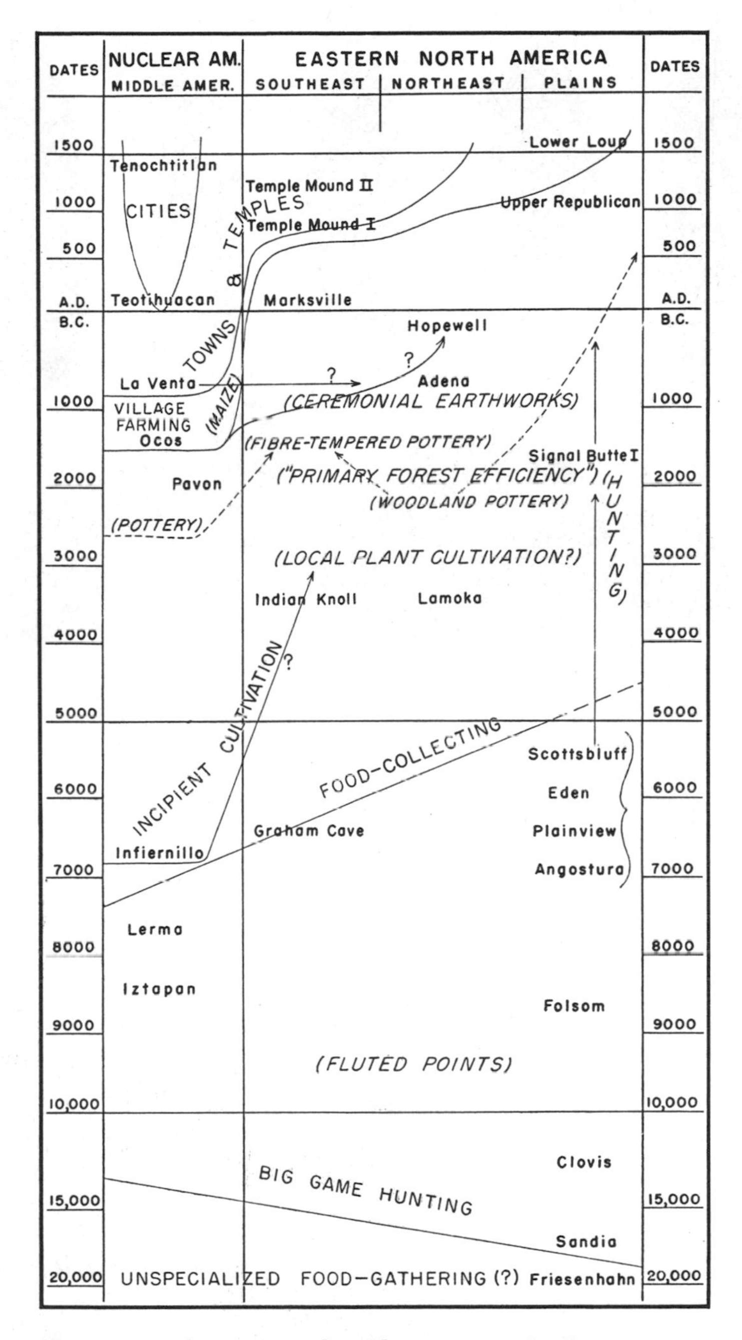

Fig. 11–3. Subsistence and settlement type levels in native America: cross section for Nuclear America and eastern North America. The first appearance of pottery is indicated by the dotted line.

Fig. 11–4. Clovis-type projectile points and associated scrapers from the Lehner site, southern Arizona. These artifacts are comparable to those found at the near-by Naco site. They are representative implements of the North American Pleistocene big-game hunters. (Courtesy Arizona State Museum.)

Fig. 11–5. Valdivia style pottery (left) and figurines (right) from coastal Ecuador. This excised ware and the crudely modeled female figurines may be among the earliest ceramic manufactures of the New World. (Courtesy Emilio Estrada.)

Fig. 11–6. Early Woodland pottery from New York State. Typical sherds of the Vinette I cord-marked ware, a ceramic that dates back 1000 B.C. or earlier. (Courtesy New York State Museum and Science Service.)

Fig. 11–7. Rocker-stamped pottery of the New World. (Top) Three rocker-stamped potsherds from the Turner site, Ohio Hopewell culture. (Right) Fragment of a zoned rocker-stamped bowl from an early level (about 800 B.C.) of the Barton Ramie site, British Honduras (Mayan territory). (Courtesy Peabody Museum, Harvard University.)

Fig. 11–8. Examples of fine Maya Classic polychrome pottery, perhaps the peak of native New World ceramic art. Note the bands of hieroglyphs used as decorative borders. (After J. M. Longyear III.)

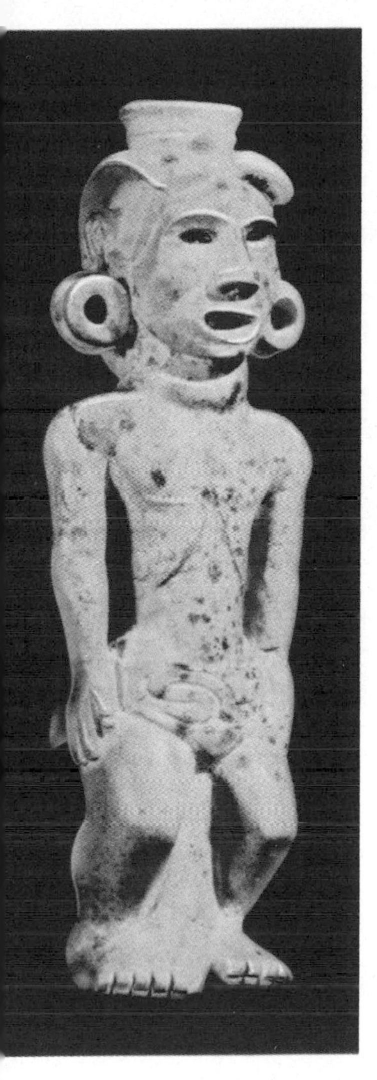

Fig. 11–9. Beautifully carved smoking pipe showing the skill with which the Adena craftsman worked small objects of stone. (Courtesy Ohio State Museum.)

Fig. 11–10. A Mayan temple of the Classic period, about A.D. 300 to 900. This is the famed "Temple of the Inscriptions," at the important ceremonial center of Palenque, Chiapas, Mexico.

Fig. 11–11. A palace-type structure at the Mayan ceremonial center of Sayil, Yucatan, Mexico. This handsome building, now largely in ruins, was built of rubble faced with cut limestone blocks and mortar. It is estimated to have contained about 100 rooms. It was probably constructed, at least in its final phases, between A.D. 600 and 900.

Fig. 11–12. A handsome masonry structure overlooking a plaza or courtyard. This building, resting upon an artificial terrace, is one of many at the Maya Classic period site of Copan in western Honduras. (Courtesy Carnegie Institution of Washington.)

Fig. 11–13. A view of the great adobe wall bordering a side of one of the huge palace and living enclosures at the Peruvian north coast site of Chanchan. The ancient urban metropolis of Chanchan consists of several such enclosures. Chanchan was in its heyday in the fifteenth century, as the capital of the Chimu kingdom. It was taken over and destroyed by the Inca armies about A.D. 1470. (Courtesy Clifford Evans, Jr.)

Chapter 12

The New American Archaeology

Joseph R. Caldwell

It is well known that the fortunes of archaeology have been greatly improved by new technical aids such as radiocarbon dating. A more important but far less celebrated advance is represented, I think, by a shift of interest in recent years toward problems of far greater generality than pertain to any single excavated prehistoric site. Part of this shift of interest to more general problems must be ascribed to the outstanding work of V. Gordon Childe and others in the Old World, but the greater reason perhaps is to be found in the close ties which most American archaeologists have maintained with general anthropology and through this, more tenuously, with the wider domain of social studies.

This juxtaposition of anthropology and archaeology in North American universities came about for the good historical reason that this continent contained living primitive cultures as well as prehistoric ones. The work of Americanists has with reason been called social-science archaeology. Not only do many Americanists have a cultural anthropological background but they find fruitful applications in archaeological thought of the studies, for example, of social anthropologists such as Julian Steward and Robert Redfield. A series of papers in a volume entitled *Seminars in Archaeology: 1955*[1] comprised the following titles: "An archaeological classification of culture contact situations"; "An archaeological approach to the study of cultural stability"; "The American Southwest: A problem in cultural isolation"; and finally, "Functional and evolutionary implications of community patterning." Had a sociologist found himself at these meetings, he would have had no trouble recognizing the problems, even if the factual data seemed a little strange. An anthropologist writes of these four seminars that they demonstrate "a growing together rather than a falling apart of archaeology and the other special disciplines of anthropology."[2]

First Approach

The understanding that is emerging as a result of shifting interests and new problems can be shown by comparing recent archaeology with the older variety. Since American archaeology is regionally specialized —Andean, Eastern, Middle American, Southwestern, and so on—and because some of these regions were ahead of others in development, I find it easier to use the older archaeology of eastern North America as a base line for the kind of contrasts that I propose to make.

Up until World War II the chief concerns of eastern archaeology—with some exceptions—seem to have been the description of archaeological sites and the description—often simply the definition—of prehistoric cultures. The latter might be presented individually or in terms of culture provinces.[3] Sometimes particular categories of material culture—for example, all the known prehistoric pottery from the eastern United States—were presented in terms of provinces.[4] Some fine work was done on the identification and methods of manufacture of prehistoric stone tools.[5]

A considerable advance was represented in the Midwest by a Linnaeanlike taxonomic system[6] that appeared just at the time it was beginning to be realized that cultural stratigraphy was present in the Eastern areas. The Midwestern taxonomic system was regarded as a necessary first step. It organized archaeological materials into categories based on degrees of likeness of the assemblages being unearthed. Unfortunately, there was a tendency to regard classification as the end of research, and some archaeologists who were obtaining long stratigraphic sequences, which in some cases showed gradual culture change, were hard put to classify these in Midwestern terms, although they continued for years to do so. In being able, now, to observe material culture changes in time and space, they already had part of the means for establishing *kinds* of historical connection, whereas the taxonomy they tried so hard to employ could only specify something about degree and could not deal with continuities.

I think it is fair to say that before World War II American archaeological studies were in a condition similar to that which F. S. C. Northrop[7] has characterized as the natural-history stage of inquiry. The emphasis was on archaeological data as things in themselves rather than on the values offered by different ways of looking at them. Moreover, it was considered, in practice, as important to excavate a site meticulously and to record every scrap of evidence that might conceivably bear on any future problem as it was to have a reason for excavating the site in the first place. One result of all this was the development of a specific kind of problem that treated classificatory entities as independent realities; one might inquire into the content of cultures known from preserved material objects, examine their temporal or spatial boundaries, or try to establish the degree of relationship among them.

A second result was the development of a specific kind of analysis to set up the comparisons required to solve problems of this nature. Types were routinely established as an economical means of describing small objects, pottery, constructions of one sort or another, and burial customs. These types were considered adequate for all comparisons that might later be made but were not designed to solve particular problems. Problems might certainly occur to one after the types had been established. Types of this kind, since they were immediately apprehensible regardless of problem, were in some quarters considered to be real entities, and rightly so.

The third result was the development of a specific kind of history—a history of material culture—which, at best, described the succession of the preserved archaeological assemblages in each culture province. At worst, such a history was confined to the area of a modern state and made unnecessarily complex by the assignment of different names in different states to cultural manifestations that, on the basis of the criteria in use, should have been assigned the same name.

The essentially dull and uninteresting character of this "culture history" was a matter of concern to some archaeologists. Walter W. Taylor[8] called for the construction of fuller cultural contexts—for attention to "the interrelationships which existed *within*" each cultural entity. Others attempted to inject a lifelike note by substituting the word *people* for *culture* whenever possible. Thus, in a semipopular book,[9] the "Savannah River Culture" became the "Savannah River People," with corresponding shifts in referential pronouns.

Transition

A trend away from local specialization was initiated in the 1940's in monographs by James A. Ford and Willey in 1941[10] and by Griffin in 1946.[11] These men made themselves familiar with a vast amount of uncollated and unpublished data that had emerged from the hundreds of excavations undertaken under various federal relief agencies. The prehistory of the eastern United States was found to be most readily susceptible to presentation in terms of a succession of pan-Eastern periods or eras, reflecting the importance of time and continuity in contemporary archaeological thought. The picture obtained from these formulations was of a steady development of material culture and of the essential unity of the East: The various prehistoric cultures assigned to each period in this vast region were usually more alike than were the temporally separated entities within any particular subarea. Hence, each of these major periods could also be regarded as a developmental stage.

In 1958, Willey and Philip Phillips[12] applied what is essentially the same kind of formulation to the whole of the New World, as a series of pancontinental stages. The theoretical foundations of their work were stated at length, some hundreds of areally based prehistoric cultures were characterized, and many special hypotheses regarding the interrelationships of these were proposed, weighed, and accepted. The result of this method was again to show the cultural interconnectedness of the area treated—in this case the entire Western Hemisphere—and to suggest that the civilizations of Mexico and Peru emerged from the same background as the other American cultures, but proceeded through additional stages leading to civilization.

American archaeology still leans heavily on the idea of areally based cultures and probably always will. We have even improved the utility of this view by the concept of tradition—a culture area having depth in that it is allowed to shift its boundaries through time. If we now suggest some new ways of thinking about areal traditions, this does not mean that we are ready to dispense with them. They do represent more or less closely one kind of natural or common-sense division among the primary materials we have to work with. Where we have improved on the older archaeology is by asking different kinds of questions of the materials, and this is directly bound up with the new interests we have noted.

The New Archaeology

The new archaeology in America is tending to be more concerned with culture process and less concerned with the descriptive content of prehistoric cultures. There are now two kinds of problem, historical and general, that can be suggested either by distinctions seen in the data themselves or by results of archaeological research in other parts of the world, or which can emerge out of other disciplines such as ethnology or philosophy, and then be brought to the data as propositions to be tested.

We may characterize our new interests in the following way. Where formerly we were concerned with the identification of things and of cultures—whether, for example, a particular artifact should be regarded as a knife or as a scraper, or whether a given archaeological assemblage should be classed with this culture or that—we have added an interest in the identification of culture processes and situations. Thus, Waldo R. Wedel's "Environment and Native Subsistence Economics in the Central Great Plains"[13] examines culture-environment connections in that area, and since that time other archaeologists, stimulated no less by Alfred L. Kroeber's "Cultural and Natural Areas of Native North America"[14] than by the fine Virú Valley project in Peru,[15] have turned their attention to the interrelations between natural ecology and human populations and settlement patterns, with respect to cultural level.

Another approach to cultural and historical processes is seen in the wealth of inferences which can be derived from changes in cultural forms seen through time—that is, through stratigraphic and constructed sequences. Whether or not changes were diffused from another region can be inferred from knowledge of whether or not they occurred earlier elsewhere. That changes are of local development can be inferred when their prototypes occur locally at an earlier time. Something about the historical situation can be inferred from rates and magnitudes of changes in cultural forms. A sudden change in a whole series of artifact forms may herald a prehistoric invasion; gradual changes in forms occurring at different times suggest a period of comparative tranquility during which cultural development was not greatly influenced by outside areas. Whereas the older Midwestern taxonomic system could establish degrees of connections among cultural assemblages, we are now finding various methods of inference which will enable us to see the kinds of connections.

Present archaeology still reflects an indiscriminate use of the notion of a prehistoric "culture," by which is sometimes meant a few artifacts of some former society and, at other times, a number of societies historically related, but perhaps in different ways and in different degrees. We are increasingly sensitive to the value of making distinctions between cultures as opposed to societies.[16] Observations that can be made about behavior are for the archaeologist mediated through cultural forms, but his inferences need not always refer back to culture. Sometimes it is better to use the concept of interaction area instead of culture area; not only is thought thus referred directly to the behavior of people instead of to a "culture," but in some cases this idea is better suited to the archaeological facts of continuous intra-areal diffusions of cultural forms.

The archaeologist's curious use of the concept of culture is a matter of convenience. He has found some use in ignoring its central reference, that is, learned behavior within a single society, in favor of some of its principal derivatives: the proclivity of cultures to occupy space, and the fact that the material elements representing one culture will differ distinctively from the material elements of another culture, with the result that a variety of culture-historical inferences can be drawn. This has consequences that have brought new understandings of old matters. The sociologist and social anthropologist tend to limit their view to single societies at a time and as seen from within. The archaeologist and historical ethnologist, whose cultures may be represented by only a few elements, if that is all they have of them, more often tend to view many societies from a distance, and the variations among these will show a geographical patterning. With the time depth available to the archaeologist, he sees, over thousands of square miles, the locally differing cultures of innumerable small societies changing in concert and in ways incompletely described by the general terms diffusion and acculturation. This phenomenon has been observed to transcend the boundaries of culture areas and linguistic groupings. To cope with it have been developed such terms as "area co-tradition," and "interaction sphere," and this is what is present when archaeologists and ethnologists speak of a cultural stage. The most hopeful approach to understanding is in terms of patterns and processes of interaction and communication. In eastern North America, where there were no great physical barriers to communication, the several periods or stages of prehistory characteristically show greater likenesses among the cultures within each stage than between the cultures of earlier and later stages.

It has long been noticed that societies lying toward the center of a group of interacting societies will in most respects change more rapidly than those lying at the margins. It can be readily argued that increased interaction among societies results in an increased rate of cultural innovation.[17] If so, we might be prepared to explain the geographical patterning of change as a result of decreasing intensity of contacts from the center to the margins of such a group of interacting societies. This idea is as old as the study of culture areas, but neither this nor the idea of marginal cultures nor of cultural "lag" explains the situation completely. Archaeological sequences of changing forms of artifacts and usages suggest that a specific kind of change takes place in the marginal societies: new traits are added to the old and both exist side by side for long periods.[18] The phenomenon is so striking that one can hardly believe it to be characteristic of all cultural change, at least to this degree. Hence the likelihood arises that it will not be so true of the centrally located societies in such a group. These should show a greater degree of replacement of old traits, a different kind of cultural evolution, perhaps related to the propensity of the more central societies to initiate cultural innovation.

Such statements about gross interactions cannot, however, become our ultimate aim. Most of their generality derives from being based on a vast mass of undigested and imperfectly understood data. There should be innumerable ways of being more specific. One example is offered in the analysis of a culture-historical situation that existed in North America roughly between 400 B.C. and A.D. 400. Between these dates we find a widely distributed manifestation called Hopewell or Hopewellian. Archaeologists have been mostly attracted to the abundant burial mounds although we now have a few excavations at habitation sites. The mounds often contain multiple burials in and outside of log tombs, and sometimes there are multiple tombs. With the dead were placed particular artifacts: pottery vessels, earthenware or finely carved stone tobacco pipes, miniature human figurines, vessels made from large *cassis* shells, finely chipped flint, chert, or obsidian projectile points. The list is long; it is enough to say that mortuary usages and grave offerings show great variety and some regional differentiation.

Variously shared at these sites and through much of eastern North America at this time is a short list of usages and mortuary-ceremonial artifacts that are of exact similarity. A figurine from the Mandeville site in Georgia can be duplicated at the Knight Mound in Illinois. A cache of thousands of chipped stone "blanks" (the initial stage in the manufacture of projectile points and other artifacts) occurs at the Baehr site in Illinois and again at the Hopewell type site in Ohio. Beaten copper panpipes in Ohio, Illinois, and Georgia are duplicated in Florida. Pottery vessels from graves in Illinois are almost indistinguishable from vessels in Ohio and Louisiana.

Artifacts and secular items from habitation sites, however, show strong regional differentiation that is especially characteristic of domestic earthenware. Therefore the Hopewellian materials cannot be said to represent a single culture even in the archaeologist's loose usage. The salient features of this situation are two: striking regional differences in secular and domestic artifacts, and an interesting short list of exact similarities in mortuary usages and artifacts over great distances. What we have is a number of societies belonging to several distinct regional traditions—and therefore cultures in their own right—that are involved in an interaction situation.

These exact similarities in mortuary matters indicate that it was in this particular aspect of culture rather than in secular aspects that interaction and communication were taking place. There was an extensive trade, for example, that brought sheet mica from West Virginia and North Carolina, copper from Lake Superior, pearls from the Midwestern rivers, *cassis* shells from Florida, and obsidian perhaps from Wyoming. This trade was devoted principally to securing elements of costume and other furnishings for the dead. When in time the mortuary practices became less lavish, some of the trade routes dried up and disappeared.

Viewing the Hopewellian situation as a period when there were notable interactions in mortuary and religious matters among a number of societies belonging to different cultural traditions rather cuts the ground out from under several hypotheses which have recently been advanced to account for the supposed decline of Hopewell "culture." We need to account instead for the decline of the Hopewellian interaction sphere and whatever reasons we might discover may well be different.

Still it must be admitted that all of this is exceedingly gross. The best work along these and related lines undoubtedly lies in the future. Even now investigations are being made that show our Hopewellian interaction sphere to be only a crude way of representing many local and particular interaction spheres within it.[19] It is also beginning to appear that these will show particular characteristics related to their occurrence in particular geographical regions as the Midwestern prairie.[20] Moreover, the eight hundred or so years we now assign to it is an exceedingly long time. There were surely many relevant events during that span, and we should be able to mark successive stages of growth and decline of communication among the constituent societies. Finally, there is reason to think that the Hopewellian interaction sphere may be an outgrowth of an earlier one.[21]

Still another basis for our changing interest stems from the idea of pattern or configuration, which has had a considerable vogue in anthropology although it is not new with that science. The archaeologist is inclined to see cultural patterns in developmental terms. A pattern represents some kind of regularity or organization. If a pattern can be recognized, the features we use to account for its presence may perhaps be stated in terms of the processes that brought it into being or perhaps in terms of the factors that operate to maintain it.

With the idea of cultural patterns and developmental patterns, modern archaeology has reached a point where many possible patterns and hypotheses can be suggested, each of which seems to propose cultural "facts" that are not necessarily mutually exclusive and that do not necessarily contradict each other but which *in the same body of materials* reflect various aspects of a many-sided reality. To take a very simple example of the way in which a given body of archaeological materials may mirror different historical facts, suppose that a stratigraphic sequence of flint projectile points is used to suggest the answer to the question of whether these points were javelin tips or arrowheads. If both types are present, it may be that the bow and arrow were replacing the javelin during this range of time. We could perhaps arrive at an answer to this problem by using a type system with criteria based upon the size and weight of the specimens. On the other hand, the question might be whether the flint was being obtained from a distance through trade, and for this we should have to examine the projectile points in the light of another type system based on kinds of flint correlated with different localities—not on sizes and weights as in the other case.

In the foregoing example it is relatively easy to see how a given body of archaeological materials represents different historical or cultural facts. In the case of cultural pattern or configuration, however, the "reality" of proposed fact is less apparent because the particular interests of the investigator, and perhaps the historical development of the science, intrude more strongly into the result. Thus, Willey and Phillips' stadial conception of New World prehistory is also concerned with a particular reality; they might have devised other conceptions of equal validity had their interests been other than what they were.

New Understandings

The views held by Julian Steward, a social anthropologist,[22] show how additional understanding has been reached by a different approach. Steward rejects "unilinear" cultural evolution, maintained at the end of the last century by ethnologists like Edward Tylor and Lewis Henry Morgan and now in part by Willey and Phillips,[23] which says that with certain allowances for diffusion, all human cultures pass historically through similar developmental stages. According to Steward's theory of "multilinear" evolution, all cultures do not pass through similar stages, but we can discern a finite number of parallel evolutions in which societies adapted to particular environments and natural resources pass through successive and distinctive levels of "sociocultural" integration. Steward's comparisons deal with societies from various parts of the world. Features of these societies are treated by Steward as types, and certain recurrent associations of important features represent "cross-cultural types."

Conclusions concerning processes involved in particular evolutionary sequences are regarded not as natural laws but as regularities or generalizations of limited range, upon which, one supposes, we may in time build further. Steward says: "Ecological adaptations can be considered as causative in the sense that a degree of inevitability in cultural adjustments is directly observable. Patrilineal bands of Bushmen, Australians, Tasmanians, Fuegeans, and others represent a type in that the ecological adaptation and level of integration are the same in all these cultures. In these and other cases, factors producing similar types such as environment, food resources, means of obtaining food, the social cooperation required, population density, the nature of population aggregates, sociopolitical controls, the functional role of religion, warfare, and other features, will have an understandable relationship to one another."

Steward's work is concerned with processes of culture change manifested in a number of distinct developmental sequences and arrives at generalizations of limited range stated in terms of cultural process, whereas, the Willey-Phillips formulation stresses the interconnectedness of the prehistoric societies of the Western Hemisphere and arrives at a series of cultural levels applying to the area.

Some of Steward's proposed cross-cultural types, such as Formative, Regional Florescent, Empire, and Conquest, are designed to show the processes leading to civilization. They are nearly parallel to the later stages of the Willey-Phillips formulation. Steward's types are now being examined and somewhat modified by archaeologists familiar with the various regions.[24] The developmental similarities of Steward's types may be stated in causal terms, because between the Old World and the New World there is not much chance that the similarities are due to historical connection.

A new approach sometimes brings a wealth of understanding. Archaeology seldom affords direct evidence of social institutions, although Childe has suggested some means by which these can be inferred, and recently William H. Sears has been able to propose a correlation between prehistoric burial mounds on the Gulf Coastal Plain with the presence or absence of strong social classes in the societies involved.[25] Now Steward provides another method for arriving at such inferences, as Fred Eggan has pointed out.[26] Archaeology usually does offer data (for example, the bones of food animals and the size and locations of sites) concerning ecological adaptation. Some social institutions can be satisfactorily inferred from this if, as Steward maintains, they are causally connected with ecological adaptations.

I have proposed[27] a conception of the development and spread of early civilizations that, like the Steward and Willey-Phillips formulations, rests on a hypothesis. The body of available data is here divided differently, and in thus shifting the focus of our interest, new cultural "facts" are created. According to this scheme, there has been, in the areas which developed civilizations as well as in those which did not, an "Archaic" culture type with certain definable developmental features. These developmental features can be used to account for the emergence of civilizations in some areas as well as for the absence of civilization in other areas. Once a civilization has developed, however, some of the processes involved in its spread are best seen in terms of a contrast between two additional culture types: "nuclear civilization" and, in the areas outside of civilization, "non-nuclear culture."

The most important developmental feature of the Archaic culture type in eastern North America was the achievement of primary forest efficiency. This was a cumulative process manifested in the development of ambush hunting, in seasonal economic cycles (transhumance), and in the discovery of new sources of natural foods. It is supposed that something like this may have occurred wherever Archaic cultures are found in forested lands. An extension of this idea leads to a definition of a "plains efficiency" for the hunters of large migratory game and a "maritime efficiency" in coastal areas. These various "efficiencies" are meant to be the logical counterparts of "primary farming efficiency"—a term originally used by Braidwood[28] to describe the economic platform upon which civilization may arise.

Plant raising was known in areas where nuclear civilization did not arise. However, we find nuclear civilization *only* in areas where food production was the economic basis for society. Perhaps the plants used had greater potentialities; perhaps growing populations or the progressive depletion of other resources, or both factors, brought about a Toynbeean challenge that was successfully met.

In the nuclear civilization culture type, it is the achievement of primary farming efficiency that permits the changes leading to civilization. In the non nuclear culture area of eastern North America, where primary forest efficiency was well established, it was this very efficiency that tended to direct subsequent economic innovation along lines previously established. Changes only represented further development of hunting-gathering systems.

While a degree of residential stability and comparative freedom from want can be achieved by peoples who live by hunting, fishing, or gathering (witness the American Indians of central California and the northwest Pacific coast), it appears that urbanization and civilization cannot appear without the development of food production on an extensive scale.

The growth potential of different economic patterns is clearly delimited in comparing the nuclear and non-nuclear culture types. The mechanics of the limiting factors can be seen in comparing each of these two with their common antecedents in the Archaic culture type.

What new understanding can be reached by viewing culture developments in the Western Hemisphere in terms of two contrasting types, nuclear civilization and non-nuclear culture? Such a view suggests one way to find connections that became established between the areas of civilization and the areas beyond, and the outward spread of civilizations can be formally examined both in time and in space. It becomes possible to ask certain questions about the spread of civilizations, and although the particular historical events may seem to be of infinite variability, it may be possible to account for these in terms of a finite number of general processes. Within the framework of the contrast between nuclear civilization and non-nuclear culture, it is relatively easy to describe certain intermediate cultural balances as of mixed descent. To do so emphasizes the role of such hybrid cultures as active agents in the spread of civilizations. Finally, it calls attention to the different developmental patterns between the spreading civilizations and the cultures that confront them. An acculturation situation consists of far more than the simple adoption of features of the greater culture by the weaker. Both are affected, and both reinterpret culture transfers in terms of their own views and interests, which we *can* see as patterned in terms of a particular historical development.

Conclusions

It is supposed that behind the infinite variability of cultural facts and behind the infinite and largely unknown detail of historical situations we shall discover the workings of a finite number of general cultural processes. This hypothesis underlies much of recent archaeological thought despite the view, often propounded, that because of level, cultural facts are much more complex than those of the physical sciences. This latter assertion does not make our task impossible. Not all cultural facts are of equal importance in determining a given pattern or trend. Certain developmental patterns must surely be overriding in their effects upon other patterns. A major historical pattern may serve to unite or in some cases to subordinate other patterns of more limited range.

I have tried in this chapter to show that different aspects of reality are reflected even in a limited body of archaeological materials. We can look beyond the immediately apprehensible characteristics of form and number to discover these, and each inference can be regarded as a particular cultural or historical "fact." But what facts they happen to be will depend on the questions we happen to have put. Our questions depend on the stage of progress of archaeology and upon the special interests of the investigator.

This may seem a long way from the apparent precision of most kinds of scientific investigation, but we may suspect that these conditions of inquiry hold for the other humanistic studies as well. If this is true, we should be aware of the difficulties, but in archaeology, at least, these difficulties are not insurmountable. The presence of many aspects of reality in a single body of materials is partly related to the typological methods we use to discover them. The kind of types we construct in each case is directly related to the question we wish to ask, and we are generally allowed to ask only one question at a time. There is an immense amount of information locked up in these materials, and our many-sided way of looking at them is probably a healthy sign.

Moreover, curious as these conditions of inquiry may seem, they do not mean that our propositions cannot be tested nor that validation cannot be secured. The pathways of archaeology are strewn with the wreckage of former theories that could no longer be supported in the light of available evidence. Even though non-opposed hypotheses can be made about the same body of materials, there are also those that can be shown to be logically inconsistent with each other and among which a choice can be made. As time goes on, tests of compendency should become increasingly specific. Archaeology, as other studies of man and society, can test its postulates. One way to disprove the Willey-Phillips postulate that all the prehistoric cultures of the New World went through similar developmental stages would be to show that an important area did not go through these stages but did go through others. I am personally inclined to think that in archaeology we shall soon be testing deductions made from general principles and that those who have maintained that this essential method of the physical sciences could not possibly be applied to historical data may have spoken too soon. If we accord the weight of a general statement to the proposition made earlier: that increased interaction among societies results in an increased rate of cultural innovation, then we can test this as a principle while being able to say something further about the Hopewellian situation in eastern North America. *If* there is a positive correlation between interaction and innovation, we can deduce as a corollary that *then* most cultural innovations in the Hopewellian situation should have taken place in the particular aspect of culture with which most of the interactions were concerned. We should expect to find more innovations in mortuary and religious usages, less in domestic and secular activities. This is in fact the case, and although it would be out of place to include a table of innovations here, a simple counting of old and new traits shows that burial mounds show much more variability, more innovations, in the usages they display.

Perhaps it will be clear from this and from what we have already said with reference to the convergence of archaeology with anthropology and social studies, that archaeology is now turning to questions of greater generality than pertain to any single excavated prehistoric site or culture. I think that our interests will become still wider. The similarities between Steward's views concerning the importance of the food quest in determining the institutions of the simpler societies and Karl Marx's production relationships, which formed the basis for his labor theory of economics, may already have occurred to the reader. V. Gordon Childe apparently found much in Marx's historical formulations to stimulate his own conceptions of prehistory.

Since archaeology expects to deal with a range of problems pertaining to former societies and often seeks the aid of other sciences to do this, it tends to make connections among various kinds of studies. Moreover, the appropriateness of archaeological data for questions that have arisen in general studies of history or art has long been recognized. Archaeological findings from the earth, viewed in terms of time, space, and cultural behavior, offer a vast body of material for inference. And as for philosophy, I think that the usefulness of archaeological data will be recognized and that closer connections with that discipline will be established. What does a stratigraphic sequence of changes in cultural forms have to say about the nature of historical causality? What does the regularity which such changes often show imply concerning historical determinism as opposed to human liberty?

If it is the wise archaeologist who now restricts his formulations to the development and persistence of civilizations, cultures, technologies, arts, and lesser matters, it must also be the very dull archaeologist who could be unconcerned with the implications of these for some of the perennial problems of Western man.

NOTES

1. "Seminars in Archaeology: 1955," *Soc. Am. Archaeol. Mem. No. 11* (1956).

2. E. H. Spicer, *Am. Antiquity*, **23**, 186 (1957).

3. H. C. Shetrone, *The Mound Builders* (Appleton, New York, 1930).

4. W. H. Holmes, *Bur. Ethnol. 20th Ann. Rept. 1898–1899* (Washington, D.C., 1903).

5. W. H. Holmes, *Bur. Ethnol. Bull. 60* (Washington, D.C., 1919).

6. W. C. McKern, *Am. Antiquity*, **4**, 301 (1939).

7. F. S. C. Northrop, *The Logic of the Sciences and the Humanities* (Macmillan, New York, 1948).

8. W. W. Taylor, *Am. Anthropologist Assoc. Mem. No. 69* (1948).

9. P. S. Martin, G. I. Quimby, D. Collier, *Indians Before Columbus* (Univ. of Chicago Press, Chicago, 1947).

10. J. A. Ford and G. R. Willey, *Am. Anthropologist*, **43**, 325 (1941).

11. J. B. Griffin, *R. S. Peabody Foundation for Archaeology Publ. No. 3* (1946), p. 37.

12. G. R. Willey and P. Phillips, *Method and Theory in American Archaeology* (Univ. of Chicago Press, Chicago, 1958).

13. W. R. Wedel, *Smithsonian Inst. Publ. Misc. Collections*, **101**, No. 3 (1941).

14. A. L. Kroeber, *Univ. Calif. (Berkeley) Publ. Am. Archaeol. and Ethnol.*, **38** (1939).

15. G. R. Willey, *Bur. Ethnol. Bull. 155* (Washington, D.C., 1953).

16. G. R. Willey and P. Phillips, *Method and Theory in American Archaeology* (Univ. of Chicago Press, Chicago, 1958).

17. This point is discussed in J. R. Caldwell, "The Origins of Civilizations," *Fulbright Series 1* (mimeo.) U.S. Comm. for Cult. Exch. with Iran. Tehran (1963).

18. The coast of the state of Georgia is discussed as a marginal area showing the persistence of old traits in J. R. Caldwell, "Trend and Tradition in the Eastern United States," *Am. Anthropological Assoc. Mem. No. 88* (1958).

19. This has come out of discussions with H. D. Winters and S. Struever.

20. J. A. Brown, personal communication.

21. See W. A. Richie's "Early Woodland Burial Cult," *N.Y. State Mus. and Sci. circular 40*, Albany, 1955.

22. J. H. Steward, *Theory of Culture Change* (Univ. of Illinois Press, Urbana, 1955).

23. G. R. Willey and P. Phillips, *Method and Theory in American Archaeology* (Univ. of Chicago Press, Chicago, 1958), pp. 70–71, assure the reader that theirs is not an evolutionary scheme. I found their arguments unconvincing, and the reader may wish to judge this matter for himself.

24. *Irrigation Civilizations: A Comparative Study*, "Social Science Monographs" (Pan American Union, Washington, D.C., 1955).

25. V. G. Childe, *Social Evolution* (Shuman, New York, 1951); W. H. Sears, *Am. Antiquity*, 23, 274 (1958). Other inferences about social organization are made by Lewis H. Binford in "Archaeological Investigations in the Carlyle Reservoir, Clinton County, Illinois," *So. Ill. Univ. Mus. Archaeological Salvage Rept.*, No. 7, 1962.

26. F. R. Eggan, in *Archaeology of Eastern United States*, J. B. Griffin, Ed. (Univ. of Chicago Press, Chicago, 1952).

27. J. R. Caldwell, "Trend and Tradition in the Prehistory of the Eastern United States," *Am. Anthropological Mem. No. 88* (1958).

28. R. J. Braidwood, *The Near East and the Foundations for Civilization* (Univ. of Oregon Press, Eugene, 1952).

Chapter 13

Microenvironments and Mesoamerican Prehistory

Michael D. Coe
Kent V. Flannery

A crucial period in the story of the pre-Columbian cultures of the New World is the transition from a hunting-and-collecting way of life to effective village farming. We are now fairly certain that Mesoamerica[1] is the area in which this took place, and that the time span involved is from approximately 6500 to 1000 B.C., a period during which a kind of "incipient cultivation" based on a few domesticated plants, mainly maize, gradually supplemented and eventually replaced wild foods.[2] Beginning probably about 1500 B.C., and definitely by 1000 B.C., villages with all of the signs of the settled arts, such as pottery and loom-weaving, appear throughout Mesoamerica, and the foundations of pre-Columbian civilization may be said to have been established.

Much has been written about food-producing "revolutions" in both hemispheres. There is now good evidence both in the Near East and in Mesoamerica that food production was part of a relatively slow *evolution*, but there still remain several problems related to the process of settling down. For the New World, there are three questions which we would like to answer.

1. What factors favored the early development of food production in Mesoamerica as compared with other regions of this hemisphere?
2. What was the mode of life of the earlier hunting-and-collecting peoples in Mesoamerica, and in exactly what ways was it changed by the addition of cultivated plants?
3. When, where, and how did food production make it possible for the first truly sedentary villages to be established in Mesoamerica?

The first of these questions cannot be answered until botanists determine the habits and preferred habitats of the wild ancestors of maize, beans, and the various cucurbits that were domesticated. To answer the other questions, we must reconstruct the human-ecological situations that prevailed.

Some remarkably sophisticated, multidisciplinary projects have been and still are being carried out elsewhere in the world, aimed at reconstructing prehistoric human ecology. However, for the most part they have been concerned with the adaptations of past human communities to large-scale changes in the environment over very long periods—that is, to alterations in the *macroenvironment,* generally caused by climatic fluctuations. Such alterations include the shift from tundra to boreal conditions in northern Europe. Nevertheless, there has been a growing suspicion among prehistorians that macroenvironmental changes are insufficient as an explanation of the possible causes of food production and its effects,[3] regardless of what has been written to the contrary.

Ethnography and Microenvironments

We have been impressed, in reading anthropologists' accounts of simple societies, with the fact that human communities, while in some senses limited by the macroenvironment—for instance, by deserts or by tropical forests[4]—usually exploit several or even a whole series of well-defined *microenvironments* in their quest for food.[5] These microenvironments might be defined as smaller subdivisions of large ecological zones; examples are the immediate surroundings of the ancient archaeological site itself, the bank of a nearby stream, or a distant patch of forest.

An interesting case is provided by the Shoshonean bands that, until the mid-nineteenth century, occupied territories within the Great Basin of the American West.[6] These extremely primitive peoples had a mode of life quite similar to that of the peoples of Mesoamerica of the fifth millennium B.C., who were the first to domesticate maize. The broadly limiting effects of the Great Basin (which, generally speaking, is a desert) and the lack of knowledge of irrigation precluded any effective form of agriculture, even though some bands actually sowed wild grasses and one group tried an ineffective watering of wild crops. Consequently, the Great Basin aborigines remained on a hunting and plant-collecting level, with extremely low population densities and a very simple social organization. However, Steward's study[7] shows that each band was not inhabiting a mere desert but moved on a strictly followed seasonal round among a vertically and horizontally differentiated set of microenvironments, from the lowest salt flats up to piñon forest, which were "niches" in a human-ecological sense.

The Great Basin environment supplied the potential for cultural development or lack of it, but the men who lived there selected this or that microenvironment. Steward clearly shows that *how* and *to what* they adapted influenced many other aspects of their culture, from their technology to their settlement pattern, which was necessarily one of restricted wandering from one seasonally occupied camp to another.

Seasonal wandering would appear to be about the only possible response of a people without animal or plant husbandry to the problem of getting enough food throughout the year. Even the relatively rich salmon-fishing cultures of the northwest coast (British Columbia and southern Alaska) were without permanently occupied villages. Contrariwise, it has seemed to us that only a drastic reduction of the number of niches to be exploited, and a concentration of these in space, would have permitted the establishment of full-time village life. The ethnographic data suggest that an analysis of microenvironments or niches would throw much light on the processes by which the Mesoamerican peoples settled down.

Methodology

If the environment in which an ancient people lived was radically different from any known today, and especially if it included animal and plant species that are now extinct and whose behavior is consequently unknown, then any reconstruction of the subsistence activities of the people is going to be difficult. All one could hope for would be a more-or-less sound reconstruction of general ecological conditions, while a breakdown of the environment into smaller ecological niches would be impossible. However, much if not most archaeological research concerns periods so recent in comparison with the million or so years of human prehistory that in most instances local conditions have not changed greatly in the interval between the periods investigated and the present.

If we assume that there is a continuity between the ancient and the modern macroenvironment in the area of interest, there are three steps which we must take in tracing the role of microenvironments.

1. Analysis of the present-day microecology (from the human point of view) of the archaeological zone. Archaeological research is often carried out in remote and little-known parts of the earth, which have not been studied from the point of view of natural history. Hence, the active participation of botanists, zoologists, and other natural scientists is highly recommended.

The modern ethnology of the region should never be neglected, for all kinds of highly relevant data on the use of surrounding niches by local people often lie immediately at hand. We have found in Mesoamerica that the workmen on the "dig" are a mine of such information.

There may be little need to thumb through weighty reports on the Australian aborigines or South African Bushmen when the analogous custom can be found right under one's nose.[8] The end result of the analysis should be a map of the microenvironments defined (here aerial photographs are of great use), with detailed data on the seasonal possibilities each offers human communities on certain technological levels of development.

2. Quantitative analysis of food remains in the archaeological sites, and of the technical equipment (arrow or spear points, grinding stones for seeds, baskets and other containers, and so on) related to food-getting. It is a rare site report that treats of bones and plant remains in any but the most perfunctory way. It might seem a simple thing to ship animal bones from a site to a specialist for identification, but most archaeologists know that many zoologists consider identification of recent faunal remains a waste of time.[9] Because of this, and because many museum collections do not include postcranial skeletons that could be used for identification, the archaeologist must arrange to secure his own comparative collection. If this collection is assembled by a zoologist on the project, a by-product of the investigation would be a faunal study of microenvironments. Similarly, identification of floral and other specimens from the site would lead to other specialized studies.

3. Correlation of the archaeological with the microenvironmental study in an over-all analysis of the ancient human ecology.

The Tehuacán Valley

An archaeological project undertaken by R. S. MacNeish, with such a strategy in mind, has been located since 1961 in the dry Tehuacán Valley of southern Puebla, Mexico.[10, 11] The valley is fringed with bone-dry caves in which the food remains of early peoples have been preserved to a remarkable degree in stratified deposits. For a number of reasons, including the results of his past archaeological work in Mesoamerica, MacNeish believed that he would find here the origins of maize agriculture in the New World, and he has been proved right. It now seems certain that the wild ancestor of maize was domesticated in the Tehuacán area some time around the beginning of the fifth millennium B.C.

While the Tehuacán environment is in general a desert, the natural scientists of the project have defined within it four microenvironments (Fig. 13–1).

1. *Alluvial valley floor,* a level plain sparsely covered with mesquite, grasses, and cacti, offering fairly good possibilities, especially along the Rio Salado, for primitive maize agriculture dependent on rainfall.

2. *Travertine slopes,* on the west side of the valley. This would have

been a niche useful for growing maize and tomatoes and for trapping cottontail rabbits.

3. *Coxcatlán thorn forest,* with abundant seasonal crops of wild fruits, such as various species of *Opuntia*, pitahaya, and so on. There is also a seasonal abundance of whitetail deer, cottontail rabbits, and skunks, and there are some peccaries.

4. *Eroded canyons,* unsuitable for exploitation except for limited hunting of deer and as routes up to maguey fields for those peoples who chewed the leaves of that plant.

The correlation of this study with the analysis, by specialists, of the plant and animal remains (these include bones, maize, cobs, chewed quids, and even feces) found in cave deposits has shown that the way of life of the New World's first farmers was not very different from that of the Great Basin aborigines in the nineteenth century. Even the earliest inhabitants of the valley, prior to 6500 B.C., were more collectors of seasonally gathered wild plant foods than they were "big game hunters," and they traveled in microbands in an annual, wet-season–dry-season cycle.[12] While slightly more sedentary macrobands appeared with the adoption of simple maize cultivation after 5000 B.C., these people nevertheless still followed the old pattern of moving from microenvironment to microenvironment, separating into microbands during the dry season.

The invention and gradual improvement of agriculture seem to have made few profound alterations in the settlement pattern of the valley for many millennia. Significantly, by the Formative period (from about 1500 B.C. to A.D. 200), when agriculture based on a hybridized maize was far more important than it had been in earlier periods as a source of food energy, the pattern was still one of part-time nomadism.[13] In this part of the dry Mexican highlands, until the Classic period (about A.D. 200 to 900), when irrigation appears to have been introduced into Tehuacán, food production had still to be supplemented with extensive plant collecting and hunting.

Most of the peoples of the Formative period apparently lived in large villages in the alluvial valley floor during the wet season, from May through October of each year, for planting had to be done in May and June, and harvesting, in September and October. In the dry season, from November through February, when the trees and bushes had lost their leaves and the deer were easy to see and track, some of the population must have moved to hunting camps, principally in the Coxcatlán thorn forest. By February, hunting had become less rewarding as the now-wary deer moved as far as possible from human habitation; however, in April and May the thorn forest was still ripe for exploitation, as many kinds of wild fruit matured. In May it was again time to return to the villages on the valley floor for spring planting.

Now, in some other regions of Mesoamerica there were already, during the Formative period, fully sedentary village cultures in existence. It is clear that while the Tehuacán Valley was the locus of the first domestication of maize, the origins of full-blown village life lie elsewhere. Because of the constraining effects of the macroenvironment, the Tehuacán people were exploiting, until relatively late in Mesoamerican prehistory, as widely spaced and as large a number of microenvironments as the Great Basin aborigines were exploiting in the nineteenth century.

Coastal Guatemala

Near the modern fishing port of Ocós, only a few kilometers from the Mexican border on the alluvial plain of the Pacific coast of Guatemala, we have found evidence for some of the oldest permanently occupied villages in Mesoamerica.[14] We have also made an extensive study of the ecology and ethnology of the Ocós area.

From this study[15] we have defined no less than eight distinct microenvironments (Fig. 13–2) within an area of only about 90 square kilometers. These are as follows:

1. *Beach sand and low scrub.* A narrow, infertile strip from which the present-day villagers collect occasional mollusks, a beach crab called *chichimeco* and one known as *nazareño*, and the sea turtle and its eggs.

2. *The marine estuary-and-lagoon system,* in places extending considerably inland and ultimately connecting with streams or rivers coming down from the Sierra Madre. The estuaries, with their mangrove-lined banks, make up the microenvironment richest in wild foods in the entire area. The brackish waters abound in catfish (*Arius* sp. and *Galeichthys* sp.), red snapper (*Lutjanus colorado*), several species of snook (*Centropomus* sp.), and many other kinds of fish. Within living memory, crocodiles (*Crocodylus astutus*) were common, but they have by now been hunted almost to extinction. The muddy banks of the estuaries are the habitat of many kinds of mollusk, including marsh clams (*Polymesoda radiata*), mussels (*Mytella falcata*), and oysters (*Ostrea columbiensis*), and they also support an extensive population of fiddler and mud crabs.

3. *Mangrove forest,* consisting mainly of stilt-rooted red mangrove, which slowly gives way to white mangrove as one moves away from the estuary. We noted high populations of collared anteater (*Tamandua tetradactyla*) and arboreal porcupine (*Coendu mexicanus*). A large number of crabs (we did not determine the species) inhabit this microenvironment; these include, especially, one known locally as the *azul* (blue) crab, on which a large population of raccoons feeds.

4. *Riverine,* comprising the channels and banks of the sluggish Suchiate and Naranjo rivers, which connect with the lagoon-estuary system not far from their mouths. Fresh-water turtles, catfish, snook, red snapper, and mojarra (*Cichlasoma* sp.) are found in these waters; the most common animal along the banks is the green iguana (*Iguana iguana*).

5. *Salt playas,* the dried remnants of ancient lagoon-and-estuary systems that are still subject to inundation during the wet season, with localized stands of a tree known as *madresal* ("mother of salt"). Here there is an abundance of game, including whitetail deer and the black iguana (*Ctenosaura similis*), as well as a rich supply of salt.

6. *Mixed tropical forest,* found a few kilometers inland, in slightly higher and better drained situations than the salt *playas.* This forest includes mostly tropical evergreens like the ceiba, as well as various zapote and fan palms, on the fruit of which a great variety of mammals thrive—the kinkajou, the spotted cavy, the coatimundi, the raccoon, and even the gray fox. The soils here are highly suitable for maize agriculture.

7. *Tropical savannah,* occupying poorly drained patches along the upper stream and estuary systems of the area. This is the major habitat in the area for cottontail rabbits and gray foxes. Other common mammals are the coatimundi and armadillo.

8. *Cleared fields and second growth,* habitats that have been created by agriculturists, and that are generally confined to areas that were formerly mixed tropical forest.

Among the earliest Formative cultures known thus far for the Ocós area is the Cuadros phase, dated by radiocarbon analysis at about 1000 to 850 B.C. and well represented in the site of Salinas La Blanca, which we excavated in 1962.[16] The site is on the banks of the Naranjo River among a variety of microenvironments; it consists of two flattish mounds built up from deeply stratified refuse layers representing house foundations of a succession of hamlets or small villages.

From our analysis of this refuse we have a good idea of the way in which the Cuadros people lived. Much of the refuse consists of potsherds from large, neckless jars, but very few of the clay figurines that abound in other Formative cultures of Mesoamerica were found. We discovered many plant remains; luckily these had been preserved or "fossilized" through replacement of the tissues by carbonates. From these we know that the people grew and ate a nonhybridized maize considerably more advanced than the maize which was then being grown in Tehuacán.[17] The many impressions of leaves in clay floors in the site will, we hope, eventually make it possible to reconstruct the flora that immediately surrounded the village.

The identification of animal remains (Fig. 13–3), together with our ecological study and with the knowledge that the people had a well-developed maize agriculture, gives a great deal of information on the subsistence activities of these early coastal villagers. First of all, we believe they had no interest whatever in hunting, a conclusion reinforced by our failure to find a single projectile point in the site. The few deer bones that have been recovered are all from immature individuals that could have been encountered by chance and clubbed to death. Most of the other remains are of animals that could have been collected in the environs of the village, specifically in the lagoon-estuary system and the flanking mangrove forest, where the people fished, dug for marsh clams, and, above all, caught crabs (primarily the *azul* crab, which is trapped at night). Entirely missing are many edible species found in other microenvironments, such as raccoon, cottontail rabbit, peccary, spotted cavy, and nine-banded armadillo.

There is no evidence at all that occupation of Salinas La Blanca was seasonal. An effective food production carried out on the rich, deep soils of the mixed tropical forest zone, together with the food resources of the lagoon-estuary system, made a permanently settled life possible. Looked at another way, developed maize agriculture had so reduced the number and spacing of the niches which had to be exploited that villages could be occupied the year round.[18]

Conditions similar to those of the Ocós area are found all along the Pacific coast of Guatemala and along the Gulf coast of southern Veracruz and Tabasco in Mexico, and we suggest that the real transition to village life took place there and not in the dry Mexican highlands, where maize was domesticated initially.[19]

Conclusion

The interpretation of archaeological remains through a fine-scale analysis of small ecological zones throws new light on the move toward sedentary life in Mesoamerican prehistory. In our terms, the basic difference between peoples who subsist on wild foods and those who dwell in permanent villages is that the former must exploit a wide variety of small ecological niches in a seasonal pattern—niches that are usually scattered over a wide range of territory—while the latter may, because of an effective food production, concentrate on one or on only a few microenvironments that lie relatively close at hand.

Fine-scale ecological analysis indicates that there never was any such thing as an "agricultural revolution" in Mesoamerica, suddenly and almost miraculously resulting in village life. The gradual addition of domesticates such as maize, beans, and squash to the diet of wild plant and animal foods hardly changed the way of life of the Tehuacán people for many thousands of years, owing to a general paucity of the environment, and seasonal nomadism persisted until the introduction of irrigation. It probably was not until maize was taken to the alluvial, lowland littoral of Mesoamerica, perhaps around 1500 B.C., that permanently occupied villages became possible, through reduction of the number of microenvironments to which men had to adapt themselves.

NOTES

1. Mesoamerica is the name given to that part of Mexico and Central America that was civilized in pre-Columbian times. For an excellent summary of its prehistory, see G. R. Willey, *Science*, **131**, 73 (1960).

2. R. S. MacNeish, *Science*, **143**, 531 (1964).

3. See C. A. Reed and R. J. Braidwood, "Toward the reconstruction of the environmental sequence of Northeastern Iraq," in R. J. Braidwood and B. Howe, "Prehistoric Investigations in Iraqi Kurdistan," *Oriental Institute, University of Chicago, Studies in Ancient Oriental Civilization No. 31* (1960), p. 163. Reed and Braidwood also convincingly reject the technological-deterministic approach of V. G. Childe and his followers.

4. See B. J. Meggers, *Am. Anthropologist*, **56**, 801 (1954), for an environmental-deterministic view of the constraining effects of tropical forests on human cultures.

5. See F. Barth, *ibid.*, **58**, 1079 (1956), for a microenvironmental approach by an ethnologist to the exceedingly complex interrelationships between sedentary agriculturists, agriculturists practicing transhumant herding, and nomadic herders in the state of Swat, Pakistan.

6. J. H. Steward, "Basin-Plateau Aboriginal Sociopolitical Groups," *Smithsonian Inst. Bur. Am. Ethnol. Bull. 120* (1938).

7. *Ibid.*

8. The pitfalls of searching for ethnological data relevant to archaeological problems among cultures far-flung in time and space are stressed by J. G. D. Clark, *Prehistoric Europe, The Economic Basis* (Philosophical Library, New York, 1952), p. 3.

9. See W. W. Taylor, Ed., "The identification of non-artificial archaeological materials," *Natl. Acad. Sci.–Natl. Res. Council Publ. 565* (1957). For a general article on the analysis of food remains in archaeological deposits see R. F. Heizer in "Application of quantitative methods in archaeology," *Viking Fund Publications in Anthropology No. 28* (1960), pp. 93–157.

10. R. S. MacNeish, *Science*, **131**, 73 (1960).

11. P. C. Mangelsdorf, R. S. MacNeish, W. C. Gallinat, *Science*, **143**, 538 (1964). We thank Dr. MacNeish for permission to use unpublished data of the Tehuacán Archaeological-Botanical Project in this article.

12. R. S. MacNeish, *Second Annual Report of the Tehuacán Archaeological-Botanical Project* (Robert S. Peabody Foundation for Archaeology, Andover, Mass., 1962).

13. The research discussed in this and the following paragraph was carried out by Flannery as staff zoologist for the Tehuacán project during the field seasons of

1962 and 1963; see K. V. Flannery, "Vertebrate Fauna and Prehistoric Hunting Patterns in the Tehuacán Valley" (Robert S. Peabody Foundation for Archaeology, Andover, Mass., in press); K. V. Flannery, thesis, Univ. of Chicago, unpublished.

14. M. D. Coe, "La Victoria, an early site on the Pacific Coast of Guatemala," *Peabody Museum, Harvard, Paper No. 53* (1961).

15. The study was carried out largely by K. V. Flannery.

16. The final report on Salinas La Blanca by Coe and Flannery is in preparation. The research was supported by the National Science Foundation under a grant to the Institute of Andean Research, as part of the program "Interrelationships of New World Cultures." The oldest culture in the area is the Ocós phase, which has complex ceramics and figurines; the palaeoecology of Ocós is less well known than that of Cuadros, which directly follows it in time.

17. P. C. Mangelsdorf, who has very kindly examined these maize specimens, informs us that they are uncontaminated with *Tripsacum*, and that probably all belong to the primitive lowland race, Nal-Tel.

18. To paraphrase the concept of "primary forest efficiency," developed by J. R. Caldwell ["Trend and Tradition in the Eastern United States," *Am. Anthropol. Assoc. Mem. No. 88* (1958)], we might think of the Cuadros phase as leaning to a "primary lagoon-estuary efficiency." We might think the same of the Ocós phase of the same region, which may date back to 1500 B.C.

19. An additional factor which may in part account for the priority of coastal Guatemala over Tehuacán in the achievement of a sedentary mode of life is the presence of an extensive system of waterways in the former region, which might have made it less necessary for local communities to move to productive sources of food. By means of canoes, a few persons could have brought the products of other niches to the village. However, our evidence indicates that the Cuadros people largely ignored the possibilities of exploiting distant niches.

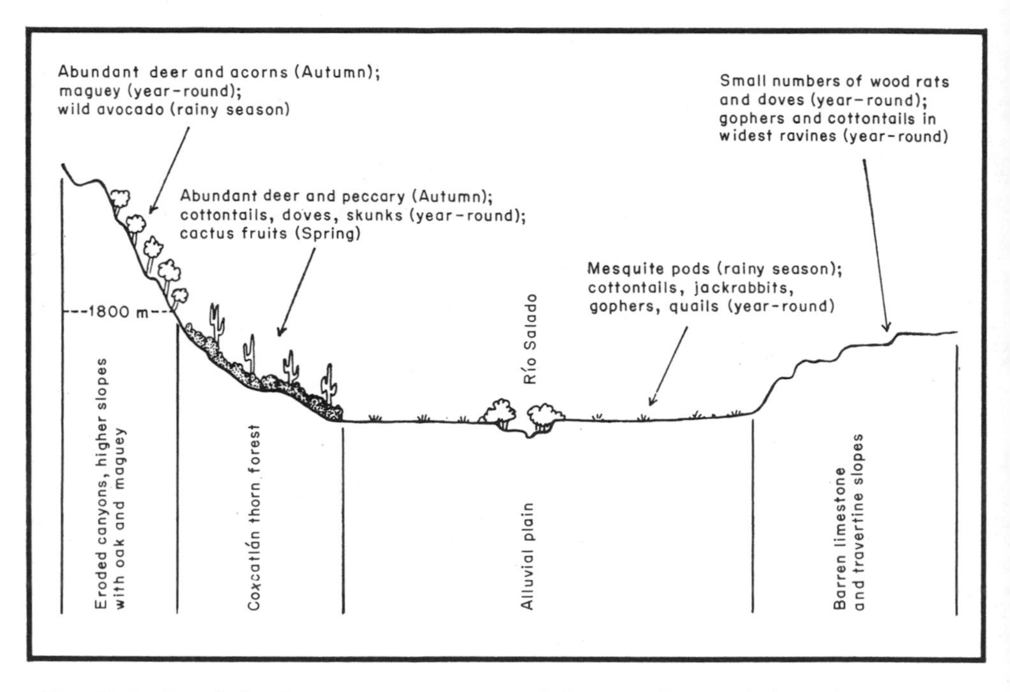

Fig. 13–1. An idealized east-west transection of the central part of the Tehuacán Valley, Puebla, Mexico, showing microenvironments and the seasons in which the food resources are exploited. East is to the left. The length of the area represented is about 20 kilometers.

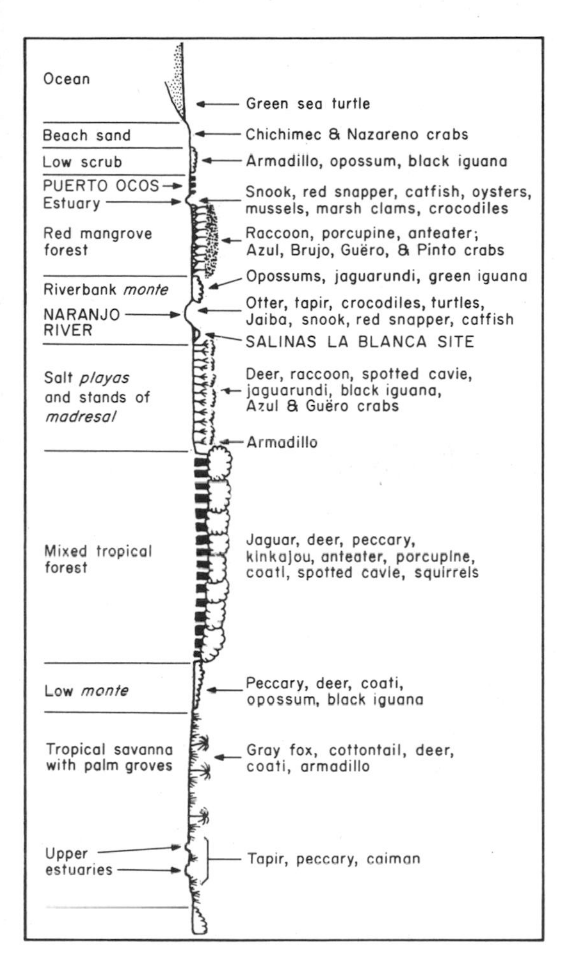

Fig. 13–2. Northeast-southwest transection of the Ocós area of coastal Guatemala, showing microenvironments in relation to the site of Salinas La Blanca. Northeast is to the right. The length of the area represented is about 15 kilometers.

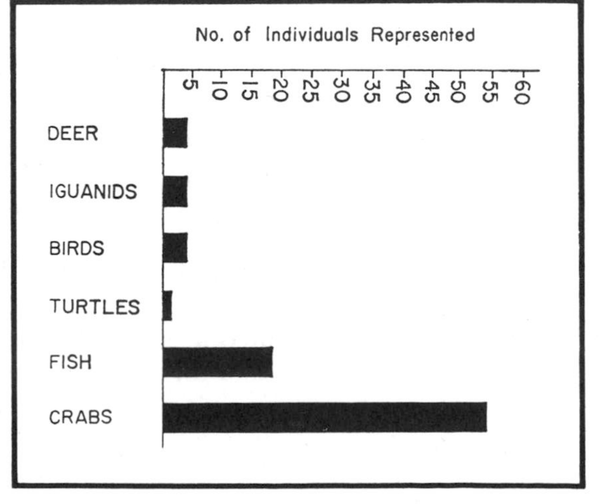

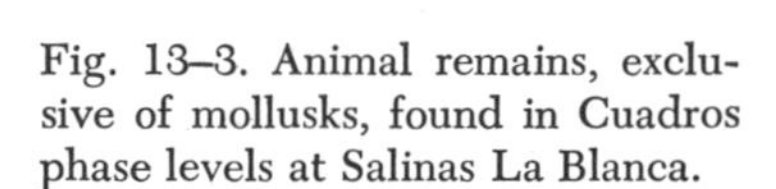
Fig. 13–3. Animal remains, exclusive of mollusks, found in Cuadros phase levels at Salinas La Blanca.

Chapter 14

Domestication of Corn

Paul C. Mangelsdorf
Richard S. MacNeish
Walton C. Galinat

The problem of the origin of corn has intrigued botanists and other students of plants for more than four centuries. The plant was unknown in any part of the Old World before 1492, while in the New World it was the basic food plant of all pre-Columbian advanced cultures and civilizations, including the Inca of South America and the Maya and Aztec of Middle America.[1] Although these facts point strongly to its American origin, some writers have continued to argue eloquently for an Old World origin. A living wild form of corn has never been discovered, despite the extensive searches for it which have been carried on in various parts of the hemisphere. The absence of a wild form has been conducive to speculation—sometimes reaching the point of acrimonious debate—about its probable nature. There has, however, been general agreement that modern corn is unique among the major cereals in its grain-bearing inflorescence (the ear), which is completely enclosed in modified leaf sheaths (the husks), the plant being thus rendered incapable of dispersing its seeds. How, then, did wild corn, which to survive in nature must have had a means of dispersal, differ from modern cultivated corn? Where did it grow? How did it evolve under domestication? These are some of the questions that comprise the corn problem.

Close collaboration in recent years between archaeologists and botanists has furnished at least partial answers to all of these questions, and has also contributed to solving the problem of the beginning of agriculture in America and the rise of prehistoric cultures and civilizations.

The first substantial contribution of archaeology to the solution of the corn problem was the finding of prehistoric vegetal material in Bat Cave in New Mexico, excavated by Herbert Dick, then a graduate student in the Peabody Museum of Harvard University, in two expeditions, in 1948 and 1950. Accumulated trash, garbage, and excrement in this cave contained cobs and other parts of corn at all levels, and these cobs and parts showed a distinct evolutionary sequence from the lower to the upper levels.[2] At the bottom of the refuse, which was some 2 meters deep, Dick found tiny cobs, 2 to 3 centimeters long, which were dated by radiocarbon determinations of associated charcoal at about 3600 B.C. Anatomical studies of these cobs led to the conclusion that the early Bat Cave corn was both a popcorn (a type with small, hard kernels capable of exploding when exposed to heat) and a pod form (a type with kernels partly enclosed by floral bracts which botanists call glumes and the layman knows as chaff).[3]

Because the Bat Cave corn was both a popcorn and a pod corn, Mangelsdorf undertook to produce a genetic reconstruction of the ancestral form of corn by crossing pod corn and popcorn and backcrossing the hybrid repeatedly to popcorn. The final product of this breeding was a pod-popcorn bearing small kernels enclosed in glumes on ears arising from the upper joints of the stalks.[4] This reconstructed ancestral form had two means of dispersal: seeds borne on the fragile branches of the tassel and seeds on ears at high positions on the stalk, which at maturity were not completely enclosed by husks. The reconstructed ancestral form served another useful purpose in showing the archaeologist approximately what to look for in seeking prehistoric wild corn.

Prehistoric Corn in Northern Mexico

A second important collection of prehistoric maize came from La Perra Cave in Tamaulipas in northeastern Mexico, excavated in 1949 by MacNeish, who was then associated with the National Museum of Canada. The specimens from this cave, as those from Bat Cave, showed a distinct evolutionary sequence from the lower to the higher levels of the accumulated refuse.[5] The earliest corn, dated 2500 B.C. by radiocarbon determination of associated wood and leaves, was identified as an early form of a still-existing race, Nal-Tel, which E. J. Wellhausen and others,[6] who have classified the present-day maize of Mexico, described as one of the four Ancient Indigenous races of Mexico. These earliest cobs were somewhat larger than the earliest cobs from Bat Cave and so gave some support to the assumption that the two radiocarbon dates involved, 3600 and 2500 B.C., might be relatively if not absolutely correct.

While excavating La Perra Cave, which is located in eastern Tamaulipas, MacNeish also made some preliminary soundings in several caves in southwestern Tamaulipas, which persuaded him that still earlier corn, perhaps even prehistoric wild corn, might be found in the lower levels of the refuse of these caves. Accordingly in 1954, with the assistance of David Kelley, then a graduate student in anthropology at Harvard, he excavated two caves, Romero's Cave and Valenzuela's Cave, in Inferniello Canyon. The earliest corn from these caves proved, disappointingly, to be not earlier than the La Perra corn but slightly later, about 2200 B.C.[7] It was, however, of a race different from the La Perra corn and showed some resemblance to the Bat Cave corn.

Of even greater interest was the discovery in Romero's Cave of a few specimens of teosinte, the closest relative of corn. Well-preserved specimens of the fruits of this plant occurred in a level dated 1400 to 400 B.C. Fragments identified as teosinte occurred in feces in a level dated 1800 to 1400 B.C. Since teosinte has not been found growing in Tamaulipas in modern times, it may be assumed either that its range is more restricted today than it was several thousand years ago or that teosinte was planted with corn as a method of improving it, a practice reported by C. Lumholtz[8] to be characteristic of certain Indians of the western part of Mexico.

While the excavations in Tamaulipas were in progress, another series of excavations was being made in caves in the states of Chihuahua and Sonora in northwestern Mexico by Robert H. Lister of the University of Colorado. In one of these caves, Swallow Cave, Lister uncovered at the lowest levels several tiny cobs similar in shape and size to the Bat Cave cobs, though slightly larger. Since it seemed inadvisable to sacrifice these to obtain radiocarbon determinations, they have not been dated. However, the fact that they occurred at a considerable depth (about two meters below the surface) and in a preceramic context suggests a substantial age. These earliest Swallow Cave cobs were identified as prototypes of Chapalote,[9] another of the Ancient Indigenous and still-existing races of corn of Mexico described by Wellhausen and others.[10]

During this same period another important discovery was made when E. S. Barghoorn and others[11] identified as pollen grains of maize some fossil pollen isolated from a drill core taken at a depth of more than seventy meters below the present site of Mexico City. This pollen was assigned to the Last Interglacial period now estimated by geologists to have occurred about 80,000 years ago. Since this period antedates the arrival of man on this continent, the pollen was thought to be that of a wild maize that once grew in the Valley of Mexico and has since become extinct. Other pollen, considered to be that of cultivated corn, occurred abundantly in the upper levels—above six meters. The earliest of these upper-level pollen grains are assigned to the later part of the post-glacial optimum and are therefore no earlier than the earliest corn from Tamaulipas or New Mexico. Although the criteria used in identifying the fossil pollen grains have been questioned,[12] more recent studies made by Barghoorn and his associates, using phase microscopy, have revealed features in which the pollen grains of corn and its relatives differ conspicuously and have confirmed the earlier identifications. There now seems to be no doubt that at least some of the fossil pollen grains were those of corn. Thus, this fossil pollen settles two important questions: it shows that corn is an American plant and that the ancestor of cultivated corn is corn and not one of corn's relatives, teosinte or Tripsacum.

On the basis of his excavations in Tamaulipas and the discovery of fossil corn pollen in the Valley of Mexico, MacNeish concluded that the evidence for the earliest domestication of maize and the beginnings of agriculture in America must be sought farther south. A reconnaissance made in Honduras and Guatemala in 1958 yielded no results of promise. Excavations in 1959 of Santa Maria Cave in Chiapas in southern Mexico uncovered corn and other vegetal material, including pollen, but none older than that which had already been found further north. Turning northward again, MacNeish made a reconnaissance of sites in Oaxaca and Puebla which led to the conclusion that the Tehuacán Valley of southern Puebla and northern Oaxaca might, because of its dry climate and ever flowing springs, offer the most promising site so far discovered for seeking prehistoric wild corn and the beginning of agriculture. A preliminary sounding in 1960, in one of the numerous caves in the cliffs surrounding the valley, uncovered cobs which were thought to be those of wild corn. Full-scale excavations conducted the following season confirmed this.

The physical features of the Valley of Tehuacán and its various cultural phases have been described elsewhere by MacNeish.[13] At first glance this valley, with its semiarid climate and its predominantly xerophytic, drought-resisting vegetation, may not seem to be a suitable habitat for wild corn, and in earlier speculation about where wild corn might have grown we did not associate it with such plants as cacti and thorny leguminous shrubs.[14] Closer examination, however, suggests that the habitat furnished by this arid valley may, in fact, have been almost ideal for wild corn. The average annual rainfall at the center of the valley is low (approximately 500 millimeters a year) and becomes somewhat higher both south and north of the center. About 90 per cent of the annual rain usually falls during the growing season, from April through October.

The other months are quite dry—in midwinter the valley is virtually a desert—and comprise a period during which the seeds of wild maize and other annual plants could have lain dormant, ready to sprout with the beginning of the summer rains and never in danger of germinating prematurely and then succumbing to the vicissitudes of winter. Thus, although the perennial vegetation of this valley, which year after year must survive the dry winter months, is necessarily xerophytic, the annual vegetation (and wild maize would have been an annual) need not be especially drought-resistant. Modern maize is not notable for its drought resistance and probably its wild prototype was not either.

The corn uncovered by MacNeish and his associates in their excavations of the caves in Tehuacán Valley is, from several standpoints, the most interesting and significant prehistoric maize so far discovered. It includes the oldest well-preserved cobs yet available for botanical analysis; the oldest cobs are probably those of wild maize. This maize appears to be the progenitor of two of the previously recognized Ancient Indigenous races of Mexico, Nal-Tel and Chapalote, of which prehistoric prototypes had already been found in La Perra and Swallow caves, respectively. The collections portray a well-defined evolutionary sequence.

Prehistoric Corn from Five Caves

Before considering the corn itself, we should say a word about the caves in which the remains of maize were uncovered. Five major caves that were excavated—Coxcatlan, Purron, San Marcos, Tecorral, and El Riego—yielded maize in archaeological levels. The caves were situated in three or four different environments, which might have had considerable bearing upon the possibility of wild corn's growing nearby and which might have affected the practice of agriculture (see Fig. 14–1).

Coxcatlan Cave, first found in 1960, was one of the richest in vegetal remains. Excavations revealed twenty-eight superimposed floors or occupational levels covering two long unbroken periods—from 10000 to 2300 B.C. and from 900 B.C. to A.D. 1500. Fourteen of the upper floors, those from 5200 to 2300 B.C. and from 900 B.C. to A.D. 1500, contained well-preserved corn cobs. The cave, a long narrow rock shelter, is situated in the southeastern part of the Valley in one of the canyons flanking the Sierra Madre mountain range (Fig. 14–1). The shelter faces north and looks out on a broad alluvial plain covered with grasses, mesquite, other leguminous shrubs, and cacti. Supplementing the meager annual rainfall is some water drainage from the nearby mountain slopes, and this would have made it possible for wild or cultivated corn to grow during the wet season. In other seasons of the year irrigation would have been necessary for corn culture.

A few miles south of Coxcatlan, in the same set of canyons, is Purron

Cave. This is a somewhat smaller rock shelter but it contains a long continuous occupation (twenty-five floors) from about 7000 B.C. to A.D. 500. It is archaeologically much poorer than Coxcatlan, and only the top twelve floors (from 2300 B.C. to A.D. 500) contained preserved remains of food plants.

El Riego Cave, situated in the north end of the Valley (Fig. 14–1) only a mile north of the modern town of Tehuacán, is a deep recess which contained an abundance of preserved specimens. Its five archaeological zones, however, do not extend far back in time and were deposited between 200 B.C. and A.D. 1500. The cave is in the travertine face of a cliff and faces south. Under these cliffs and flowing out from them are the famous Tehuacán mineral springs, and the soils in front of the cave are fertile and well watered. Because of the fertile soils and the abundant water there is an oasislike vegetation around the cave. This is an excellent area for agriculture, and it may even have originally supported a vegetation too lush for wild corn to compete with.

The last two caves, San Marcos and Tecorral, occur in a steep canyon in the west side of the Valley (see Fig. 14–1). They are small shelters situated side by side, facing east. Tecorral contained three floors and only the top floor (about A.D. 1300) had a few corn cobs. San Marcos, however, was very different; although small, it yielded five superimposed floors with an abundance of preserved maize and other remains. The top four floors have been dated, by the carbon-14 method, at about 4400 B.C., 3300 B.C., 1100 B.C., and A.D. 300, respectively, and the earliest one is estimated to have been laid down about 5200 B.C. The shelters look out over broad alluvial terraces covered by grass and small thorny trees. Plants collected from this canyon bottom reveal a number of endemics —species not found elsewhere in the valley. The surrounding travertine-covered canyon walls and hilltops, however, have a vegetation like that of the Sonoran Desert. The area receives water in the rainy season and much of it floods the lower terraces. All occupations found in the caves were from the rainy seasons. Agriculture would have been possible in the rainy season with or without irrigation. The alluvial terraces would have furnished an almost ideal habitat for wild corn.

In all, 23,607 specimens of maize were found in the five caves; 12,857 of these, or more than half, are whole or almost intact cobs. There are, in addition to the intact cobs, 3,273 identified cob fragments and 3,880 unidentified cob fragments. Among the remaining specimens are all parts of the corn plant: 28 roots, 513 pieces of stalks, 462 leaf sheaths, 293 leaves, 962 husks, 12 prophylls, 127 shanks, 384 tassel fragments, 47 husk systems, 6 midribs, and 600 kernels. There are also numerous quids, 64 representing chewed stalks, and 99, chewed husks.

The prehistoric cobs from the five caves can be assigned to six major and five minor categories. The frequency polygons of Fig. 14–2 show graphically the time of first appearance of a type of corn, the corresponding cultural periods, and the relative prominence (in terms of percentages) of the number of identified cobs for each of these categories. The polygons show patterns similar to those exhibited by artifacts, and for good reason—man's cultivars are artifacts as surely as are his weapon points or pottery. A brief description of the types of maize represented by these categories follows.

Prehistoric Wild Maize

The earliest cobs from the El Riego and Coxcatlan cultural phase, dated 5200–3400 B.C., are regarded as being those of wild corn for six reasons. (i) They are remarkably uniform in size and other characteristics and in this respect resemble most wild species. (ii) The cobs have fragile rachises as do many wild grasses; these provide a means of dispersal that modern corn lacks. (iii) The glumes are relatively long in relation to other structures and must have partially enclosed the kernels as they do in other wild grasses. (iv) There are sites in the Valley, such as the alluvial terraces below San Marcos Cave, which are well adapted to the growth of annual grasses, including corn, and which the competing cacti and leguminous shrubs appear to shun. (v) There is no evidence from other plant species that agriculture had yet become well established in this valley, at least in the El Riego phase or the earlier part of the Coxcatlan phase. (vi) The predominating maize from the following phase, Abejas, in which agriculture definitely was well established, is larger and more variable than the earliest corn.

This combination of circumstances leads to the conclusion—an almost inescapable one—that the earliest prehistoric corn from the Tehuacán caves is wild corn. We shall assume here that it is.

The intact cobs of the wild corn vary in length from 19 to 25 millimeters (Fig. 14–3A). The number of kernel rows is usually eight but a few cobs with four rows were found. None of the earliest cobs have kernels, but the number of kernels that they once bore can be determined by counting the number of functional spikelets. These vary from 36 to 72 per cob. The average number of kernels borne by the earliest intact cobs of wild corn from San Marcos Cave was 55.

The glumes of the spikelets are relatively long in relation to other structures and are soft, fleshy, and glabrous (lacking in hairs). On some cobs the glumes are rumpled, probably as a result of the forcible removal of the kernels. The cobs have the general aspect of a weak form of pod corn.

The spikelets are uniformly paired and are attached to a slender, soft, and somewhat fragile stem (technically known as a rachis) in which the cupules, or depressions in the rachis, are shallow and almost glabrous, bearing only sparse, short hairs.

Most of the wild-type cobs were apparently once bisexual, bearing pistillate (female) spikelets in the lower regions and staminate (male) spikelets above. Of fifteen apparently intact cobs from San Marcos Cave, ten had stumps at the tip where a staminate spike had presumably been broken off. In this respect the Tehuacán wild maize resembles corn's wild relative, Tripsacum, which regularly bears pistillate spikelets below and staminate spikelets above on the same inflorescences.

The uniformly paired spikelets and relatively soft tissues of the rachis and glumes provide further proof, in addition to that furnished by the fossil pollen of the Valley of Mexico, that the wild ancestor of cultivated corn was corn and not one of its relatives, teosinte or Tripsacum.

The wild maize declined somewhat in prominence in the Abejas phase, where it comprises 47 per cent of the cobs. It persisted, however, as a minor element of the corn complex until the middle of the Palo Blanco phase, dated at about A.D. 250.

What caused the wild corn finally to become extinct? We have for some years assumed that two principal factors may have been involved in the extinction of corn's ancestor. (i) The sites where wild corn grew in nature might well be among those chosen by man for his earliest cultivation. (ii) Wild corn growing in sites not appropriated for cultivation but hybridizing with cultivated corn, after the latter had lost some of its essential wild characteristics, would become less able to survive in the wild. Of these two causes of extinction the second may have been the more important. Corn is a wind-pollinated plant, and its pollen can be carried many miles by the wind. It is virtually inevitable that any maize growing wild in the Valley would have hybridized at times with the cultivated maize in nearby fields, which was producing pollen in profusion. Repeated contamination of the wild corn by cultivated corn could eventually have genetically "swamped" the former out of existence.

There is now good archaeological evidence in Tehuacán to suggest that both of these assumed causes of extinction were indeed operative. The alluvial terraces below San Marcos Cave, where wild corn may once have grown, now reveal the remains of a fairly elaborate system of irrigation, indicating that the natural habitat of wild corn was replaced by cultivated fields. Abundant evidence of hybridization between wild and cultivated corn is found in the prehistoric cobs. We have classified 252 cobs as possible first-generation hybrids of wild corn with various cultivated types and 464 cobs as backcrosses of first-generation hybrids to the wild corn.

Is there a possibility that wild corn may still be found in some remote and inaccessible locality in Mexico or elsewhere? We suspect not. Whatever wild corn may have persisted until the sixteenth century was almost certainly rapidly extinguished after the arrival of the European colonists with their numerous types of grazing animals: horses, burros, cows, sheep, and—worst of all—the omnivorous and voracious goats. To all of these animals young corn plants are a palatable fodder, one that is to be preferred to almost any other grass.

Wild Corn Reconstructed

A well-preserved early cob, an intact husk system consisting of an inner and outer husk from the Abejas phase in the San Marcos Cave, and a piece of staminate spike from the Ajalpan phase of the same cave provide the materials for a reconstruction of the Tehuacán wild corn. This is illustrated in actual size in Fig. 14–4. An ear with only two husks was probably borne high on the stalk and its husks opened at maturity, permitting dispersal of the seeds. Other early specimens show that the plants lacked secondary stalks, technically known as tillers; the leaf sheaths were completely lacking in surface hairs; the kernels were somewhat rounded and were either brown or orange.

In its lack of tillers, its glabrous leaf sheaths, its rounded kernels, and the color of its pericarp, the Tehuacán wild corn differs quite distinctly from a third Ancient Indigenous race of Mexico, Palomero Toluqueño, described by Wellhausen and others.[15] This finding suggests that the latter may have stemmed from a different race of wild corn growing in another place. Fossil corn pollen from the Valley of Mexico suggests still a third locality where wild corn once occurred. It is becoming increasingly apparent that cultivated corn may have had multiple sites of origin, of which southern Mexico is only one, but the earliest one so far discovered.

The corn which we have called "early cultivated" is similar to the wild corn except in size (Fig. 14–3*B*). It has the same long glumes and the same soft, somewhat fragile rachises. It is probably a direct descendant of the wild corn, sightly modified through growing in a better environment. Initially the better environment may have been nothing more than that produced by the removal, by man, of other vegetation competing with wild corn growing in its natural habitat. Later the corn was actually planted in fields chosen for the purpose. Still later it was irrigated.

Exactly when the maize was first cultivated in Tehuacán Valley is difficult to determine. Two cobs classified as early cultivated appeared in the Coxcatlan culture, dated 5200 to 3400 B.C., but we cannot tell whether the cobs represent the upper or lower part of this phase. Since remains of the bottle gourd and two species of squashes (*Cucurbita moschata* and *C. mixta*), as well as tepary beans, chili peppers, amaranths, avocados, and zapotes, occurred at this phase, they may have been at least an incipient agriculture and it is not unreasonable to suppose that maize, too, was being cultivated. However, two cobs scarcely furnish conclusive evidence on this point.

What we can be certain of is that during the Abejas phase, dated 3400 to 2300 B.C., this corn was definitely a part of an agricultural complex that included, in addition to maize, the following cultivars: the bottle gourd, *Lagenaria siceraria*; two species of squashes, *Cucurbita moschata* and *C. mixta*; *Amaranthus* spp.; the tepary bean, *Phaseolus acutifolius*, and possibly also the common bean, *P. vulgaris*; Jack beans, *Canavalia enciformis*; chili peppers, *Capsicum frutescens*; avocados, *Persea americana*; and three varieties of zapotes. Among the 99 cobs representing this phase, 45 (almost half) were classified as early cultivated. Thereafter this type gradually decreased in relative frequency, becoming extinct before the beginning of the Venta Salada phase at A.D. 700.

Other Prehistoric Types

Making its first appearance in the Abejas phase but represented there by a single cob, and becoming well established in the Ajalpan phase, is a type we have called "early tripsacoid" (Fig. 14–5A). The term *tripsacoid* is one proposed by Edgar Anderson and R. O. Erickson[16] to describe any combination of characteristics that might have been introduced into corn by hybridizing with its relatives, teosinte or Tripsacum. In both of these species the tissues of the rachis and the lower glumes are highly indurated and the lower glumes are thickened and curved. Archaeological cobs showing these characteristics are suspected of being the product of the hybridization of maize with one of its two relatives.

Since neither teosinte nor Tripsacum is known in Tehuacán Valley today and since neither is represented in the archaeological vegetal remains, we suspect that the tripsacoid maize was introduced from some other region, possibly from the Balsas River basin in the adjoining state of Guerrero where both teosinte and Tripsacum are common today.

The introduced tripsacoid is a corn somewhat similar to the race (Nal-Tel) described by Wellhausen and others[17] but smaller. This introduced corn evidently hybridized with both the wild and the early domesticated corn in the Tehuacán Valley to produce hybrids with characteristics intermediate between those of the parents. First-generation hybrids, in turn, backcrossed to both parents to produce great variability in both cultivated and wild populations (figs. 14–5*B* and 14–6).

The introduced tripsacoid maize, together with its various hybrids, was the most common maize during the Ajalpan and Santa Maria phases from 1000 to 200 B.C. Thereafter it declined in frequency until it became almost extinct in the Venta Salada phase. But this complex apparently gave rise to two still-existing races of Mexico, Nal-Tel and Chapalote, and to a prehistoric type which we have called "late tripsacoid."

The earliest cobs from the Tehuacán caves were, because of their shape and glabrous cupules, thought to be prototypes of Chapalote, one of the Ancient Indigenous races described by Wellhausen and others. This race is today found only in northwestern Mexico in the states of Sinaloa and Sonora. Archaeologically it is the predominating early corn in all sites excavated in northwestern Mexico and the southwestern United States.

Some of the later cobs from the caves, because their cupules were beset with hairs, a characteristic of the early Nal-Tel of La Perra Cave, seem to resemble this race more than they resemble Chapalote. Also, since Nal-Tel is found today in southern Mexico, it would not be surprising to find its origin there.

Actually Nal-Tel and Chapalote are quite similar in their characteristics, the principal conspicuous difference between them being in the color of the kernel, which in the former is orange and in the latter chocolate brown. Our hope that the first kernels to appear among the remains would enable us to distinguish between the two races was not realized. Of the kernels occurring in the Santa Maria phase, about half were brown and the other half orange, and both brown and orange kernels were also found in the later levels. Being unable, by any single criterion or combination of characters, to distinguish the cobs of the two races, we have designated this category the Nal-Tel-Chapalote complex (Fig. 14–7).

It was this corn, more than any other, that initiated the rapid expansion of agriculture that was accompanied by the developments of, first, large villages and, later, secular cities; the practice of irrigation; and the establishment of a complex religion. If it is too much to say that this corn was responsible for these revolutionary developments, it can at least be said that they probably could not have occurred without it. Perhaps it is not surprising that present-day Mexican Indians have a certain reverence for these ancient races of corn, Nal-Tel and Chapalote, and continue to grow them although they now have more productive races at their command.

A type which we have designated "late tripsacoid" corn differs from the early tripsacoid primarily in size. It comprises principally the more tripsacoid cobs of the Nal-Tel-Chapalote type and were it not that it includes some tripsacoid cobs of a slender popcorn, it could be considered part of the Nal-Tel-Chapalote complex to which it is closely related and with which it is contemporaneous. If the tripsacoid cobs resembling Nal-Tel or Chapalote are considered along with these late tripsacoid cobs, this complex seems even more likely to have been the basic maize of the Tehuacán cultures from 900 B.C. to A.D. 1536, representing about 65 per cent of all the corn at the end of the Venta Salada phase.

A type called "slender pop" has very slender cylindrical cobs, many rows of grain, and small rounded kernels, yellow or orange. This may be the prototype of a Mexican popcorn, Arrocillo Amarillo, one of the four Ancient Indigenous races described by Wellhausen and others.[18] This race, which is now mixed with many others, occurs in its most nearly pure form in the Mesa Central of Puebla at elevations of 1,600 to 2,000 meters, not far from the Tehuacán Valley and at similar altitudes. Appearing first in the Santa Maria phase between 900 and 200 B.C., the slender pop increased rapidly and steadily in frequency, comprising 20 per cent of the cobs in the final phase.

Judged by its cobs alone, the slender pop might be expected to be less productive than Nal-Tel or Chapalote, and its increased prominence deserves an explanation. A plausible one is that, although the ears are small, the stalks may have been prolific, normally bearing more than one ear. The present-day race to which it bears some resemblance and to which it may be related is prolific, usually producing two or three ears per stalk.

Minor categories include cobs and kernels, which appear in later levels, and which are recognized as belonging to several of the modern races of Mexico described by Wellhausen and others.[19] They occur much too infrequently to be of significance in the total picture of food production but they are important in showing that these modern Mexican races were already in existence in prehistoric times. The only previous evidence of this was the fact that casts of ears appear on Zapotec funerary urns.

Other Parts of the Corn Plant

In all, 3,597 specimens of parts of the corn plant, other than cobs, were found in the five caves. These specimens confirm the conclusions reached from the study of the cobs. There has been no change in the basic botanical characteristics of the corn plant during domestication.

Then, as now, corn was a monoecious annual bearing its male and female spikelets separately, the former predominating in the terminal inflorescences and the latter in the lateral inflorescences, which, as in modern corn, were enclosed in husks. Then, as now, the spikelets were borne in pairs; in the staminate spikelets one member of the pair was sessile, the other pediceled. The only real changes in more than 5,000 years of evolution under domestication have been changes in the size of the parts and in productiveness.

The importance of these changes to the rise of the American cultures and civilizations would be difficult to overestimate. There is more foodstuff in a single grain of some modern varieties of corn than there was in an entire ear of the Tehuacán wild corn. A wild grass with tiny ears—a species scarcely more promising as a food plant than some of the weedy grasses of our gardens and lawns—has, through a combination of circumstances, many of them perhaps fortuitous, evolved into the most productive of the cereals, becoming the basic food plant not only of the pre-Columbian cultures and civilizations of this hemisphere but also of the majority of modern ones, including our own.

Summary

Remains of prehistoric corn, including all parts of the plant, have been uncovered from five caves in the Valley of Tehuacán in southern Mexico. The earliest remains, dated 5200 to 3400 B.C., are almost certainly those of wild corn. Later remains include cultivated corn and reveal a distinct evolutionary sequence that gave rise ultimately to several still-existing Mexican races. Despite a spectacular increase in size and productiveness under domestication, which helped make corn the basic food plant of the pre-Columbian cultures and civilizations of America, there has been no substantial change in 7,000 years in the fundamental botanical characteristics of the corn plant.[20]

NOTES

1. P. C. Mangelsdorf and R. G. Reeves, "The origin of Indian corn and its relatives," *Texas Agr. Expt. Sta. Bull. No. 574* (1939).

2. P. C. Mangelsdorf, R. G. Reeves, and C. E. Smith, Jr., in *Botanical Museum Leaflets, Harvard Univ.*, **13**, 213 (1949).

3. P. C. Mangelsdorf, *Science*, **128**, 1313 (1958).

4. *Ibid.*

5. P. C. Mangelsdorf, R. S. MacNeish, W. C. Galinat, in *Botanical Museum Leaflets, Harvard Univ.*, **17**, 125 (1956).

6. E. J. Wellhausen, L. M. Roberts, E. Hernandez, in collaboration with P. C. Mangelsdorf, *Races of Maize in Mexico* (Bussey Institution, Harvard University, Cambridge, 1952).

7. R. S. MacNeish, *Trans. Am. Phil. Soc.*, **48**, pt. 6, 1 (1958).

8. C. Lumholtz, *Unknown Mexico* (Scribner, New York, 1902).

9. P. C. Mangelsdorf and R. H. Lister, in *Botanical Museum Leaflets, Harvard Univ.*, **17**, 151 (1956).

10. E. J. Wellhausen, L. M. Roberts, E. Hernandez, in collaboration with P.C. Mangelsdorf, *op. cit.*

11. E. S. Barghoorn, M. K. Wolfe, K. H. Clisby, in *Botanical Museum Leaflets, Harvard Univ.*, **16**, 229 (1954).

12. E. B. Kurz, J. L. Liverman, H. Tucker, *Bull. Torrey Botan. Club*, **87**, 85 (1960).

13. *Science* 143: 533 *et seq.*

14. P. C. Mangelsdorf and R. G. Reeves, "The origin of Indian corn and its relatives," *Texas Agr. Expt. Sta. Bull. No. 574* (1939).

15. E. J. Wellhausen, L. M. Roberts, E. Hernandez, in collaboration with P. C. Mangelsdorf, *op. cit.*

16. E. Anderson and R. O. Erickson, *Proc. Natl. Acad. Sci. U.S.*, **27**, 436 (1941).

17. E. J. Wellhausen, L. M. Roberts, E. Hernandez, in collaboration with P. C. Mangelsdorf, *op. cit.*

18. *Ibid.*

19. *Ibid.*

20. The archaeological excavations were supported by a grant from the National Science Foundation; the botanical studies, by a grant from the Rockefeller Foundation. Both grants are acknowledged with appreciation and thanks.

Fig. 14–1. The principal physical features of the Valley of Tehuacán, Mexico, and the approximate locations of the five caves in which remains of prehistoric corn were uncovered. The insert shows the locations of other archaeological sites which have yielded evidence on the origin and evolution of corn.

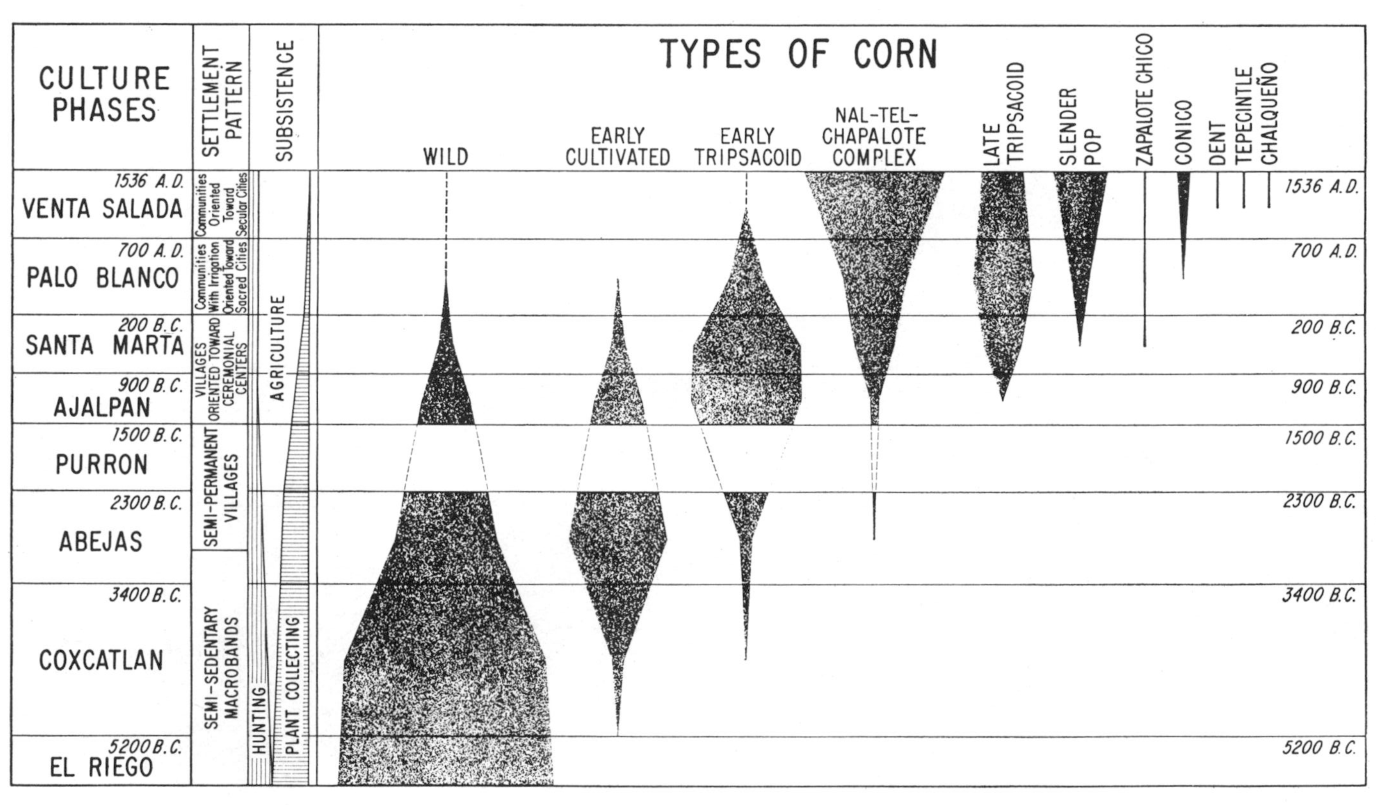

Fig. 14–2. Frequency polygons, in terms of percentages of number of cobs identified, showing changes in the types of corn in the Valley of Tehuacán from 5200 B.C. to A.D. 1536. Specimens of prehistoric corn were almost totally lacking for the Purron culture phase, which is recognized by other types of artifact.

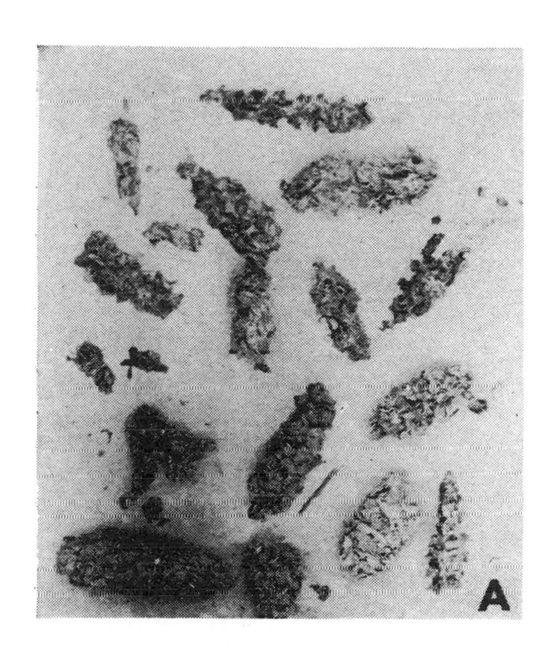

Fig. 14–3. (A) Cobs of wild corn from San Marcos Cave, representing the Coxcatlan culture phase, dated 5200 to 3400 B.C. These cobs are characterized by uniformity in size of the intact cobs, relatively long glumes, and fragile rachises. (B) Cobs of early cultivated corn from San Marcos Cave, representing the Abejàs culture phase, dated 3400 to 2300 B.C. These are larger and more variable than the cobs of wild corn but are similar to them in having long glumes and fragile rachises. (Actual size.)

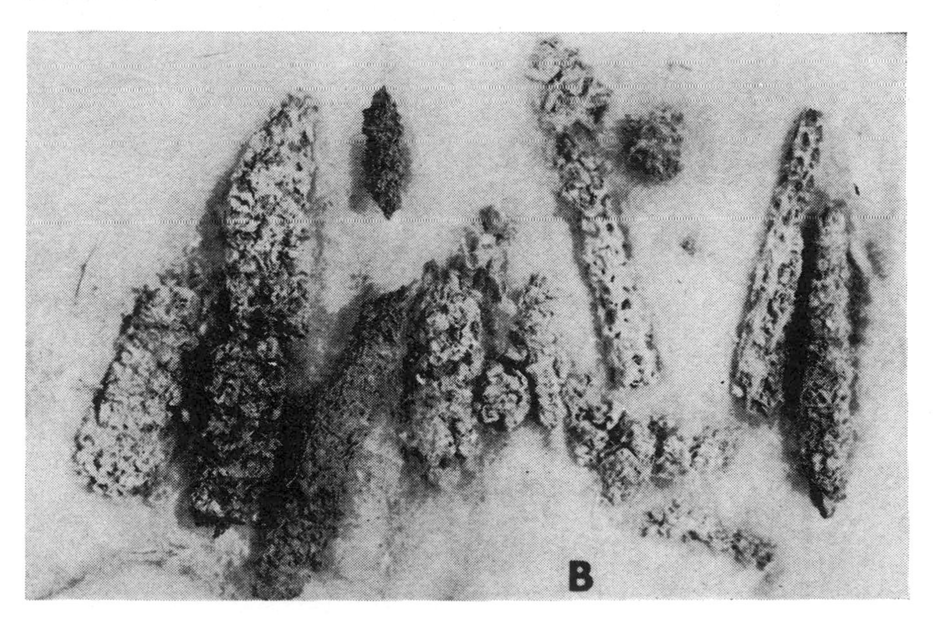

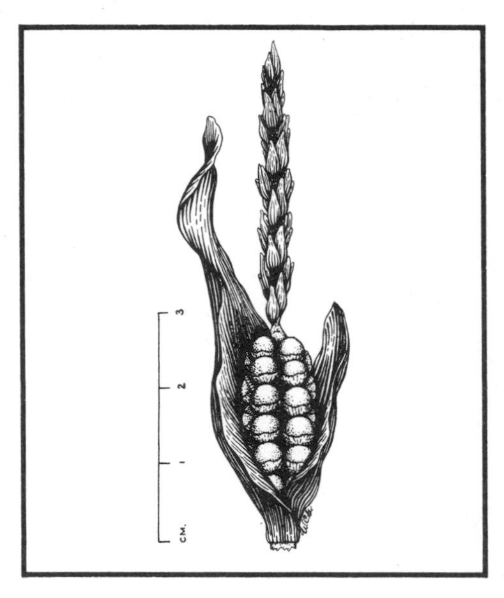

Fig. 14–4. Artist's reconstruction of wild corn based on actual specimens of cobs, husks, a fragment bearing male spikelets, and kernels uncovered in the lower levels of San Marcos Cave. The husks probably enclosed the young ears completely, but opened up at maturity, permitting dispersal of the seeds. The kernels were round, brown or orange, and partly enclosed by glumes. (Actual size.)

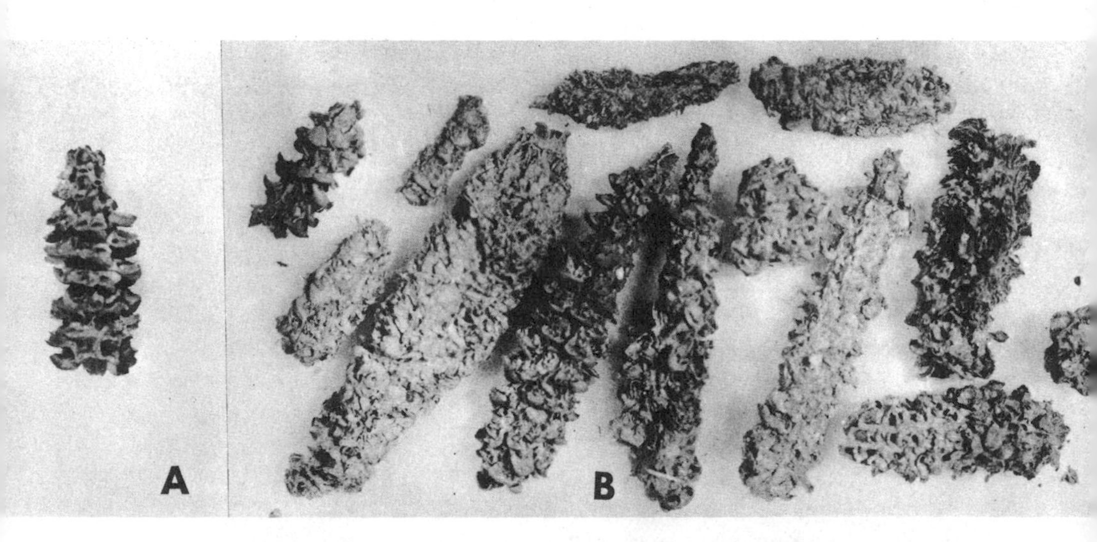

Fig. 14–5. (A) Cob of early tripsacoid corn. This is characterized by stiff indurated glumes. It is thought to be the product of hybridization of corn with one of its relatives, teosinte or *Tripsacum*. (B) Various hybrid combinations resulting from the crossing of Tehuacán early cultivated corn with the early tripsacoid corn and back-crossing to both parents. (Actual size.)

Fig. 14–6. Cobs from a single cache in San Marcos Cave, showing the great variation which followed the hybridization of the Tehuacán corn with the early tripsacoid corn. The two small upper cobs are wild-type segregates. (Actual size.)

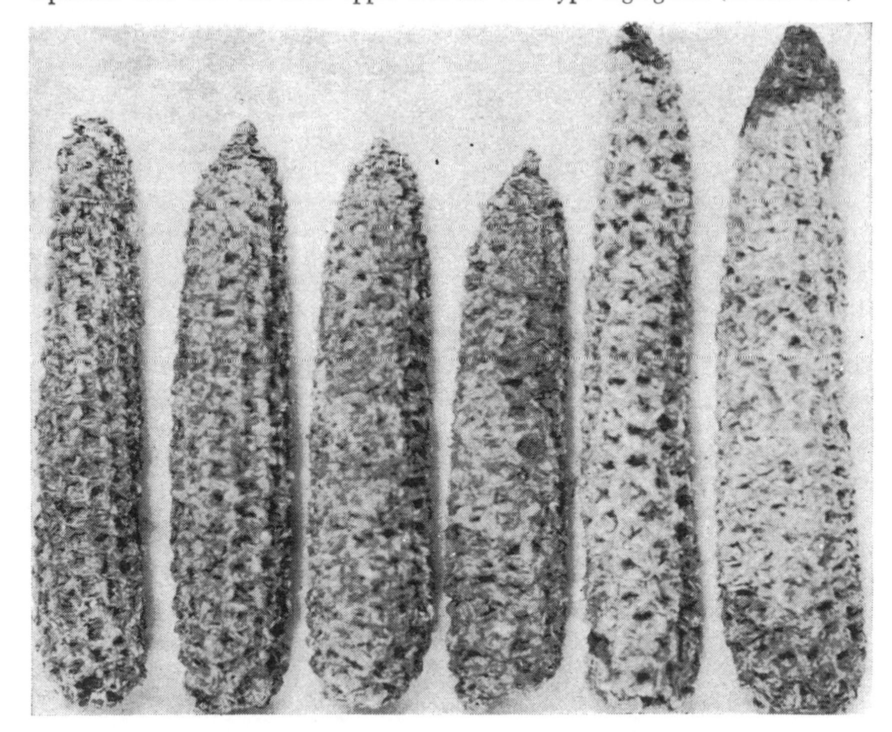

Fig. 14–7. Cobs of the Nal-Tel-Chapalote complex from San Marcos Cave representing the Palo Blanco phase, dated A.D. 200 to 700. It was this corn, the product of hybridization between the Tehuacán corn and the early tripsacoid, which, in providing an adequate food supply, contributed to the rise of an advanced culture and civilization in Tehuacán Valley. (Actual size.)

Chapter **15**

Prehistory of the West Indies

Irving Rouse

The islands of the West Indies are of interest to prehistorians because they lie like a series of stepping stones between the northeastern part of South America and the peninsulas of Florida and Yucatan, projecting from North and Middle America, respectively (Fig. 15–1). From which of these three mainland regions did the Indians reach the islands? When did various groups of Indians first arrive? Did they continue through the islands into other mainland regions? And did certain customs and beliefs spread to or from the islands without an accompanying displacement of the population? Research on the prehistory of the West Indies is designed to answer these questions.[1,2,3]

The islands have also attracted attention because they were the scene of the first significant contacts between the Old and New Worlds. The earlier Norse encounters with the Eskimo had contributed nothing to the development of Western civilization, but when Columbus discovered the New World in the West Indies, he set in motion a chain of events which led to the adoption by Europeans of a number of new crops, such as maize and tobacco, and new artifacts, such as hammocks and canoes, which we now consider our own. Even the names for these crops and artifacts are taken from the West Indian languages. It is of some interest, therefore, to determine how they reached the islands.

Natural and Cultural Setting

The Lesser and the Greater Antilles form the backbone of the West Indies. The Lesser Antilles consist mainly of small, volcanic islands, which curve to the north and west from the mouth of the Orinoco River in eastern Venezuela (Fig. 15–1). The Greater Antilles, composed of much larger, mainly sedimentary islands, extend westward from the northern end of the Lesser Antilles toward Florida and Yucatan. From east to west, the principal islands of the Greater Antilles are Puerto Rico, Hispaniola (which is now divided between the Dominican Republic and Haiti), Jamaica, and Cuba.

Lesser island groups include the Turks and Caicos Islands and the Bahamas, scene of Columbus's first landfall; both groups stretch north from Haiti and Cuba along the southeast coast of Florida. A series of islands just off the coast of Venezuela are often also considered part of the West Indies. Only the easternmost of these concern us here: Margarita, Cubagua, and Coche, which belong to Venezuela, and Trinidad, a former British possession.

Trinidad is closest to the mainland and, indeed, was attached to it until about 6000 B.C., well after the time of the Indians' arrival in South America.[4] There is a gap of 145 kilometers between Trinidad and Grenada, the southernmost of the Lesser Antilles, but the islands of the Lesser and Greater Antilles are closer together, most within sight of one another. Cuba is 195 kilometers from Yucatan and 145 kilometers from Key West, the closest part of Florida.

The prevailing winds and currents proceed westward from the Guianas past Trinidad to Margarita, Cubagua, and Coche islands. Some currents are deflected northward from Trinidad into the Lesser Antilles, and this northward movement is reinforced by the water pouring out of the mouth of the Orinoco River. When the Orinoco River is in flood it muddies the sea out past Trinidad, and some of its debris is carried into the Lesser Antilles.[5]

The winds and currents likewise proceed mainly from east to west through the Greater Antilles. The straits of Yucatan and Florida serve to channel the flow of the currents, swinging them back to the northeast between Florida and the Bahamas, where they unite to form the Gulf Stream. Both straits lack large rivers, the waters of which would cut across the currents as the Orinoco does in South America.

These factors have favored movement, first of animals and later of man, out into the Antilles from South America, rather than from either Middle or North America.[6] So also has the existence of two large, sheltered gulfs at the southern end of the island chain: Paria, between Trinidad and the mainland, and Cariaco, south of Margarita, Cubagua, and Coche islands. Here, as we shall see later, the Indians acquired the seafaring skills they needed to move out into the Antilles.

That the Indians did, in fact, move out from South America is indicated by the distribution of their culture in the time of Columbus. Eastern Venezuela, the adjacent part of the Guianas, and the West Indies form a single, Caribbean culture area, the native inhabitants of which were closely interrelated in language and culture (Fig. 15–2).

Unfortunately for Columbus, who had hoped to impress the King and Queen of Spain with the importance of his discoveries, the Indians of the Caribbean area had not attained civilization. This was limited to two areas on the Pacific side of the hemisphere: Mesoamerica, comprising the modern countries of Mexico, Guatemala, and Honduras, and the Central Andes area in the present nations of Peru and Bolivia. In these two areas the Indians had cities, monumental architecture, empires or kingdoms, advanced scholarship and scientific knowledge, extensive commerce and industry, and other manifestations of civilization.[7]

The Mesoamerican and Andean civilizations were separated by a region of lesser development, extending south along the Pacific coast from Nicaragua to Ecuador and east along the Caribbean shore as far as western Venezuela. This has become known as the Intermediate area, because of its position between the two civilizations (Fig. 15–2). Unlike them, it had only towns and villages, constructed of perishable materials; chiefdoms were its largest political units; and commerce, industry, and scientific knowledge were all poorly developed. The Indians of the Intermediate area are notable mainly for their skill in working precious metals, in which they surpassed the Mesoamericans, and for their art, expressed in the form of grave objects and religious figures.[8]

In the time of Columbus the Caribbean area was at the same general level of development as the Intermediate area, differing from it only in details. The staple crop was manioc rather than maize, though the latter frequently served as a secondary crop; metallurgy was poorly developed, despite Columbus's claims to the contrary; burial received little attention; and religious figures were rare. There were a few notable exceptions which, as we shall see, are likely to have been the result of influences from the Intermediate area or Mesoamerica. A game played with a rubber ball was probably also derived from one or the other of these sources.

To fill out the picture of historic distribution of Indian culture, let us take a brief look at the situation south and north of the Caribbean area (Fig. 15–2). Amazonia, to the south, had a simpler version of Caribbean culture, lacking most of the latter's influences from the Intermediate area and Mesoamerica. The eastern United States contained a different and more advanced form of culture, characterized by maize agriculture and temple mounds, both of which seem to have diffused directly from Mesoamerica by way of the Gulf Coast.[9]

Ethnic and Linguistic Groups

Three major groups of Indians inhabited the Caribbean area in the time of Columbus: Marginal (Archaic) peoples, the Arawak, and the Carib. The principal Marginal tribes were the Warrau, who lived in the delta of the Orinoco River, and the so-called Ciboney, who inhabited the western part of Cuba (from Havana to the Yucatan Channel), islets off the coast of Cuba, and the long southwestern peninsula of Haiti. They were apparently remnants of an earlier population that had been pushed back into these peripheral positions by later migrants. They lived by hunting and fishing, without agriculture; inhabited small, temporary camps, which were often in caves; and had simple forms of social organization and religion.[10]

It is important to note that the southern part of Florida, up to and including Cape Kennedy, was likewise occupied by hunting and fishing peoples when the Spaniards arrived. One problem of West Indian archaeology is to determine the relationship, if any, between the Marginal peoples of Florida and those of the Caribbean area. Were the former derived from the latter, or vice versa? Or, as the archaeology seems to indicate, did the two groups develop separately, each being pushed back into its historic, peripheral position by more advanced people, moving in from the south and north, respectively?

The Arawak were also widely dispersed. One group lived on the South American mainland and on the islands immediately offshore, where their settlements were sometimes interspersed among those of the Carib. By far the greater part of the Arawak, numbering in the millions according to the conquistadores, were in the Greater Antilles, the Turks and Caicos Islands, and the Bahamas. These island Arawak were skilled agriculturists (though they also obtained much food by fishing), lived in settled villages ruled by hierarchies of chiefs, made good pottery, and had a relatively elaborate religion, centering around the worship of deities known as *zemis*. The mainland Arawak were somewhat less highly developed.[11, 12]

Both groups spoke languages of the great Arawakan family, which was widespread throughout Amazonia and extended into the Intermediate area as well but did not occur in either Middle or North America. Hence, we may assume that the Arawak invaded the West Indies from the south, pushing the Marginal tribes back into the peripheral positions they occupied in historic time. Columbus was able to use the same interpreters wherever he went among the island Arawak, and this suggests that the Arawak invasion of the Antilles cannot have begun very long before his time, else there would have been stronger dialectal differences.

The island Arawak told Columbus that they were subject to raids by man-eating Indians known as Carib.[13] Intrigued, Columbus directed his second voyage into the Lesser Antilles, where the Carib lived, and confirmed the Arawak report. The Carib were agriculturists and pottery makers, but, unlike the Arawak, they paid more attention to warfare than to religion and ate captives in order to absorb their fighting ability.

(Our word cannibal is a corruption of *Caribal*, the Spanish form of Carib.) The chiefs were not hereditary, as they were among the Arawak, but were chosen for their ability in warfare.[14]

The Carib claimed to have arrived in the Lesser Antilles only a few generations before Columbus, having come from South America, where many Carib still lived. They had conquered the previous Arawak inhabitants of the Lesser Antilles and, so they said, killed off the men but married the women. Apparently the women's language prevailed over that of the men, for the descendants of the island Carib speak an Arawakan language.[15] The Cariban languages were confined to South America proper, where they were as widespread as Arawakan, occurring in Amazonia and the Intermediate area as well as in the Caribbean area.

There can be little doubt, then, that both the Arawak and the Carib, if not the Ciboney, entered the West Indies from South America. It remains for archaeology to determine when and how they came and to work out the extent of their contacts, if any, with the Indians of Mesoamerica and the eastern United States.

Chronology and Cultures

Extensive stratigraphical excavations in the Caribbean area made it possible in the 1940's to set up a relative time scale consisting of five periods, which were arbitrarily numbered from I to V. These were subsequently grouped into three epochs, and a fourth was added at the beginning, as shown in Table 15–1.

Following a common practice in prehistory and history, we have named each epoch after its most advanced people. The original Palaeo-Indians were hunters; this epoch is characterized by artifacts of chipped stone. The Meso-Indians turned to fishing if they lived along the coast, and probably to the gathering of wild vegetable foods if they lived in

TABLE **15–1**

Relative time scale for Caribbean cultures.

EPOCHS	PERIODS	DURATION
Palaeo-Indian		15000–5000 B.C.
Meso-Indian	I	5000–1000 B.C.
Neo-Indian	II	1000 B.C.–A.D. 300
Neo-Indian	III	A.D. 300–1000
Neo-Indian	IV	A.D. 1000–1500
Indo-Hispanic	V	A.D. 1500 on

the interior. They made a greater variety of artifacts, including the first shell tools (on the coast) and the first pottery (in the interior). The Neo-Indians were agriculturists and pottery makers, and the Indo-Hispanic people, as the term implies, were Indians who had become more or less influenced by European civilization. The Indo-Hispanic people are beyond the scope of this discussion.

With the development of radiocarbon analysis in the 1950's, it has become possible to determine the duration of the epochs and periods (see Table 15–1, col. 3). A total of sixty-five dates has been obtained for the Caribbean area proper. These are well distributed over all the epochs except the first; the duration assigned this first epoch is based upon evidence from outside the Caribbean area.[16, 17]

A number of cultures have been distinguished for each of the epochs and periods. Those of the Palaeo- and Meso-Indian epochs are termed complexes, since each one has been defined in terms of all the types of artifacts represented in its sites. The Neo-Indian cultures, on the other hand, are called "styles," in recognition of the fact that they must be defined primarily in terms of pottery, since relatively few nonceramic artifacts are found in the Neo-Indian sites.

Most of the complexes and styles can be fitted into lines of development, in each of which it appears that the original complex or style gave rise to a second, the second to a third, and so on. Such lines of development are termed "series."[18] The simplest form of series consists of a succession of complexes or styles which the people of a single locality developed with the passage of time. In other cases the people of the original locality seem to have migrated, sometimes over long distances, changing from one complex or style to another as they went. In still other cases it is probable that the people of the original locality influenced people of a second locality, those of the second locality passed the influences on to a third, and so on; and that a series of new complexes or styles was thus produced by a process of acculturation rather than migration. All the series are widely distributed in either time or space, or in both.

In accordance with standard archaeological practice, each complex or style is named after a type site. Each series is similarly named after a typical complex or style, by addition of the suffix *-oid* to the name of the complex or style. For example, the style found in a group of sites just above the delta of the Orinoco River is termed Barrancas, since the modern town of that name is situated directly on the type site, and this style is assigned to the Barrancoid series, so-called because the Barrancas style is the type member of the series.

The complexes and styles are shown in the chronological charts of figures 15–3 to 15–5. The Meso-Indian complexes appear below the heavy black line on each chart and the Neo-Indian styles above it. Various kinds of shading are used to indicate how the complexes and styles are related to form the series.

The charts of figures 15–3 and 15–5, which represent the mainland part of the Caribbean area and the Greater Antilles, respectively, can be considered fairly reliable, since they are based upon extensive stratigraphic excavations; but the chart of Figure 15–4 is only a first approximation, for excavation has lagged in the Lesser Antilles, to which it refers. It is included only to give some idea of the present state of our knowledge, and it will undoubtedly have to be modified when current research in the Lesser Antilles is completed.

It may be seen that Period I was purely Meso-Indian. Periods II to IV were marked by gradual encroachment of the Neo-Indians upon Meso-Indian territory, a process which had not been completed by the time of Columbus. This substantiates the theory concerning the fate of the Ciboney Indians presented earlier—that they had been gradually pushed back by later migrants into the peripheral positions in which Columbus found them. It follows that some of the Meso-Indian series and complexes must be ancestral to the Ciboney, but we do not know which ones were. Neither can we say for sure which of the Neo-Indian series and styles were Arawak and which Carib. Therefore, in the following survey of the archaeology, epoch by epoch, the series, complexes, and styles are discussed per se, and the problem of tribal identifications is treated separately at the end.

Palaeo-Indian Epoch

The nearest known remains of the Palaeo-Indian epoch are in western Venezuela, just outside the limits of the Caribbean area. Here, José M. Cruxent has succeeded in distinguishing a single, Joboid series and several other complexes, which cannot yet be assigned to series. Similar remains should eventually turn up in the mainland part of the Caribbean area and on the island of Trinidad, which was attached to the mainland during the Palaeo-Indian epoch; but it is doubtful that the West Indies were inhabited at the time, since the Palaeo-Indians were not seafarers. Moreover, they did not eat sea foods, so far as we know, and remains of the large, now extinct land mammals—for example, the mastodon, horse, and sloth—upon which they apparently relied for food, do not occur on any of the islands except Trinidad.[19, 20]

Let me briefly summarize the finds in western Venezuela as a background for considerations of the Meso-Indians, who did reach the Antilles. Cruxent distinguishes a succession of Camare, Las Lagunas, El Jobo, and Las Casitas complexes within the Joboid series. These are based upon his finds in the valley of the Rio Pedernales in the state of Falcón, east of Lake Maracaibo. They are associated, respectively, with the uppermost, upper middle, lower middle, and lower terraces formed by the river. In the Camara sites, Cruxent found only choppers and scrapers of quartzite, which, he suggests, may have served to make wooden spears for use in hunting mammals (Fig. 15–6*A*). The Las Lagunas sites also contained large bifacially worked blades, which could have been hafted in heavy thrusting spears (Fig. 15–6, *E–G*). The subsequent, El Jobo complex yielded lanceolate projectile points, small enough to fit in darts (Fig. 15–6, *B–D*). The final, Las Casitas complex had in addition a few stemmed points with triangular blades.[21]

Three dates, obtained by the radiocarbon method, of about 15,000 to 13,000 years ago have been obtained, two from Muaco, on the coast near the mouth of the Rio Pedernales, and the third from Rancho Peludo, in the Maracaibo Basin farther west, where a presumed Palaeo-Indian deposit underlies successive occupations by Meso- and Neo-Indians.[22] The Muaco site belongs to the Joboid series, but the deposit at Rancho Peludo apparently represents, instead, a Manzanillo complex, which is characterized by chopping tools of fossil wood.[23]

Meso-Indian Epoch (Period I)

The Meso-Indian epoch began with the retreat of the Pleistocene ice sheets in the Northern Hemisphere. This caused a rise in the sea level and separation of Trinidad from the mainland. However, the sea cannot have reached its present level until relatively late in the epoch, for a number of the Meso-Indian sites along the shore are now partially under water.[24]

The beginning of the epoch was also marked by extinction of the large land mammals upon which the Palaeo-Indians had relied as a main source of food. Either these animals were overhunted or they failed to adapt to the gradual drying up of the climate that took place in western Venezuela at the time.[25] The Indians of the Guiana highlands continued to hunt the surviving game, as evidenced by the presence of stemmed projectile points in the Canaima complex of that area, but elsewhere the Meso-Indians must have de-emphasized hunting, for they ceased to produce projectile points of stone. Evidently they turned to new sources of food—fish and shellfish along the shore and vegetable foods in the interior. These foods, which had been relatively little used during the Palaeo-Indian epoch, are hallmarks of the Meso-Indian epoch.

The Meso-Indian sites along the shore consist of large refuse heaps containing not only shells but also fish bones and the remains of echinoderms. Most of those investigated in eastern Venezuela belong to a single, Manicuaroid series, which is best represented on Cubagua Island at the large midden of Punta Gorda, four meters high. Excavation in this midden has revealed a stratigraphic succession of three complexes —Cubagua, Manicuare, and Punta Gorda (Fig. 15–3)—marked by an increasing variety of shell artifacts. The Cubagua complex has only hammers, cups, and a disk (Fig. 15–7, *H*, *I*, *L*, *G*); to these the Manicuare complex adds beads, pendants, and gouges (Fig. 15–7, *M*, *Q*, *E*, *J*, *K*), and the Punta Gorda complex, celts (Fig. 15–7, *F*) and points. The gouges are most distinctive; each consists of a triangular section from the outer whorl of a conch shell that has been ground along the base to form a bit. The shell points appear to be copies of bone points, which are common to the three complexes; both may have been hafted in fishhooks or on harpoon heads (Fig. 15–7, *B–D*). Also distinctive are stone pebbles which have been pointed at either end, possibly for use as sling stones (Fig. 15–7, *O*).

To be able to colonize Cubagua and Margarita islands, where remains are most abundant, the Manicuaroid Indians must have developed considerable seafaring ability. We may suppose that they used dugout canoes, which they hollowed out with their shell gouges. Dates obtained by the radiocarbon method indicate that the series had begun by at least 2500 B.C. and that it lasted until shortly after the time of Christ, which is the date of trade pottery in the latest, Punta Gorda complex.[26]

Remains of the Manicuaroid series have been found eastward along the coast of Venezuela only as far as the present city of Carúpano (Fig. 15–3). The Meso-Indian remains farther east, on the peninsula of Paria and the island of Trinidad, cannot be assigned to the series. For example, the Ortoire complex of Trinidad lacks the distinctive Manicuaroid artifacts of shell and stone and is characterized instead by tiny chips of stone, of unknown use.[27]

There is no evidence that either the Manicuaroid or the Ortoire people followed the course of the currents out into the Lesser Antilles, although they seemingly had the seafaring ability to do so. Indeed, no trace of any Meso-Indians has yet been found in the Lesser Antilles, except on St. Thomas in the Virgin Islands, farthest away from South America, and there the remains are very different from those on the mainland. They belong to a Krum Bay complex, characterized by bifacially chipped and ground stone artifacts.[28]

By contrast, Meso-Indian remains have been discovered on most islands of the Greater Antilles, but here, too, evidence of a progression from east to west, in the direction of the prevailing winds and currents, is lacking (Fig. 15–5). Each island has its own complexes, the various complexes being unrelated in the form of series, and each is characterized by its own types of artifact—pebble grinders, faceted on the edges rather than the sides, in the case of the María la Cruz (Loiza

Cave) complex of Puerto Rico; plain blades, worked only on the edges, in the case of the Marban, Cabaret, and Couri complexes of Hispaniola; and shell gouges in the Guayabo Blanco and Cayo Redondo complexes of Cuba.[29, 30] The Couri and Cayo Redondo complexes also have a series of ground stone and shell artifacts, several of which are decorated with rectilinear incised designs (figs. 15–8 and 15–9).

Another puzzling point about the Meso-Indian complexes of the Greater Antilles is that there are resemblances to complexes in different parts of the mainland. The edge grinders of María la Cruz link that complex with Cerro Mangote in Panama and with several complexes of central and western Venezuela, notably El Heneal.[31, 32] The plain blades of Marban, Cabaret, and Couri are in the Central American tradition of flint working; their most elaborate type, a stemmed projectile point belonging to the Couri complex (Fig. 15–8, *A*), is duplicated among the Maya of Yucatan, in Central America, and nowhere else.[33] The shell gouge of Guayabo Blanco and Cayo Redondo is found also in the Manicuaroid series of Venezuela and in the preceramic sites on the St. Johns River in Florida.[34] Rectilinear designs resembling those on Couri and Cayo Redondo artifacts likewise occur in Florida.

How are we to explain the irregularities in distribution of the Meso-Indian complexes in the Antilles and the diversity of their resemblances to mainland complexes? These may simply be due to gaps in our knowledge, but there is an alternative possibility. It happens that the mammalian fauna of the Antilles shows a similar irregularity of distribution and diversity of resemblances. George Gaylord Simpson has accounted for this irregularity and diversity by theorizing that mammals accidentally floated out to the Antilles on natural rafts—that they were cast into the sea by the great rivers of northern South America, such as the Orinoco, and were then blown to different islands.[35] Man may have first reached the Antilles in a similar manner; that is, Meso-Indian families traveling in canoes along the shore or to nearby islands such as Cubagua may have been caught in storms, blown out to sea, and, if lucky enough to survive, deposited haphazardly on the shores of different islands.[36]

A special problem is presented by the occurrence of shell gouges, not only in the Manicuaroid complex of Venezuela, but also in Cuba and Florida. If all three of these occurrences are related—and they may not be, since there is so great a gap between the first and the other two—do they indicate diffusion from South America northward or from North America southward? There is disagreement about this problem; Cruxent and I, for example, have favored a Venezuelan origin, whereas Ricardo E. Alegría has argued for diffusion from Florida to Cuba.[37, 38]

The time of arrival of Meso-Indians in the Antilles is also in doubt. The Cuban complexes are reportedly associated with extinct forms of ground sloths and monkeys, but this means nothing, since the Indians are likely to have caused the extinction.[39] The earliest date obtained for the area by the radiocarbon technique is 2190 ± 160 B.C., for the Marban complex, but this may not be reliable, since the next earliest date is only 450 ± 175 B.C., for the Krum Bay complex.[40, 41] A date of A.D. 990 ± 60 for the Cayo Redondo complex attests to the survival of Meso-Indians alongside the Neo-Indians in the Greater Antilles.[42]

Neo-Indian Epoch: Period II

The Meso-Indian epoch corresponds to Period I of our relative time scale (figs. 5–3 to 15–5). The Neo-Indian epoch includes the three remaining periods, II–IV. It is convenient to discuss these separately.

The most important problem of Period II is that of the origin of Neo-Indian culture. Here we must turn from the Meso-Indians of the coast and islands to the Meso-Indians of the interior. Unable to exploit the sea, the inland Indians subsisted mainly on fruits and vegetables, which they at first gathered wild and eventually learned to cultivate. The original cultivation was too rudimentary to supply more than a small part of the Indians' food, but the techniques and crops were gradually improved to the point where agriculture replaced hunting and gathering as the principal means of subsistence. At this point, we may say, the Indians crossed the threshold between Meso- and Neo-Indian culture and the Neo-Indian epoch began.

No traces of the transition from Meso- to Neo-Indian culture have yet been found in the Caribbean area, though some have recently turned up at various places in the Mesoamerican, Intermediate, and Central Andean areas.[43] This may mean that Neo-Indian culture diffused full-blown into the Caribbean area from the west, but it is more likely that there was some sort of local development, since the earliest Neo-Indian culture of the Caribbean area differs from anything to the west.

For an idea of what this local development may have been like, let us turn to the site of Rancho Peludo in western Venezuela, which has the nearest known occurrence of the Meso- to Neo-Indian transition. The presumed Palaeo-Indian deposit at the site has already been discussed. Overlying it is refuse first of Meso- and then of Neo-Indians. There is pottery in both layers, as in the other sites that contain the later stages of the Meso- to Neo-Indian transition. The pottery includes flat, circular griddles of the type used in the time of Columbus to bake bread prepared from the flour of the manioc plant. These indicate that the Meso-Indians of Rancho Peludo cultivated manioc, though, to judge from the scarcity of griddles at the site, it cannot have formed an important part of their diet. A series of six dates places the time of this deposit between 2000 and 500 B.C.[44]

The earliest known pottery-bearing sites of the Caribbean area proper must be considered fully Neo-Indian, since griddles are so numerous as to suggest that manioc had already become the principal item of the diet. These sites are situated along the middle and lower parts of the Orinoco River and belong to a series of styles known as Saladoid, for which we have three dates at the beginning of the first millennium B.C. Saladoid pottery is thin, hard, and so well made that there must be a long tradition of pottery making behind it. (It is better made than any of the later pottery in the Caribbean area.) Other traits that distinguish it are bowls shaped gracefully in the form of inverted bells; white-on-red painted designs; crosshatching in red paint; simple incised designs; and tabular lugs (Fig. 15–10). Its origin is a complete mystery.[45]

At the type site of Saladero, just above the delta of the Orinoco River, Saladoid refuse is overlaid by refuse of the Barrancoid series; dates for the two series partially overlap. Barrancoid pottery is much thicker, heavier, and coarser than Saladoid pottery. Bowls tend to have vertical sides and thick, flanged rims. Elaborate incised and modeled-incised designs occur on the flanges, on vessel walls, and on lugs.[46] There is reason to believe that the Barrancoid people intruded into the lower part of the Orinoco Valley from the west—Barrancoid remains are common in the Valencia basin of central Venezuela (Fig. 15–1)—and that they split the Saladoid people into two parts, one group remaining in the middle part of the Orinoco Valley and the other passing out through the delta to the island of Trinidad and the peninsula of Paria (Fig. 15–3).

When the Saladoid people arrived in the Trinidad-Paria region they came into contact with the Meso-Indian fishermen who had survived there. They probably pushed some of them back into the delta of the Orinoco River (the modern Warrau Indians may be the descendants of this group) and absorbed others, teaching them the arts of horticulture and ceramics. But the acculturation cannot have been all one way; we may suppose that the Saladoid people acquired a taste for sea foods from the Meso-Indians, since shells and fish bones are common in their sites, and that they also learned seafaring from those Indians.

Several centuries elapsed before the Saladoid people put this seafaring ability to use in colonization. At about the time of Christ they began to expand in the directions of the prevailing winds and currents—that is, westward along the coast of Venezuela to the region of Cumaná and thence out to the islands of Cubagua and Margarita (Fig. 15–3); and northward through the Lesser Antilles to Puerto Rico, the first island of the Greater Antilles (figs. 15–4 and 15–5). Along the coast of Venezuela they met and replaced the Meso-Indians of the Manicuaroid series, as is evidenced by the presence of Saladoid trade sherds in the latest Manicuaroid sites, and in the Virgin Islands and Puerto Rico they displaced the Meso-Indians of the Krum Bay and María la Cruz complexes, respectively (figs. 15–3—15–5). They had reached Carúpano, halfway along the Venezuelan coast to Cumaná, by about A.D. 1, according to radiocarbon dating, and were on Margarita Island by A.D. 300. A series of ten dates places their movement out through the Lesser Antilles to Puerto Rico at about A.D. 200.[47]

There is a remarkable uniformity of pottery and other artifacts throughout this vast area that can only be explained by postulating a regular expansion of the Neo-Indians at the expense of the previous Meso-Indian inhabitants. The pottery continues the tradition of the original Saladoid pottery on the Orinoco River but with two important modifications: modeled-incised lugs have become common, and crosshatching is incised, not applied in red paint (Fig. 15–11). Both these changes can be ascribed to Barrancoid influence, resulting from the contact between the two series on the lower Orinoco River. The incised crosshatching provides a particularly good time marker for the movement of Neo-Indians out into the islands, since this trait lasted for only two to four centuries, to judge from the results of radiocarbon dating.[48]

Neo-Indian Epoch: Period III

The Neo-Indians continued during Period III to expand into the Greater Antilles at the expense of the Meso-Indians, and by the end of the period they had pushed the Meso-Indians back into the peripheral positions they occupied at the time of Columbus (Fig. 15–5). The Neo-Indians of the Lesser Antilles still made Saladoid pottery, but in the Greater Antilles there had been a change, and by Period III*b* there were three new series of pottery—Ostionoid, Chicoid, and Meillacoid.

The development of the Ostionoid series was foreshadowed in Puerto Rico by a gradual loss of decoration during periods II*b* and III*a*. Hacienda Grande, the earliest ceramic style found in Puerto Rico (Fig. 15–5), possesses the full range of Saladoid decoration. Cuevas, the next style, has lost modeled-incised lugs and incised crosshatching but may still be considered Saladoid because it retains the bell shape of the bowl and the white-on-red painting. These traits, too, are gone in the subsequent Ostiones style, and with their loss, we may say, the Ostionoid series had begun.[49] This series is characterized by a smooth finish, more-or-less straight-sided or incurving-sided bowls, plain tabular lugs, and simple red painting (Fig. 15–12).

It was apparently people of the Ostionoid series who introduced Neo-Indian culture to the Dominican Republic, Haiti, and Jamaica (Fig. 15–5). The series survived until the end of Period III, but only in Puerto Rico. Late in the period it gave rise to the Chicoid series in the Dominican Republic and to the Meillacoid series in Haiti and Jamaica. Meillacoid Indians then expanded into Cuba, into the Turks and Caicos islands, and possibly also into the Bahamas, thereby completing the Neo-Indian colonization of the Antilles.[50]

Both the Chicoid and the Meillacoid series of pottery continue in the Saladoid-Ostionoid tradition so far as materials and shapes are concerned, but they differ in decoration, and therefore we may presume that the two series developed locally. The Chicoid series is distinguished by a renewed interest in modeled-incised lugs and incision (Fig. 15–13). Its designs look Barrancoid, and it may be no accident that, at the time the Chicoid series was arising in the Dominican Republic, Barrancoid people were expanding into Trinidad and the northwestern part of British Guiana (Fig. 15–3). There is no evidence of direct contact between the two groups, but Barrancoid decorative traits do occur throughout the Lesser Antilles and on the latest Ostiones pottery of Puerto Rico, whence they might have contributed to the formation of Boca Chica, the original Chicoid style in the Dominican Republic.[51, 52] One wonders, though, why the Barrancoid traits did not take hold as strongly in the Lesser Antilles and Puerto Rico as in the Dominican Republic.

The Meillacoid potters broke with the tradition of smooth surfaces that had previously prevailed in the Greater Antilles and developed a new set of techniques that roughened the surfaces: appliqué work, both on lugs and on vessel walls; punctation; and incising, done in such a way that the edges of the grooves are jagged (Fig. 15–14). The lugs look like crude copies of Chicoid lugs,[53] but the incised designs may well have been acquired from the Meso-Indians, for they resemble the latter's designs engraved on shell, stone, and wooden artifacts (Fig. 15–8).

Neo-Indian Epoch: Period IV

While there is no evidence of major population movements during Period IV, a number of significant changes in pottery did take place. The original Saladoid and Barrancoid series finally came to an end.[54] They were replaced on the mainland and nearby islands by the Guayabitoid, Dabajuroid, and Arauquinoid series, which need not concern us here (Fig. 15–3). In the Lesser Antilles, Saladoid pottery gave way to cruder material, as yet undefined except on St. Lucia, where Marshall McKusick has distinguished a succession of Choc and Fannis styles (Fig. 15–4). Both are characterized by griddles with legs. Choc pottery has monochrome painting in red, and the Fannis vessels are decorated principally with finger impressions on the rim.[55]

In the Greater Antilles the Ostiones series also came to an end, through expansion of the Chicoid series. This appears to have been primarily a matter of acculturation rather than migration; the original,

Boca Chica potters of the Dominican Republic influenced the people both to the east and to the west of them, causing the development of local versions of the Chicoid series in Puerto Rico and the Virgin Islands, to the east, and in Haiti, eastern Cuba, and the Turks and Caicos Islands, to the west.[56] The Meillacoid series survived only in Jamaica and central Cuba, and Meso-Indian culture, in southwestern Haiti and western Cuba. This produced the situation found by Columbus when he discovered the New World at the end of Period IV (Fig. 15–5).

Columbus found the Arawak of the Greater Antilles to be a gentle and religious people. Evidences of their religion first appear in the archaeological sites of the late Period III and increase to a climax at the end of Period IV. They center in the Dominican Republic and Puerto Rico, whence they may have spread east and west along with Chicoid pottery.

Intrigued by the Arawak religion, Columbus commissioned a priest, Ramón Pané, who accompanied him on his second voyage, to make a study of it on the island of Hispaniola. Pané's account, which has been called the first anthropological research in the New World, explains many features of the archaeological record of Period IV. Pané informs us that the Indians worshipped deities known as *zemis*, which were either human or animal in form, and that they were accustomed to portray these deities on their household utensils and implements. Apparently he was referring to the human and animal lugs of Chicoid pottery (Fig. 15–13); to effigy vessels, which are found principally in the Dominican Republic; and to comparable carvings on stone celts, axes, and pestles (figs. 15–15 and 15–16, left). Pané states that *zemis* were also portrayed on amulets, and these occur archaeologically in stone, bone, or shell.

From various Spanish sources we learn that the Arawak villages contained plazas at the ends of which were temples devoted to the worship of *zemis*. These temples have not survived, for they were made of perishable materials, but there are still traces of the plazas—rectangular or oval areas, leveled by digging wherever necessary and lined with rough stone slabs, some of which bear pictures of *zemis*. The plazas were also used as ball courts. They have yielded large carved stone "collars" that resemble the stone yokes worn about the waist by ball players in Mexico (Fig. 15–17, left). These and other bizarre artifacts, including three-pointed stones and elbow stones (figs. 15–16, right and 15–17, right) are not explained in the Spanish sources. They must have had some ceremonial significance, since all bear carvings of *zemis*.[57, 58]

Fortunately for us, the Arawak also worshipped *zemis* in caves, where perishable materials are better preserved. Among the objects found there are tubes, statues, and stools of wood. According to Pané, the native priests sniffed tobacco or another narcotic through the tubes as part of their ritual (Fig. 15–18). He states that the priests placed the snuff on top of *zemis*; many of the statues found in caves have platforms on top for this purpose (Fig. 15–19). The stools were used by chiefs as a sign of rank. The one illustrated (Fig. 15–20) is inlaid with gold disks and decorated with the head of a *zemi*. Inlaying of shell is more common, and plaster was also used.

Inspiration for the cult of *zemis* may have come from the mainland, where many of its elements are widespread. Mesoamerica is the most likely source of influences, since it is closest and has yielded the most detailed resemblances, such as effigy celts, stone yokes, and inlaying, but some authorities favor the longer route of diffusion from the Intermediate area by way of Venezuela and the Lesser Antilles.[59, 60] Other elements of the cult appear to have had a local origin; for example, the large, elaborately carved three-pointed stones of Period IV can be traced back to small, plain, three-pointed stones of Period III, and these in turn may go back to even smaller three-pointers of shell, made during Period II.[61] The fact that the cult reached its highest development in the central part of the Antilles—that is, in the Dominican Republic and Puerto Rico—and shades off as one moves westward toward Mesoamerica and southward toward the Intermediate area, also indicates local development.

Summary and Conclusions

We have traced the prehistory of the West Indies through three epochs, Palaeo-, Meso-, and Neo-Indian. During the Palaeo-Indian epoch, beginning about 15000 B.C., hunting peoples colonized the Caribbean mainland but did not continue into the islands, apparently because they had no interest in sea food and lacked the ability to travel by sea. By the Meso-Indian epoch, about 5000 B.C., the large land mammals upon which the Palaeo-Indians had relied for food had become extinct, making it necessary for the Indians to develop new sources of food. Along the coast they turned to fishing and the gathering of shellfish, and in the process acquired enough seafaring ability to colonize the nearby islands. I have suggested that some of them, from different parts of the mainland, may have been accidentally blown out into the Greater Antilles. This seems the best way to explain the irregular distribution of Meso-Indians on the islands, the diversity of the mainland resemblances, and, in particular, the lack of Meso-Indian remains in the Lesser Antilles. The Meso-Indians appeared on the coast soon after 5000 B.C., had reached Cubagua Island by 2500 B.C., and may have arrived in the Greater Antilles by 2000 B.C.

Lacking maritime resources, the Meso-Indians of the interior came to rely upon vegetable foods. They must originally have gathered these wild, but eventually they learned to cultivate them. The Neo-Indian epoch begins at the point where agriculture had become efficient enough to serve as the principal means of subsistence. This point was reached about 1000 B.C. by Indians of the Saladoid series, who lived on the lower Orinoco River. Soon afterward, Indians of the Barrancoid series intruded from the west and pushed some of the Saladoid people through the delta of the Orinoco River to the coast of Venezuela. There they came into contact with the surviving Meso-Indians, learned fishing and seafaring from them, and gradually expanded at their expense. The subsequent prehistory of northeastern Venezuela and the West Indies is primarily one of encroachment of the Neo-Indians on the Meso-Indians, until the latter had finally been driven back into the peripheral positions they occupied at the time of Columbus. The Neo-Indians reached Puerto Rico by A.D. 200 and were in Cuba by A.D. 1000.

Two other sets of events have been discussed. We have traced the development of Neo-Indian pottery from the Saladoid series to (i) cruder, as yet undefined ceramics in the Lesser Antilles, and (ii) a succession through the Ostionoid series to the Chicoid and Meillacoid series in the Greater Antilles, and we have discussed the rise of a cult of *zemis* in the Greater Antilles. Both Chicoid pottery and the cult of *zemis* show evidences of influence from the mainland, yet both seem to have arisen first in the Dominican Republic and to have spread westward from there to eastern Cuba and eastward as far as the Virgin Islands.

We have seen that there is an unbroken continuity from the later Meso-Indians of the Greater Antilles, who survived the Neo-Indian invasion, to the Ciboney (Marginal) Indians of Columbus's time. There is a second, likewise unbroken, continuity among the Neo-Indians of the Greater Antilles, beginning with the original Saladoid series and extending through the Chicoid and Meillacoid series to the Insular Arawak. Between them, these two continuities account for all the Indians who inhabited the Greater Antilles, the Turks and Caicos Islands, and the Bahamas in the time of Columbus.

The situation in the Lesser Antilles is not so clear. Since the earliest known remains are those of the Saladoid series, we may assume, pending the discovery of contrary data, that the first inhabitants were the ancestors of the Insular Arawak on their way out to the Greater Antilles. Whether or not another Arawak group subsequently invaded the Lesser Antilles is not certain, though linguistic evidence suggests that one did;[62] nor are we able to say when the Carib reached the islands. McKusick[63] has correlated the arrival of the Carib with the shift, during Period IV on St. Lucia, from the Choc to the Fannis style (Fig. 15–4), but pottery may not be a good indicator of this event. As I mentioned earlier, when the Carib conquered the Lesser Antilles they killed the Arawak men and married their women, and it was the women's language that survived. Since the women were the potters, their ceramics should also have survived.

Religion may be a better indicator than ceramics of the arrival of the Carib. Evidence of the earlier stage of the cult of the *zemis*, in the form of small, plain, three-pointed stones, occurs throughout the Lesser Antilles, but evidence of the later stage, in which large, sculptured three-pointers, stone collars, and other distinctive artifacts were made, is lacking.[64] This suggests that the Carib invasion may have taken place at the end of Period III—that is, before the development of either the Choc or the Fannis style (Fig. 15–4).

From this unsatisfactory discussion of the prehistory of the Ciboney, Arawak, and Carib, let us turn in conclusion to the problem of their relationships with the Indians of Middle and North America. Investigations have brought out a number of similarities with Middle America. Hahn, for example, thinks that certain features of late Ciboney stonework, such as stone balls and disks, may have diffused from that direction.[65] I am more impressed by resemblances in Arawak culture, such as hammocks, the ball game, and the elements in the cult of *zemis* discussed earlier. A prehistoric Maya record of a Carib raid upon Yucatan should also be mentioned.[66]

Relationships with Florida and the rest of the southeastern United States are more difficult to find. William Sturtevant has reviewed this problem from the standpoint of ethnology and has concluded that the similarities seen by previous writers are superficial and cannot be taken as evidence of valid relationships.[67] On the archaeological level, perhaps the best evidence consists of similarities in designs between Ciboney stone and shell work and the pottery of the Glades Indians of southern Florida.[68] It is not clear whether this similarity is due to origin of the Ciboney Indians in Florida or to subsequent contacts between the Ciboney and Glades Indians.

NOTES

1. See, for example, C. Gower, "The Northern and Southern Affiliations of Antillean Culture," *Mem. Am. Anthropol. Assoc. No. 35* (1927).

2. See S. Lovén, *Origins of the Tainan Culture, West Indies* (Elanders, Göteborg, 1935).

3. See W. C. Sturtevant, "The Significance of Ethnological Similarities between Southeastern North America and the Antilles," *Yale Univ. Publ. Anthropol. No. 64* (1960).

4. T. van Andel and H. Postma, *Verhandel. Koninkl. Ned. Akad. Wetenschap.*, **20**, No. 5 (1954).

5. O. G. Ricketson, Jr., in *The Maya and Their Neighbors* (Appleton, New York, 1940), pp. 18–26; A. de Hostos, in *Anthropological Papers* (Government of Puerto Rico, San Juan, 1941), pp. 30–53.

6. G. G. Simpson, "Zoogeography of West Indian Land Mammals," *Am. Museum Novitates No. 1759* (1956).

7. G. R. Willey, *Science*, **131**, 73 (1960).

8. I. Rouse, in "Courses towards Urban Life," R. J. Braidwood and G. R. Willey, Eds., *Viking Fund Publ. Anthropol. No. 32* (1962), pp. 39–54.

9. See, for example, A. L. Kroeber, *Anthropology* (Harcourt, Brace, New York, 1948), pp. 815–24, 834–36.

10. J. Wilbert, *Kölner Z. Soziol. Sozialpsychol.*, **10**, 272 (1958); I. Rouse and P. García Valdes, in "Handbook of South American Indians," J. H. Steward, Ed., *Smithsonian Inst. Bur. Am. Ethnol. Bull. 143* (1948), vol. 4, pp. 497–505.

11. S. Lovén, *op. cit.*

12. I. Rouse, in "Handbook of South American Indians," J. H. Steward, Ed., *Smithsonian Inst. Bur. Am. Ethnol. Bull. 143* (1948), vol. 4, pp. 507–65.

13. *The Journal of Christopher Columbus*, C. Jane, Trans. (Potter, New York, 1960), pp. 92–95, 146–52.

14. I. Rouse, in "Handbook of South American Indians," J. H. Steward, Ed., *Smithsonian Inst. Bur. Am. Ethnol. Bull. 143* (1948), vol. 4, p. 547.

15. D. M. Taylor, "The Black Carib of British Honduras," *Viking Fund Publ. Anthropol. No. 17* (1951), pp. 41–54.

16. I. Rouse and J. M. Cruxent, *Venezuelan Archaeology* (Yale Univ. Press, New Haven, 1963), p. 155.

17. I. Rouse, *Final Technical Report on NSF-G24049: Dating of Caribbean Cultures* (mimeographed, available from the author, Yale University, New Haven, Conn., 1963). The National Science Foundation's support of the project is gratefully acknowledged. The dates have been corrected for the recently adopted convention of using a base line of A.D. 1950.

18. I. Rouse and J. M. Cruxent, *Venezuelan Archaeology* (Yale Univ. Press, New Haven, 1963), p. 23.

19. G. G. Simpson, *op. cit.*

20. I. Rouse, "The Entry of Man into the West Indies," *Yale Univ. Publ. Anthropol. No. 61* (1960).

21. I. Rouse and J. M. Cruxent, *Venezuelan Archaeology* (Yale Univ. Press, New Haven, 1963), p. 27.

22. I. Rouse and J. M. Cruxent, *Am. Antiquity*, **28**, 537 (1963).

23. J. M. Cruxent, *ibid.*, **27**, 576 (1962).

24. I. Rouse and J. M. Cruxent, *Venezuelan Archaeology* (Yale Univ. Press, New Haven, 1963), p. 39.

25. J. Royo y Gómez, *Soc. Vertebrate Paleontol. News Bull. No. 58* (1960), pp. 31–32.

26. I. Rouse and J. M. Cruxent, *Venezuelan Archaeology* (Yale Univ. Press, New Haven, 1963), p. 44.

27. I. Rouse, "The Entry of Man into the West Indies," *Yale Univ. Publ. Anthropol. No. 61* (1960), p. 10.

28. R. P. Bullen and F. W. Sleight, "The Krum Bay Site—a Preceramic Site on St. Thomas, American Virgin Islands," *William L. Bryant Foundation, American Series, No. 5* (1964).

29. R. Alegría, H. B. Nicholson, G. R. Willey, *Am. Antiquity*, **21**, 113 (1955); J. M. Cruxent, personal communication.

30. I. Rouse, "Culture of the Ft. Liberté Region, Haiti," *Yale Univ. Publ. Anthropol. No. 24* (1941); "Archeology of the Maniabón Hills, Cuba," *Yale Univ. Publ. Anthropol. No. 26* (1942).

31. I. Rouse and J. M. Cruxent, *Venezuelan Archaeology* (Yale Univ. Press, New Haven, 1963), p. 46.

32. C. R. McGimsey III, *Am. Antiquity*, **22**, 151 (1956).

33. W. R. Coe II, *ibid.*, **22**, 280 (1957).

34. I. Rouse, "The Entry of Man into the West Indies," *Yale Univ. Publ. Anthropol. No. 61* (1960), p. 20.

35. G. G. Simpson, *op cit.*, p. 6.

36. I. Rouse, "The Entry of Man into the West Indies," *Yale Univ. Publ. Anthropol. No. 61* (1960), p. 23.

37. I. Rouse and J. M. Cruxent, *Venezuelan Archaeology* (Yale Univ. Press, New Haven, 1963).

38. R. E. Alegría, in *Miscelanea de Estudios dedicados a Fernando Ortiz* (Havana, Cuba, 1955), vol. 1, pp. 43–62.

39. I. Rouse, "The Entry of Man into the West Indies," *Yale Univ. Publ. Anthropol. No. 61* (1960), p. 21.

40. R. E. Algería, in *Miscelanea de Estudios dedicados a Fernando Ortiz* (Havana, Cuba, 1955), vol. 1, pp. 43–62.

41. R. P. Bullen and F. W. Sleight, *op. cit.*

42. P. G. Hahn, thesis, Yale University, unpublished.

43. See, for example, R. S. MacNeish, *Second Annual Report of the Tehuacán Archaeological-Botanical Project* (R. S. Peabody Foundation, Andover, Mass., 1962); G. Reichel-Dolmatoff, *Rev. Colombiana Antropol.*, **10**, 349 (1961); J. B. Bird, in "A Reappraisal of Peruvian Archaeology," *Soc. Am. Archaeol. Mem. No. 4* (1948), pp. 21–28.

44. I. Rouse and J. M. Cruxent, *Venezuelan Archaeology* (Yale Univ. Press, New Haven, 1963), p. 48.

45. J. M. Cruxent and I. Rouse, "An Archeological Chronology of Venezuela," *Pan American Union Social Sci. Monographs No. 6* (1958–59), vol. 1, pp. 223–33.

46. I. Rouse and J. M. Cruxent, *Venezuelan Archaeology* (Yale Univ. Press., New Haven, 1963).

47. I. Rouse, *Final Technical Report on NSF-G24049: Dating of Caribbean Cultures, op. cit.*

48. I. Rouse and J. M. Cruxent, *Venezuelan Archaeology* (Yale Univ. Press, New Haven, 1963), p. 121.

49. I. Rouse, R. E. Alegría, M. Stuiver, "Recent Radiocarbon Dates for the West Indies," unpublished; I. Rouse, *Porto Rican Prehistory*, vol. 18, pt. 3, of *Scientific*

Survey of Porto Rico and the Virgin Islands (New York Acad. of Sciences, New York, 1952).

50. I. Rouse, *Southwestern J. Anthropol.*, **7**, 248 (1951).

51. I. Rouse and J. M. Cruxent, *Venezuelan Archaeology* (Yale Univ. Press, New Haven, 1963).

52. An alternative theory is suggested in note 16: that the Chicoid people may have elaborated their Barrancoid traits from those which the ancestral Saladoid people had obtained on the mainland before the time of Barrancoid intrusion into Trinidad.

53. Rouse, Alegría, Stuiver, *op. cit.;* Rouse, *Puerto Rican Prehistory, op. cit.*

54. I. Rouse and J. M. Cruxent, *Venezuelan Archaeology* (Yale Univ. Press, New Haven, 1963).

55. M. McKusick, thesis, Yale University, unpublished; C. Jesse, *J. Barbados Museum and Hist. Soc.*, **27**, 49 (1959).

56. I. Rouse, *Southwestern J. Anthropol.*, **7**, 248 (1951).

57. I. Rouse, in "Handbook of South American Indians," J. H. Steward, Ed., *Smithsonian Inst. Bur. Am. Ethnol. Bull. 143* (1948), vol. 4, pp. 507–565.

58. G. F. Ekholm, in S. K. Lothrop et al., *Essays in Pre-Columbian Art and Archaeology* (Harvard Univ. Press, Cambridge, 1961), pp. 356–71.

59. S. Lovén, *op. cit.*

60. J. H. Steward, *Southwestern J. Anthropol.*, **3**, 85 (1947).

61. I. Rouse, in S. K. Lothrop et al., *Essays in Pre-Columbian Art and Archaeology* (Harvard Univ. Press, Cambridge, 1961), pp. 342–55.

62. D. Taylor and I. Rouse, *Intern. J. Am. Linguistics*, **21**, 105 (1955).

63. M. McKusick, personal communication.

64. I. Rouse, in S. K. Lothrop et al., *Essays in Pre-Columbian Art and Archaeology* (Harvard Univ. Press, Cambridge, 1961), pp. 342–55.

65. P. G. Hahn, *op. cit.*

66. H. Berlin, *Rev. Mex. Estud. Antropol.*, **4**, 141 (1940); I. Rouse, "Mesoamerica and the West Indies," "Handbook of Middle American Indians" (Tulane Univ. Middle American Research Institute, New Orleans, in press).

67. W. C. Sturtevant, *op. cit.*

68. R. P. Bullen, "Similarities in Pottery from Florida, Cuba, and the Bahamas," *Actas del XXXIII Congreso Internacional de Americanistas, San José, Costa Rica* (1959), vol. 2.

Fig. 15–1. Map of the Caribbean Sea and surrounding lands.

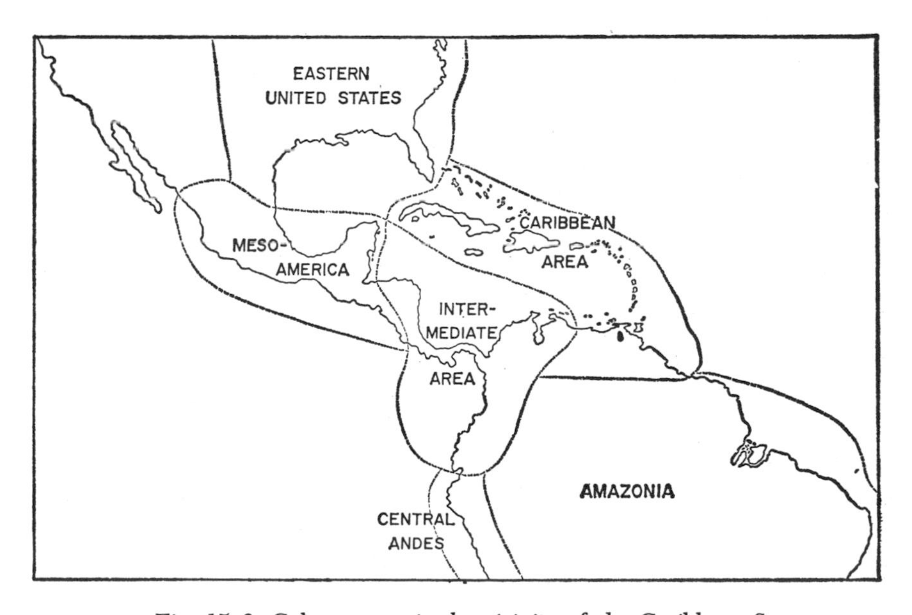

Fig. 15–2. Culture areas in the vicinity of the Caribbean Sea.

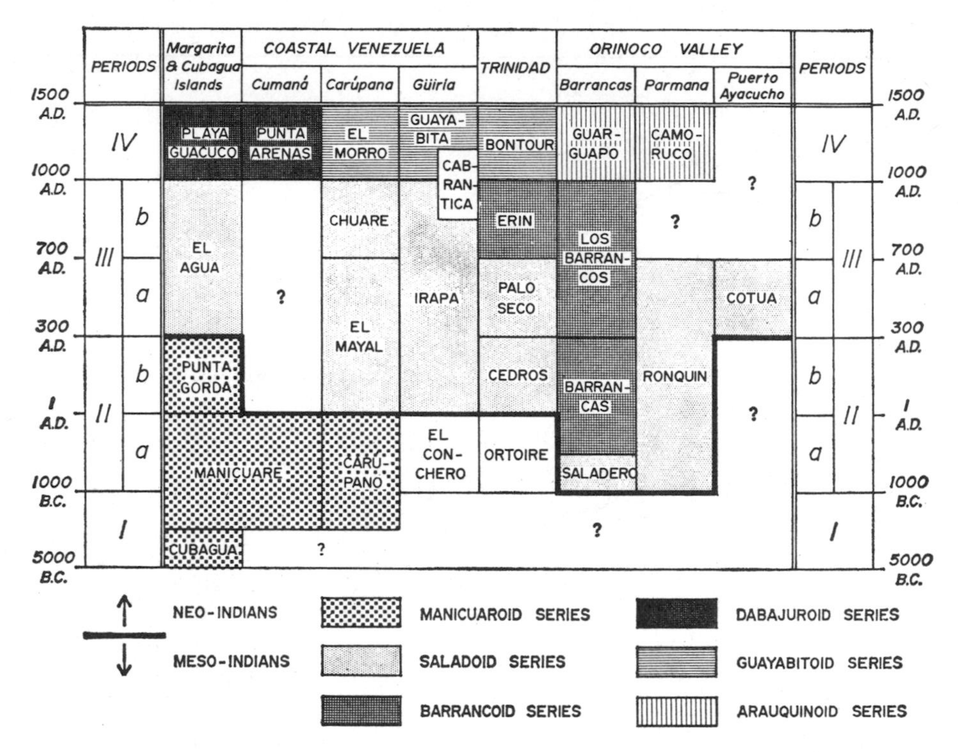

Fig. 15–3. Chronology of eastern Venezuela and the islands immediately offshore.

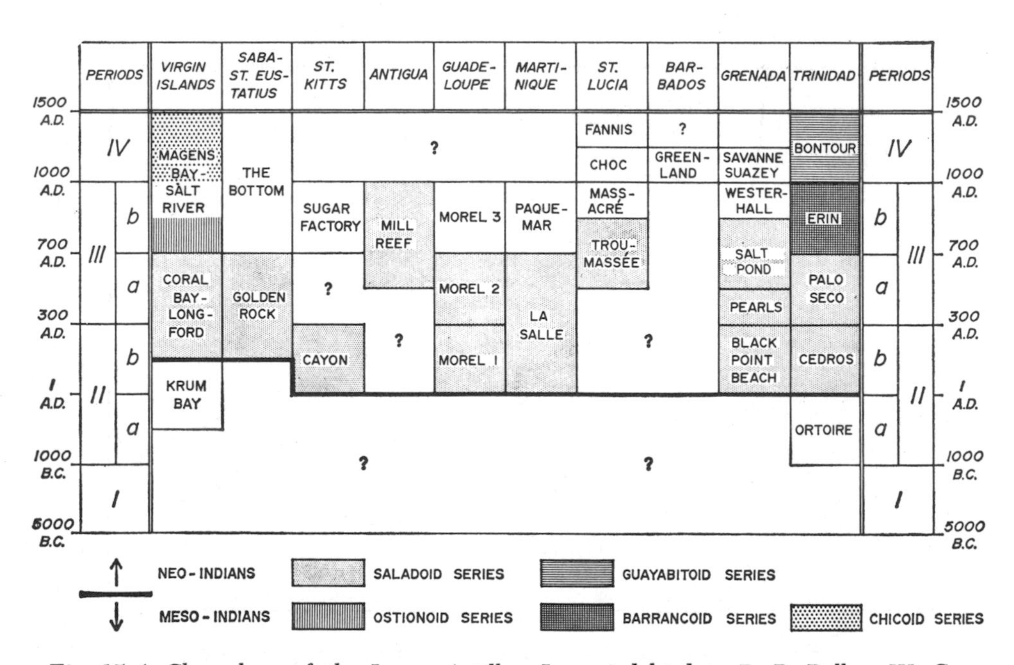

Fig. 15–4. Chronology of the Lesser Antilles. I am indebted to R. P. Bullen, W. G. Haag, C. Hoffman, M. McKusick, and F. Olsen for information upon which this figure is based, though they are in no way responsible for my placement of the cultures.

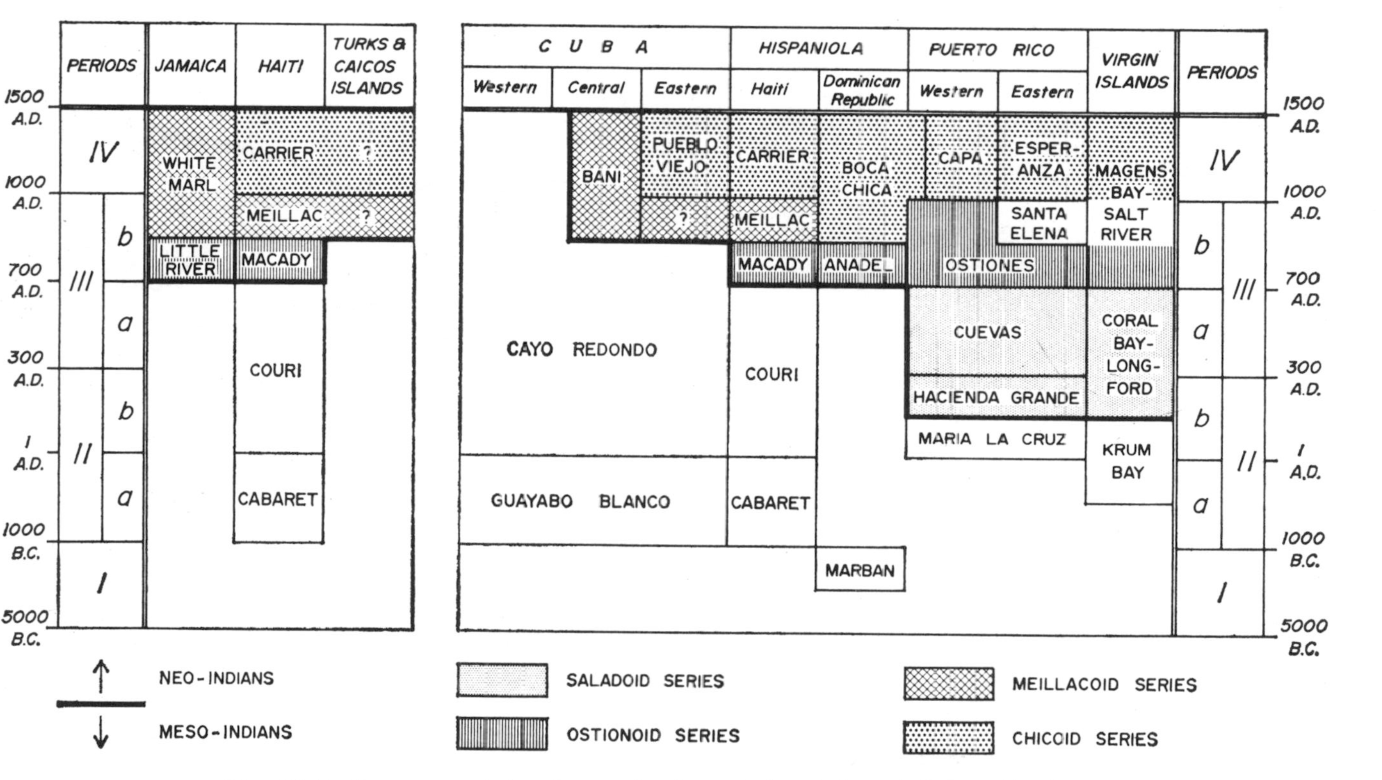

Fig. 15–5. Chronology of the Greater Antilles and the Turks and Caicos Islands.

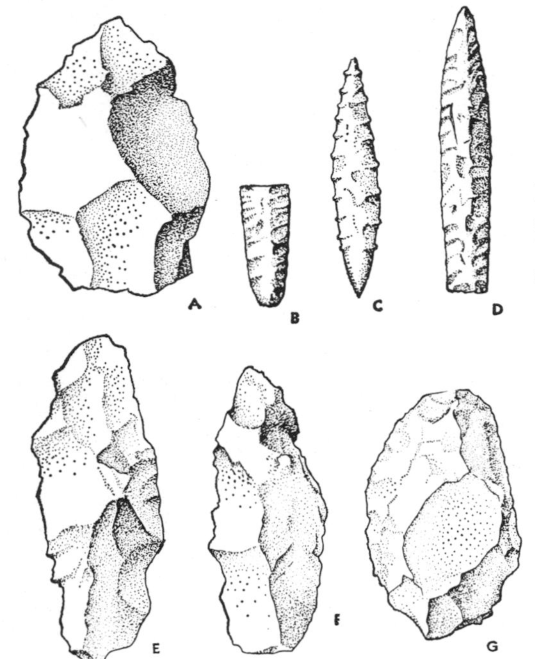

Fig. 15–6. Some artifacts of the Joboid series, western Venezuela. (Courtesy Yale University, Department of Anthropology.)

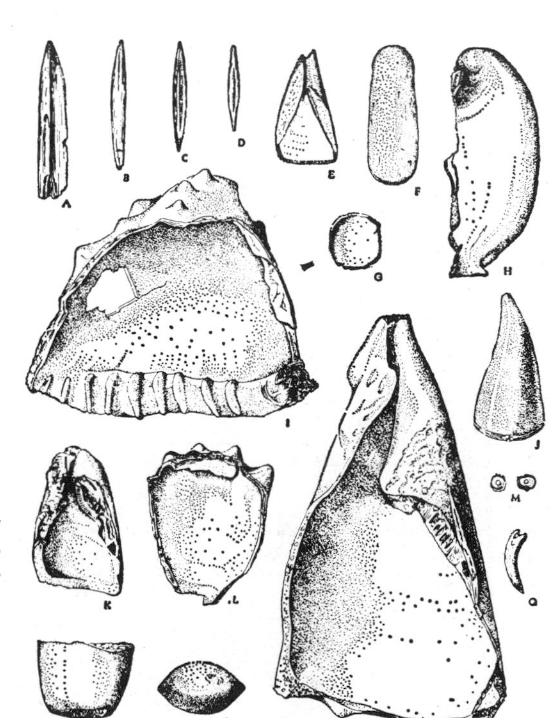

Fig. 15–7. Artifacts of the Manicuaroid series, eastern Venezuela. (Courtesy Yale University, Department of Anthropology.)

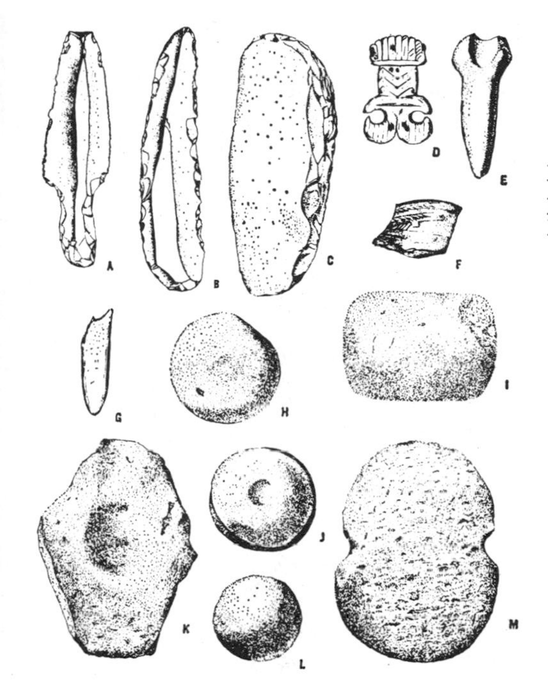

Fig. 15–8. Artifacts of the Couri complex, Haiti. (Courtesy Yale University, Department of Anthropology.)

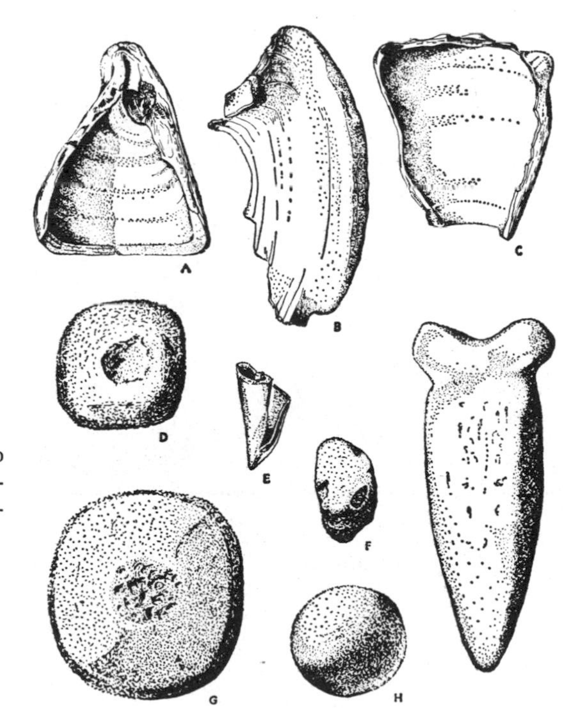

Fig. 15–9. Artifacts of the Cayo Redondo complex, Cuba. (Courtesy Yale University, Department of Anthropology.)

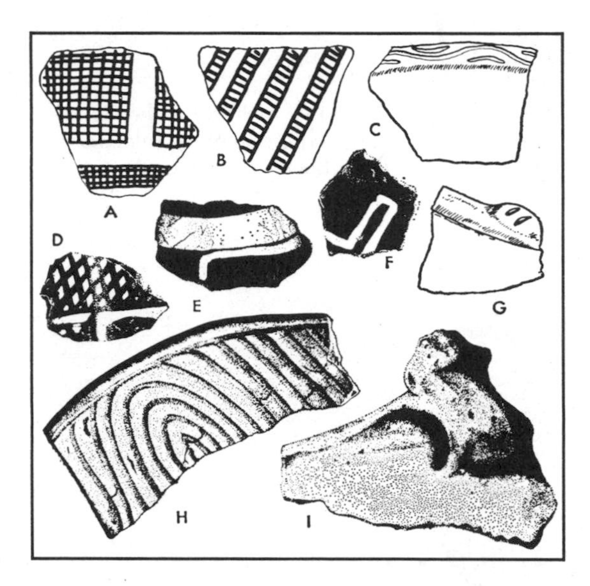

Fig. 15–10. Pottery of Saladero, type style of the Saladoid series, eastern Venezuela. (Courtesy Yale University Press.)

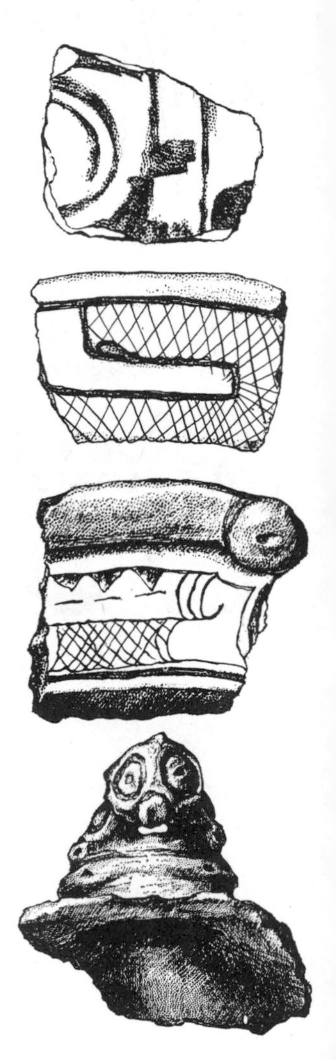

Fig. 15–11. Pottery of the Saladoid series from layer 1, Morel site, Guadeloupe.

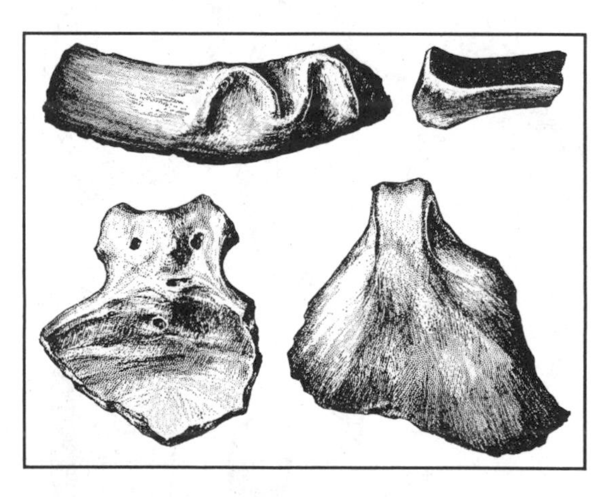

Fig. 15–12. Pottery of Ostiones, type style of the Ostionoid series, Puerto Rico.

Fig. 15–13. Pottery of Boca Chica, type style of the Chicoid series, Puerto Rico.

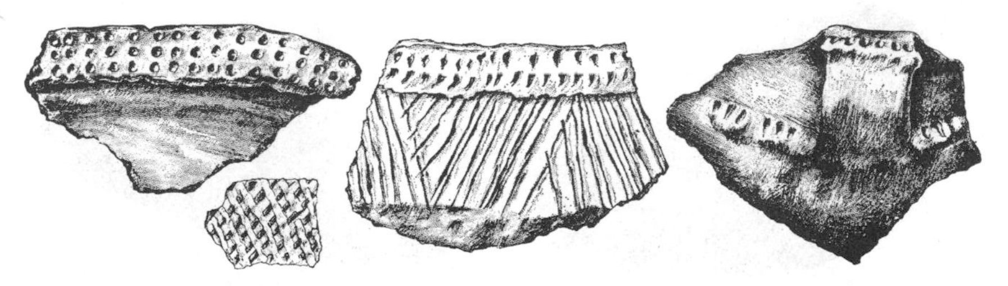

Fig. 15–14. Pottery of Meillac, type style of the Meillacoid series, Haiti.

Fig. 15–15. Ceremonial stone celts and figure, Greater Antilles. (From T. A. Joyce, *Central American and West Indian Archaeology* [Macmillan, London, 1916], plate xxiii.)

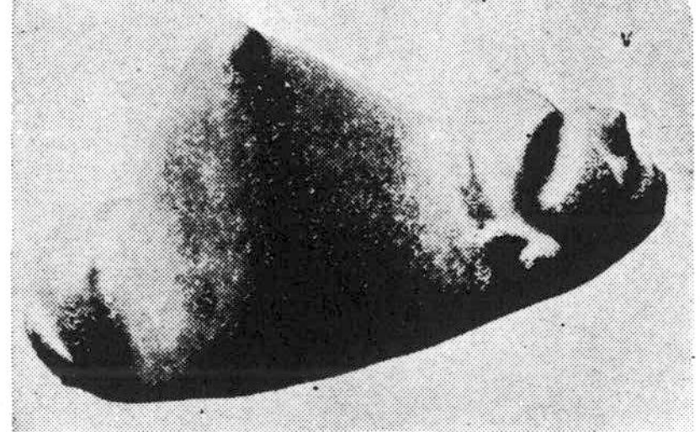

Fig. 15–16. (Left) Ceremonial stone ax, Puerto Rico. (Right) Three-pointed stone, Puerto Rico. (From H. A. Lavachery, *Les artes antiguas de América en el Museo Arquéologico de Madrid* [Sikkel, Antwerp, 1929].)

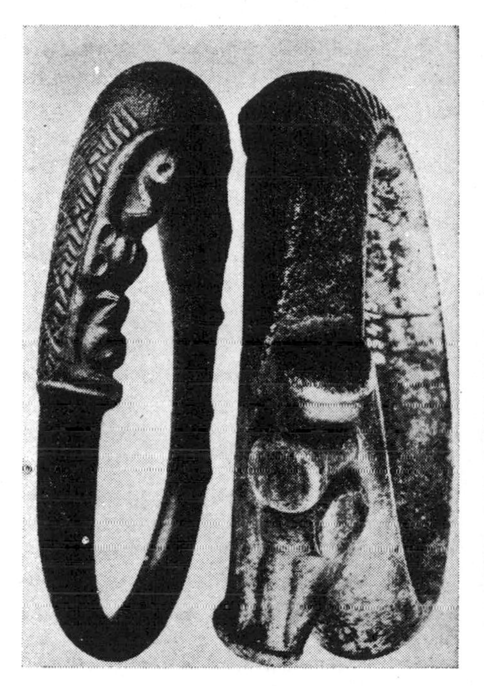

Fig. 15–17. (Left) Stone collars, Puerto Rico. (From S. Lovén, *Origins of the Tainan Culture, West Indies* [Elanders, Gothenburg, 1935], plate xviii.) (Right) Elbow stone, Puerto Rico. (From H. A. Lavachery, *Les artes antiguas de América en el Museo Arquéologico de Madrid* [Sikkel, Antwerp, 1929].)

Fig. 15–18. Snuffing tube of wood, Haiti. (From E. Mangones and L. Maximilien, *L'art précolombien d'Haiti* [L'Imprimerie de L'Etat, Port-au-Prince, 1941], plate L.)

Fig. 15–19. Wooden figure of a *zemi*, Greater Antilles. (From T. A. Joyce, *Central American and West Indian Archaeology* [Macmillan, London, 1916], plate xxi.)

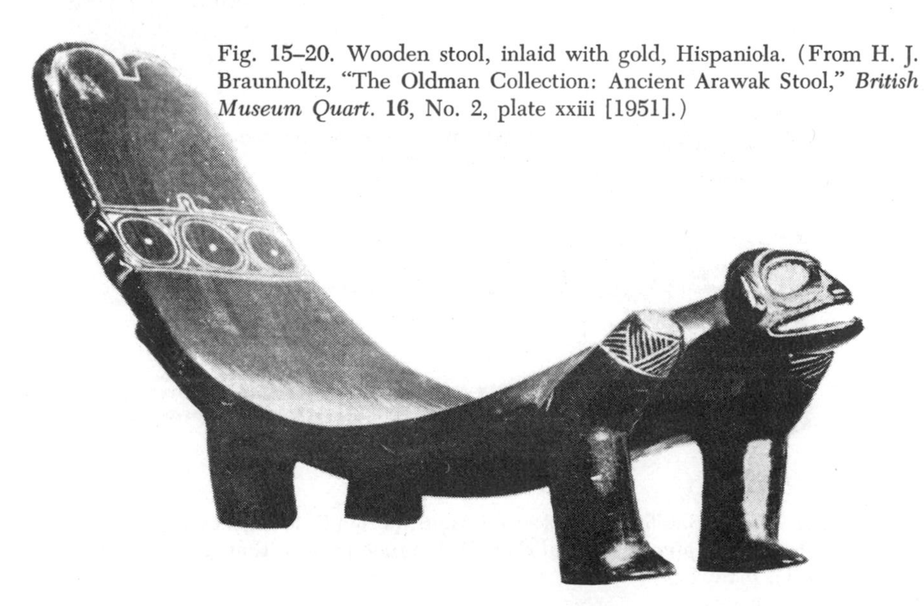

Fig. 15–20. Wooden stool, inlaid with gold, Hispaniola. (From H. J. Braunholtz, "The Oldman Collection: Ancient Arawak Stool," *British Museum Quart.* **16**, No. 2, plate xxiii [1951].)

Part III

CITIES AND CIVILIZATIONS

Chapter 16

The Hasanlu Project

Robert H. Dyson, Jr.

The aim of any archaelogical project is to explore the unknown. In the case of the Hasanlu project carried out over the past five years by the University of Pennsylvania Museum, the Metropolitan Museum of Art of New York City, and the Archaeological Service in Iran, the "unknown" is the entire pre- and protohistoric period of southern Azerbaijan, Iran. While exploration has proceeded in a general way on this overall problem, a more specific objective has been the elucidation of the chaotic and mysterious period that corresponds to the first appearance of the Medes and Persians in western Iran—that is to say, the first half of the first millennium B.C. In pursuing this objective the project (it is hoped) in part exemplifies archaeology as systematic scholarship, for it employs the natural sciences on key problems as working tools that contribute toward a better evaluation of the evidence and a keener judgment of its significance.[1] The following brief review indicates the steps by which this objective is being pursued and the degree to which success has been achieved.[2, 3]

Geographical Setting

The initial framework for any archaeological problem is the geographical setting of the site to be studied. This setting determines the limitations and the opportunities—the nonavailability or availability of local raw materials, water, and land, as well as the degree of isolation from, or contact with, foreigners that results from the presence of travel barriers and accessible trade routes. In the case of Tepe Hasanlu, the site chosen for exploration, these factors may be summarized briefly.

The site, which is a high mound, is located in the center of a basin-like valley adjoining the Qadar River in the general region known as

"Solduz," just southwest of the great salt lake Urmia (more recently called Rezaiyeh) (Fig. 16–1). The valley is separated from the lake-shore by a low range of hills. The slopes surrounding the valley are devoid of forest, and little rain falls except during the late spring. During the remainder of the year agriculture is made possible through the use of canal water brought several miles from the Qadar River and discharged into low mud flats, where temporary lakes are formed. These lakes are a favorite haunt of quantities of wild ducks and other water birds. In particularly dry years, when there is little snow in the mountains to feed the Qadar, the lakes dry up. The bottom of the valley is filled with gray clay, sand, and gravel. Ledges of white limestone and more rarely red sandstone project from some of the nearby hills. Poplar trees and small groves of fruit trees, such as apple and apricot, are grown in the villages. In the winter the snow blows down into the valley, and wild boars from the mountains root through the fields in search of food.

The present inhabitants raise grain (wheat and barley, chiefly), some vegetables, and many grapes. They raise a few chickens, more water buffalo (for milk and for use as draft animals), and even more sheep. The landlords also keep horses. So long as water is available and the administration of its distribution functions effectively, the land yields a rich harvest. The area is shielded from its neighbors to the north by the salt lake, a thick and viscous body of water subject to sudden high winds. To the west the high wall of the Zagros Mountains rises to about 14,000 feet above sea level, the valley itself being at about 4,900 feet.[4] To the south lie the mountains of Kurdistan. The isolation of the area is broken, however, by the trade routes that cross from Iraq (ancient Assyria) in the west, via the Rowanduz and Kel-i-shin passes, to the east (central Iran) along the Qadar River, and from the Caucasus and Rezaiyeh in the north to Kurdistan in the south, along the western shore of the lake and southward through a series of interconnecting mountain valleys. This geographical position thus provides the first element in the cultural dynamics of the area: a partial isolation with a consequent tendency toward local cultural development, yet full exposure to intrusions from the four directions, the north and east being somewhat more open than the west.[5]

Historical Background

The known history of Azerbaijan province documents the shifting influences, particularly of a political and military nature, which traveled these geographical routes from time to time. Since the ninth century B.C. this area has been successively controlled by Assyrians, Urartians, Assyrians, Scythians, Medes, Achaemenian Persians, Greeks, Parthians, Sassanians, tribesmen of the caliph Uthman, Abbasids, Buwayhids, Seljuk Turks, Ildijiz Atabeks, Mongols, Ilkhans, Jalayrs, Turkomans of the Black Sheep, Turkomans of the White Sheep, Safavids, Ottoman Turks, Kajars, Russians, and Pahlavis. The dates of these periods of control and the directions from which influences came are given in Table 16–1.[6]

Aside from providing an indication of the nature of the relationship with surrounding areas, the available history also provides us with a more detailed frame of reference for the period in which our archaeological problem falls—the early first millennium B.C. This period is vague, owing to the fact that it is seen only indirectly, through the records of the Assyrian and Urartian kings.[7,8] Because this history is described in

TABLE **16–1**

Influences on the present Azerbaijan province of Iran, over twenty-eight centuries.

INVADER	DATE	DIRECTION FROM WHICH INVADERS CAME
	B.C.	
Assyrians	~850	West
Urartians	~800	North
Assyrians	714	West
Scythians	650–625	North
Medes	625–550	Southeast
Achaemenian Persians	550–331	Southeast
Greeks	331	West
Parthians	150	East
	A.D.	
Sassanians	226–651	Southeast
Tribesmen of caliph Uthman, Abbasids, Buwayhids	651–1055	West
Seljuk Turks, Ildijiz, Atabeks	1136–1225	North and east
Mongols, Ilkhans	1256–1349	North and east
Jalayrs	1340–1411	West
Turkomans of the Black Sheep	1378–1469	North and east
Turkomans of the White Sheep	1469–1478	North and east
Safavids	1500–1750	East
Ottoman Turks	1514–1603, 1724–1732	North
Kajars	1750–1925	East
Russians	1915–1917, 1941–1946	North
Pahlavis	1925 to the present	East

foreign annals rather than local records we refer to this period in Iran as "protohistoric."

The records indicate that the area to the south of Lake Urmia was in general occupied by tribal groups loosely organized at first under a hereditary leader aided by his nobles. Gradually, under Assyrian influence, this governing structure appears to have become more formal and to have acquired the trappings of a true monarchy. The capital, Izirtu, was probably near modern Saqqiz. The number of villages and towns listed in the annals of the conquering Assyrian kings indicates quite clearly that while part of the population may have been pastoral, living in tents (especially in the high summer pastures), there was, nevertheless, a substantial settled population as well. It seems probable that the pattern of life was somewhat similar to that of present-day Kurds, among whom part of the population moves to the high pastures during the summer and returns to spend the winter in the villages of the lower valleys. The annals refer to the burning of small towns as well as the burning and pulling down of fortified citadels.

From these pieces of evidence we may educe a settlement pattern involving a scattering of fortified citadels located in the fertile valley areas or in commanding positions in the hills, each with a number of small associated villages and pasture lands. The economy would have been partly agricultural and partly one of husbandry, with luxury goods largely imported from the more advanced centers of Assyria and Urartu (the area around Lake Van in eastern Turkey). Some local manufacturing of goods would have been done by artisans, who often no doubt copied the more fashionable styles from abroad. Most of these goods would have been concentrated in the hands of the nobles. The latter, to judge by their names and the names of their towns, spoke a dialect of Hurrian, a language family that spread widely over the Near East in the late second millennium B.C. During the part of their history about which we are informed (from about 850 B.C. onward), these people, the Mannaeans, were constantly on the defensive against their most powerful Assyrian and Urartian neighbors to the west and north. Their history is one of alliance and counteralliance, attack and counterattack, all in an effort to retain their independence. During the first millennium, Indo-European speaking tribes began to settle in the area (possibly in Solduz). By 600 B.C. the Mannaeans had been absorbed into the newly established kingdom of the Medes centered on Hamadan to the southeast.

Stratigraphic Sequence

The next step in the investigation was the stratigraphic excavation of the mound in order to obtain a series of associated groups of artifacts and architecture. By digging downward from the surface and removing one soil layer at a time in selected areas, the sample cultural remains of the different periods of time have been recovered to a depth of 27 meters, beginning with the latest "Islamic" material and going backward to Neolithic levels of the early sixth millennium B.C. This sequence of cultures has been established on the high Citadel Mound in the center of the site and at three small neighboring sites. The lower mound (8½ meters) surrounding the Citadel, called for convenience the "Outer Town," proves to be much younger; its lowest level, which rests on virgin soil, corresponds with an intermediate level in the much longer sequence of the Citadel Mound. Apparently at a certain point in time (which we estimate to be around 2200 B.C.) the central mound had become inconveniently high (through its growth upward due to incessant leveling and rebuilding of mud-brick houses), and people began settling around its base.

The major excavations to date have been carried out in the southwestern quadrant of the Citadel Mound. This area has been cleared through four major periods of construction. These architectural periods, numbered I through IV from the youngest to the oldest, correspond to ceramic phases (that is, groups of soil layers containing fragments of the same distinctive pottery) which were first recognized in trial soundings or test pits. Period I is known to be Islamic by its plain pottery and small bricks, as well as by the local tradition that standing walls were still visible less than one hundred years ago. Below this level lies the "Mystery" phase (Period II), a period of dull, plain buff-and-red wares, stone cist graves, and the foundation of a large building. Below this lies a stratum (Period III) of small rooms set against the inside of a reused fortification wall. The pottery includes some red burnished ware and finely made buff pottery painted with small hanging triangles, hence the working name, Triangle Ware phase. And finally (Period IV), the Gray Ware phase—the period of the building of the fortification wall, with its towers, and the large buildings with pillared halls lying within (Fig. 16–2). A cross-section (Fig. 16–3) shows the sequence of these periods as seen on an unexcavated face or balk of the site. Below lie the remains of an even earlier, Button-base phase (Period V), named after the burnished gray vessels with tiny bases. Burials belonging to both the Button-base and the Gray Ware phases are found in a cemetery area around the northern edge of the Outer Town. The relative positions of the graves show that material from the Button-base phase is lower, and therefore earlier in date. Given all of this information, the problem remains of how to relate this stratigraphic sequence to the major problem of the early first millennium B.C.—the clarification of cultural conditions in the country at the time of the arrival of the Medes and Persians—and to the local framework of known geography and history.

Relative Chronology

The first step taken in this direction was the comparison of the sequence at Hasanlu to the sequences already known at adjacent sites. By this process of typological comparison the general relative positions of the individual phases were estimated, and some idea as to their historical position was obtained through the known history of the compared sites. Thus, for example, the finely made burnished gray ware vessels of the type seen in Figure 16–4, with loop handle and small disk base, were compared with similar forms known from Tepe Sialk A in central Iran; from Tepe Giyan I, to the south in Luristan; and from Geoy Tepe B, immediately to the north near Rezaiyeh.[9, 10] Such similarities relate the ceramics of this phase to the general tradition of central and western Iran in the late second millennium B.C. This correlation is supported by the presence of a few button-base vases of buff pottery painted with simple lines. These forms belong to a tradition found in Mesopotamia during the latter part of the second millennium B.C. at the sites of Nuzi, Assur, and elsewhere.[11] The pottery of this phase thus relates the local culture to cultures in the two directions from which influence would be expected, given the geography, and confirms the period of that influence as being about 1200–1000 B.C.

The objects found in the burned ruins of Period IV are even more revealing as to relative date. The pottery includes rare glazed vessels of late Assyrian type, along with other items obviously of Assyrian inspiration if not of Assyrian origin. Glazed wall tiles, beads of paste, small stamp seals in the form of crouching lions, fragments of ivory with incised rampant goats and rams flanking a sacred tree, a piece of gold foil with the figure of a man in the act of lustrating, and a fragment of a carnelian bead bearing part of a prayer in Akkadian cuneiform are among the items that show a close relationship with the Assyria of the ninth century B.C. Figure 16–5 shows a sacred-tree scene of Assyrian style cut into the surface of a vase fragment of an artificial material often called "Egyptian blue."[12] Other objects of this material were also found: beads, a small wall tile, cylinder seals.[13] Similar items are known from Egypt to Assyria, and subsequently the material was found in quantity in Iran in the Achaemenian period. Its introduction to Iran remains one of the interesting problems facing archaeologists who investigate this period and area.

Most striking of all the finds, for purposes of comparison, are cylinder seals of late Assyrian type showing archers and griffons, which link Hasanlu to the west. The common pottery of the period, on the other hand, is burnished gray-black and continues an earlier tradition of firing in a reduced atmosphere. The identifying shape is a vessel with a long pouring spout, always found one to a grave but also found in ruined buildings. These vessels pour very well and are decorated in numerous ways by incising, grooving, ribbing, and so on. A similar type has been found in the cemetery of Sialk B in central Iran. This form, and smaller objects in metal, such as horse trappings, relate Hasanlu IV to Sialk B to the east, in contrast to the western, Assyrian-style objects.

The artifacts thus again illustrate influences from both east and west. Significantly, however, the Sialk B material is composed of a mass of additional elements foreign to Hasanlu, including elaborately painted vases suggestive of Anatolian connections. At the same time Hasanlu IV is not specifically Assyrian. Clearly, therefore, a somewhat locally specialized culture is represented by this period, which we estimate as about 1000 to 800 B.C.

In the period following the great fire which ended Hasanlu IV, all of the older types of pottery have disappeared, along with the building style. The few fragments of pottery recovered, and one trilobate copper arrowhead, show some relationship to the artifacts reported from the Persian village at Susa,[14] far to the south. This typological similarity would appear to indicate that in this period, the Triangle Ware phase, we are about to enter, or have entered, the Achaemenian period, the first fully historical period in Iran.[15] The degree of typological change suggests either a lapse of time or a change of population between the Gray Ware phase and the Triangle Ware phase.

Absolute Chronology

While these typological comparisons allow us to fit our major periods into the known structure of relative chronology, they do not provide any precision in terms of actual calendar dates, owing to the absence of written records. True, the objects from the burned buildings of the Gray Ware phase do correspond closely to similar items in Assyria dating to the reign of Shalmaneser III in the middle of the ninth century B.C., and the Persian village remains are dated by tablets to between 600 and 400 B.C. But it would be useful to have an independent check on these interpretations. At this point physics comes to the rescue of archaeology through the method of radiocarbon analysis. From the samples of ash, charred grain, and structural timber collected during the excavations a series of assays have been made by Elizabeth K. Ralph of the University of Pennsylvania's carbon-14 laboratory. The results of tests run to date are summarized in Figure 16–6.[16] Four stratigraphically placed samples are from the Button-base phase (Period V), ten are from the Gray Ware phase (Period IV), and two are from the early and two from the late Triangle Ware phase (periods IIIB and

IIIA respectively). The dates obtained for the Button-base phase range from 1340 to 971 B.C. with a median date of 1155 B.C., which falls directly within the estimated date range obtained by relative chronology. The dates for the Gray Ware phase are more problematical. It may be seen on the chart that they consistently fall earlier than the typological correlation in the ninth century B.C., the average of eight structural timber samples being 1001 ± 20 B.C. At first sight this result appears at variance with the typological result. It is not, however, necessarily unexpected. The remains of the buildings excavated at this level include a number of poorly built walls and blocked doorways as well as numerous areas which had been robbed of paving *before* the fire. This evidence indicates that there was a passage of time between the initial construction (as measured by the carbon samples) and the fire (as measured by the artifacts) and two samples of food destroyed by the fire (P576, P577). A change in function and perhaps in political control is indicated for the Citadel during the period, and it is quite possible that the builders of the Citadel fortifications were not the defenders of it in its last days. Another possibility is that some of the wooden beams used had been salvaged from earlier buildings and thus yield dates somewhat earlier than those of the buildings in which they were found.

A further question remains, involving the hiatus between periods III and IV which is indicated by the presence of secondary erosion deposits underlying walls of the later period. Two radiocarbon dates help here: one from an underground storeroom dug into the ruins of Period IV and filled with charred grain; the other charcoal from a charred floor area connected with an auxiliary fortification wall on the west slope. These samples give an average of 622 ± 38 B.C. and perhaps indicate a local burning. This event could be the dividing line between the early buildings of the period (IIIB) and their later renovations (IIIA). Two samples (P420 and P582) pertain to the end of Period IIIB and indicate that we are now well into the historical Achaemenid period.

In another line of chronological investigation, use is made of the new technique for measuring the age of glass through the effects of weathering on its surface.[17] Whether or not the glass beads and small fragments at present available will prove useful in this regard is a question now being reviewed by Robert H. Brill of the Corning Glass Museum.

Local Population

As indicated in the section on the historical background, some of the local ninth-century population probably spoke a Hurrian form of language, along with other scattered groups in the Zagros Mountains, who reached as far south as the country of Elam around Susa as well as westward through eastern Turkey and northern Mesopotamia. A continuity of local population from the Button-base through the Gray Ware phase at Hasanlu is suggested by a similarity in burial customs (the earlier burials have a drinking vessel as a standard item in the grave; the later burials, a spouted pouring jar), the lack of any appreciable time gap in either the visible stratigraphic record or the radiocarbon dates, and the continuing use of the same cemetery areas.

Pottery of the underlying Period VI and the rare painted vessels of Period V are derived from the second millennium B.C. pottery of northern Mesopotamia. It is thus possible to suggest tentatively that Hurrian may have already been spoken at Hasanlu as early as 1200 B.C., although whether by the whole population or by only selected individuals remains problematical (one grave found had *only* Mesopotamian "Hurrian" period tumblers in it).

Men and women of the Gray Ware phase are depicted on a number of objects found: a silver beaker, ivory fragments, a fragment of sculptured wood, a bronze tetrapod stand (Fig. 16–7), and metal fragments. They are shown as having large fleshy noses. The men wore their hair long and had full beards. The women, seen only on the Hasanlu gold bowl, braided their hair and wrapped it around their heads. It is possible that the bowl was produced during the earlier, Button-base phase, in view of some of its stylistic connections.[18] People are shown occasionally wearing sandals, dressed in long robes or short kilts fastened at the waist with belts, the women with necklaces, the men with fillets around their heads. Small fragments of textiles (studied by Harold Burnham of the Royal Ontario Museum's textile department) were preserved by charring and appear to include both fine and coarse weaves. The possibility that some of them at least were dyed is presented by the identification of one of the shells found as *Murex brandaris* Linné imported from the Mediterranean. This mollusk was commonly used for the production of "royal purple."[19] In medieval times Azerbaijan was also famous for textiles dyed red from the *kirmiz* insect (hence *crimson* and *carmine*), which feeds on the oak trees in the area.[20] It is possible that this source of dye was already known, but we have no material evidence to prove it.[21] Sargon of Assyria lists, among the spoils collected on his famous eighth campaign in this area from the sacking of the city of Musasir (somewhat west of Hasanlu in the mountains) in the eighth century B.C., "130 multicolored garments and tunics of linen; of blue wool and of wool woven in the scarlet of the countries of Urartu and Kilhu."[22]

As for the actual physical characteristics of the population, the skeletal remains were studied by William Bass of the University of Kansas. A preliminary review of the individuals excavated in the 1957 and 1958 seasons (a total of 28) shows that five of the six males from periods IV and V are dolichocranic, or longheaded, as is the adult male from Period I (Islamic). The five males from Period IV have cranial indices ranging from 66.32 to 75.84. The latter measurement is of the one slightly mesocranic individual, who also has a mesorrhinic or medium nasal index of 48.00. The one other individual on whom the nasal index could be measured proved to be leptorrhinic (index of 45.83). These individuals thus would seem to fit into the general prehistoric population presently known from western Iran at Hasanlu, Geoy Tepe, Tepe Sialk, and Tepe Giyan.[23, 24] It is notable that at both Tepe Sialk B in central Iran and at Shah Tepe I on the shore of the Caspian Sea, the population, in contrast to that of Hasanlu, became markedly brachycranic or even hyperbrachycranic in the ninth to eighth century B.C.

The remains of twenty-two additional individuals were added to this study group from the last two seasons' work and provided a wider basis for more detailed study. Almost all periods are represented by at least one individual. Skeletal remains of the earlier periods were multiplied by the work of T. Cuyler Young, Jr., during the 1961 season; hence it should ultimately be possible to make some long-range observations on the local population.

Local Economy

The population at Hasanlu was supported by an economy that, it would seem, was not unlike that of the Assyrians (stripped of its imperial aspects). The fields around the Citadel-town yielded crops of hulled barley (evidently of more than one type), bread wheat (*Triticum vulgare*), and millet (*Panicum miliaceum*). Hans Helbaek[25] of the Danish National Museum, who identified the plant remains, comments that the barley probably contains both two-row and six-row spikes with a dense-spike type among the six-row barley, but no detailed study has yet been made. Among the barley grains, he reports, are some larger than any he has seen before. Iron hoes and sickle blades were among the tools used in the field. The vineyards of Hasanlu, then as now, yielded grapes (*Vitis vinifera*), which were undoubtedly dried in the sun for raisins as well as used for the making of wine. The husbandry practiced was also similar to that of today; cattle, sheep, and goats are most abundant among the animal remains. That horses were used is shown by the presence of one skeleton and of bits and harness bosses and buttons. Horses are shown on a silver beaker, pulling a chariot as well as being led riderless among the prisoners following the chariot.[26, 27] On the Hasanlu gold bowl, on the other hand, are depicted what appear to be asses rather than horses.[28] Boars' tusks show that wild pig was hunted then as today. We may guess that the game birds in the lakes were also the target of hunters, who possibly used the small pyramidal bone arrowheads which are occasionally found. To date, no remains of wheeled vehicles have been found, but their use is indicated by the chariot scene on the silver beaker and by the three chariots on the gold bowl. Moreover, wheeled vehicles were obviously needed to transport the large slabs of building stone (some as much as three meters long) from the rock ledges in the hills ten or more miles away, and to bring in the harvested grain from the surrounding fields. A road of hard packed clay flanked by stone-lined drains and "pavements" leads up the west slope of the mound.

The size and elaborate nature of the structures so far excavated on the Citadel indicate that they belonged to the ruling hierarchy rather than to private individuals. Indeed, Hasanlu was not much of a town in the ordinary sense of the word, although a few scattered private houses appear to have stood around the Outer Town area. One of these, the Artisan's House, excavated for us in 1959 by George F. Dales, Jr., consisted of a large square room entered through two small vestibules. Outside in the courtyard stood a small kiln. The building itself was of sun-dried mud-brick set on rock foundations; it was probably only a single story high. The house had been burned in the sacking, and its collapsed debris showed that a large number of heavy pottery storage jars had been stored on what must have been a flat roof. Inside the house, scattered on the floor, were sherds of fine black table ware. The occupation of the owner was indicated by the presence of a crude medium-sized crucible for pouring molten metal, together with fragments of open-face and two-piece molds for casting ingots, axes, and jewelry. We are thus assured that some craftsmen lived in the area, as well as farmers and herders.

Certainly some of these craftsmen were the potters who produced the fine burnished gray-black pottery by firing it in a reducing atmosphere. That it is possible to produce similar fine pottery from local washed clay was demonstrated experimentally by Frederick Matson during a visit to the site in 1960. Matson also made a technical study of the composition of the frit and glazes of Hasanlu at the ceramic laboratory at Pennsylvania State University. Eric Parkinson at the University Museum has tested the glazes on several different types of objects—tiles, vases, beads, and so on—to ascertain whether or not the same type of glaze was used on different objects; in preliminary examination he has found no significant differences between the glazes used.

Several of the graves found contained the bodies of adult males with their weapons—bronze daggers and spears, iron maceheads, and in one instance flat bronze arrowheads—indicating that professional soldiers also occupied the Citadel (although all of the male population undoubtedly served in this capacity when needed). Other, fallen warriors have been found entombed in the remains of the burned buildings. Some of these men appear to have been wearing garments of colored leather or textile, as areas of red have been found underlying armbones. The red often lies beneath a thin layer of powdery yellow. An attempt to analyze this material by the Museum chemist, Eric Parkinson, ended in failure,[29] as no organic material remained. The conclusion reached was that "the colors may have been due to yellow or red ochre, possibly mixed with copper oxide."

Among these warriors may well have been some of the ruling nobles. However, to distinguish nobles from ordinary soldiers is, at the present state of our knowledge, impossible. Equally difficult is identification of any of the remains as those of priests. In this respect it is of interest to note that most of the forty-four skeletons found burned and crushed on the floor of the great pillared hall of Burned Building II in the 1960 excavations were of preadolescent and early adolescent females, as indicated by their costumes. Since the building may have served a religious function (another point difficult to prove conclusively without excavation in other quarters of the Citadel), it is possible that these poor souls, trapped in the flaming building, may have been somehow connected with religious activities in the Citadel. On the other hand, they may only have been children of the ruling families. Problems of interpretation such as these make archaeology a scholarly rather than a purely scientific pursuit.

The Citadel itself (Fig. 16–8) was completely surrounded by a massive fortification wall consisting of a heavy foundation of roughly fitted limestone blocks. A section of it built in period III is preserved in places to a height of 2.60 meters, and has a superstructure of large mud-bricks (39 by 39 and 36 by 36 by 13 cm). The base of the original wall was 3.20 meters wide, and the estimated height (calculated with Sargon's descriptions as a guide) would have been about 9 meters. At spaced intervals around the walls were narrow stone reinforcement piers. Between every two of these stood a defensive tower. A similar plan was used at the Urartian Citadel of Karmir Blur. To date, four structures have been cleared inside the walls: two buildings with pillared halls (burned buildings I and II), a two-room Bead House, and a one-room South House. The two major burned buildings are separated by a narrow paved street, South Street. All of the open-paved areas are provided with drains to carry off rainwater through underground stone-lined channels. Unfortunately the terminal area for these drains was destroyed by later digging so that we do not know whether the runoff was collected. Drains from the high northwest area flow under a street leading out through the West Gate, and down the slope. The wall foundations and paving were all of limestone slabs. The wooden columns, which were a half meter in diameter, rested on stone base blocks and were of *Populus* sp.[30]

Among the other wooden specimens identified were the remains of a small wooden bowl (*Crataegus* sp., or hawthorn), a shaped fragment with a hole through it (*Cupressus* sp., or cypress), the wooden core (*Populus* sp.) of a bird figure covered with an overlay of copper plates and the shafts of two bronze maceheads (one of *Buxus sempervirens* or common box tree, the other of *Malus* sp. or *Pyrus* sp., apple, pear, or crab apple). The *Buxus* identification is of particular interest, since Sargon lists as part of the loot he carried off from Musasir, "sticks of ivory, ebony, of boxwood with [their] knobs [maceheads?]."[31]

Foreign Trade

While much that has been found at Hasanlu in the Gray Ware levels is of local manufacture, as is shown by the quite distinctive styles in decoration (Fig. 16–7), many of the raw materials for these products had to be imported. This is particularly true of metal ores such as copper, tin, silver, and gold. At the same time the imitation locally of foreign styles shows that the craftsmen had at their disposal actual imported objects, or had traveled widely enough to have seen the originals. The glazed wall tiles, for example, are provincial copies of the tiles seen in the palaces of the kings of Assyria.[32] On some of the ivory and metal fragments are scenes in imitation of standard Assyrian scenes.

The imitation of decorative style is relatively easy to trace. Not so the source of the metals used. The process of tracking sources is slow, owing to the need to analyze metal objects and samples of ores. In one instance we have pretty good evidence of the importation of ore. The beads and buttons that, in the field, we took to be cast native silver proved, upon analysis by Eric Parkinson in the University Museum laboratory, to be almost pure antimony, containing a little copper, calcium, magnesium, and possibly strontium. The discovery is of particular interest in that antimony and antimony bronze were common in the Tiflis and Kuban areas of the Caucasus after the tenth century B.C., especially in the graves of the Redkinlager on the Aksatfa, a tributary of the Kura River.[33, 34] The metal is brittle and easily pulverized. It was apparently cast in small open molds to make the buttons and beads found at Hasanlu. It seems quite probable that these were imported from the Caucasus, although antimony is also known to occur near Takht-i-Suleiman in Afshar province, southeast of Azerbaijan.

The star-shaped maceheads found among the ruins may also indicate foreign contact. These special forms are of cast bronze and represent one among a number of types, the others being somewhat simpler. The star form is of particular interest, as it is also reported from the Caucasus area in the Gandša-Karabulag culture, which is dated slightly earlier than Hasanlu IV.[35]

Whether the four star-shaped maceheads found to date at Hasanlu represent imports, locally manufactured articles, or weapons of the attackers we cannot determine. In an effort to shed some light on this question, Parkinson made a spectroscopic analysis of four maceheads, including one that was star-shaped. He reports that "in each sample copper and tin were predominant. Silicon, probably in very small amount, was identified in all except one . . . where its presence was doubtful, and each probably had traces of calcium and magnesium. The star-shaped macehead (Has 60-943) showed two lines which were attributable to zinc, which the others lacked. . . . The indication of zinc . . . may be significant, but it should be noted that only the two principal lines of zinc were identified, so that in any case it could have been present only in very small amount."[36] It may be added that zinc was lacking not only in the other three maceheads but also in a dagger from the Button-base phase.[37] From the qualitative spectrographic analysis of the dagger (No. 58-4-11), the following estimates for the base metal were derived: copper and tin (the major components), over 5 per cent; arsenic, 0.1 to 0.9 per cent; silicon, iron, nickel, and antimony, 0.01 to 0.09 per cent; and magnesium, lead, and silver, 0.001 to 0.009 per cent. Three other Luristan specimens analyzed at the same time from the collection of W. O. D. Pierce also failed to contain any trace of zinc. Nor is zinc reported in the analyses of Luristan objects published by Desch in *A Survey of Persian Art*.[38] According to R. J. Forbes[39] (Fig. 57), zinc is found at Tabriz, and farther north at Kara Dagh and elsewhere in the Caucasus. The next step in this study will be to examine the other star-shaped maceheads to determine whether they too contain zinc. Clearly, much work remains to be done on metal in this period before any solid conclusions can be reached.[40]

Conclusion

The aim of this chapter has been to indicate some of the work in progress, both in the field and in the laboratory, in connection with the Hasanlu project. Since this work is still going on, no definitive conclusions can be reached in regard to the many special studies which our raw data have made possible.

Nevertheless, in terms of the initial aim of elucidating a period which, before this project was undertaken, was virtually unknown in this area, we feel that important preliminary results have been achieved. A general stratigraphic sequence has been built up through careful excavation and documentation in the field. This sequence provides a guide for approximately 5,000 years of pre- and protohistoric time. It has been successfully related to both relative and absolute chronology in a general way. A surface survey in the area has confirmed the historically indicated pattern of local citadels and scattered villages in the early first millennium B.C. The contents of burials and buildings excavated on the Citadel and in the Outer Town at Hasanlu illuminate the way of life of the period. The picture that emerges coincides well with the inferences already drawn from indirect historical references. Information of a technical nature has been, and is being, made available to the fields of botany, zoology, ceramics, physics, physical anthropology, and metallurgy, among others; this information will contribute fundamentally to the known history of technology and to the improvement of research methods within these fields. Most important of all, the study of Hasanlu contributes toward our own understanding of the immense ranges of experience already explored by the human mind even in protohistoric times. In the pursuit of these objectives, in the pursuit of archaeology, certainly scholarship and science go hand in hand.

NOTES

1. For a general commentary on tactics and strategy in archaeology, see R. E. M. Wheeler, *Archaeology from the Earth* (Oxford Univ. Press, London, 1954).

2. Previous explorations have been reported as follows: M. T. Mustafavi, *Nagsh-o-Negar Magazine*, No. 6 (1959) (translated for the project by P. Barzin); A. Stein, *Old Routes of Western Iran* (Macmillan, London, 1940); and A. Hokimi and M. Rad, *Guzárishhá-yi bástanshinási I* (Archaeological Museum, Teheran, 1950) (translated for the project by P. Barzin). Preliminary reports on recent excavations have appeared as follows: R. H. Dyson, Jr., *Illustrated London News*, **236**, 132 (1960); R. H. Dyson, Jr., *ibid.*, **239**, 534 (1961); R. H. Dyson, Jr., *Archaeology*, **13**, 118 (1960); Anon., *Life*, **46**, 50 (1959); V. E. Crawford, *Metropolitan Museum of Art Bull.*, **1961**, 84 (November, 1961).

3. R. H. Dyson, Jr., *Illustrated London News*, **236**, 250 (1960).

4. These data were kindly supplied by Dr. Erich F. Schmidt. Reference is made to Tepe Hasanlu under the name of Khasani in his *Flights over Ancient Cities of Iran* (Univ. of Chicago Press, Chicago, 1940), p. 69. The exact location of the site is recorded in Schmidt's files as 37°N and 45°27′E.

5. As far as we know there is no detailed geography dealing specifically with this area of Iran. As a rule, Azerbaijan is treated under Iran in general, and in a general way in travelers' accounts and general geographies. A brief summary may be found in the *Encyclopaedia Britannica* (1954). The geography of the medieval period is

reviewed in G. Le Strange, *The Lands of the Eastern Caliphate* (Cambridge Univ. Press, Cambridge, 1905).

6. W. L. Langer, *An Encyclopedia of World History* (Houghton Mifflin, Boston, 1956); R. Roolvink et al., *Historical Atlas of the Muslim Peoples* (Djambatan, Amsterdam, Netherlands, 1957); W. R. Shepherd, *Historical Atlas* (Barnes and Noble, New York, 1956).

7. D. D. Luckenbill, *Ancient Records of Assyria and Babylonia* (Univ. of Chicago Press, Chicago, 1926); G. A. Melikishvili, *Vestnik Drevnei Istorii,* **1**, 57 (1949) (translated for the project by M. Van Loon).

8. F. Thureau-Dangin, *Une Rélation de la Huitième Campagne de Sargon* (Geuthner, Paris, 1912).

9. T. Burton Brown, *Excavations in Azerbaijan, 1948* (Murray, London, 1951).

10. G. Conteneau and R. Ghirshman, *Fouilles du Tépé-Giyan près de Néhavend, 1931 et 1932* (Geuthner, Paris, 1935); R. Ghirshman, *Fouilles de Sialk près de Kashan 1933, 1934, 1937* (Geuthner, Paris, 1938, 1939).

11. B. Hrouda, *Istanbuler Forschungen* (Mann, Berlin, 1957), vol. 19.

12. F. R. Matson, in E. F. Schmidt, *Persepolis* (Univ. of Chicago Press, Chicago, 1957), vol. 2, pp. 134–35.

13. Samples of this material are being analyzed by Dr. F. R. Matson at Pennsylvania State University.

14. R. Ghirshman, *Mémoires de la Mission Archéologique en Iran, No. 36* (1954).

15. G. C. Cameron, *History of Early Iran* (Univ. of Chicago Press, Chicago, 1936).

16. These dates are calculated (using the 5730 year half-life adopted by the Carbon-14 Conference in July 1962) from the University of Pennsylvania Radiocarbon Dates VI, R. Stuckenrath, Jr., *Radiocarbon*, 1963.

17. R. H. Brill, *Archaeology*, **14**, 18 (1961).

18. E. Porada, *Expedition*, **1**, 19 (1959).

19. The shell identifications were made by Dr. T. Abbott of the Academy of Natural Sciences, Philadelphia. Other shells came from the Red Sea (*Engina mendicaria* Linné, *Columbella fulgurans* Lamarck), Red Sea or Indian Ocean (*Charonia tritonia* Linné, *Conus* sp., *Olivia* sp.), Red Sea or Persian Gulf (*Murex virgineus* Röding), Persian Gulf (*Strombus decorus* subsp. *persicus* Swainson), and Mediterranean (*Conus mediterraneus* Brugineve, *Nassarius gibbosulus* Linné). The last-named shell is also found in the Black Sea.

20. G. Le Strange, *The Lands of the Eastern Caliphate* (Cambridge Univ. Press, Cambridge, 1905), p. 57.

21. R. J. Forbes, *Studies in Ancient Technology* (Brill, London, 1950), vol. 4, pp. 102–3.

22. F. Thureau-Dangin, *op. cit.*

23. T. Burton Brown, *op. cit.*

24. G. Morant, *Biometrika*, **30**, 130 (1938); H. Vallois, in R. Ghirshman, *Fouilles de Sialk près de Kashan 1933, 1934, 1937* (Guethner, Paris, 1939), vol. 2, p. 178.

25. H. Helbaek, personal communication.

26. R. H. Dyson, Jr., *op. cit.*

27. Anon., *Archaeology*, **12**, 171 (1959).

28. E. Porada, *op. cit.*

29. E. Parkinson, personal communication.

30. Wood samples from the buildings were identified through the courtesy of B. Francis Kukachka, acting chief of the Forest Products Laboratory, U.S. Department of Agriculture, Madison, Wis.

31. F. Thureau-Dangin, *op. cit.*, p. 53.

32. W. Andrae, *Colored Ceramics from Assur* (Kegan Paul, Trench, Trubner, London, 1925), fig. 41.

33. R. J. Forbes, in *History of Technology* (Oxford Univ. Press, London, 1954), vol. 1, p. 588.

34. R. J. Forbes, *Metallurgy in Antiquity* (Brill, London, 1950), p. 263.

35. F. Hançar, *Eurasia Septentrinalis Antiqua*, **9**, 62 (1934), Fig. 17; C. F. A. Schaeffer, *Stratigraphie Comparée et Chronologie de l'Asie Occidentale* (Oxford Univ. Press, London, 1948). Recently a number of these maceheads have been excavated by E. Negahban near Rudbar in northern Iran for the Archaeological Service of Iran.

36. E. Parkinson, personal communication.

37. The analysis of the dagger was made by Lucius Pitkin, Inc., through the courtesy of W. O. D. Pierce.

38. A. U. Pope, Ed., *A Survey of Persian Art* (Oxford Univ. Press, London, 1938), vol. 1, p. 278.

39. R. J. Forbes, in *History of Technology* (Oxford Univ. Press, London, 1954), vol. 1, fig. 57.

40. J. E. Burke of the Research Laboratory of the General Electric Company, Schenectady, N.Y., has agreed to help in some of the work that involves bronze.

Fig. 16–1. Map of northwestern Iran, showing Hasanlu and some related sites.

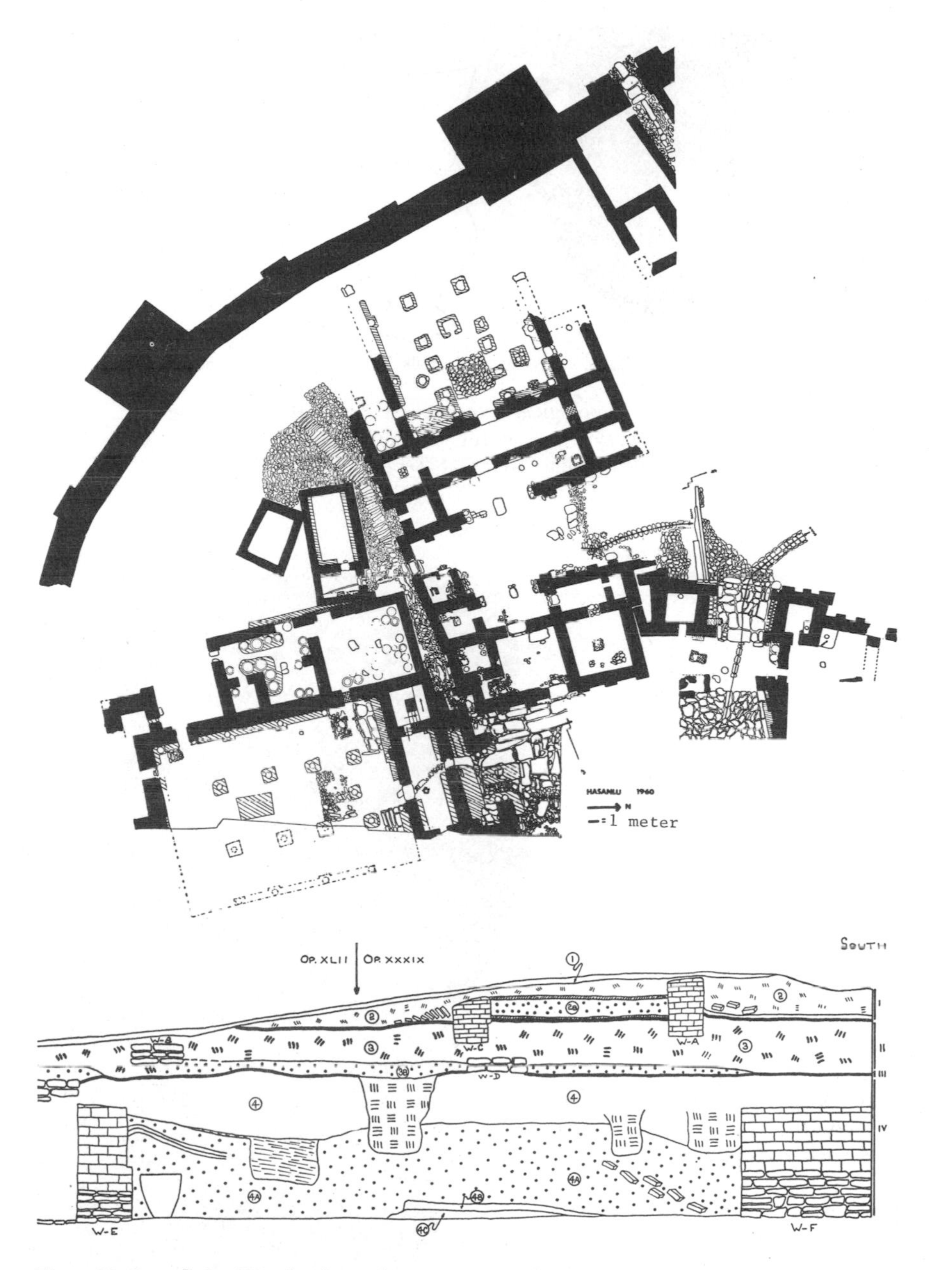

Figs. 16–2 and 3. (Top) Plan of excavations of period IV (9th century B.C.) of the Hasanlu Citadel at the end of the 1960 season. (Bottom) Section through periods I to IV on the Hasanlu Citadel, areas XLII and XXXIX. I: 1, turf; 2, collapse of walls W-A and W-C; 2B, trash. II: 3, collapse of wall W-B. III: 3B, ashy occupation, wall W-D. IV: 4, erosion deposits from collapsed Burned Building II; 4A, burned bricky collapse of Burned Building II, walls W-E and W-F. Width of section, 14.60 meters.

Fig. 16–4. Typological comparison of pottery, linking a burnished-gray-ware vessel of the Button-base phase at Hasanlu with similar items at Tepe Sialk and Tepe Giyan.

Fig. 16–5. "Egyptian blue" vase fragment. A stylistic link with the Assyrians is provided by motifs on decorated objects such as this.

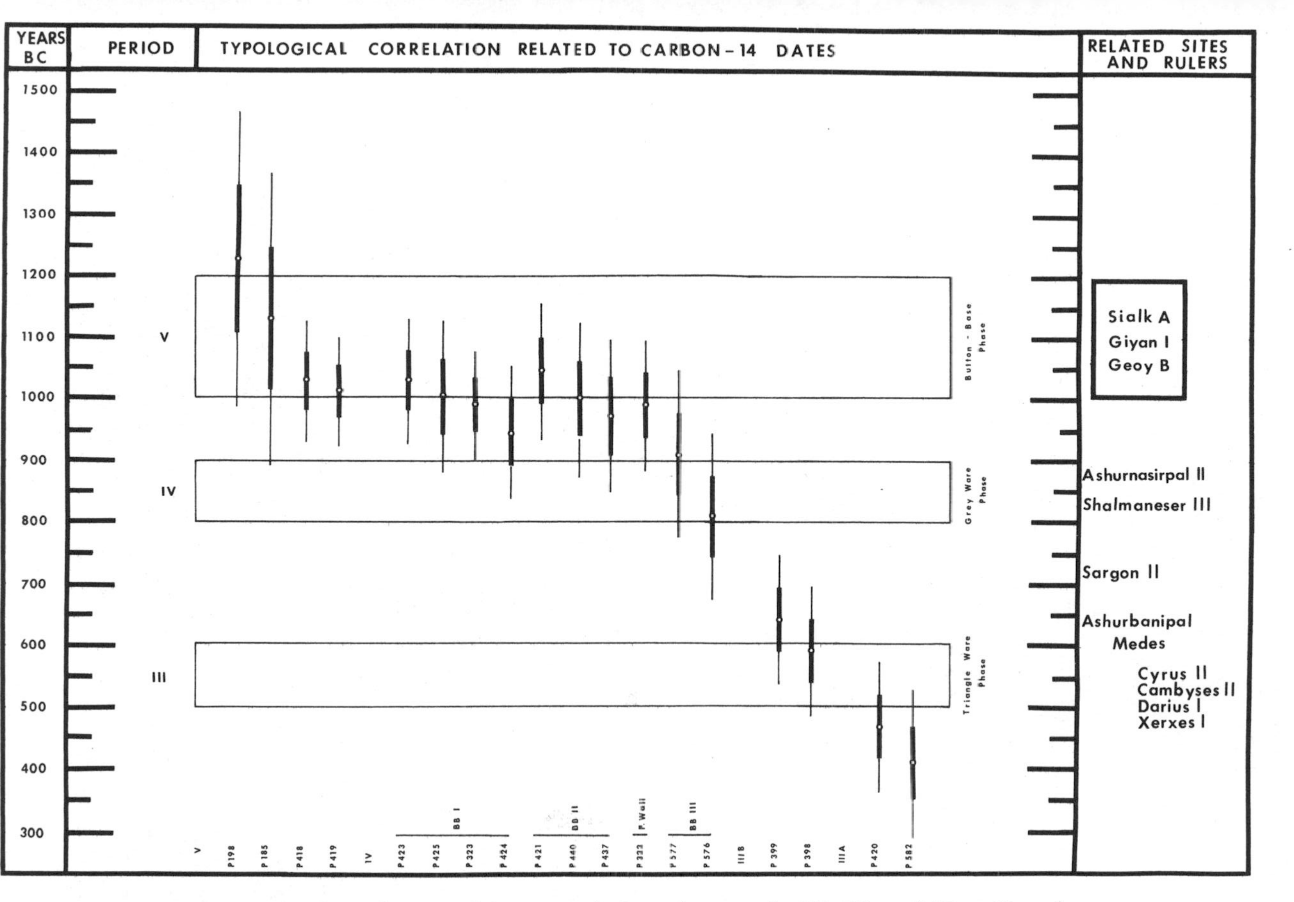

Fig. 16–6. Chart showing the estimated dates for periods III, IV, and V at Hasanlu, based on typological comparisons as related to a series of radiocarbon dates run by the University of Pennsylvania's carbon-14 laboratory.

Fig. 16–7. One leg of a bronze tetrapod stand after being cleaned electrolytically at the University of Pennsylvania Museum.

Fig. 16–8. View of the excavated area at Hasanlu, looking south. The original surface level is marked by the top of the pinnacle in the center. The level of period IV corresponds to the paved area, steps, and courtyard in center background, which forms the center of Burned Building I. A covered stone drain is in center foreground.

Chapter 17

Agriculture and Urban Life in Early Southwestern Iran

Robert M. Adams

Modern agricultural development programs promise to transform the levels of economic well-being of much of southwestern Asia. Among the many integrated steps comprising such programs are the application of chemical fertilizers, shifts to more intensive and better-balanced systems of crop rotation, comprehensive drainage schemes, and above all the acquisition of storable water surpluses behind high dams for greatly extended summer irrigation. Where these technical benefits are made widely available through accompanying social and economic reforms, the means clearly are now at hand by which the prevailing vicious circle of rural underemployment and poverty can be broken.

The frequent dependence of these programs on foreign technical direction, and the unprecedented investments of scarce capital they require of the nations undertaking them, combine to underscore the break that the new programs represent with the recent past. Yet in another sense they focus attention on the past as well. How do the current plans and promises compare, we may ask, with earlier achievements in the same area, whether under the hegemony of ancient Ur, or the Persian Empire of Cyrus and Darius, or the "golden age" of the Abbasid Caliphate? Fortunately, a minor by-product of the present widespread concern over development is the opportunity it sometimes affords to study these questions.

The Khuzestan Region

This has been the case in Khuzestan (Fig. 17–1). Geologically, this area represents only an extension of the great Mesopotamian alluvial plain into southwestern Iran, but its inhabitants have often participated closely in—and sometimes have briefly exercised a determinative influence upon—the political history of the alluvial zone as a whole. Moreover, the initial appearance here of agricultural villages and, several millennia later, of an urban civilization centered in the ancient Elamite capital of Susa, proceeded step by step with similar developments in Mesopotamia proper. Nevertheless, new resources of data such as soil surveys, hydrographic reports, and aerial photographs call attention to significant environmental features in Khuzestan that are not found elsewhere in lower Mesopotamia. This suggests in turn that gross descriptive categories like "semiarid steppeland" and "dependence on large-scale irrigation agriculture" may be as inadequate for a deeper historical understanding as they are for the contemporary planner. In brief, the ambitious development program currently under way in Khuzestan thus has stimulated and made possible a reappraisal of historic patterns of human subsistence and settlement within that region. By combining the results of archaeological surface reconnaissance with historical and documentary records, we can briefly and tentatively chart the changing conditions of human occupance in the area.[1]

The Mesopotamian Plain was earlier regarded as a trough slowly filling with alluvial soil carried down from the mountains to the north and east; it was assumed that there had been a regular withdrawal of the Persian Gulf before the advancing deltas of the major rivers. According to a more recent view that is supported by much convincing evidence of local geological uplift and subsidence, the entire basin is in fact a complex and unstable geosyncline which probably has tended to settle about as rapidly as it has filled. On the other hand, the testimony in the itineraries of Assyrian and Greek travelers (particularly that of Nearchus, admiral of Alexander's fleet on its return from India) continues to support the assertion that marshy but navigable lagoons, tidewater mudflats, and perhaps even open sea, extended well to the north of their present limits as late as the first millennium B.C.[2] Reports of the topography of the time are somewhat confused and contradictory, but the mouth of the Karun River, for example, apparently lay not much more than twenty-seven kilometers below what must have been a precursor of the modern town of Ahwaz, on a large inland lake that was in turn connected with the sea. Moreover, archaeological explorations of the lower plain, although still unsystematic and very limited in scope, thus far at least have failed to disclose any evidence of widespread occupation there antedating the Christian Era. Only in Sassanian or Islamic times, as will be shown presently, did these lower plains become a major focus of settlement and agricultural activity—and probably then as a consequence not merely of an advancing shoreline but of broad administrative and social changes affecting the methods of land use and the incentives to its exploitation.

Above the old shoreline there is a much longer and fuller record upon which this discussion will necessarily concentrate. The last anticlinical outlying fold of the Zagros Mountains crosses the Khuzestan Plain from the northwest just above Ahwaz, but except at the strategic crossing and rapids on the Karun River at Ahwaz itself there is nothing to attest a significant occupation prior to Alexander's conquests for perhaps fifty kilometers farther north. In this intermediate zone the presence of the sea hardly can be adduced as an explanation for the paucity of human settlement after Pleistocene times at the latest. However, modern conditions of widespread salinity, poor drainage, and extensive dune formation may account for the relative neglect of this area for almost 3,000 years after Susa had crossed the threshold of urbanism not far to the north. Before the arrival of the Greeks, and probably for several centuries afterward as well, it would appear that land had not been a directly limiting factor upon population but instead was available in sufficient quantities to be utilized selectively where conditions were optimal. The exceptional grain yields of a hundred-fold and more which Strabo reports for Susa may have their explanation in this abundance of unused land, for he adds that furrows in that district were widely spaced to maximize sprouting.

The band of upper plains in which Susa lies presents a quite different picture than either the lower or intermediate zones. Increased surface gradients and widespread underlying gravel deposits provide sufficient natural drainage over most of the area to minimize the problems of salinization and waterlogging that usually attend irrigation agriculture. These conditions also must have been advantageous during the early stages in the development of irrigation, since they permit an adequate level of flow during the winter growing season with relatively short and easily maintained canals. In addition, the pebbly soils in the upper portion of this region receive natural subirrigation from underground springs, while rainfall on the steep slopes that bound the plain is carried out onto it for some distance by numerous winter and spring freshets. Under these conditions rich natural pasturelands tend to form (when not destroyed by overgrazing), containing a wide variety of leguminous herbs and grasses. According to the medieval Arab geographers these meadows were covered with wild narcissus; moreover, they provided the chief winter grazing lands for large groups of nomadic Bakhtiari tribesmen until well into the present century. To be sure, the attractive conditions that prevailed from October through April were balanced by the oppressive summer heat and desiccation for which Khuzestan is notorious; but cooler hill slopes and upland valleys with good forage that could no longer be found on the parched plains have always been available during the summers within easy marching distance into the mountains. It is, in fact, erroneous to consider the upper plains as a zone of occupance distinct from the surrounding uplands. Both together constitute a single natural ecosystem, whose seasonal alternation of resources provides as strong an inducement to migratory stockbreeding as to intensive, settled agriculture.

More important even than all these factors is the volume of precipitation that the upper plains receive. Under Mesopotamian conditions, the 200-millimeter isohyet is generally regarded as the absolute lower limit within which dry farming is possible. However, there is a substantial "zone of uncertainty" above this limit within which permanent settlement will be avoided even though sporadic catch-crop cultivation may be practicable. In Khuzestan the essential point is that permanent settlement based on dry farming is roughly coterminous with the upper plains, extending to the north and northeast of a line slightly below what is shown in Figure 17–2 as the 300-millimeter isohyet. In an area now under development above this line to the north of the Shaur River (see Fig. 17–3) a recent agricultural appraisal showed average net wheat yields of 410 kilograms per hectare (about six bushels per acre) from dry farming as compared with 615 kilograms per hectare on irrigated land. In short, irrigation on the upper plains around Susa is a valuable adjunct in the cultivation of the basic cereal crops during the traditional winter growing season, but it is by no means an indispensable condition for the practice of agriculture.[3]

Early Village Settlement

The earliest major phase of settlement on the upper plains of Khuzestan is illustrated in Fig. 17–3. This map records all of the known sites where prehistoric painted pottery has been obtained in surface collections; thus it covers the long span from perhaps as early as 5500 or 6000 B.C. to about 3500 B.C. However, all but a few of the sites shown (116 out of 130) were occupied during the latest part of this span, so that the map can be regarded as a representation of the distribution of settlements particularly during the relatively brief "Susa A" period soon after the beginning of the fourth millennium B.C. The hallmarks of this period, incidentally, are the highly stylized and beautifully decorated pottery beakers and other vessels that, since their excavation from basal levels at Susa by an impressive series of French expeditions beginning in the nineteenth century, have graced the pages of many histories of art.[4]

The most obvious feature of the pattern shown in Fig. 17–3 is its density—a grid of villages fully comparable to that of the present day in spacing, and in some cases extending into areas no longer permanently settled. All of these sites are small, most of them covering one or two hectares or even less, but the considerable heights of the mounds that mark their ruins suggest that they were occupied relatively continuously and for long periods.

It must be stressed that the vigorous growth of settlement evident here at an early time-level almost certainly does not stem from any priority of the region in the basic practices of cultivation and stockbreeding. The potential domesticates are thought to have occurred naturally only at somewhat higher elevations, and in any case these animals must have been at home all along the lower slopes of the mountains overlooking the Mesopotamian Plain (wherever favorable soils and microclimates created suitable niches) and not merely in the Persian uplands surrounding Khuzestan. Instead, the high early density of settlement in this region perhaps can be traced to the exceptionally favorable circumstances the region would have offered for the transition from dry farming to irrigation agriculture, a transition that began only after the initial food-producing revolution had been consummated elsewhere. Chief among these locally favorable factors, of course, was the adequacy of precipitation for dry farming while rudimentary irrigation systems were being developed, together with the suitable slopes and soils on the upper plain and the numerous small, easily diverted watercourses.

Some further light is shed on the time of introduction of irrigation techniques by comparing the patterns of settlement for the different component periods into which the long prehistoric span can be divided. For the earliest of these, the Susiana *a* period, only thirty-four sites are known. It is noteworthy that these uniformly small mounds occur almost exclusively in the northern part of the upper plains, where also the highest annual rainfall is encountered. Moreover, a number of them, like those west of the Karkheh River and on the rolling downlands near the modern town of Dizful, are so located that irrigation would not have been possible in their vicinities without the great irrigation works that are known only to have been constructed much later. From these observations it would appear that, in the earliest known phase of settled village life in Khuzestan, agriculture depended mainly or even exclusively on rainfall.

In the immediately following periods of the Susiana prehistoric sequence, probably equivalent to the Halaf and early Ubaid horizons in Mesopotamia and falling within several centuries after 5000 B.C., the number of known village sites jumps to 102, and the sites assume a distribution which then continued with little change for at least a millennium. This pattern differs from its predecessor not only in the sharply increased number of sites but also in the geographical extension and concentration of the sites much farther to the south. On the other hand, the new lower limit of settlement still conforms roughly to that existing today for villages partly dependent on dry farming. From this it seems reasonable to conclude that, while some irrigation practices had been introduced and had led to a greatly expanded population, they were still quite localized and probably were regarded merely as an adjunct to farming practices oriented in the main toward rainfall.

Also worth noting is the fact that these later prehistoric sites were not uniformly spaced but in many cases tended to form clusters, some of them being grouped along the margins of shallow fossil valleys that are still traceable at intervals on the alluvial land surface. Considering these valleys together with the general distribution of sites, it is apparent that fundamentally different river regimes must have obtained for at least the Karkheh and Diz rivers along this part of their courses. Instead of the single, rapidly degrading channels that they occupy today, these rivers seem to have divided themselves into numerous bifurcating and rejoining channels of a presumably aggrading character. While no clear over-all pattern of the more important contemporary watercourses can be distinguished from the position of settlements—in part because alternative sources of domestic water were so readily available that villages were not closely bound to the major stream levees—Fig. 17–3 suggests that during this period a large part of the drainage from the present Karkheh watershed may have run southeastward along what is now the bed of the Diz River. In addition there are suggestions that some of the minor outwash channels reaching the plain between the Diz and Karun rivers may have carried a larger, or at least less strictly seasonal, flow than they do at present. Quite possibly both tectonic activity and the effects of human settlement (principally wood cutting and overgrazing) have contributed to the subsequent far-reaching changes.

Virtually nothing is directly known at present of the subsistence practices of the established early village range in Khuzestan, although an investigation of the transition from food gathering to food production has recently been undertaken in the higher valleys to the north.[5] Widely distributed flint sickle-blades, hard-baked clay sickles, and stone hoes attest the existence of a specialized tool kit associated with cultivation by the late fifth millennium. At the same time, chipped flint arrowheads, carved gazelle bones, and stylized but plausible representations of bowmen and of the pursuit of wild game with dogs provide fragmentary suggestions of the continued importance of the hunt. By the time of the proto-Elamite tablets, around the end of the fourth millennium, signs representing orchards as well as fields can be distinguished, and the existence of a plow also can be demonstrated. The latter implies the domestication of bovids or equids, and contemporary representations (including animals with mounted riders and others pulling chariots or carts) and administrative records of both are found. However, it is not certain how much earlier than the introduction of the tablets the plow and these domesticates can be assumed to have made their appearance in Elam. The specific identification of animals from their highly stylized representation on the pottery of the Susa A period is difficult, but at least it is clear that fish, birds, and members of the *Capra* and *Gazella* genera are very numerous, while bovids and equids are decidedly rare. Conceding the unsatisfactory state of the evidence, it might be tentatively concluded that hoe cultivation constituted the dominant form of food production at that time, and that livestock was largely limited to sheep and goats.

Growth of Towns and City-States

The pattern of subsistence and settlement described above reached its apogee during the Susa A period, roughly equivalent to the terminal Ubaid and early Uruk horizons in Mesopotamia and falling within the first half-millennium after 4000 B.C. By this time (if not somewhat earlier) a few sites began to stand out as larger centers among the numerous small villages. The contemporary remains at Susa itself are too deeply buried for their extent to be plotted, but a small number of other mounds then covered four or five hectares and so perhaps can be classified as small towns. This process of differentiation intensified further during the Susa B and C periods that followed. Exclusive of Susa, individual centers are known which extended over as much as twenty hectares, and we may infer from its thousands of proto-Elamite account tablets that Susa was at least as large as any of the contemporary unexcavated towns around it.

On the other hand, the total number of settlements declined by almost two-thirds, for only thirty-nine are known in the surveyed area which were occupied during the half-millennium before 3000 B.C. Without more extensive excavations it will be impossible to determine all of the factors involved in this decisive shift. Some instability of population is suggested by the sharp cultural break that is especially evident in ceramics, but its character remains elusive. Moreover, there is contrary evidence for considerable continuity of settlement in the fact that less than a sixth of these sites were newly founded during the Susa B and C periods. In part at least, the newly emerging pattern must have consisted of the drawing together of the population into larger, more defensible political units, some of which—on the analogy of the better-known Mesopotamian sequence—began at this time to attain truly urban status and to be enclosed by massive walls. At the same time, large areas that formerly had been settled and cultivated were well beyond the radius of easy communication from any of the reduced number of towns that remained and hence must have reverted to pastureland or other nonintensive use. Thus, to continue with the reconstruction advanced earlier, we may visualize the growth of large towns and urban centers like Susa as proceeding hand-in-hand with changes in the exploitation of the surrounding hinterlands. Small-scale irrigation, the planting of intensively cultivated gardens and orchards, and plow cultivation were all adopted increasingly within limited enclaves around the towns. Elsewhere, however, considerable tracts were abandoned to shifting cultivators or nomadic herdsmen, who would have left few material remains for the archaeologist and no written records for the historian.

The trends set in motion before 3000 B.C. seem to have continued throughout the third and into the second millennium. To judge from the strength occasionally displayed by rulers of Susa, like Puzur-Inshushinak who campaigned successfully as far afield as Opis (probably not far below Baghdad) and Kirkuk in the late third millennium, Susa must have been a substantial and prosperous city for at least part of this time.

Still, even a calculation based on the larger, later area of its ruins and the unusually high assumed density of 400 persons per hectare within its walls suggests that ancient Susa did not exceed—and quite possibly never approached—a maximum of 40,000 inhabitants. This population, it should be pointed out, is about at the level which could be adequately provided for with irrigation of no more than the area between the Karkheh and Diz rivers, within less than a day's walk from the capital to the furthest of its supporting fields. As for other towns in the region, their number and average size declined somewhat further during the Susa D or protohistoric period and only began to increase again very slowly afterward. As late as 2000 B.C., in spite of the greatly intensified administrative and economic relations with southern Mesopotamia that were maintained under the hegemony of the Third Dynasty of Ur, not more than two towns seem to have existed in the surveyed area which covered as much as ten or fifteen hectares. In short, well into the second millennium B.C. Susa was the only known settlement on the upper Khuzestan Plain west of the Karun River that might be called a city or even a large town, while substantial parts of that plain were not regularly settled at all.

Thus it would appear that Elamite military prowess did not derive from a large, densely settled peasantry occupying irrigated lowlands in what is often loosely considered the heart of Elam. Instead, the enclave around Susa must have been merely one component in a more heterogeneous and loosely structured grouping of forces. Other towns and settled districts that were strung out along somewhat higher valleys like those of the Middle Karkheh, Saimarreh and Kashgan rivers to the northwest of the Khuzestan Plain, and perhaps even seminomadic tribal groups with no major centers, often must have been of roughly equivalent political importance. A reflection of this distinctly hybrid geographical character probably is to be found in the successive roles played by towns like Awan, Simash, Anzan, and Madaktu, all probably to be sought in the higher valleys and each with at least comparable prestige and importance to that of Susa in external political relations and royal titulary during the whole of Elam's pre-Achaemenian history.

Second millennium texts from Susa attest the presence in considerable quantities of a number of domesticated plants and animals. While the specialized transactions recorded in the texts may be unrepresentative of the broader subsistence picture, wheat and barley obviously were the most important crops. In addition, dates, chick-peas, and lentils, as well as considerable quantities of sesame, are mentioned. Most of the sesame probably was converted to oil, while some of the barley was used for beer. The most numerous category of livestock seems to have been sheep, and both sheep and cattle in some cases were specified to have been fed on barley. Goats and donkeys also were present.

The details of the later political history of Elam are not important here.[6] A dynasty of "kings of Anzan and Susa," vigorous contemporaries of Hammurabi and his successors in the First Dynasty of Babylon, seems to have been followed by a long and uncertain interregnum and then by a powerful but short-lived "empire." By the time of the latter, the total number of occupied settlements on the upper plains between the Karkheh and Karun rivers had slowly increased to forty-eight, less than half of what it had been in prehistoric times but more than twice its total at any time during the third millennium (Fig. 17–4). In this process the disparity was somewhat alleviated between urban Susa and a surrounding region with no more than widely scattered small towns and villages. Eight other towns each now occupied more than ten hectares and one of them, Chogha Zanbil or the ancient Dur Untashi, covered about 1 square kilometer and thus stands comparison as a city with Susa itself.[7] Relative to the available land, however, these changes did not basically alter Khuzestan's earlier aspect as a still lightly settled region in which towns and their regularly cultivated hinterlands formed only widely separated enclaves.

The Assyrian Invaders

Elamite power culminated during the reign of Shilhak-Inshushinak (about 1165–1151 B.C.). Afterward Elam and Babylon were increasingly drawn together to resist the growing strength of Assyria. In spite of this alliance, the Assyrian annals describe campaigns which gradually subdued the desert tribes west of Elam, ravaged its border districts, and even descended upon its seacoast in assaults launched from across the Persian Gulf. As the consolidation of Assyrian rule continued to be resisted in Babylonia, these attacks increased in what has been called their "calculated frightfulness." Ultimately, under Assurbanipal (668–626 B.C.), the struggle was carried not merely into the intervening mountain valleys to the northwest but directly into Khuzestan itself, and most or all of the royal strongholds there were stormed and sacked. The Assyrian king boasts of the slaughter of his Elamite enemies and the burning of their cities, of having carried off population and livestock "more numerous than grasshoppers," and even of the scattering of salt over the devastated province.

For once, the Assyrian version does not seem to have been greatly exaggerated. Virtually every town of the period that was visited during the archaeological survey was found to have been abandoned at a time roughly corresponding with these campaigns, and to have remained unoccupied for a long time afterward. More than a century elapsed before Darius undertook to restore Susa as an Achaemenian capital (about 521 B.C.), and in spite of the ambitiousness of his constructions there the level of population in the surrounding region apparently failed to approach what it had been at the outset of the Elamite-Assyrian rivalry. To Herodotus, two generations later, the plains around Susa and even the city itself were merely part of Cissian territory. The Cissians were described by the Greeks as rude and warlike mountaineers whom the Persian kings had placated with annual tribute in order to prevent infestation of the plains with brigandage, and their gradual infiltration and resettlement of the plain is further evidence that much arable land had been left empty in the wake of Assurbanipal's armies. Under these conditions it would have been unlikely for the Achaemenians to initiate an extensive program of agricultural development, in spite of their concern with fostering commerce and their renown as builders.[8] And in fact only along the eastern margin of the surveyed region, on the right bank of the Karun River below Shustar, is there evidence suggesting that a group of small agricultural villages may have been linked by a new canal dug during the time of Darius or one of his successors.

Emergence of a New Pattern

Although there are numerous accounts of the conquests of Alexander, the confrontation of Greek and Persian that continued in Khuzestan under his Seleucid heirs (about 311–140 B.C.) is virtually unreported. Accordingly, the enduring effects of those conquests upon patterns of subsistence and settlement can only be discerned in very general terms.

The absence of detailed, local sources is particularly unfortunate in that many decisive changes must have had their origins in the intensive cultural interchange that went on during this period, although these changes do not come clearly into focus until five hundred or more years later.

Among the most crucial effects of Greek influence was a renewed emphasis on city building. This entailed in some cases the actual foundation of important urban centers, like one of the numerous Alexandrias which continued on into Parthian times as the kingdom of Charax. Situated on the lower Khuzestan Plain near the mouth of the Tigris, its location implies some effort at systematic settlement and cultivation of the marshes along the lower edge of the alluvium, an effort of which we are otherwise largely ignorant. Other Greek cities were strategically located so as to pacify mountain peoples like the Cissians, in turn providing the security which permitted a renaissance of settled life around cities like Susa on the plains as well. Another Macedonian practice was to implant garrisons of soldier-colonists in the major existing towns, surely stimulating the growth in Asia of juridical and institutional concepts upon which the independence and prosperity of the Greek *polis* had been based. Finally, mention might be made of the diffusion of particular cultural activities like grape cultivation, for Strabo says of Khuzestan that "the vine did not grow there until the Macedonians planted it." The cohabitation of Greeks and Persians must have had many other, less simple and overt, effects upon agricultural techniques, but they escaped the notice of the chroniclers and hence remain largely for speculation.

Sources on the Parthian period (about 140 B.C.–A.D. 226) are, if anything, more fragmentary and less informative than those on the preceding Seleucids.[9] The impressive showing of Parthian armies against the Romans in the west implies at least a periodically effective internal administration and a reasonably adequate level of economic well-being, but Khuzestan was too distant from the frontier for Roman accounts to furnish many details. From an inscription in Susa we learn that the Greek garrison there still retained its corporate identity, and that it recorded its gratitude to the governor of the province for initiating certain irrigation works. Possibly the installations referred to include several impressive networks of canals which can be shown from aerial photographs to have antedated the still greater networks constructed early in the Sassanian period. At any rate, both kinds of data make it clear that a considerable program of canal building was under way before the end of Parthian times. Similarly, the remains of Parthian towns located during the archaeological reconnaissance suggest a substantial increase in the extent and density of settlement, although their full area is often masked by the massive Sassanian ruins that almost always overlie them. Such general indications as these, however, are at best vague and unsatisfactory. Only new historical sources and extended, patient excavations will illuminate more fully the changing practices and conceptions of land use that prepared the way for the Sassanians.

As in other parts of the Mesopotamian plain, one of the most striking observations of recent archaeological reconnaissance has been the immense and variegated impact of the Sassanians (A.D. 226–637) upon the Khuzestan landscape.[10] In a way which seemingly had no parallel in earlier periods, vast efforts were devoted to comprehensive programs of irrigation extending over virtually the entire arable surface. This entailed bold and imaginative planning and administration, a whole series of technical innovations, and above all the investment of state funds on what must have been an unprecedented scale. The Sassanian effort differs from its modern counterpart in that it aimed to increase agricultural output (and thereby, of course, state revenues) primarily by extending the area of cultivation and only secondarily by introducing a more intensive agricultural regime and increasing labor productivity. Still, since this choice was dictated by the existing socioeconomic system and level of technology, it provides a distinction more apparent than real. For Iran at least, we are justified in regarding the Sassanian administrators as the spiritual ancestors of the modern teams of developers, and in hoping that the latter are as successful by contemporary standards as the Sassanians must have seemed in their own time.

Some elements of the Sassanian program have long been known. Great weirs were constructed of stone and brick across the Karkheh River at what is now Pa-i-Pol, the Diz River at Dizful, and the Karun River at Shustar and Ahwaz. Although the available flow varied widely and was markedly reduced in summer due to the absence of water storage, radiating canal networks from these strategic locations at least could provide more reliable winter irrigation than had existed heretofore. The remains of these dams are still identified locally as "Roman," and there seems little reason to doubt the statements of medieval Arab historians and geographers that a central role in their construction was played by seventy thousand Roman legionnaires who, together with the Emperor Valerian, had been captured by King Shapur I at Edessa. With positions reversed since the time of Macedonian conquests, soldiers from the west again played a vital part in Khuzestan's agricultural development.[11]

A fuller view of the Sassanian program is made possible by linking the study of aerial photographs with ground reconnaissance. Figure 17–5 illustrates the layout of the major branches in the Sassanian canal system, even though the system is now obscured by long-continued erosion, modification and re-use. Several aspects deserve brief mention. In the first place, the readiness of the Sassanian engineers to cut through ridges and other natural obstacles can be documented in many places. By brute force, as it were, they undertook to impose a unified system of canalization upon a broken topography that always before had been irrigated (where it was irrigated at all) in relatively small, unrelated segments. In one particularly illuminating case water diverted from the Diz River was conducted southeastward by canal near the upper limits of the plain to lands on the right bank of the Karun, which were too high to be irrigated directly from the latter. This bold reshaping of basic drainage patterns is fully consistent with the approach followed in the design of the system as a whole.

Another feature of the general pattern shown in Fig. 17–5 is the attention that was obviously devoted to providing irrigation supplies not only for large, topographically suitable, highly productive areas but also for small and marginal tracts which hardly seem to have warranted the investment that was made in them. This determinedly full utilization of land does not suggest a grandiose project whose potentialities were never fully realized but, on the contrary, a system that operated so successfully that it was ultimately extended to the fullest possible limits.

A final point, related to the previous ones, is that at least the major canals in this system apparently were designed and executed under a series of comprehensive plans. This is evident both from the regularly branching patterns of minor distributary canals and from the directness of most of the larger channels. By contrast, the small-scale private irrigation networks that have been introduced in the Near East in recent years are characterized by extremely broken and erratic patterns of canalization in which the absence of central planning is immediately apparent. There were, of course, periodic changes in the scope and objectives of the Sassanian system, so that Fig. 17–5 presents a composite rather than an actual layout. It has been possible to distinguish some of the major phases within this composite, beginning with the initial construction of the weirs and the first of their offtakes early in the Sassanian period. Not surprisingly, the main trunk canal leading southeast from the Karkheh weir and the vented tunnel proceeding southwest from the Karun at Gutwand—both major undertakings with a large water-carrying capacity—appear to date from the latter part of the period. Probably they are to be attributed to the reign of Chosroes I (A.D. 531–579), whose military successes and administrative reforms placed him in control of unprecedented state revenues for costly enterprises of this kind.[12]

Apart from the well-planned and well-executed construction of the weirs themselves, the principal innovation that appears in this great system is the extensive use of tunnels with periodic vent holes, not only as subsurface conduits through ridges and other topographic obstacles to ordinary canal construction but also as collectors for ground water. Knowledge of the technique of constructing these tunnels may well go back to Achaemenian times or earlier (although probably not earlier than the introduction of cheap iron tools), but their first extensive application in Khuzestan came only under the Sassanians. Another interesting innovation, suggestive of the technical ingenuity that originally must have been applied at many points, was the construction of an inverted siphon (a ruined example of which survives below the modern town of Gutwand) to carry a large canal across a seasonal watercourse.

While the introduction of an extensive network of irrigation canals is the most tangible surviving evidence of the development of the area during the Sassanian period, the full range of measures that were taken was far more extensive and complex. A greatly increased stress upon commercial crops and handicraft industries, centering on the manufacture of fine silks, satins, brocades, and cotton and woolen textiles, accompanied the resettlement of prisoners here after successful western campaigns. Some of the rich variety of orchard products for which Khuzestan was praised by early Arab writers, including plums, pears, melons, pomegranates, olives, and citrus fruits, must have been introduced at the same time. Date palms, already native, were spread so widely that later it was claimed there was no place in Khuzestan without them. Probably from the east came sugar cane, implying a new emphasis on year-round irrigation within the limits of the available summer supplies of water. By Arab times at least, Khuzestan's annual tax payments included twenty metric tons of refined sugar and most or all of the sugar that was traded throughout the eastern Caliphate (at a normal market price of about $3.30 per kilogram, sufficient to maintain a couple in modest circumstances for a month) is said to have come from this province; much additional cane was described as unsuitable for refining, so that the raw stalks were consumed locally. Rice, another summer crop, had already been noted in Khuzestan by companions of Alexander, but only later became an important item in the diet. By the tenth century rice-flour is reported so much a staple in Ahwaz that people sickened and died if forced to eat bread made of wheat-flour instead.

Taken together, the newly introduced crops signify more than simply an increase in the *number* of cultivated plants known to Khuzestan's agriculturalists. Most of them required new and highly specialized cultivation procedures (elaborate Arab accounts of those for sugar fortunately have come down to us), provisions for more intensive irrigation than previously had been necessary, and in some cases the investment in semi-industrial processing equipment before the natural harvest could be utilized. Moreover, as a group the new crops imply the growth of a market economy at the expense of the subsistence economy that had prevailed previously. Thus their appearance suggests a qualitative change in the orientation, structure, and technological complexity of agriculture as a whole. Still, limitations on the available water supply must have restricted cultivation of at least four-fifths of the arable land on the upper plains to the winter production of wheat and barley under the aboriginal system of alternate years in fallow.

Since both sugar and rice became important crops in the districts served by the new dams, part of the original purpose of the weirs may have been the encouragement of summer cultivation. On the other hand, numerous later references to lifting devices along the major streams, particularly undershot waterwheels, may indicate that summer cultivation was largely independent of the great canal networks radiating from the weirs. Probably there was considerable variation in how the problem of summer cultivation was met. Above Jundi Shapur vented tunnels were dug as an alternative source of water for the main canals after the dam on the Diz River had been built. Their installation may have been related either to an increasing need for summer water, or merely to the need for assuring winter irrigation supplies during periods when the weir was inoperative due to washouts.

There is both archaeological and documentary evidence that the agricultural and commercial development of Khuzestan was accompanied by the multiplication of urban settlements on an unprecedented scale. Royal concern for the area is shown by the construction there of two new capital cities, Jundi Shapur by Shapur I (A.D. 241–271) and Iran-shahr-Shapur by Shapur II (A.D. 306–380), each with a planned rectangular layout comprising several square kilometers within impressive outer walls. In addition, the ancient mound at Susa was extensively rebuilt, and Shustar became an important fortified town. Between these walled major centers, as Fig. 17–5 records, a great number of sprawling (presumably undefended) towns and villages sprang up—not a few of them approaching the maximum size that even Susa had achieved in earlier antiquity. The total population of at least the upper plains during Sassanian times thus appears to have exceeded by several times what it had ever been previously, and in fact to have reached a level that has not been equalled since.

Although details are lacking, something similar probably happened on the lower plains as well. At any rate, Ahwaz (originally Hurmuz-Shahr) was founded by Ardashir I (A.D. 226–241), the first of the

Sassanian kings, and other towns still farther out on the lower plains are known to have been occupied at the time of the initial Arab conquest. Given the poorer soils and the virtual absence of leguminous weeds on the alluvium below Ahwaz, the extension of agriculture into this area required greater innovations than merely the introduction of irrigation canals. Drainage systems comparable to those so important in development programs today apparently were never constructed, perhaps suggesting that irrigation water in the canals was usually inadequate and had to be sparingly applied. But at least by Arab times there are references to the importation by barge of plant-ash and night soil for fertilizer from the great metropolitan center of Basra in lower Iraq.

This chapter is concerned primarily with the material conditions of life, but it is worth noting in passing that the benefits of Sassanian policy extended into other realms as well. Shapur II is credited with having founded a university in Jundi Shapur which became widely noted in the ancient world as a center of astronomical, theological, and medical learning. A marked religious tolerance generally prevailed as Sassanid policy, and to the large and thriving Jewish communities dating from the neo-Babylonian period were added Nestorian congregations fleeing Byzantine persecution. To the emerging syncretistic traditions of science and theology, thus there was added the polyglot confusion of Pahlavi, Greek, and Syriac in the marketplace, symbols of a cosmopolitanism for which the early medieval world provided few equals. This was a milieu in which the products of Classical learning were valued and preserved, and from which they were handed on in time to the Arabs (and ultimately to the West) with the rise of the Abbasid Caliphate. As the vigorous commercial orientation of many Khuzestan towns implies, it was also a milieu that stimulated practical enquiry. The process of refining sugar is said to have been worked out at Jundi Shapur, only a short distance from the site where, after a lapse of many centuries, a new sugar plantation and refinery were put in operation by the Iranian Government. A Great Pharmacopoeia, probably the first of its kind ever to be issued officially, also was a product of the Jundi Shapur hospital and medical school. Although appearing only in the ninth century, it surely embodies a tradition of scholarship there that was deeply rooted in the Sassanian period.[13]

The Medieval Breakup

The initial effects of conquest by invading Arab armies in A.D. 639 were relatively mild. Resistance was mainly confined to the important fortified towns, and while the defenders in some cases held out and were put to the sword, they more often capitulated and resumed a life little different from what it had been earlier. Nevertheless, documentary sources and the results of archaeological survey converge to indicate that the agricultural economy failed to return quickly to its previous levels, and in fact went into a discontinuous but cumulative decline that has only been reversed in the modern period.

Taxes submitted to the central government provide perhaps the simplest and most clear-cut index to this process, although it is a very imperfect index at best. To begin with, direct comparisons between the Sassanian and Islamic periods are obscured by changes in the breadth of application of land and poll taxes as a result of religious conversions.[14] A further difficulty is that even within the Islamic period collections probably decreased more rapidly than economic well-being, as the central government's means of coercion were attenuated by unsettled conditions. But since the Muslims took over the Sassanian tax system substantially as they found it, it does not seem unwarranted to assume that at least some degree of correspondence held in the long run between the volume of tax receipts, on the one hand, and the volume of agricultural produce and activity from which those revenues had been derived on the other.

With due allowance for their defects, the trends in state revenues are very striking. In the late Sassanian period, tax receipts in Khuzestan had reached 50 million dirhems, equivalent to $5 million or more at current price levels—and something on the order of 12 times more than the annual tribute exacted from approximately the same area by the Achaemenian kings a thousand years earlier. While they fluctuated widely afterward, receipts never again reached this figure, and within three centuries or so after the Arab conquest they had been reduced to less than 40 per cent of it. Four centuries later they were reported to have been only the equivalent of about 6 per cent of the Sassanian amount—even ignoring the effects of the debasement of the currency. In the mid-nineteenth century, before the consolidation of modern Iran had begun, collections still remained at approximately the same level.

The progressive economic decline suggested by these figures—and corroborated by other accounts and archaeological data—obviously is only a local manifestation of processes which were at work through much of the Islamic world. Moreover, declining commerce and agriculture were an integral part of a series of interdependent changes affecting the whole fabric of society, and they hardly can be understood without reference to this broader context within which they occurred. The destructive long-term consequences of the introduction of tax farming as a general practice, the increasingly corrupt and inefficient character of the Abbasid Caliphate after the ninth century A.D., the replacement of citizen-soldiers by bands of power-seeking mercenaries, the breakup of the former domain of the Caliph into unstable local polities, and finally the appearance of great conquering armies like the Mongols who swept over the whole area, all are important components of any full account of the events in a particular district.[15] But having admitted this broader historical context, here we can only describe some of its consequences in concrete, local terms.

New Patterns

From a comparison of Sassanian and early Abbasid settlement patterns on the upper plains (figs. 17–5 and 17–6), it is apparent that by the time of the latter a considerable retraction had taken place outside of the cities and their immediate environs. Substantial areas of formerly dense occupation now were more or less abandoned. And while a few new towns were founded (the most important of them being 'Askar Mukram, where Khuzestan's sugar crop was brought for refining), surface reconnaissance of the ruins suggests that more commonly there was a reduction in the occupied quarters of even the more populous centers. At Jundi Shapur (Fig. 17–7), for example, the zone of continuing Islamic occupance seems to have been confined to roughly the central third of the great walled rectangle that Shapur I originally laid out.

While large areas of good land on the upper plains were being abandoned, it is interesting to note that some new lands of marginal quality appear to have been irrigated for the first time. Within the surveyed area a good example is furnished by the Shu'aibiyah district immediately north of the confluence of the Diz and Karun rivers. This is a poorly drained and moderately saline tract that had been very sparsely settled previously and which even today receives only some speculative, tractor-based rainfall farming from very few permanent settlers. Yet Figure 17–6 shows that early in the Islamic period a fairly large trunk canal and numerous offtakes were placed in operation across this district, possibly even requiring the installation of some sort of diversion structure in the bed of the Diz River. What explanation can there be for the simultaneous taking-up of unproductive land on the one hand and the abandonment of larger areas of much better soil on the other?

Too little is known of the above example for the question to be answered on the basis of evidence from that district alone, but the same practice of large-scale development of marginal lands had a wider occurrence. Another example of the practice is reported from the Masrukan district, on the opposite bank of the Karun, where one of the early Arab generals is said to have turned his attention to the irrigation of lands which previously had been uncultivated waste. A different, more illuminating case is provided by the events leading up to a great slave rebellion centering in lower Khuzestan and the lower Tigris marshes, as they were set down in surprising fullness by the great contemporary historian, Tabari. The social content of the rising was a militant protest over the intolerable conditions of servitude on great latifundia or landed estates, where masses of slaves (15,000 are mentioned in one district) apparently were employed in the physical removal of the saline surface crust which had prevented cultivation over great areas of the lower plains. In the sequel, the Zanj (after Zanzibar, for most of the slaves were from East Africa) held out for fourteen years of frequently heavy struggle (A.D. 869–883), in the course of which many of the towns of Khuzestan were repeatedly and heavily damaged.[16]

To generalize from these examples, the development of marginal lands probably stemmed at least in part from the availability of unprecedented numbers of unfree laborers, who could be economically supervised only in great gangs (one of 500 men is reported). Since the initial Arab conquest had not immediately led to the wholesale expulsion of the indigenous population from the more productive land, this placed a premium on the development of large undivided tracts, if necessary even on poor soils where potential yields were relatively low and where salinization would become a problem after a few years. In addition, the rapidly shifting winds of court intrigue in Baghdad encouraged grandiose, speculative, usually short-lived undertakings; for example, a case is known in which a minister out of favor at the court spent ten million dirhems merely to obtain restitution of his former Khuzestan holdings. From the paucity of accompanying settlement and the absence of any substantial canal levees, it is clear that the example in Shu'aibiyah is one of those that ended quickly. So, in fact, did most of the agriculture on the lower plains. It was only the firmly rooted peasantry around towns like Shustar and Dizful in the north that continued to cultivate their lands through the turmoil of the later Middle Ages.

A succession of reports by Arab travelers and geographers makes it possible for the history of many individual towns in Khuzestan to be pieced together,[17] providing an "urban" perspective on the history of the region to complement the more "rural" view obtained from the study of changing canal patterns. Of Jundi Shapur, for example, we learn that already by the tenth century it was suffering from inroads by nomadic Kurds, while to Yakut, a generation before the sack of Baghdad by the Mongols (A.D. 1258), it was a ruin whose glories lay in the past. Susa still was a thriving mercantile and textile center in the tenth century, and as late as 1170 a widely traveled rabbi reported that 7,000 Jews were numbered among its inhabitants. Shortly afterward it was precipitately but temporarily abandoned in one of a continuing series of local military actions, and then perhaps was reoccupied on a declining scale until its final destruction by Tammerlane late in the fourteenth century. Basinna, in the tenth century, was a smaller but thriving town, the fine workmanship of whose veils and tapestries is said to have promoted their export to the farthest ends of the earth. Still commercially active in the time of Yakut, it disappears from view within a century or so afterward. Its name is no longer known in the area, and the location given in Fig. 17–6 is provisional.

Ahwaz, the chief city which gave its name to the whole province in Arab times, suffered particularly heavily under Zanj assaults but subsequently was partly rebuilt. The trade upon which the importance of the city was based had begun to bypass it, however, and by the mid-twelfth century the greater part of it was reported to be in ruins. The great weir at Ahwaz must have gone out of service not long afterward, and with it the canals radiating from the city onto the marginal lands of the lower alluvium. By the nineteenth century the lower Karun's banks below Ahwaz were entirely uncultivated and virtually without permanent settlement. Ahwaz itself may have survived uninterruptedly through the succeeding centuries, but by the 1870's it had shrunk to a small village. The center of political power in the province, meanwhile, inevitably moved northward toward the remaining concentrations of settled population. Shustar, already so noted a textile center that its brocades once draped the walls of the Ka'ba in Mecca, became the capital of the province after the Mongol conquest. It only finally relinquished that position, first to Dizful and subsequently to Ahwaz, in the nineteenth century after a particularly devastating epidemic. The return of Ahwaz to its present prominence stems from the opening of steamship navigation on the Karun, and still more from the development of Khuzestan's oil resources after World War I.

It is important to note that the withdrawal of settled life from the lower plains was accompanied by a deterioration of agriculture on the upper plains as well. While Susa could exist as a prosperous enclave for long periods 3,000 years earlier, such enclaves now were frequent prey to overwhelming external forces. The growth and coalescence of numerous other power centers all over the Near East, and the corresponding evolution of new and increasingly predatory sociopolitical forms, seem to have made the small, independent city-state a helpless anachronism by the later Middle Ages.

The fate of Khuzestan's vaunted sugar production illustrates one of the new threats to a local agricultural economy. With proper care this crop thrived on the upper plains, as it is doing again today. Yet, in spite of the continuity of agricultural settlement around centers like Shustar and Dizful, even the memory of its cultivation had disappeared before the current development program was launched. The exact circumstances behind the total disappearance of so well-adapted a crop are obscure, but a plaintive account of contemporary events in Egypt provides a likely parallel. To increase state revenues some of the Mameluke sultans imposed strict monopolies on the production and exchange of certain commodities, and it was said that under Barsbai (A.D. 1422–38) the cost of sugar rose so high as a result that even victims dying of plague were unable to obtain their customary syrups as palliatives. Surely in a similar artificial constriction of the sugar crop for some such purpose as this lies at least the precondition for the abandonment of sugar cane cultivation in Khuzestan altogether. Damascus merchants who were victims of this monopoly are said to have imported sugar from Khuzestan in 1433, but thereafter the records are silent. From the seventeenth century onward Indian sugar was one of the most lucrative products imported into Iran by European traders.

A different kind of illustration of the declining potentialities of the upper plains as the prosperous, continuously cultivated zone they had become in Sassanian times is provided by archaeological reconnaissance. Outside of the few large towns, we can trace the withering away of all the smaller settlements that had grown up at a distance from the main streams and hence had come to depend on the great Sassanian canal network for irrigation water. One by one they were abandoned, the last holdouts abandoning hope for water from the accustomed sources and desperately seeking to develop tiny enclaves of intensive cultivation by tapping small seasonal watercourses. It is uncertain whether their inhabitants ultimately moved into the larger fortified towns for better protection against marauding nomads, or instead became nomads themselves.

Conditions were not much pleasanter even in Shustar and Dizful. European visitors to them both in the early and middle nineteenth centuries are impressively united in their descriptions of prevalent disease, corruption, poverty, abandoned living quarters, stagnant commerce, and declining agriculture. Thus only a thin thread of never-quite-extinguished urban life ties the bustling Khuzestan towns of today to their more remote and prosperous past.

In the face of the great physical and economic changes contemplated by present planners, practical discoveries from this long record of human settlement that can be directly applied to our contemporary needs are few and minor. But if valid general insights ever can be sought in the history of so small an area, two may be suggested here. The first is that at least the immediate opportunities and impediments to the enhancement of man's economic well-being seem to have lain more often in his social institutions than in the presence or absence of particular items of material equipment. The second is that the myth of the "changeless Orient" is ready for burial.

NOTES

1. I conducted field reconnaissance from December 1960 through March 1961, under joint sponsorship by the Oriental Institute of the University of Chicago and the Development and Resources Corporation of New York. Thanks are due to the Khuzestan Development Service, operating instrumentality of the D.R.C. in Iran, for unstinted cooperation and material support. In particular, I am grateful to Leo L. Anderson, its chief representative, for assistance in understanding many complex problems of agricultural development—present as well as past. The cooperation of the Iran Antiquities Service, and especially of Professor E. O. Negahban and M. M. Moshirpour, also is gratefully acknowledged.

2. See J. De Morgan, "Etude géographique sur lar Susiane" [Délégation en Perse. Mission archéologique en Iran. *Mémoires,* vol. 1 (Paris, 1900)]; G. M. Lees and N. L. Falcon, "The geographical history of the Mesopotamian plains," *Geograph. J.,* **118** (1952).

3. For the most comprehensive of many descriptions of the Khuzestan landscape just prior to the modern period, see A. H. Layard "A description of the province of Khuzistan" [*J. Roy. Geograph. Soc. London,* **16** (1846)]. On soils and the modern agricultural regime, see "Report to the Government of Iran on the development of land and water resources in Khuzestan" [*F.A.O. Report 553* (Rome, 1956)]; "Unified report on the soil and land classification survey of Dizful Project, Khuzistan, Iran" [F.A.O. (Teheran, 1958)]; Plan Organization, Government of Iran "Report on the Diz Irrigation Project: agricultural and civil-technical analyses; cost and benefit appraisal" [Nederlandsche Heidemaatschappij (Arnhem, 1958)].

4. The classification of pre- and protohistoric periods followed here is that of L. Le Breton "The early periods at Susa, Mesopotamian relations" [*Iraq,* 19 (1957)]. For a critical introduction to the voluminous (and sometimes contradictory) reports on excavations at Susa see H. W. Eliot, "Excavations in Mesopotamia and Western Iran" [*Peabody Museum Spec. Publ.* (Cambridge, 1950)]. An up-to-date and full bibliography on archaeological and textual reports from Khuzestan is available in L. Vanden Berghe, "Archéologie de l'Iran ancien" [*Documenta et Monumenta Orientis Antiqui,* **6** (Brill, Leiden, 1959)]. An authoritative overview of the prehistory and history of Iran as a whole through Sassanian times is provided by R. Ghirshman, *Iran* (Pelican, Harmondsworth, Middlesex, 1954).

5. R. J. Braidwood, B. Howe, C. A. Reed, "The Iranian prehistoric project," *Science,* **133,** 2008 (1961).

6. See G. G. Cameron, *History of Early Iran* (University of Chicago Press, Chicago, 1936); R. Mayer, "Die Bedeutung Elams in der Geschichte des alten Orients" [*Saeculum,* **7** (1956)]. Documentation given here and elsewhere is confined to the most general and pertinent references. Thanks are due to Professor E. Reiner for advice at numerous points on dealing with Elamite materials.

7. R. Ghirshman, "The Ziggurat of Tchoga-Zanbil," *Sci. American,* **205** (Jan. 1961).

8. See A. T. Olmstead, *History of the Persian Empire* (University of Chicago Press, Chicago, 2nd impression, 1959).

9. N. C. Debevoise, *A Political History of Parthia* (University of Chicago Press, Chicago, 1938).

10. T. Jacobsen and R. M. Adams, "Salt and silt in ancient Mesopotamian agriculture," *Science*, **128**, 1251 (1958).

11. G. van Roggen, "Notice sur les anciens travaux hydrauliques en Susiane," *Mém. Délégation en Perse*, 2e sér., **7** (Paris, 1905).

12. A. Christensen, *L'Iran sous les Sassanides* (Copenhagen, ed. 2, 1944); T. Nöldeke, *Geschichte der Perser und Araber zur Zeit der Sasaniden aus der arabischen Chronik des Tabari* (Brill, Leiden, 1879).

13. C. Elgood, *A Medical History of Persia and Eastern Caliphate* (Cambridge, 1951); D. L. E. O'Leary, *How Greek Science Passed to the Arabs* (Routledge and Kegan Paul, London, 1949).

14. D. C. Dennett, Jr., *Conversion and the Poll Tax in Early Islam* (Harvard Univ. Press, Cambridge, 1950); A. K. S. Lambton, *Landlord and Peasant in Persia* (Oxford Univ. Press, London, 1953).

15. A recent general account is in P. K. Hitti, *History of the Arabs* (Macmillan, London, ed. 6, 1956).

16. T. Nöldeke, *Sketches from Eastern History* (Black, London, 1892).

17. The observations of many Arab historians and geographers on Khuzestan are conveniently made available in P. Schwarz, *Iran im Mittelalter nach dem arabischen Geographen*, vol. 4 [Quellen und Forschungen zur Erdund Kulturkunde (Leipzig, 1921)]. The same material is much more concisely treated in G. Le Strange, *The Lands of the Eastern Caliphate* (Cambridge Univ. Press, Cambridge, 1905). I acknowledge the advice on specific points of Professor M. Mahdi and Professor N. Abbott.

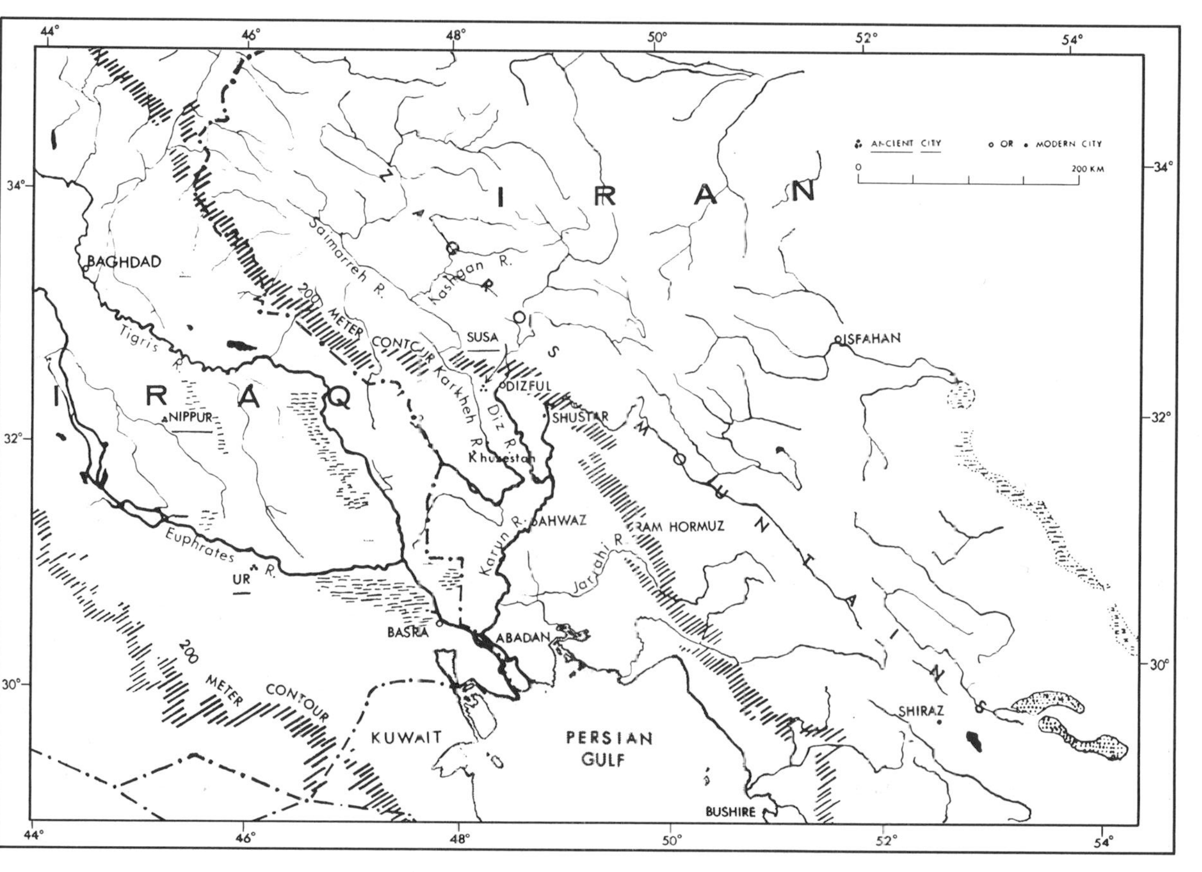

Fig. 17–1. Location of Khuzestan, the Persian Gulf, Iran, and Iraq.

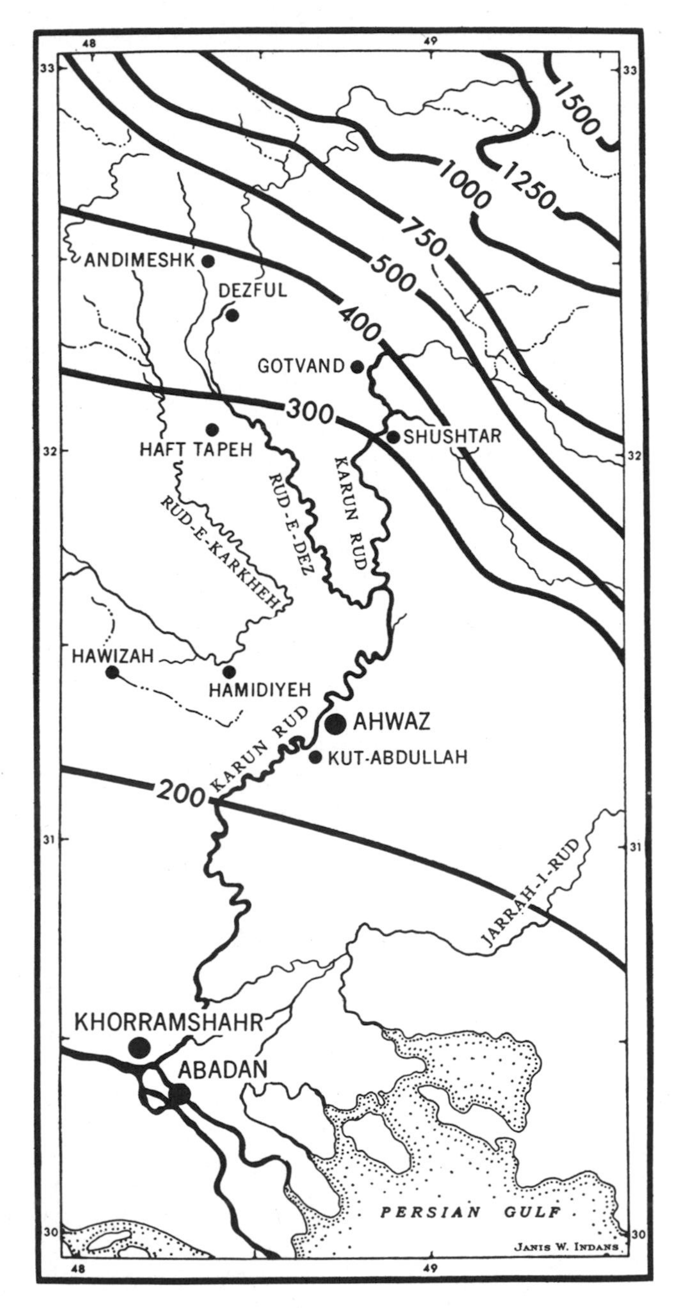

Fig. 17–2. Rainfall in Khuzestan. The isohyets give the approximate annual precipitation, in millimeters.

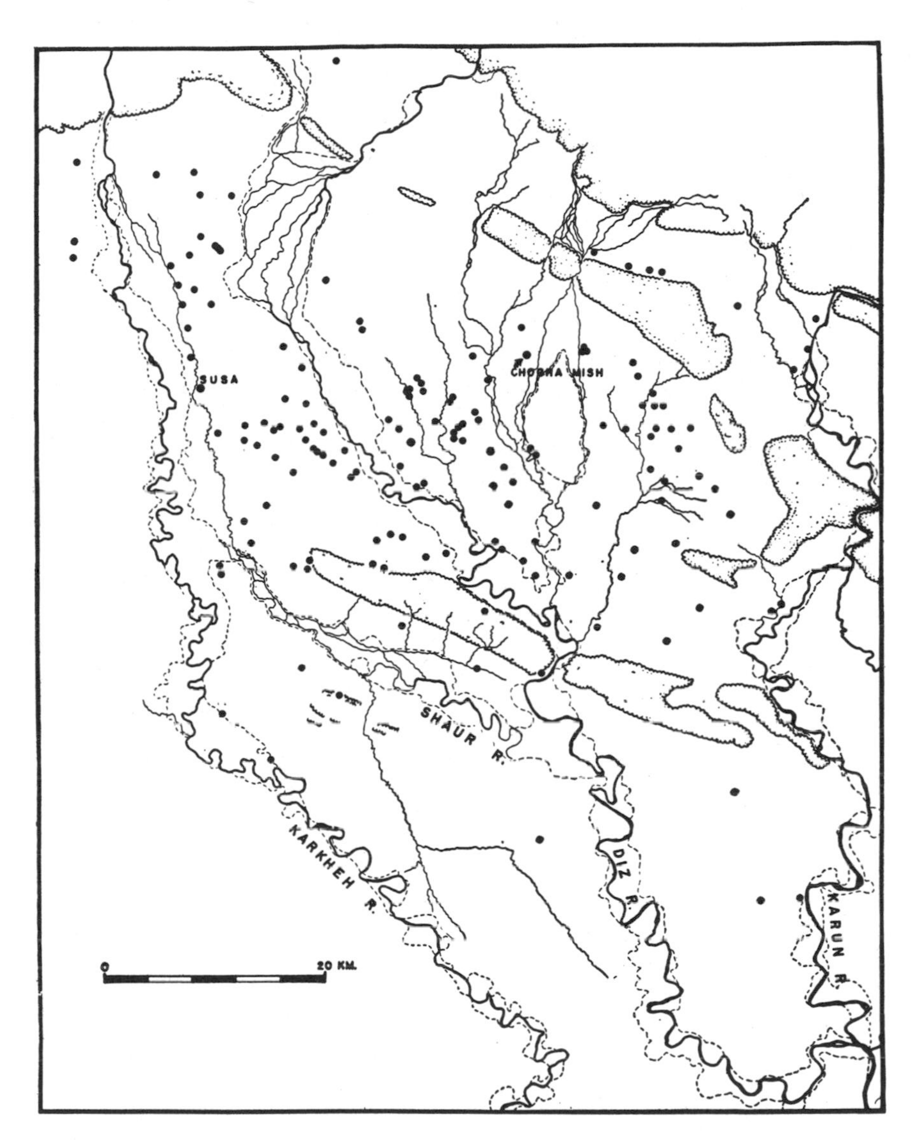

Fig. 17–3. Prehistoric agricultural settlement in the upper Khuzestan plains prior to about 3500 B.C. Numerous villages and a few small towns formed a widespread, relatively dense network over the area. Linear groupings of sites generally conform to former channels of natural watercourses; irrigation was of limited scope and secondary importance. Villages (here defined as settlements occupying less than 4 hectares, or slightly less than 10 acres) and small towns are shown much enlarged.

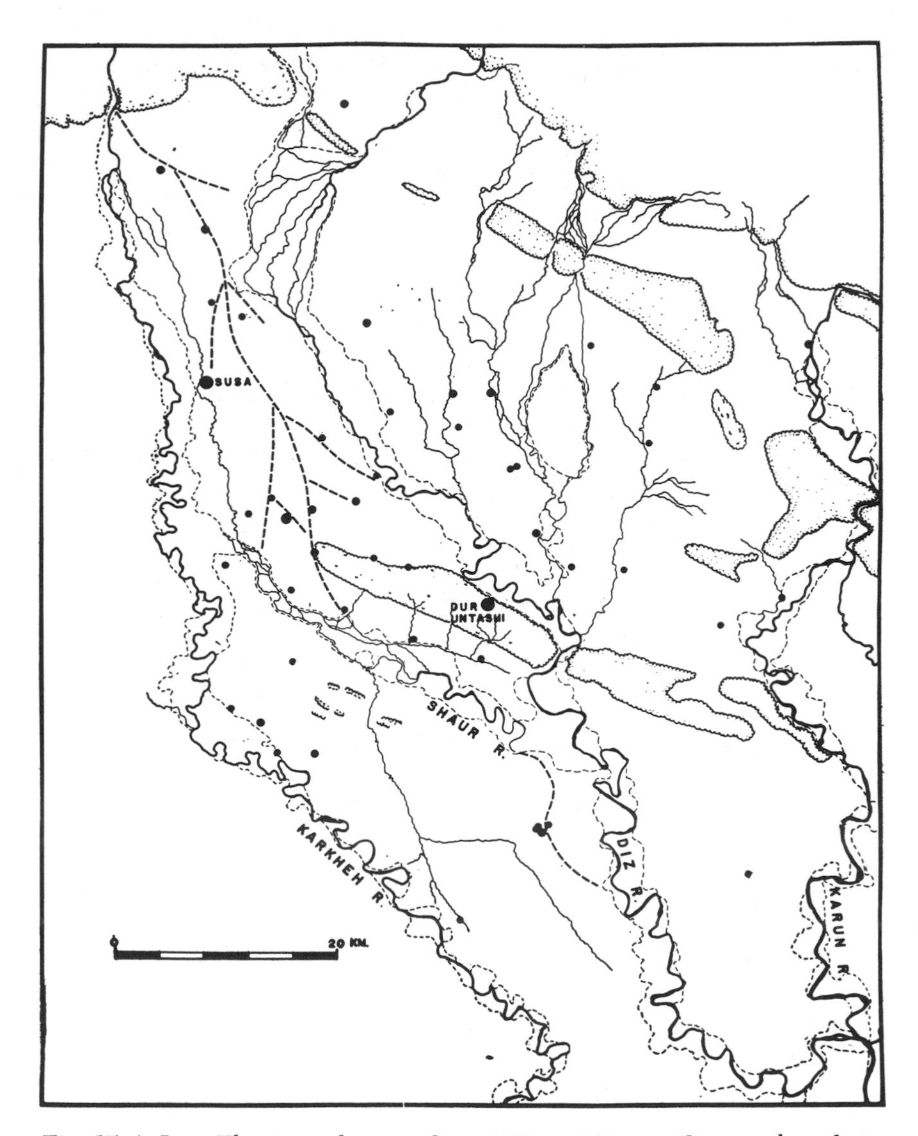

Fig. 17–4. Late Elamite settlement, about 1200 to 640 B.C. This was the culminating phase of a slow process of town growth, although the number of individual sites did not approach the prehistoric total. Only the two urban centers are drawn approximately to scale. Except for the area around Susa, settlements tended to lie along natural watercourses, and probably only local irrigation was practiced. Susa lay in the center of a larger enclave, which may have been continuously cultivated, but elsewhere there still must have been many tracts of good land used only seasonally for grazing. Because of the overlying deposits, the canal system shown around Susa is a largely speculative reconstruction.

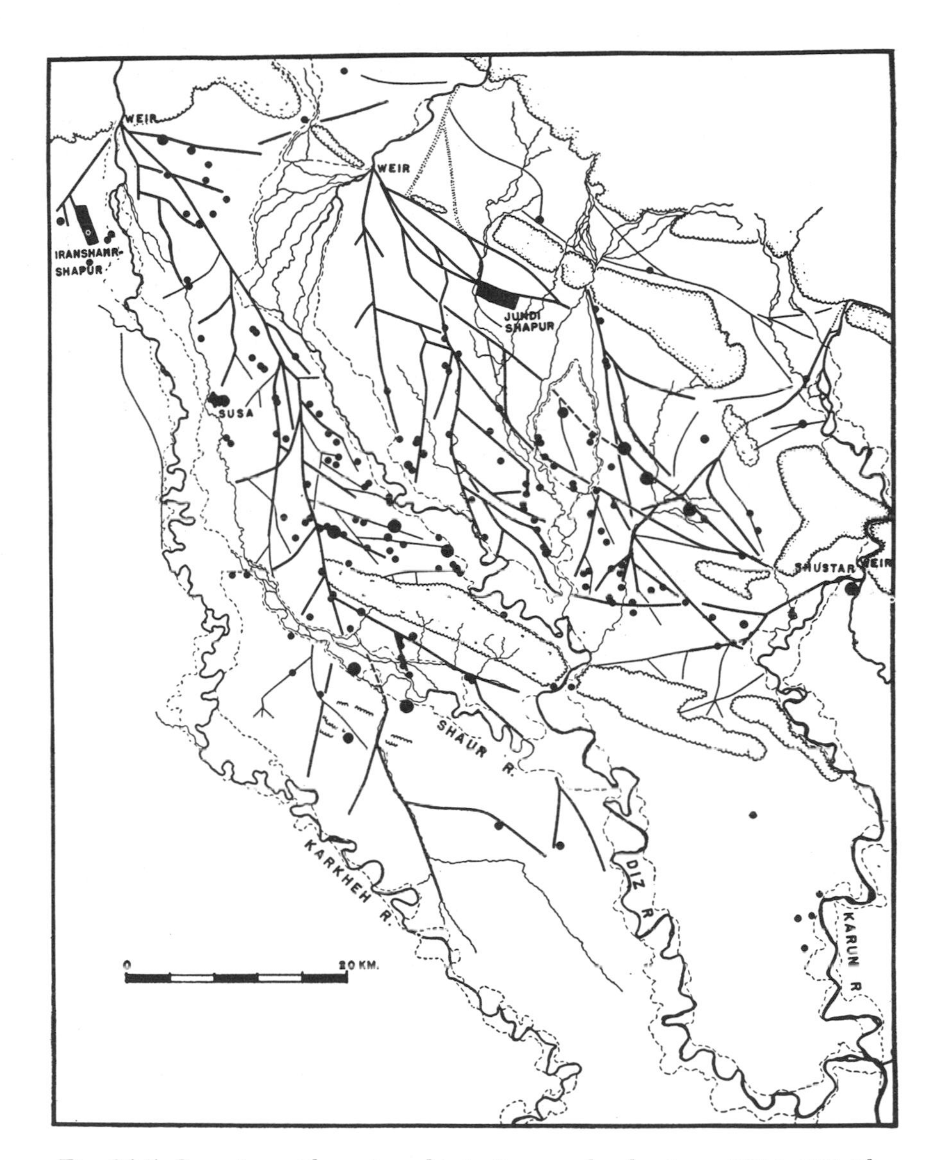

Fig. 17–5. Sassanian settlements and irrigation canals, about A.D. 226 to 639. The extension of the irrigation system to virtually the widest possible limits was accompanied by new and more intensive irrigation techniques, state-directed urbanization, and a population maximum. The three largest cities are drawn to scale, but nine others are shown only slightly enlarged. Note that what is today the Shaur River was apparently at that time a major branch of the Karkheh River. Dotted lines represent vented tunnels used as water conduits.

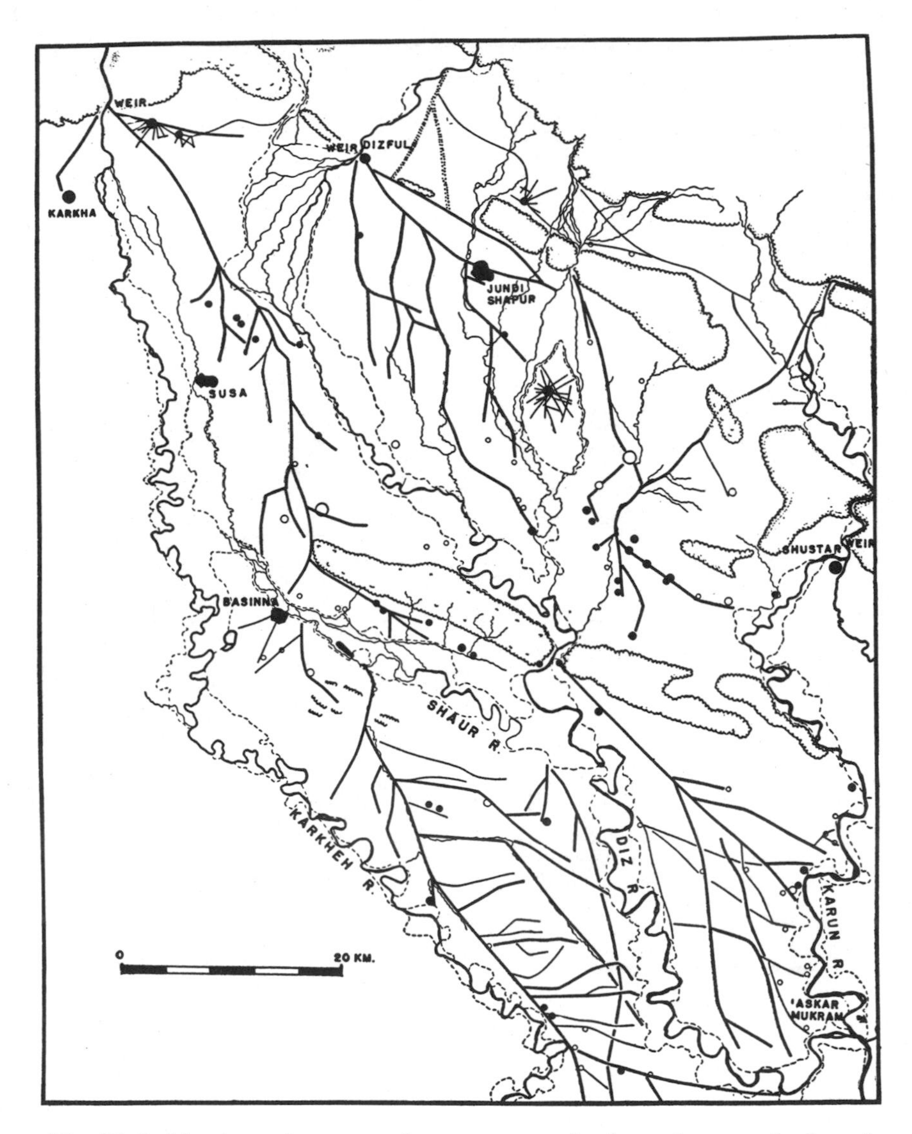

Fig. 17–6. Islamic settlements and irrigation canals, from the seventh through approximately the ninth century A.D. Sites shown as hollow rings were apparently much reduced or abandoned by A.D. 900, while those shown as solid circles continued for varying periods. New, but generally short-lived, irrigation enterprises were constructed in the less productive lower (southern) part of the area, while on the upper plains the Sassanian system gradually deteriorated. The small-scale canal networks that are shown around some of the surviving towns must have been constructed only after the weirs and main supply canals had ceased to function; hence, they may be somewhat later than the period covered by the map as a whole.

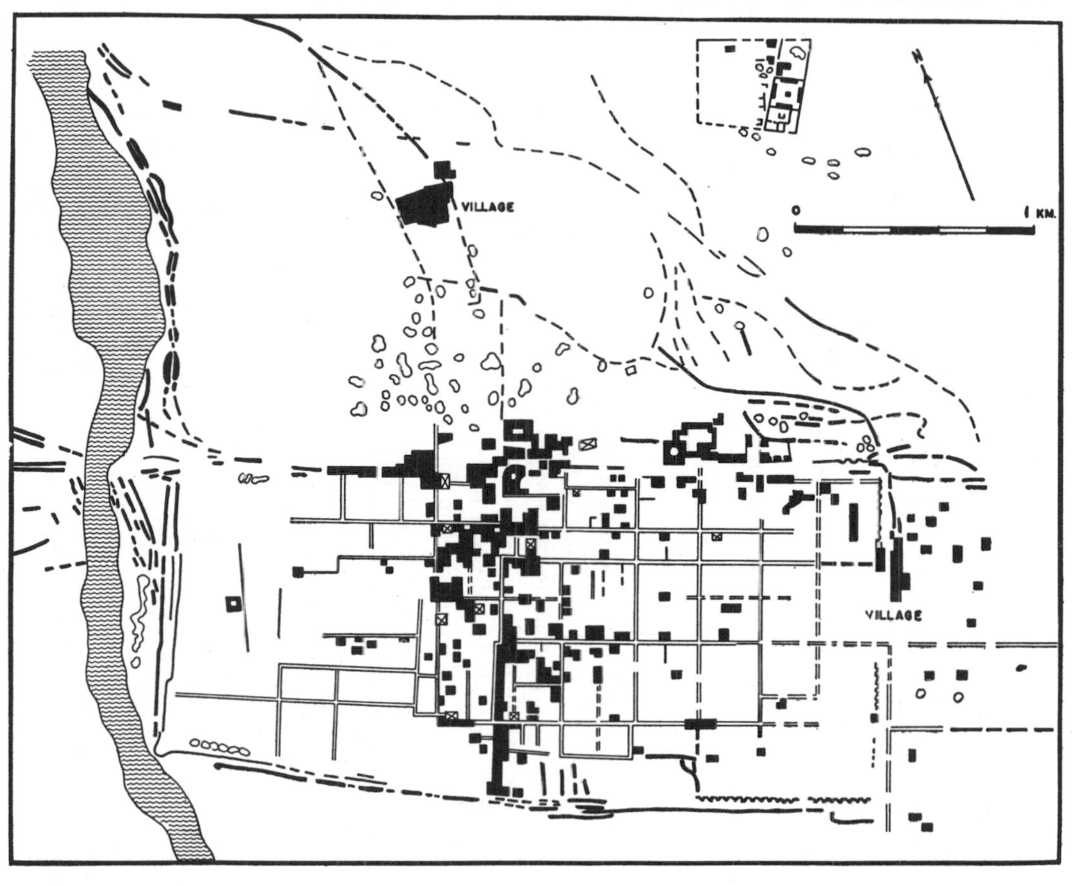

Fig. 17–7. The city plan of Jundi Shapur, as reconstructed mainly from aerial photographs.

Chapter **18**

Salt and Silt in Ancient Mesopotamian Agriculture

Thorkild Jacobsen
Robert M. Adams

Under the terms of a farsighted statute, 70 per cent of the oil revenues of the Iraqi Government are set aside for a program of capital investment that is transforming many aspects of the country's predominantly agricultural economy. As compared with the subsistence agriculture, which largely has characterized Iraq's rural scene in the past, new irrigation projects in formerly uninhabited deserts are pioneering a rapid increase in land and labor productivity through crop rotation, summer cultivation in addition to the traditional winter-grown cereals, and emphasis on cash crops and livestock.

But these and similar innovations often have disconcerting effects in a semiarid, subtropical zone—effects that cannot be calculated directly from the results of experiment in Europe and America. At the same time, old canal banks and thickly scattered ruins of former settlements testify to former periods of successful cultivation in most of the desert areas now being reopened. The cultural pre-eminence of the alluvial plains of central and southern Iraq through much of their recorded history provides still further evidence of the effectiveness of the traditional agricultural regime in spite of its prevailing reliance on a simple system of fallow in alternate years. Accordingly, the entire 6,000-year record of irrigation agriculture in the Tigris-Euphrates flood plain furnishes an indispensable background for formulating plans for future development.

At least the beginnings of a comprehensive assessment of ancient agriculture recently were undertaken on behalf of the Government of Iraq Development Board. In addition to utilizing ancient textual sources from many parts of Iraq, which today are widely scattered in the world's libraries and museums, this undertaking included a program of archaeological field work designed to elucidate the history of irrigation and settlement of a portion of the flood plain that is watered by a Tigris tributary, the Diyala River.[1] Here we cannot report all the diverse findings of the project and its many specialists, but instead will outline some aspects of the general ecological situation encountered by agriculturalists in the Mesopotamian alluvium that seem to have shaped the development of irrigation farming. And, conversely, we hope to show that various features of the natural environment in turn were decisively modified by the long-run effects of human agencies.

Historical Role of Soil Salinization

A problem that recently has come to loom large in Iraqi reclamation planning is the problem of salinity. The semiarid climate and generally low permeability of the soils of central and southern Iraq expose the soils to dangerous accumulations of salt and exchangeable sodium, which are harmful to crops and soil texture and which can eventually force the farmer off his land.

For the most part, the salts in the alluvial soils are presumed to have been carried in by river and irrigation water from the sedimentary rocks of the northern mountains. In addition, smaller quantities may have been left by ancient marine transgressions or borne in by winds from the Persian Gulf. Besides the dominant calcium and magnesium cations, the irrigation water also contains some sodium. As the water evaporates and transpires it is assumed that the calcium and magnesium tend to precipitate as carbonates, leaving the sodium ions dominant in the soil solution. Unless they are washed down into the water table, the sodium ions tend to be absorbed by colloidal clay particles, deflocculating them and leaving the resultant structureless soil almost impermeable to water. In general, high salt concentrations obstruct germination and impede the absorption of water and nutrients by plants.

Salts accumulate steadily in the water table, which has only very limited lateral movement to carry them away. Hence the ground water everywhere has become extremely saline, and this probably constitutes the immediate source of the salts in Iraq's saline soils. New waters added as excessive irrigation, rains, or floods can raise the level of the water table considerably under the prevailing conditions of inadequate drainage. With a further capillary rise when the soil is wet, the dissolved salts and exchangeable sodium are brought into the root zone or even to the surface.

While this problem has received scientific study in Iraq only in recent years, investigation by the Diyala Basin Archaeological Project of a considerable number and variety of ancient textual sources has shown that the process of salinization has a long history. Only the modern means to combat it are new: deep drainage to lower and hold down the water table, and utilization of chemical amendments to restore soil texture. In spite of the almost proverbial fertility of Mesopotamia in antiquity, ancient control of the water table was based only on avoidance of overirrigation and on the practice of weed-fallow in alternate years. As was first pointed out by J. C. Russel, the later technique allows the deep-rooted *shoq* (*Proserpina stephanis*) and *agul* (*Alhagi maurorum*) to create a deep-lying dry zone against the rise of salts through capillary action. In extreme cases, longer periods of abandonment must have been a necessary, if involuntary, feature of the agricultural cycle. Through evapotranspiration and some slow draining they could eventually reduce an artificially raised water table to safe levels.

As to salinity itself, three major occurrences have been established from ancient records. The earliest of these, and the most serious one, affected southern Iraq from 2400 B.C. until at least 1700 B.C. A milder phase is attested in documents from central Iraq written between 1300 and 900 B.C. Lastly, there is archaeological evidence that the Nahrwan area east of Baghdad became salty only after A.D. 1200.

The earliest of these occurrences particularly merits description, since it sheds light on the northward movement of the major centers of political power from southern into central Iraq during the early second millennium B.C. It seems to have had its roots in one of the perennial disputes between the small, independent principalities that were the principal social units of the mid-third millennium B.C. Girsu and Umma, neighboring cities along a watercourse stemming from the Euphrates, had fought for generations over a fertile border district. Under the ruler Entemenak, Girsu temporarily gained the ascendancy, but was unable to prevent Umma, situated higher up the watercourse, from breaching and obstructing the branch canals that served the border fields. After repeated, unsuccessful protests, Entemenak eventually undertook to supply water to the area by means of a canal from the Tigris; access to that river, flowing to the east of Girsu, could be assured without further campaigning against Umma to the northwest. By 1700 B.C. this canal had become large and important enough to be called simply "the Tigris," and it was supplying a large region west of Girsu that formerly had been watered only by the Euphrates. As a result, the limited irrigation supplies that could be drawn from the latter river were supplemented with copious Tigris water. A corresponding increase undoubtedly occurred in seepage, flooding, and overirrigation, creating all the conditions for a decisive rise in ground-water level.

Several parallel lines of evidence allow the ensuing salinization to be followed quantitatively:

1. Beginning shortly after the reign of Entemenak, the presence of patches of saline ground is directly attested in records of ancient temple surveyors. In a few cases, individual fields, which at that time were recorded as salt-free, can be shown in an archive from 2100 B.C. to have developed conditions of sporadic salinity during the three hundred intervening years of cultivation.

2. Crop choice can be influenced by many factors, but the onset of salinization strongly favors the adoption of crops which are more salt-tolerant. Counts of grain impressions in excavated pottery from sites in southern Iraq of about 3500 B.C., made by H. Helbaek, suggest that at that time the proportions of wheat and barley were nearly equal. A little more than 1,000 years later, in the time of the ruler Entemenak at Girsu, the less salt-tolerant wheat accounted for only one-sixth of the crop. By about 2100 B.C. wheat had slipped still further, and it accounted for less than 2 per cent of the crop in the Girsu area. By 1700 B.C., the cultivation of wheat had been abandoned completely in the southern part of the alluvium.

3. Concurrent with the shift to barley cultivation was a serious decline in fertility which for the most part can be attributed to salinization. At about 2400 B.C. in Girsu a number of field records give an average yield of 2,537 liters per hectare—highly respectable even by modern United States and Canadian standards. This figure had declined to 1,460 liters per hectare by 2100 B.C., and by about 1700 B.C. the recorded yield at nearby Larsa had shrunk to an average of only 897 liters per hectare. The effects of this slow but cumulatively large decline must have been particularly devastating in the cities where the needs of a considerable superstructure of priests, administrators, merchants, soldiers, and craftsmen had to be met with surpluses from primary agricultural production.

The southern part of the alluval plain appears never to have recovered fully from the disastrous general decline that accompanied the salinization process. While never completely abandoned afterward, cultural and political leadership passed permanently out of the region with the rise of Babylon in the eighteenth century B.C., and many of the great Sumerian cities dwindled to villages or were left in ruins. Probably there is no historical event of this magnitude for which a single explanation is adequate, but that growing soil salinity played an important part in the breakup of Sumerian civilization seems beyond question.

Silt and the Ancient Landscape

As with salt, the sources of the silt of which the alluvium is composed are to be found in the upper reaches of the major rivers and their tributaries. Superficially, the flatness of the alluvial terrain may seem to suggest a relatively old and static formation, one to which significant increments of silt are added only as a result of particularly severe floods. But in fact, sedimentation is a massive, continuing process. Silt deposited in canal beds must be removed in periodic cleanings to adjoining spoil banks, from which it is carried by rain and wind erosion to surrounding fields. Another increment of sediment accompanies the irrigation water into the fields themselves, adding directly to the land surface. In these ways, the available evidence from archaeological sounding indicates that an average of perhaps ten meters of silt has been laid down at least near the northern end of the alluvium during the last 5,000 years.

Of course, the rate of deposition is not uniform. It is most rapid along the major rivers and canals, and their broad levees slope away to interior drainage basins where accumulated runoff and difficult drainage have led to seriously leached soils and seasonal swamps. However, only the very largest of the present depressions seem to have existed as permanent barriers (while fluctuating in size) to cultivation and settlement for the six millennia since agriculture began in the northern part of the alluvium. More commonly, areas of swamp shifted from time to time. As some were gradually brought under cultivation, others formed behind newly created canal or river levees that interrupted the earlier avenues of drainage.

As the rate of sedimentation is affected by the extent of irrigation, so also were the processes of sedimentation—and their importance as an agricultural problem—closely related to the prevailing patterns of settlement, land-use, and even sociopolitical control. The character of this ecological interaction can be shown most clearly at present from archaeological surveys in the lower Diyala basin, although other recent reconnaissance indicates that the same relationships were fairly uniform throughout the northern, or Akkadian, part of the Mesopotamian Plain.[2] To what degree the same patterns occurred in the initially more urbanized (and subsequently more saline) Sumerian region farther south, however, cannot yet be demonstrated.

The methods of survey employed here consisted of locating ancient occupational sites with the aid of large-scale maps and aerial photographs, visiting most or all of them—in this case, more than 900 in a 9,000-square-kilometer area—systematically in order to make surface collections of selected "type fossils" of broken pottery, and subsequently determining the span of occupation at each settlement with the aid of such historical and archaeological crossties as may be found to supplement the individual sherd collections.[3] It then can be observed that the settlements of a particular period always describe networks of lines that must represent approximately the contemporary watercourses necessary for settled agricultural life. For more recent periods, the watercourses serving the settlements often still can be traced in detail as raised levees, spoil banks, or patterns of vegetation disturbance, but, owing in part to the rising level of the plain, all of the older watercourses so far have been located only inferentially.

A number of important and cumulative, but previously little-known, developments emerge from the surveys. By comparing the over-all pattern of settlement of both the early third and early second millennium B.C. (Fig. 18–1) with the prevailing pattern of about A.D. 500 (Fig. 18–2) these developments can be seen in sharply contrasting form. They may be summarized conveniently by distinguishing two successive phases of settlement and irrigation, each operating in a different ecological background, and each facing problems of sedimentation of a different character and magnitude.

The earlier phase persisted longest. Characterized by a linear pattern of settlements largely confined to the banks of major watercourses, it began with the onset of agricultural life in the Ubaid period (about 4000 B.C.) and was replaced only during the final centuries of the pre-Christian Era. In all essentials the same network of watercourses was in use throughout this long time-span, and the absence of settlement along periodically shifting side branches seems to imply an irrigation regime in which the water was not drawn great distances inland from the main watercourses. Under these circumstances, silt accumulation would not have been the serious problem to the agriculturalist that it later became. The short branch canals upon which irrigation depended could have been cleaned easily or even replaced without the necessary intervention of a powerful, centralized authority. Quite possibly most irrigation during this phase depended simply on uncontrolled flooding through breaches cut in the levees of watercourses (as the lower Mississippi River) flowing well above plain level.

It is apparent from the map in Fig. 18–1 that large parts of the area were unoccupied by settled cultivators even during the periods of maximum population and prosperity that have been selected for illustration therein. An extended historical study of soil profiles would be necessary to provide explanations for these uninhabited zones, but it is not unreasonable to suppose that some were seasonal swamps and depressions of the kind described above, while others were given over to desert because they were slightly elevated and hence not subject to easy flooding and irrigation. Still others probably were permanent swamps, since it is difficult to account in any other way for the discontinuities in settlement that appear along long stretches of some watercourses. One indication of the ecological shift that took place in succeeding millennia is that permanent swamps today have virtually disappeared from the entire northern half of the alluvium.

Considering the proportion of occupied to unoccupied area, the total population of the Diyala Basin apparently was never very large during this long initial phase. Instead, a moderately dense population was confined to small regional enclaves or narrow, isolated strips along the major watercourses; for the rest of the area there can have been only small numbers of herdsmen, hunters, fishermen, and marginal catch-crop cultivators. It is significant that most of the individual settlements were small villages, and that even the dominant political centers in the area are more aptly described as towns rather than cities.[4]

An essential feature of the earlier pattern of occupation, although not shown in a summary map like Fig. 18–1, is its fluctuating character. There is good historical evidence that devastating cycles of abandonment affected the whole alluvium. The wide and simultaneous onset of these cycles soon after relatively peaceful and prosperous times suggests that they proceeded from sociopolitical, rather than natural, causes, but at any rate their effects can be seen clearly in the Diyala region. For example, the numerous Old Babylonian settlements shown in Fig. 18–1 had been reduced in number by more than 80 per cent within 500 years following, leaving only small outposts scattered at wide intervals along watercourses that previously had been thickly settled. An earlier abandonment, not long after the early Dynastic period that is shown in gray in Fig. 18–1, was shorter-lived and possibly affected the main towns more than the outlying small villages. Village life in general, it may be observed, remains pretty much of an enigma in the ancient Orient for all "historical" periods.

Under both ancient and modern Mesopotamian conditions, a clear distinction between "canals" and "rivers" is frequently meaningless or impossible. If the former are large and are allowed to run without control they can develop a "natural" regime in spite of their artificial origin. Some river courses, on the other hand, can be maintained only by straightening, desilting, and other artificial measures. Nevertheless, it needs to be stressed that the reconstructed watercourses shown in Fig. 18–2 followed essentially natural regimes and that at least their origins had little or nothing to do with human intervention. They were, in the first place, already present during the initial occupation of the area by prehistoric village agriculturalists who lacked the numbers and organization to dig them artificially. Secondly, the same watercourses persisted for more than three millennia with little change, even through periods of abandonment when they could not have received the maintenance that canals presuppose. Finally, the whole network of these early rivers describes a "braided stream" pattern that contrasts sharply with the brachiating canal systems of all later times, which are demonstrably artificial.

Specific features of the historic geography of the area are not within the compass of this chapter, but it should be noted that the ancient topography differed substantially from the modern. Particularly interesting is the former course of the Diyala River, flowing west of its present position and joining the Tigris River (apparently also not in its modern course) through a delta-like series of mouths. A branch that bifurcated from the former Diyala above its "delta" and flowed off for a long distance to the southeast before joining the Tigris has been identified tentatively as the previously unlocated "River Dabban" that is referred to in ancient cuneiform sources.

The pattern of occupation illustrated in Fig. 18–2 began to emerge in Achaemenian times (539–331 B.C.), after nearly 1,000 years of stagnation and abandonment. Perhaps the pace of reoccupation quickened with the conquest of Mesopotamia by Alexander, but the density of population reached during much older periods was attained again, and then surpassed, only in the subsequent Parthian period (about 150 B.C.–A.D. 226). New settlements large enough to be described as true cities, on the other hand, were introduced to the area for the first time by Alexander's Macedonian followers—demonstrating, if doubt could otherwise exist, that the onset of urbanization depends more on historical and cultural factors than on a simple increase in population density.

A central feature of this second phase of settlement is the far more complete exploitation of available land and water resources for agriculture. There is some evidence that the irrigation capacity of the Diyala River was being utilized fully even before the end of the Parthian period, and yet both the proportion of land that was cultivated and the total population rose substantially further, reaching their maxima in this area, for any period, under the Sassanian dynasty (A.D. 226–637) that followed. A rough estimate of the total agricultural production in the area first becomes possible with records of tax collections under the early Abbasids, perhaps three hundred years after the maximum limits of expansion shown in Fig. 18–2 had been reached. From a further calculation of the potentially cultivable land it can then be shown that (with alternate years in fallow and assuming average yields) virtually the entire cultivable area must have been cropped regularly under both the Sassanians and early Abbasids.

Increased population, the growth of urban centers, and expansion in the area of cultivation to its natural limits were linked in turn to an enlargement of the irrigation system on an unprecedented scale. It was necessary, in the first place, to crisscross formerly unused desert and depression areas with a complex—and entirely artificial—brachiating system of branch canals, which is outlined in Fig. 18–2. Expansion depended also on the construction of a large, supplementary feeder canal from the Tigris which, with technical proficiency that still excites admiration, and without apparent regard for cost, brought the indispensable, additional water through a hard, conglomerate headland, across two rivers, and thence down the wide levee left by the Dabban

River of antiquity. Enough survives of the Nahrwan Canal, as the lower part of this gigantic system was called, even to play a key part in modern irrigation planning. Excavations carried out by the Diyala Basin archaeological project at one of several known weirs along the 300-kilometer course of this canal provided a forceful illustration not only of the scale of the system but also of the attention lavished on such ancillary works as thousands of brick sluice gates along its branches. In short, we are dealing here with a whole new conception of irrigation which undertook boldly to reshape the physical environment at a cost that could be met only with the full resources of a powerful and highly centralized state.[5]

In spite of its unrivaled engineering competence, there were a number of undesirable consequences of the new irrigation regime. For example, to a far greater degree than had been true earlier, it utilized long branch canals that tended to fill rapidly with silt because of their small-to-moderate slope and cross-sectional area. Only the Nahrwan Canal itself —and that only during the first two centuries or so of its existence— seems to have maintained its bed without frequent and costly cleaning. Silt banks left from Parthian, Sassanian, and Islamic canal cleaning are today a major topographic feature not only in the Diyala region but all over the northern part of the Mesopotamian alluvium; frequently they run for great distances and tower over all but the highest mounds built up by ancient towns and cities. Or again, while massive control installations were essential if such a complex and interdependent system was to operate effectively, they needed periodic reconstruction at great cost (six major phases at the weir excavated by the Diyala project) and practically continuous maintenance. Moreover, the provision of control works of all sizes acted together with the spreading networks of canal branches and subbranches to reduce or eliminate flood surges that otherwise might have contributed to the desilting process.

None of these consequences, to be sure, vitiated the advantages to be obtained with the new type of irrigation *so long as there remained a strong central authority committed to its maintenance.* But with conditions of social unrest and a preoccupation on the part of the political authorities with military adventures and intrigues, the maintenance of the system could only fall back on local communities ill equipped to handle it. These circumstances prevailed fairly briefly in late Sassanian times, leading to a widespread but temporary abandonment of the area. After an Islamic revival, they occurred again in the eleventh and twelfth centuries A.D., accompanied by such storm signals of political decay as the calculated breaching of the Nahrwan during a military campaign. On this occasion there was no quick recovery; it still remains for the modern Iraqis to re-establish the prosperity for which the region once was noted.

A closer look at the role of sedimentation along the Nahrwan during the years of political crisis under the later Abbasids is given in Fig. 18–3. In the first illustrated phase, in late Sassanian times, irrigation water was drawn from the Nahrwan at fairly uniform intervals and applied almost directly to fields adjoining its course. During a second phase, roughly coinciding with the rise of the Abbasid caliphate, irrigation water tended to be drawn off farther upstream from the field for which it was destined. This is best exemplified by the increasing importance of the weir as a source for branch canals serving a considerable area. For some distance below the weir the level of the Nahrwan apparently no longer was sufficient to furnish irrigation water above the level of the fields.

By the time of the final phase, soon after A.D. 1100, practically all irrigation in the very large region below the weir had come to depend on branch canals issuing from above it; it is worth noting that two of the largest and most important of these branches simply paralleled the Nahrwan along each bank for more than twenty kilometers. The same unsuccessful struggle to maintain irrigation control is shown by the shrinkage or disappearance of town and city life along the main canal and the depopulation of the initial five to ten kilometers along each major branch issuing from it, while lower-lying communities at the distal ends of the branches continued to flourish.

This cumulative change in the character of the system probably was a consequence of both natural and social factors. On the one hand, silt deposition had raised the level of the fields by almost one meter over a five-hundred-year period. Since the natural mechanisms for maintaining equilibrium between the bed of a watercourse and its alluvial levee were largely inoperative in such a complex and carefully controlled system, this rise in land surface may have reduced considerably the level of water available for irrigation purposes. At the same time, inadequate maintenance and subsequent siltation of the Nahrwan's own bed in time sharply reduced its flow and surely also reduced the head of water it could provide to its branches. But whatever the responsible factors were, the result was an especially disastrous one. At a time when the responsibility of the central government for irrigation was eroding away, and when population had been reduced substantially by warfare and by prolonged disruption of the water supply, the heavy burden of desilting branch canals remained constant or even increased for the local agriculturalist. If the accumulation of silt was no more than a minor problem at the beginning of irrigation in the Diyala Basin 5,000 years earlier, by the late Abbasid period it had become perhaps the greatest single obstacle that a quite different irrigation regime had to deal with.

With the converging effects of mounting maintenance requirements on the one hand, and declining capacity for more than rudimentary maintenance tasks on the other, the virtual desertion of the lower Diyala area that followed assumes in retrospect a kind of historical inevitability. By the middle of the twelfth century most of the Nahrwan region already was abandoned. Only a trickle of water passed down the upper section of the main canal to supply a few dying towns in the now hostile desert. Invading Mongol horsemen under Hulagu Khan, who first must have surveyed this devastated scene a century later, have been unjustly blamed for causing it ever since.

NOTES

1. The Diyala Basin Archaeological Project was conducted jointly by the Oriental Institute of the University of Chicago and the Iraq Directorate General of Antiquities, on a grant from the First Technical Section of the Development Board. It was directed by one of us (T. J.), with the other (R. M. A.) and Sayyid Fuad Safar, of the Directorate General of Antiquities, as associate directors. Excavations were under the supervision of Sayyid Mohammed Ali Mustafa, also of the Directorate General of Antiquities. Field studies of palaeobotanical remains were undertaken in association with the project by Dr. Hans Helbaek, of the National Museum, Copenhagen, Denmark. Intensive study of the cuneiform and Arabic textual sources on agriculture was made possible through the collaboration of scholars of many countries. Especial thanks for assistance to the field program in Iraq, and for advice in the interpretation of its results, are due to Mr. K. F. Vernon, H. E. Dr. Naji al-Asil, Dr. J. C. Russel, and Sayyid Adnan Hardan.

2. R. M. Adams, *Sumer*, in press.

3. A preliminary application of this approximate methodology to conditions prevailing in Iraq was introduced by one of us (T. J.) in the Diyala Basin in 1936–37, and the results of that earlier survey have been incorporated in the present study. Fortunately for the archaeologist, there is sufficient disturbance from routine community activities (for example, foundation, well, and grave digging, and mud-brick manufacture, and so forth) for some traces of even the earliest of a long sequence of occupational periods to be detected on a mound's surface.

4. Partial town plans for the political capital of the region at Tel Asmar (ancient Eshnunna) and for two other slightly smaller centers are available from extensive Oriental Institute excavations carried out in the Diyala region between 1930 and 1937. See P. Delougaz, *The Temple Oval at Khafajah* [Oriental Inst. Publ., 53 (Univ. of Chicago Press, Chicago, 1940)]; P. Delougaz and S. Lloyd, *Pre-Sargonid Temples in the Diyala Region* [Oriental Inst. Publ., 58 (Univ. of Chicago Press, Chicago, 1942)]; and H. Frankfort, *Stratified Cylinder Seals from the Diyala Region* [Oriental Inst. Publ., 72 (Univ. of Chicago Press, Chicago, 1955)], plates 93–96. For recent general overviews of the history and culture of the earlier periods, see A. Falkenstein "La cité-temple Sumérienne" [*Cahiers d'Histoire Mondiale*, 1 (1954)] and T. Jacobsen, "Early political developments in Mesopotamia" [*Z. für Assyriologie* (N.F.), 18 (1957)].

5. General accounts of political, social, and cultural conditions in Mesopotamia during the Persian dynasties and under the Caliphate are to be found in R. Ghirshman, *Iran* (Pelican, Harmondsworth, Middlesex, 1954) and P. K. Hitti, *History of the Arabs* (Macmillan, London, ed. 6, 1956).

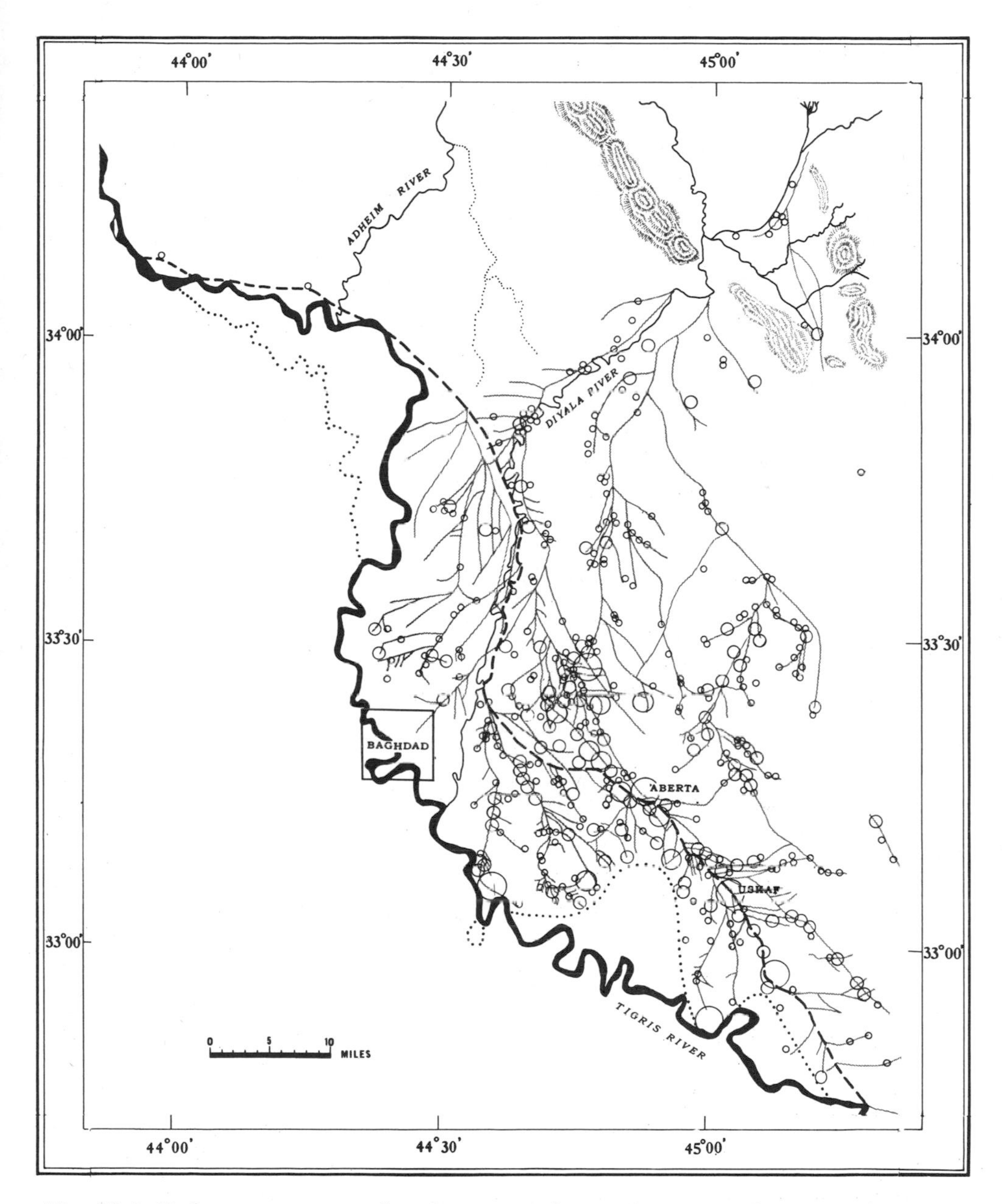

Fig. 18–1. Early watercourses and settlements in the Diyala region. The system shown in gray was in use during the Early Dynastic period, about 3000–2400 B.C. Sites and watercourses shown in black, slightly displaced so that the earlier pattern will remain visible, were occupied during the Old Babylonian period, about 1800–1700 B.C. In this and subsequent figures, size of circle marking an ancient settlement is roughly proportional to the area of its ruins. Modern river courses are shown in gray.

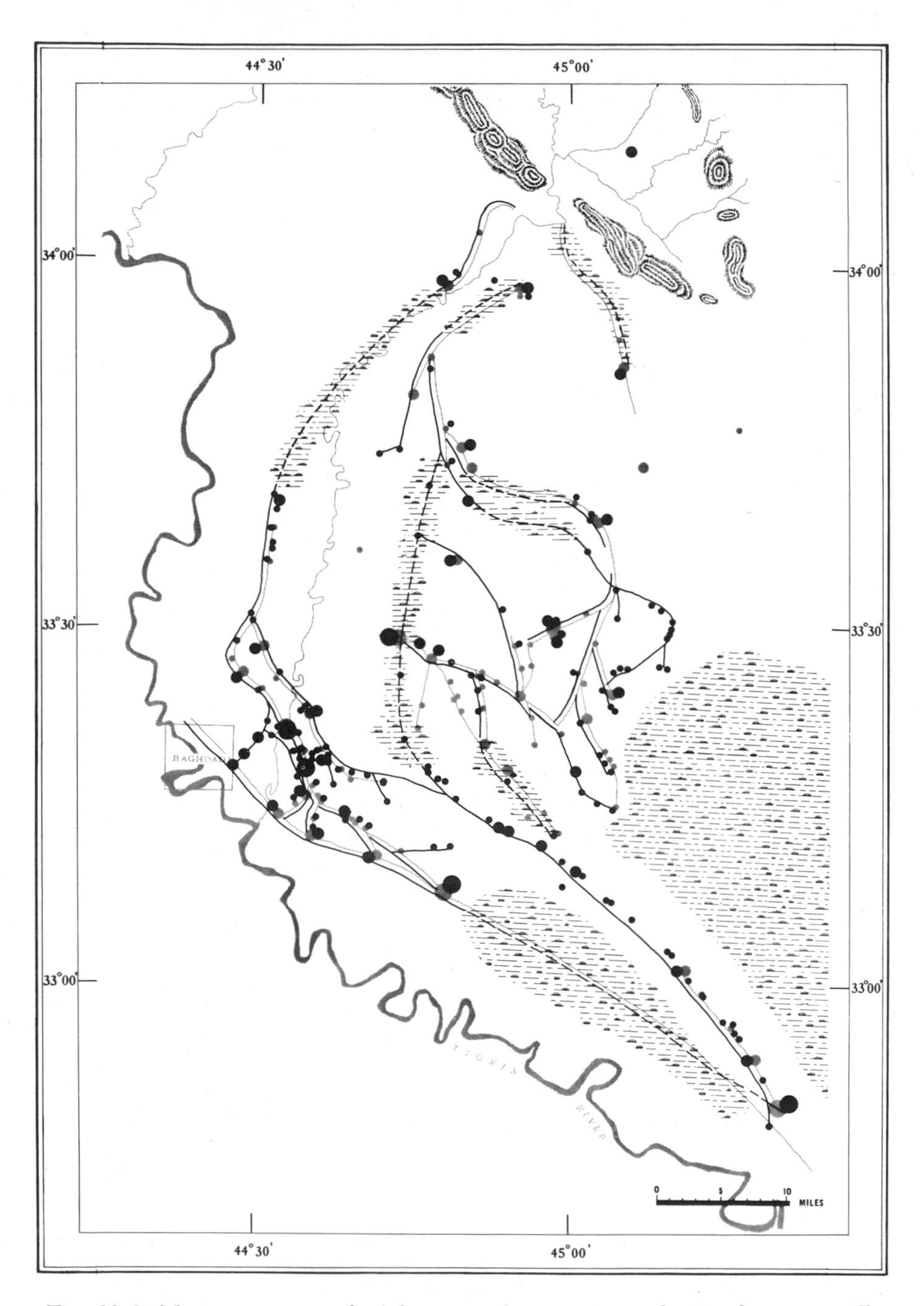

Fig. 18–2. Maximum extent of settlement and irrigation in the Diyala region. All canals shown by lines with minute serrations were in use during the Sassanian period, A.D. 226–637. However, expansion to the full limits came only with construction of the Nahrwan Canal (shown as a dashed black line) late in the period. Settlements shown as black circles are also of Sassanian date. The different course probably followed in places by the Tigris River during the Sassanian period is suggested by black dotted lines.

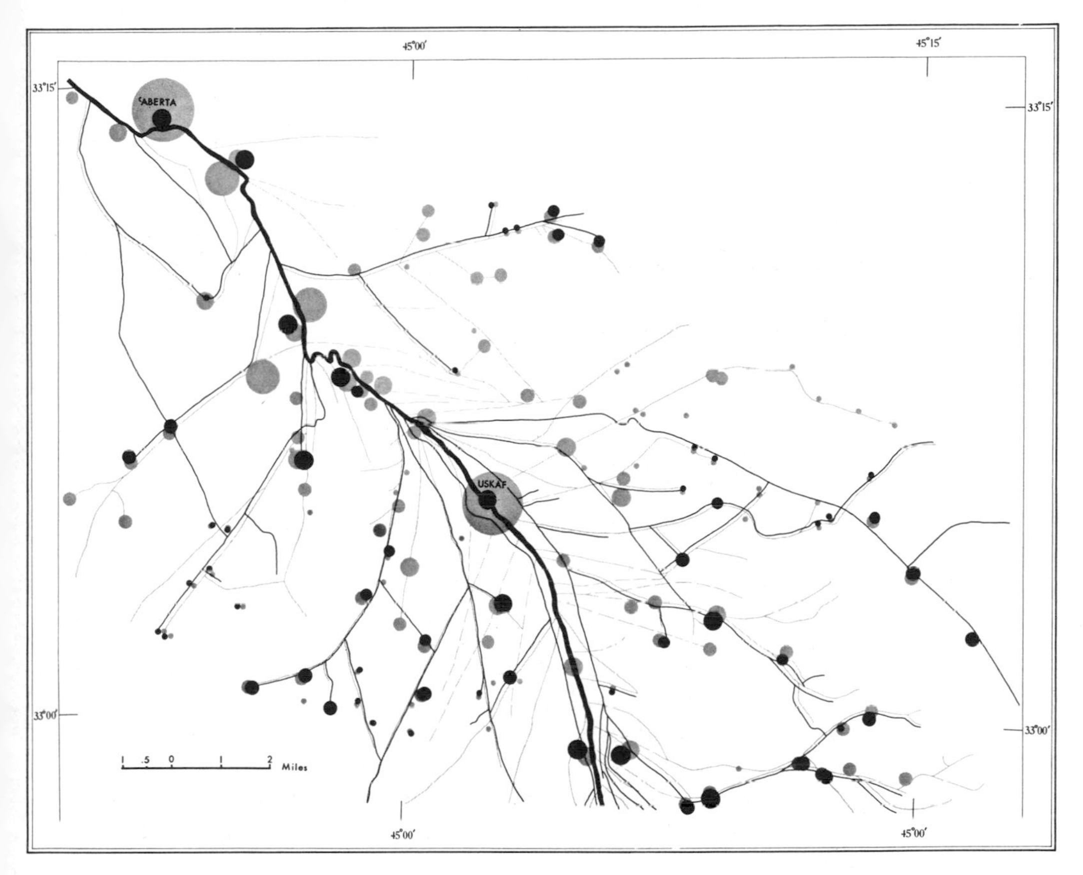

Fig. 18–3. Branch canal sequence along the Nahrwan. Branches shown as dashed gray lines date to the later Sassanian period (about A.D. 500–637). Settlements shown as gray circles and branch canals shown as continuous gray lines belong to the Early Islamic and Samarran periods, prior to about A.D. 900. Settlements and branch canals shown in black are those in use during the final phase of irrigation in the lower Nahrwan district, about A.D. 1100. The weir excavated by the project was located at the junction of numerous branch canals northwest of the city of Uskaf.

Chapter 19

Ancient Agriculture in the Negev

M. Evenari
L. Shanan
N. Tadmor
Y. Aharoni

The Negev desert of Israel, with its numerous, clearly visible traces of ancient civilizations dating back at least four to five thousand years, has attracted the attention of many scientists. Since E. H. Palmer[1] in 1871 described the general character of these civilizations as well as the intriguing agricultural remnants that he observed in the area, the Negev has become a field of research for many phases of science.

We have been working as a team in the Negev desert for five years with the specific aim of solving the enigma of the once flourishing agricultural civilizations in a now barren desert. This team covers the fields of botany, archaeology, ecology, hydrology, and water engineering, and this combination of experience and interests enabled us to correlate widely differing fields of observation. In this chapter we present some of our conclusions as to how the ancient civilizations maintained a thriving agriculture in the desert and also indicate their possible application in the future.

Description of the area. The Negev is shaped like a triangle (Fig. 19–1). Its base line stretches in the north from an imaginary line drawn from Gaza on the Mediterranean Sea, through Beersheba, to Ein Gedi on the Dead Sea. Its two sides stretch from Gaza and from Ein Gedi down to Eilat on the Gulf of Aqaba. The 12,500 square kilometers of the Negev can be divided into the six following subregions: (i) the coastal strip; (ii) the lowlands and foothills; (iii) the central highlands; (iv) the sedimentary southern Negev, mostly consisting of rolling gravel plains; (v) the crystalline southern Negev representing the northeast corner of crystalline Sinai; and (vi) the Wadi Araba depression.

The physiographic and climatic conditions vary from subregion to subregion, and the various civilizations naturally adapted their agricultural projects to these differing features. The densest settled areas have been discovered in the lowlands and the highlands, and since most of our investigations have been concentrated in these subregions, we will describe them briefly.

The lowlands and foothills. This subregion is a strip about 10 to 25 kilometers wide, bounded by the coastal region on the west and the central highlands on the east and covering about 150,000 hectares. The morphological structure is made up mostly of Eocene limestone hills separating wide rolling plains, with the elevations ranging from 200 to 450 meters above sea level. This area contains the ancient towns of Nessanah, Sbeita, Ruheibeh, and Khalassah. A number of large wadis, whose sources are in the highlands, cut through the plains and drain toward the Mediterranean Sea. The hillsides are generally covered with a very shallow, gravelly, saline soil possessing an immature profile. The flora is dominated by the Zygophylletum dumosi association.[2]

On the other hand, the Quaternary aeolian-fluviatile loess soils of the plains are relatively deep (2 to 3 meters) and only slightly saline. The Haloxylonetum articulati association is typical for these areas.

The highlands. This subregion covers some 200,000 hectares and contains the ancient towns of Mamshit (Kurnub) and Avdat (Abde). It is composed of a series of parallel anticlines, and the elevations vary between 450 and 1,000 meters above sea level. The anticlines are composed of Cenomanian Turonian limestones and cherts.

Between the high ridges, the main wadis drain to the Mediterranean and Dead seas. Adjacent to the wadis lie relatively narrow alluvial plains, and near the watershed divides where the wadis have not cut down to a stable base level, there are a number of expansive plains.

There are two principal plant habitats common to the area. On the rocky slopes (80 to 90 per cent of the area) where the soil cover is shallow, gravelly, and saline, the Artemisietum herbae albae association prevails with transitions to the Zygophylletum dumosi.

On the loessian plains and in the wadi bottoms where loess has accumulated, the vegetation consists of sparsely distributed low shrubs of the Haloxylonetum articulati association.

Rainfall conditions. The rainfall records of our area have not been kept systematically for any long period of time. But even the few short records that have been published show that we deal here with that typical pattern of rainfall which is so characteristic for all deserts.

A study of Table 19–1 shows that the variations between maximum and minimum annual amounts of precipitation are large and that most of the rain falls in quick short showers of less than ten millimeters. The difference between the average and median annual values should be noted, as, for agriculture, the median and not the average is significant.

The average number of rainy days with daily totals of precipitation of 0 to 3, 3 to 10, and more than 10 millimeters is another important figure, as it touches on the problem of the minimum "effective" rainfall.[3]

Agricultural history of the Negev. The Northern Negev was, in historical times, first settled during the Chalcolithic Period (fourth millennium B.C.). But up to now, no trace of this period has been found in the central and southern Negev.[4]

During Middle Bronze I (twenty-first to nineteenth century B.C.), the Negev was quite densely populated. The next period of sedentary settlement dates from the end of the tenth century B.C. to the beginning of the sixth—that is, the period of the Judaean Kingdom (Israelite periods II–III, or Iron Age II). However, the time between about 200 B.C. and A.D. 630 represents the longest and most flourishing period of almost continuous settlement in the Negev. During this time, the Nabataeans and Romans (about 200 B.C. to A.D. 330) and the Byzantines (A.D. 330 to

TABLE 19–1

Rainfall data (in millimeters).

ITEM	STATION		
	ASLUG	AUJA	MAMSHIT
No. of years of record	13	14	9
Highest total annual recorded	170	285	171
Lowest total annual recorded	52	25	58
Average annual	100	89	98
Median annual	86	65	80
Av. no. of days per year with total of 0–3 mm	9	5	5
Av. no. of days per year with total of 3–10 mm	7	3	6
Av. no. of days per year with total of more than 10 mm	2	2	2

630) ruled the area.[5] After the Arab conquest, from the seventh century A.D. up to our time, the Negev was occupied only by nomadic Bedouins.[6]

As far as the *agricultural* history of the Negev is concerned, our own surveys and excavations of Israelite farms and settlements[7] and the surveys of Nelson Glueck[8] have shown that the Israelite Period III settlers already carried out desert agriculture based on flood-water irrigation. We may mention that this is corroborated by the Bible (II Chron. 26:10), where it says of King Uzziah, who ruled the Negev down to Elath, "also he built towers in the desert, and hewed out many cisterns . . . for he loved the land." But it may even be that the Middle Bronze I people practiced runoff desert agriculture, as the Negev is full of their settlements.[9] During the Nabataean-Roman-Byzantine period, desert agriculture reached its peak of development. After the Arab conquest, the ancient desert agriculture slowly disintegrated, and the Bedouins of the area at best merely utilize dilapidating old systems for patch cultivation.

Outside of the Negev of Israel, ancient desert agriculture is known from the following areas: (i) North Africa (Algeria, Tunisia, Libya,[10, 11, 12, 13, 14]), Syria,[15] and Transjordan,[16] where it flourished under Roman domination; (ii) Southern Arabia;[17] (iii) North America, where it was practiced and is being practiced by the Indians;[18] and (iv) South America, where it was carried out by the pre-Hispanic civilizations.[19]

Ancient desert techniques of water utilization for agriculture. There are relatively large areas in the desert where the soils are suitable for cultivation, and the only requirement is water. This is true for most of the not too steep wadis, the flood plains, and the depressions where loess soils have accumulated to a depth of one to two meters. The key to establishing sedentary agriculture in a desert is, therefore, maximum utilization of the meager rainfall.

For this reason, our work has been concentrated on studying the techniques used by the ancient civilizations to collect and exploit the meager water resources of the area. The techniques that we have so far studied in detail, and that are presented in this article, can be divided into the three following main categories: (i) exploitation of runoff from small watersheds (up to 100 hectares in size); (ii) exploitation of runoff from large watersheds (up to 10,000 hectares in size); and (iii) chain-well systems.

Exploitation of Runoff from Small Watersheds

The exploitation of runoff from small watersheds[20] is by far the most interesting of all the methods utilized by the ancients, since it made possible the very intensive development of the area.

The basic principle of the method was simple but nevertheless required a good understanding of the sciences of hydrology, soils, and meteorology. Table 19–1 shows that most of the rainfall in the desert falls in relatively light showers—three to ten millimeters at a time. These meager amounts of rainfall are generally regarded as ineffective—that is, they wet a very shallow depth of soil, which dries by evaporation before plants can utilize the moisture. However, the loess soils of the area have a characteristic of forming a crust when wet. This crustal formation was studied by D. Hillel,[21] who has shown it to be an intrinsic feature of the Negev loess soil: the aggregated structure of the soil surface is destroyed by a wetting or slaking process. The crust decreases the water-intake rate of the soil and so increases the rate of runoff.

This phenomenon was observed by the ancients and exploited to the maximum. The loessial hillsides, which became more or less impermeable after wetting, were utilized as catchment basins to produce runoff for subsequent utilization in nearby fields. The desert farmer's aim was to prevent a penetration of rain on the slopes and so produce maximum runoff, whereas the farmer in more humid lands aims to soak all of the rain into the soil and so minimize runoff. The desert farmer directed the runoff from a large area on the slope to a small cultivated area in the bottomlands, and in this way he was able to collect sufficient water to ensure a crop even under adverse desert conditions.

This ingenious type of runoff agriculture we define as runoff farming, and the cultivated units to which it was related we call runoff farms.

Each runoff farm consisted of the farm area proper, containing the cultivated fields, and the surrounding catchment basin. The cultivated area was subdivided into terraces by low terrace walls. The function of the terrace walls was to retain the flood water on the field, where it could soak into the soil and be stored for subsequent use by the crops. A number of terraced fields were surrounded by a stone wall, constituting a distinct unit.[22] Within the area bounded by the wall there is very often a farmhouse or a watchtower. The hillsides surrounding the farm served as a catchment area from which water conduits channeled the runoff water onto the fields. Once the water was inside the farm, drop structures, ditches, and dividing boxes gave the farmer complete mastery over the distribution of the water. Figures 19–2 and 19–3 illustrate this very well. Figure 19–2 represents a system near Avdat. The whole catchment area comprises about seventy hectares and is artificially divided into a number of smaller catchment basins by several conduits, each leading to a specific terraced field in the narrow valley. Some of the conduits begin high on the plateau and collect runoff from there.

The seventy hectares of watershed of this system supplied water to about 2.2 hectares of cultivated land.

Figure 19–3 represents a number of runoff farms in the Shivtah area.

Each farm received its runoff water from its own small wadi and from the many conduits that collected water from the small catchment basins on the hillside adjoining each farm.

About a hundred runoff farms, together with their catchment basins, have been studied in detail.

Each farm unit formed an entity comprising a catchment basin and cultivated land. The larger the catchment basin, the more the water yield and the greater the corresponding area that could be irrigated. The ancient farmers often extended their water-collecting conduits to the plateaus high above their fields in order to increase the available water supply, and sometimes conduits were led around the hillsides so as to increase artificially the natural drainage area of the runoff farm. These catchments were therefore "water rights," and each runoff farm possessed a water right on a definite portion of the slope. These water rights, which generally vary in size from ten to one hundred hectares, were no less important a part of the runoff farm than the cultivated land itself. The man who owned water rights on the slopes could always build himself a farm, but not vice versa.

The farmland and its catchment on the slope are thus a mutually balanced system of land and water. All the precious water collected from the slope was used. If there was any surplus water on the farm, the cultivated area was extended by adding a new terrace downstream. It was probably only in exceptionally rainy years that surplus water passed over the lowest spillway of a runoff farm and flowed to the next terrace. On the other hand, permanently "dry" terraces were of no avail, and the farmer only built a new terrace if his expectations of getting it wet were reasonably good. Catchment and cultivated area are thus seen as a clearly defined unit—an integral part of an over-all plan of watershed subdivision.

The conduits generally collected water from a relatively small area, sometimes as small as 0.1 to 0.3 hectare and generally not larger than 1.0 to 1.5 hectares. The result was that the over-all runoff was always divided into small streams of water, preventing the occurrence of large flash floods. Such controlled flows are suited to the dry stone structures of the ancients; moreover, only such small flows could be handled by a farmer and allow him to control the flow during the flood period. Flows from even one hectare of catchment might reach a high peak intensity for short periods. For example, with a peak rain intensity of 30 millimeters per hour and a 60 per cent (an extreme figure) runoff for a short period during a single rain storm, one hectare of slope might yield a peak flow of 180 cubic meters per hour, if only for a few minutes. This requires a ditch with a cross section of 0.05 to 0.10 square meter (depending on gradient), a requirement which readily fits observed ditch dimensions.

The farmer could therefore not allow the waters to collect from a larger area, as the resulting peak flow would have been unmanageable and would have destroyed his terrace structures. The over-all runoff was thus effectively broken up into small streams.

The crucial question that arises is the amount of runoff the ancient farmers received per unit area. Actual field measurement of runoff already initiated on our two reconstructed farms, discussed below, will have to be made for at least ten years before a reliable estimate can be made. We approached this question indirectly by analyzing the ratio

$$R = \frac{\text{area of catchment basin of ancient farm}}{\text{area of cultivated area of ancient farm}}$$

About a hundred farms in the Avdat, Shivtah, and Auja areas show that this ratio varies between 17:1 and 30:1, with an average value of about 20:1. This means that between 20 and 30 hectares of catchment area were needed to irrigate 1 hectare of cultivated field. Present-day agricultural experience has shown that flow of at least 3,000 to 4,000 cubic meters per hectare has to be applied as supplementary irrigation in order to insure any crop in this desert area. (Each 1,000 cubic meters of water per hectare will wet about a meter of soil depth.) Taking these figures as a basis, we calculated that if 20 hectares on the slopes supplied the 3,000 to 4,000 cubic meters of water, every hectare of catchment supplied 150 to 200 cubic meters of water per year. If each hectare of catchment supplies 150 to 200 cubic meters of water (which is equivalent to 15 to 20 millimeters of rainfall), we can safely conclude that the coefficient of runoff was at least 15 to 20 per cent of the total annual precipitation.

These runoff farms formed an important part of the desert settlements throughout the ages, and primitive but nevertheless well-defined runoff farms have been found dating from the tenth to the eighth centuries B.C.[7] This form of intensive sedentary agriculture was probably continuous throughout all the civilizations, reaching its peak in the Roman-Byzantine era. Interesting and conspicuous features related to the runoff farms are the gravel mounds and gravel strips (see figs. 19–4, 19–5, 19–6) and stone mounds and strips. These man-made structures cover thousands of acres and are common in the vicinity of the ancient cities of Avdat and Shivtah. The gravel mounds are low heaps of gravel artificially arranged in long rows with a more or less uniform distance between the mounds. The strips are of the same material. Mounds and strips are often intermingled and form all kinds of intricate patterns. They are only found on hammadas[23] covered by small gravel and are made by raking together the gravel.

The stone mounds and strips are built of much bigger stone fragments and are typical for those areas where for geological reasons the slopes are covered with big stone fragments and not with gravel. Gravel mounds, gravel strips, stone mounds, and stone strips are found exclusively on slopes leading to farms or cisterns.[24, 25]

Since Palmer[26] first discovered these structures, all authors dealing with them agree that they are related to agriculture, and the following theories have been proposed concerning their function:

1. Palmer was told by his Bedouins that the Arabic name for these structures is *teleilât el 'anab* or *rujum el Kurum*—that is, "grape mounds" or "vineyard heaps." "These sunny slopes," he concluded, "would have been admirably adapted to the growth of grapes and the black flint surface would radiate the solar heat, while these little mounds would allow vines to trail along them and would still keep the clusters off the ground."

A number of authors,[27] and lately P. Mayerson,[28] follow Palmer's theory. In our opinion, the slopes can never have been used for growing grapes because there is either no soil at all or only a very shallow superficial soil cover that is highly saline (2 to 5 per cent total soluble salt). The naturally occurring plant associations on these slopes indicate the most difficult growing conditions for plants. As we have shown, the amount of rain water these slopes receive is insufficient for growing grapes, and since the ancient farmers never used all the good loess soils available, there was no reason for them to cultivate the worst soil to be found in all the desert.[29]

2. Some authors[30, 31, 32] believe that the function of the mounds was to condense dew. But experiments have shown that no dew can be collected in the mounds, and that the water relations of the soil below the mounds do not differ from those of the surrounding soil.[33]

3. Y. Kedar[34, 35] put forward the theory that the main function of the mounds was to increase soil erosion from the slopes in order to accumulate more soil in the wadi bottoms ("accelerated erosion"), as in his opinion the main hindrance to agriculture was lack of suitable soil and not lack of water. There are a number of objections to this theory. First, there is and was plenty of good loess soil in the valley bottoms and flood plains close to the ancient agricultural systems. As today, lack of water and not lack of cultivable soil was the main problem of the ancients. It is hard to believe that the ancients, who knew so much about water spreading, would have endangered their elaborate systems by intentionally introducing silting, the arch enemy of any water-spreading system. Furthermore, some of the gravel-mound areas lead to water cisterns, where the accumulation of silt by erosion is most undesirable. The gradient of some of the collecting ditches varies from 0.5 to 1 per cent. If they had been designed to carry silt, a much steeper gradient would have been necessary. But the main objection lies in a simple calculation. According to Kedar's experimental figures,[36] an ancient farmer

would have had to wait patiently for about twenty to fifty years after building an elaborate structure in the wadi before sufficient soil had accumulated to justify the planting of a crop.[37]

4. We have proposed[38] that mounds and strips were established in order to increase the amount of surface runoff and gain more water for the fields below.[39] Hillel[40] has shown that the infiltration capacity of the prevailing soil of the region decreases markedly with the formation of a characteristic surface crust through the physical slaking of the upper layer during the wetting-drying cycle. This increases the runoff. Crust formation is prevented by the presence of a protective surface of gravel. Therefore, by clearing the slopes, the soil surface was exposed, crust formation was enhanced, and runoff was increased. This resulted in greater water yields from the slopes.

Thus, mounds were only a by-product of clearing the surface of stones. Strips sometimes fulfilled an additional function in channeling the water from the slopes to the fields. This is especially obvious in connection with the stone strips and conduits (see Fig. 19–2).

The fact that the mound and strip areas are always connected to the fields by channels (Fig. 19–2) is in conformity with this theory. Apparently this ingenious system was not restricted to the ancient desert agriculture of the Negev.[41]

Exploitation of Runoff from Large Watersheds

For purposes of our work the term *large watersheds* is taken to mean watersheds greater than about a hundred hectares in size. The hydrology of these large watersheds differs from that of small catchment areas. In the small catchment, runoff may begin after a small amount of rain (3 to 6 millimeters) has fallen, while on the other hand, a rainfall of at least 10 to 15 millimeters is required to cause a flow in the wadi of a large watershed.[42, 43] Furthermore, the percentage of runoff from a small catchment basin may be as high as 20 to 40 per cent of the annual rainfall, but in the larger watershed it would not be greater than 3 to 6 per cent. The small watersheds produce relatively small streams that can be handled easily by simple structures, whereas the flash-flood flows of the large wadis can destroy even the strongest of engineering structures. These factors led to the development of systems of water exploitation that differed both in form and extent from those described above for small watersheds. The Mamshit system is one of the best preserved. The ancient town of Mamshit is situated on a range of Turonian Cenomanian hills overlooking the Tureiba Plain. Just south of the town, Wadi Kurnub cuts a narrow gorge through the Hatira anticline and enters the Tureiba Plain. At the point where the gorge enters the Tureiba Plain, the drainage basin has an area of about 27 square kilometers, and it was below this point that the flood waters from the large watershed were exploited.

Figure 19–7 is a map of this system and Fig. 19–8 is an aerial photograph of the area. A large diversion channel, about 400 meters in length, leads the diverted waters of Wadi Kurnub at a 2:1000 gradient to the flood plain. The original diversion dam has been completely destroyed but need only have been a simple rock structure to have raised the water level 30 to 50 centimeters in order to control the lower flood plain. This diversion channel leads the water to a series of broad terraces that are all in good condition. The terraces are more or less level in the transverse direction but have a slight gradient (2:1000 to 4:1000) in the direction of flow of the water. This arrangement made it possible to irrigate the area either in large basins or in small plots. The excess water from each terrace flowed to the next lower terrace through well-built drop structures.

The total area of the cultivated terraces is about 10 to 12 hectares. Agricultural experience has shown that about 3,000 to 4,000 cubic meters of water per hectare should be applied each year in order to insure an agricultural crop. This means that the watershed supplied about 40,000 to 50,000 cubic meters per year to the cultivated terraces. This represents less than 2 per cent of the annual rainfall on the large watershed and could be expected every year as runoff. This quantity of water could have been carried by the diversion canal in six to ten hours, according to the depth of flow (which probably did not exceed 40 to 60 centimeters).

A detailed examination of the area disclosed that the most ancient system was established when the wadi flowed in a shallow depression in the flood plain and before it had cut through the alluvial soils. The first walls were built primarily as stabilizing structures for the shallow depression, and only subsequently were they extended, in order to spread the water across the flood plain. Some of these walls can still be found on the opposite side of the wadi, showing that they predate the gully stage of Wadi Kurnub. The most ancient potsherds found in the vicinity belong to the Middle Bronze and Iron ages, but there is still no certain evidence that this first system antedates the Nabataean period.

These stabilizing walls assisted in the deposition of alluvial silt in the terraces, and so their level was gradually raised. At some period, either through natural flood conditions or because the inhabitants abandoned the area for historical reasons, the wadi destroyed the stabilizing walls, and an ever-deepening gully was cut through the system. The next users of the area were therefore faced with an entirely different problem: the runoff water no longer flowed in a shallow depression but concentrated in a wadi, one or two meters below the flood plain. They therefore had to base their system on a diversion structure that raised the water out of the wadi bottom and directed it to a diversion canal, which in turn led the water to the old terraces. The remains of this system stand out clearly on the aerial photographs and are the easiest to find in the field. Close inspection also revealed a number of diversions in the lower reaches of the wadi. These indicate that the elevations of the terraces were continually rising because of a silting process, and that new diversion structures at higher elevations had to be built in order to control these new elevations.

Chronologically, the next system that is clearly discernible in the field seems to have been constructed when the diversion channel had become so silted that the whole system based on a diversion channel may have had to be abandoned. This system is based on a completely different principle. The main area with the diversion channel was not used, and only the lower terraces (about three to four hectares) were irrigated. This area was developed as a runoff farm and received its water from the relatively small watershed (3,500 dunams) adjoining the area, not from Wadi Kurnub.

The system in Nahal Lavan (Wadi Abiad) (see Fig. 19–9) is much more complicated but nevertheless shows similar lines of development. Nahal Lavan is the largest wadi in the vicinity of the ancient town of Shivtah and drains from the high plateau of the Matrada through a large area of barren rocky Eocene hills. The torrential floods which have poured off these hillsides have cut a deep wadi through the alluvial plain. In the upper reaches, the plain is narrow (100 to 200 meters wide), but in the lower reaches the flood plain is more than a kilometer in width. Today the wadi is a gravel-bed watercourse typical of the area. All along these alluvial flood plains are remnants of ancient walls and terraces, some of the walls reaching a height of four to five meters. These high walls (Fig. 19–10) attracted our attention, and a specific area covering 200 hectares was studied in detail.[44] The drainage area of Nahal Lavan at this point is about 53 square kilometers.

A close examination of the area disclosed again the superimposition of many systems. For a long time it was difficult to unravel the intricacies of each period or even to differentiate between the systems. Only toward the end of the survey did we realize that the capacity and size of the spillways, canals, and drop structures give the key to understanding the area. The spillways, which served as drop structures to carry the water from one terrace to the next lower one, can be classified into three distinct categories: (i) spillways with crest lengths of 30 to 60 meters, capable of handling flows in the range of 10 to 30 cubic meters per second (see Fig. 19–11); (ii) spillways with a crest length of 3 to 8 meters, capable of handling flows in the range of 1 to 5 cubic meters per second; and (iii) small spillways up to 1 meter wide for flows of less than 1 cubic meter per second.

Using this criterion as a starting point, we were able to differentiate between three different types and stages of development.

The earliest use of the area was found in the lower reaches of the area surveyed, where well-constructed stone spillways with a 30- to 60-meter opening are the common form of structure. However, these spillways were not connected to any stone walls, and it seemed as though these structures were all that remained of some ancient system —that is, that the stone walls had been dismantled and only the structures had been left standing. However, a special helicopter reconnaissance flight revealed that these wide stone spillways were connected to faint lines in the fields. Inspection of these lines disclosed them to be the remains of earth embankments that had stretched across the flood plain. The complete extent of this flood-plain spreading system was not surveyed, but it was clear that it was in use long before Nahal Lavan became a deep gravel-bed watercourse. Some of the spillways are capable of handling a flood flow of up to 100,000 cubic meters an hour (see Fig. 19–11). The topographic situation of this system indicated that it was in use when Nahal Lavan was a shallow depression and that the earth embankments were built in order to spread the runoff waters across the wide flood plain. The wide stone spillways were used to control or direct the water as it passed from a higher to a lower elevation.

In the upper reaches of the surveyed area, a second system based on diversion canals and structures (capable of handling one to five cubic meters a second) was discovered. Some of these main diversion canals are more than a kilometer in length and five to ten meters wide, and most possess a gradient of 4 to 5 per cent. All lead to diversion structures which served to divide the canal flow into as many as seven secondary canals leading to leveled terraces. Some of these terraces are in good condition, but most of them are badly eroded by gullies which join Nahal Lavan five meters or more below the level of the terraced fields. Each diversion canal serves an area of about two to four hectares.

Detailed investigation of the walls of the diversion canals and the terraced walls associated with them showed that these systems were also built in stages. Figure 19–10 shows one of the diversion canal walls and the three distinct periods of construction. Excavation alongside the terraced walls showed similar periods of construction. These observations indicated that the diversion system silted up during its operation and that the settlers were continually faced with the problem of raising the elevations of the terraced walls as well as the diversion structures. Potsherds in the area dated from the Nabataean-early Roman period.

The next use of the area was again as runoff farms connected to adjoining small watersheds (Fig. 19–9). These farms adapted the existing structures and stone walls of the diversion systems to their needs and did not exploit the runoff from Nahal Lavan. Potsherds in the vicinity of these farm units generally dated to the Byzantine era.

We were originally under the impression that diversion systems of this type were widely used by the ancient civilizations. Although we have traveled widely in the area and have studied hundreds of aerial photographs, we have now come to the conclusion that this method was used only in very special restricted areas, and furthermore no diversion canal has been found that served more than three to five hectares.

All the systems studied showed a remarkable similarity in their development. This development is characterized by three stages each related to the erosion that was taking place in the flood plains and wadis associated with the large watersheds.[45] This development can be divided into three stages, as follows:

Stage 1: Flood-plain development. The major wadis were originally wide shallow depressions meandering in alluvial plains. Cultivation of these depressions necessitated the construction of stone walls in order to stabilize the cultivated fields. These walls were subsequently extended so as to spread the water over larger sections of the flood plain.[46]

The main spillways of this system were characterized by wide openings (30 to 60 meters) for handling the whole flood flowing in the depression. The embankments in some cases were built of earth.

This flood-plain development period dates back at least to the Nabataean period and may be earlier.

Stage 2: Diversion systems. At some stage, these flood-plain spreading systems were abandoned and the system deteriorated through lack of maintenance. During or subsequent to this abandonment, the wadi cut a deep gully through the flood plain. The next settlers in the area utilized the technique of raising the water from the wadi with the aid of a diversion structure and leading the water by means of a channel to the flood plain. These diversion channels generally served small areas, and in most cases the new settlers utilized the remnants of the previous flood-plain development walls and structures. During this period the wadi continued to erode, and at the same time silt from the large watershed (or from the eroding banks of the wadi itself) was deposited in the terraced fields. This silt raised the level of the fields until a stage was reached that first necessitated raising the walls and later required the building of a new diversion structure higher up the wadi to find the diversion canal.

The period of construction of these diversion systems must have been one in which the science of engineering was well developed, since all the structures required sound knowledge of hydrology and hydraulics. Furthermore, this period must have been one in which a central authority controlled the whole system and had the legal authority to distribute the flows during the short flood period that occurred in the ephemeral wadis. In both the Roman and Byzantine periods these conditions existed, and the Roman and Byzantine potsherds found in the area probably relate to this diversion-system period.

Stage 3: Runoff farms. The diversion system may have become unmanageable because of the silting problem, or serious flood conditions may have destroyed the main diversion features, and the system was abandoned. The next system no longer relied on the main wadis but utilized the small watersheds adjoining the area in order to obtain the required runoff water. These runoff farms adapted existing walls and structures to their new requirements and generally utilized only part of the original diversion-system area.[47]

Chain-Well Systems

In the Middle East and Central Asia, chain-well systems ("artificial springs") have been used since ancient Persian times and are still widely used today. Their construction and operation have been fully described in the literature.[48]

While well digging was a common method of exploiting shallow ground-water resources in the ancient civilizations in Palestine, the more intricate chain-well systems have only been found in Jordan and the Arava Rift Valley.[49, 50] As the mean annual rainfall in the valley is only about 40 millimeters and this amount of rainfall is without agricultural value, the chain-well systems must have been the main source of irrigation water.

Chain-well systems have been located at three oases in the Wadi Arava Rift Valley. The largest and most intricate is near the Ein Ghadian (Yotvata) oasis. Other systems were discovered near Ein Zureib and near Ein Dafieh (Ein Evrona).[51] Since these systems are hardly discernible on aerial photographs and are difficult to discover from the air or even in the field, it is likely that a thorough investigation of the Wadi Arava would disclose many other systems.

A chain-well system is composed of three essential parts: (i) one or more wells (sometimes called "mother-wells") dug down to the water table; (ii) an almost horizontal underground tunnel leading the water, at small gradient, to the soil surface and ending in an open ditch; and (iii) vertical shafts connecting the tunnel to the ground surface. These shafts facilitate the construction of the tunnel and the disposal of excavated material in a molelike fashion and also provide access and ventilation to the tunnel for maintenance purposes. The surplus excavated material is deposited near the shafts, forming a circular mound around the shaft opening.

The oasis of Ein Ghadian, which was examined in detail, is presented as a typical example of a chain-well system of the Arava Valley. The oasis itself is of the playa type, and the central part, where the water table is 1 to 1.5 meters deep, is saline and sterile. It was natural that this oasis, the largest on the western side of the Arava valley floor, was constantly settled. There are several remnants of ancient settlements extending from Middle Bronze to Roman-Byzantine times. Ein Ghadian was also the first station on the Roman road from Eilath to the north of Palestine.[52]

Figure 19–12 shows a part of one chain-well system at Ein Ghadian as seen from the air; Fig. 19–13 shows details of one of the systems.

The chain-well systems vary in length; some are three to four kilometers long, others seem to be only a few hundred meters long. The vertical shafts are spaced at distances of about 15 to 25 meters, center to center. In most systems, only relatively few of the original circular mounds and shafts are still intact, owing to the obliterating action of winter flash floods. In those sections where the danger of destruction by floods was greatest, remnants of stone protection walls are found on the upstream side of the line of shafts.

All systems apparently begin in the gravelly wadi-fans on the western edge of the Arava depression and may possibly be connected to a definite fault line. The tunnel part of the system always seems to terminate in an elevated earth ridge on which there is a thick growth of *Eragrostis bipinnata* ("love grass"). These ridges are probably the old irrigation channels.

The systems and their channels lead to the northwestern edge of the Ein Ghadian playa, which is covered by stands of *Eragrostis bipinnata* rooted in the water table. Closer inspection reveals that the individual tussocks of *Eragrostis bipinnata* form regular checkerboard patterns. It is possible that these stands indicate the area of ancient irrigation. Each tussock would then represent an irrigated basin, or possibly the point where a palm tree was rooted. However, if the water level in ancient times was different, the irrigable area would, of course, have changed correspondingly.

Practical application. Many authors,[53] investigating the area, have suggested that the ancient and forgotten civilizations of the Negev could teach a practical lesson for the future. We, too, felt that some of the principles on which the ancient civilizations developed their desert agriculture could be applied today. The written records of ancient agriculture in the desert are limited principally to the Nessanah documents.[54] But even the little information given in these publications encouraged us in this line of thinking.

However, the first question that we had to decide was whether there had been any climatic changes during this period of time. We are of the opinion that there has been no major climatic change in the area—that is, that the Negev has always been a desert with an average annual rainfall of about a hundred millimeters. If there had been a more humid climate in ancient times, there would have been no need to develop this ingenious desert agriculture based on maximum water conservation; necessity was the mother of invention. However, we are also of the opinion that there were definite variations in the average annual rainfall. The twenty-year moving average may have fluctuated between 70 and 150 millimeters, but these differences would probably have evened out on a 100- to 200-year moving average.

We then decided to reconstruct two ancient runoff farms, one near Shivtah and one near Avdat. In doing so, our aims were (i) to collect exact data about rainfall and runoff and, if possible, develop an analytical relationship between them, and (ii) to find out what, if any, agricultural crops and fruit trees could be grown by utilizing *only* the runoff from small watersheds.

After a careful survey, both farms were reconstructed with all their terraces, walls, and channels.

The Shivtah farm. The runoff farm shown in Fig. 19–14 was chosen for reconstruction.[55] The reconstruction was started in the summer of 1958. Where a channel of the wadi led water into the farm, a weir and an automatic flood-recording gauge were built. A meteorological station was erected near the farm, and nine automatic and simple rain gauges were distributed over the whole catchment area. In February, 1959, after the first flood, 250 fruit trees and vines were planted (grape, almond, apricot, peach, plum, carob, olive, pomegranate, and fig). During the summer of 1959 the young trees received small amounts of additional irrigation in order to insure their establishment. From then on, they received only runoff water from the small watersheds. The results by August, 1960, were very encouraging, since during this short period the young saplings grew from a height of 40 to 50 centimeters into trees 2 to 2.50 meters high, despite the fact that both years were severe drought years, the season 1959–60 being the driest one since meteorological measurements were established in Palestine-Israel. During the rainy season 1960–61 we planted another 50 fruit trees.

The Avdat farm. This reconstructed farm lies at the foot of the hill on which the ancient city of Avdat is situated. Its reconstruction was started in July, 1959, after a careful topographic and soil survey.

The soil (as in the case of the Shivtah farm) is the typical aeolian-fluviatile loess of the Negev. It is uniform over the whole farm area with the exception of small strips in the upper part and along the sides of the farm. These parts will not be used for our agricultural experiments. The loess is uniformly 1.50 to 2.50 meters deep. A farmhouse (see Fig. 19–15) containing a laboratory, a kitchen, two sleeping rooms for the staff, and one sleeping room for visiting scientists was built on the hill overlooking the farm.[56] Near the house there is a meteorological station more complete than that at the Shivtah farm. Two automatic and seventeen simple rain gauges were distributed over the whole area. Eight weirs and automatic runoff recording gauges were set up as described for the Shivtah farm.

After the first rain of 16 to 22 millimeters, in November, 1959, a heavy flood wetted the whole farm area down to a depth of 1 to 2.5 meters. Barley of the Beacher variety was sown. It sprouted quickly and was harvested in May, 1960. On selected parts of the area, the yield was 125 kilograms per dunam (or 500 kilograms per acre) (Fig. 19–16). This is a quite astonishing yield for this most severe drought year, with only 40 millimeters of rain, when thousands of dunams of barley in the more northern area of Israel, with 80 millimeters and more of rain (but without additional runoff), failed utterly.

On the basis of the encouraging results of the first year, our agricultural committee has drawn up the following plan, which is now being carried out.[57]

1. Field crops: 80 plots (3 by 25 meters) are established. The water-distribution system to these plots is so arranged that the plots get equal, known quantities of runoff water and there is full control of this distribution of the flood runoff. The plan provides for different field crops to be sown according to the time of year when the first flood occurs and according to the depth of penetration.

2. Pastures: 10 plots are established as nursery areas. An additional 5 to 7 dunams are used as observation areas which receive only partially controlled quantities of runoff water.

3. Orchards: On 10 to 12 dunams, 200 fruit trees and vines are planted (pistachio, cherry, peach, apricot, and grape).

NOTES

1. E. H. Palmer, *The Desert of Exodus* (Cambridge, England, 1871).

2. For phytogeographical, phytosociological, and ecological data about the Negev, see H. Boyko, *Palestine J. Botany Rehovot Ser.*, **7**, 17 (1949); D. Zohary, *Palestine J. Botany, Jerusalem Ser.*, **6**, 27 (1953); M. Zohary, *ibid.*, **4**, 24 (1947); M. Zohary and G. Orshan, *Végetatio,* **5–6**, 341 (1954); M. Zohary, *Geobotany* (Sifriath Hapoalim, 1955) (in Hebrew).

3. The question of the minimum "effective" rainfall is a most important one for all desert areas. N. H. Tadmor and D. Hillel [Israel Agr. Research Sta. Rehovot, paper No. 38 (1957)] suggest that rainfall in amounts of less than 10 to 15 mm is "ineffective"—that is, has no or little effect on vegetation and runoff. Y. Kedar [*Econ. Quart.*, **5**, 444 (1958) (in Hebrew)] estimates that 50 per cent of the average yearly rainfall in the Negev is effective. However, our own first measurements showed that rainfalls much smaller than 10 to 15 mm are effective and start runoff.

4. For the history of the Negev, see the many publications of N. Glueck and especially his book *Rivers in the Desert* (Farrar, Straus, and Cudahy, New York, 1959). As for prehistoric times, there is much evidence of prehistoric settlement in the Negev, perhaps even reaching back to the Palaeolithicum [see N. Glueck, *Bull. Am. Schools Oriental Research*, **142**, 17 (1956)]. Concerning the Chalcolithicum, we may have to change our opinion, as N. Glueck [*Biblical Archaeologist*, **22**, 82 (1959)] reports that he found chalcolithic sites in the Central Negev.

5. An excellent historical sketch on the Nabataeans has lately been written by J. Starcky [*Biblical Archaeologist*, **18**, 84 (1955)]; see also M. Evenari and D. Koller, *Sci. Am.*, **194**, 39 (1956).

6. N. Glueck [*Biblical Archaeologist*, **22**, 82 (1959)] writes of this period: "The Byzantine period in the Negev came to an end . . . as a result of the Mohammedan conquest. Darkness and disintegration and reversion to desert have characterized its history since then."

7. M. Evenari, Y. Aharoni, L. Shanan, N. H. Tadmor, *Israel Exploration J.*, **8**, 231 (1959); Y. Aharoni, M. Evenari, L. Shanan, N. H. Tadmor, *ibid.*, **10**, 23, 97 (1960).

8. N. Glueck has stressed this point in his surveys of the Negev published in many issues of the *Bulletin of the American Schools of Oriental Research*; see also N. Glueck, *Rivers in the Desert*. See also F. M. Cross and J. T. Milik [*Bull. Am. Schools Oriental Research*, **142**, 5 (1956)], who reported Iron Age II sites and desert agriculture from the wilderness of Judaea (Wadi Buqeah, near Qumran). According to their description, they found what we call "runoff farms."

9. The Middle Bronze I period presents two main problems. To what ethnic group did the people of this period belong? N. Glueck [*Biblical Archaeologist*, **18**, 2 (1955); *Bull. Am. Schools Oriental Research*, **149**, 8 (1958)] calls this period in the Negev the "Abrahamitic age." But the scholars are not yet agreed on the date of Abraham's wandering through the Negev. Even if Glueck's date is right, the people of this period cannot be identified with Abraham's people. The second question arises in connection with the occupation of the people in the Middle Bronze I period. Were they cattlemen or agriculturists or both? Glueck [*Bull. Am. Schools Oriental Research*, **138**, 7 (1955)] calls them "Tillers of the soil" and states in many of his publications that they practiced agriculture. However, no investigator has yet related ancient fields to any of these settlements.

10. R. Calder, *Man Against the Desert* (Allen and Unwin, London, 1951).

11. J. Baradez, *Fossatum Africae* (Arts et Métiers Graphiques, Paris, 1949).

12. Carton, *Rec. notes et mém. soc. archael. Constantine*, **43**, 193 (1909).

13. O. Brogan, *Illustrated London News* (22 Jan., 1955); M. Renaud, *Rev. agr. Afrique du Nord*, **56**, 689 (1958).

14. Carton, in his excellent paper, was perhaps the first to recognize clearly the main principles of ancient desert agriculture—that is, the use of runoff from sterile hills and the storing of the runoff water in the soil. He writes: "Il s'agit ici d'ouvrages ayant pour but *non pas l'irrigation proprement dite, mais l'inondation ou la submer-*

sion" (italics ours). The main aim of the system was "de faire pénétrer lentement et profondément l'eau dans le sol." He was the first, too, to point out "l'ingénuité et la prévoyance des Anciens qui, au lieu d'énormes barrages-réservoirs, coûteux et dangereux, destinés à l'irrigation, avaient préféré réserver l'eau dans ces immenses réservoirs-souterrains. . . ." We cite him verbatim because his paper is not easily available. Though this desert agriculture flourished most in Roman times, Carton is of the opinion that it may be older than the times of Carthage and Rome and dates perhaps back to the old Berber population.

15. A. de Poidebard, "La trace de Rome dans le désert de Syrie," (Librairie orientale, P. Geuthner, Paris, 1934); S. Mazloum in R. Mouterde and A. de Poidebard, *Le limes de Chalcis* (Librairie orientale P. Geuthner, Paris, 1945).

16. N. Glueck, *Ann. Am. Schools Oriental Research,* **14** (1934); **17–19** (1939); **25–28** (1951).

17. F. Stark, *Geograph. J.,* **93,** 1 (1939); H. St. J. B. Philby, *Sheba's Daughters* (Methuen, London, 1939); W. Phillips, *Quataban and Sheba* (Harcourt, Brace, New York, 1955); R. L. Bowen, Jr., in *Archaeological Discoveries in South Arabia* (Johns Hopkins Press, Baltimore, 1958). Special mention must be made of the enormous ancient irrigation dam of Marib—possibly the biggest ever built in ancient times—constructed in the eighth century B.C., which broke down in the sixth century A.D. [E. Glaser, *Reise nach Marib* (Hölder, Vienna, 1913); A. Grohmann in *Encyclopedia of Islam* (Brill, Leiden, 1913)]. The oldest irrigation dam, which apparently broke immediately after it was finished, was found in Egypt, dating back to the Third or Fourth Dynasty (about 3000 B.C.) [see B. Hellström, *Houille blanche,* **1952,** No. 3, 424 (1952)].

18. E. F. Castetter and W. H. Bell, *Pima and Papago Agriculture* (Univ. of New Mexico Press, Albuquerque, 1942); *Yuman Indian Agriculture* (Univ. of New Mexico Press, Albuquerque, 1951).

19. G. de Reparaz, *El programa de estudios de la zona arida Peruana* (UNESCO, 1958).

20. See L. Shanan, N. Tadmor, M. Evenari, *Ktavim,* **9,** 107 (1958); **10,** 23 (1960).

21. D. Hillel, *Bull. Israel Agr. Research Sta. Rehovot,* **63,** 1 (1959) (in Hebrew with English summary).

22. These farm fences apparently served two purposes. They are a symbol of property [Y. Kedar, *Israel Exploration J.,* **7,** 178 (1957)] and, at the same time, a control structure. Most of them run around the farm at the base of the slopes preventing undesired material from the slopes from being carried into the fields and permitting the runoff water to enter the fields only at the places desired [see also N. Glueck, *Bull. Am. Schools Oriental Research,* **149,** 8 (1958); **155,** 2 (1959).

23. In an earlier paper [M. Evenari and G. Orshansky, *Lloydia,* **11,** 1 (1948)], hammadas were described as follows: "Hammadas are slightly rolling gravelly desert plains whose surfaces are strewn with vari-sized stone fragments and pebbles. Such fragments are brown or black, encased in the so-called 'Schutzrinde' of the German authors, regardless of whether the core itself is composed of chalk, granite, flint, or schist. The brown and black surface of the pebbles shines brightly as it is covered by the 'desert lacquer'. . . . This black lacquer gave rise to the Arab legend that these stones were scorched by heavenly fires."

24. Y. Kedar, *Bull. Israel Exploration Soc.,* **20,** 31 (1956) (in Hebrew with English summary).

25. Kedar was the first to point out the difference between stone and gravel mounds, according to the lithological material available.

26. E. H. Palmer, *op. cit.*

27. A. Musil, *Arabia Petraea* (Hölder, Vienna, 1907); T. Wiegand, *Sinai* (Gruyter, Berlin and Leipzig, 1920); C. A. Woolley and T. E. Lawrence, "The wilderness of Zim," *Palestine Exploration Fund Annual* (1914–15).

28. P. Mayerson, *Bull. Am. Schools Oriental Research*, **153**, 19 (1959).

29. Most of the arguments against this theory are discussed in N. H. Tadmor, M. Evenari, L. Shanan, D. Hillel, *Ktavim*, **8**, 127 (1957). N. Glueck [*Bull. Am. Schools Oriental Research*, **149**, 8 (1958); **155**, 2 (1959)] and Y. Kedar [*Bull. Israel Exploration Soc.*, **20**, 31 (1956); *Geograph. Rev.*, **123**, 179 (1957)] also refute Palmer's (and Mayerson's) theory. There are only two points of Mayerson's which merit attention additional to that given by Glueck (1959). First, Mayerson presents a photograph of some stone heaps which were not on a slope but in a wadi bottom and uses this single observation as an argument against Kedar, Glueck, and ourselves. He fell victim to an error, as the rubble piles depicted in his photograph from Wadi Isderiyeh are the leftovers of a relatively recent excavation made by the Mandatory Government of Palestine for a telephone cable which was laid along Wadi Isderiyeh. Second, Mayerson agrees that the vineyards planted on the slopes could not have existed on the available rainwater, but he believes that they were hand-irrigated from water stored in cisterns. A very simple calculation, already partly made by Kedar, shows that this is an impossibility. Kedar calculates that there are about 80 mounds per hectare (this is an underestimate; Mayerson talks about 600 per hectare) and that about 2,300 hectares are covered by these mounds in the vicinity of Avdat. We estimate that each vine planted on or near a mound would require at least 0.5 cubic meter of additional water per year. The ancient farmers would, therefore, have had to supply 92,000 cubic meters as additional irrigation. Kedar has calculated that all the cisterns in the vicinity do not contain more than 4,000 cubic meters altogether. The discrepancy between the figures is even more enormous if we assume that the people used some of the water from the cisterns for domestic purposes, and for cattle, as the Bedouins do today.

30. R. Calder, *op. cit.*

31. H. Boyko, *Proc. UNESCO Symposium on Plant Ecol.* (1955), pp. 1–8.

32. A. Reifenberg, *The Struggle between the Desert and the Sown* (Mossad Bialik, Jerusalem, 1955).

33. N. H. Tadmor, M. Evenari, L. Shanan, D. Hillel, *Ktavim*, **8**, 127, 151 (1957).

34. Y. Kedar, *Bull. Israel Exploration Soc.*, **20**, 31 (1956).

35. Y. Kedar, *Geograph. Rev.*, **123**, 179 (1957).

36. Y. Kedar, in *Study in the Geography of Erets Israel* (1959), vol. 1, pp. 122–124 (in Hebrew).

37. D. Sharon, in an excellent experimental study [see *Study in the Geography of Erets Israel* (1959), vol. 1, pp. 86–94], came to the following conclusions concerning soil erosion from the slopes: (i) The mounds were made by clearing the slopes of their dense stone cover. (ii) The soil beneath the mounds is undisturbed, but the clearance of ground between the mounds disturbed the natural equilibrium, exposed the slope to erosion, and through differential action on the soil-stone mixture led to the reformation of the stone cover between the mounds and the restoration of equilibrium. This explains why today the slopes between the stones are again covered by a stone pavement. As the difference in height between the old and "new"

level of stone pavement is 10 to 15 cm only, only this amount of soil can have been washed down the slopes during the centuries (actually less, as the 10 to 15 cm contained a considerable number of stones, now left on the slopes). (iii) There were originally two types of strips—strips built of stones and strips made of soil. The latter type is the more frequent. As the soil from the soil strips has been eroded, only the original stone cover lying originally beneath the soil strips remains today. In Sharon's opinion, the function of the strips was to direct the water down slope. Therefore they were made of soil, as in this way their impermeability was greatly increased in comparison with that of strips built of stone. All this tallies well with our own findings.

38. N. H. Tadmor, M. Evenari, L. Shanan, D. Hillel, *Ktavim*, **8**, 127, 151 (1957).

39. N. Glueck [*Bull. Am. Schools Oriental Research*, **149**, 8 (1958)] came to the same conclusions.

40. D. Hillel, *Bull. Israel Agr. Research Sta. Rehovot*, **63**, 1 (1959).

41. G. Caton Thompson and E. W. Gardner [*Geograph. J.*, **93**, 32 (1939)] report "evenly spaced stone-rubble heaps" tied up with ancient fields from Hadhramaut, and W. J. H. King [*ibid.*, **39**, 133 (1912)] reports similar findings from Libya. Photographs in the book of Baradez (note 11) seem to show areas of mounds and strips on the slopes near ancient fields in Algeria. A most interesting observation was made by B. Hellström [*Roy. Inst. Technol. Stockholm, Inst. Hydraulics, Bull. No. 46* (1955)]. In the desert between Cairo and Alexandria he found numerous sand walls called today *kurum* (compare our *rujum el kurum*), dating from Roman times. His explanation is as follows: "When it was raining, the water ran quickly downwards along the sides of the walls. . . . The walls were constructed for the sole purpose of irrigating surrounding cultivated areas by means of the discharging water. . . . The areas along the walls were used for vineyards."

42. M. G. Jonides, "Report on the water resources of Transjordan and their development," *Publ. Govt. Transjordan* (London, 1939).

43. A. Schori and D. Krimgold, *Internal Rept. Dept. Agr. Israel* (1959) (in Hebrew).

44. Y. Kedar [*Israel Exploration J.*, **7**, 178 (1957)] studied part of this area. However, he did not differentiate between the various periods of development in the area, and hence he shows the system as having been built and operated all at one time.

45. Y. Kedar indicates a different erosion cycle (see notes 24, 35). We feel that he did not notice the superimposition of earlier structures on later ones.

46. Examples of this flood-plain development were mentioned in our work on the Matrada plain and Sahel-El Hawa (note 7) and are common in all those flood plains where the main wadi has not yet eroded down below the flood plain. In these areas the process of active head gully growth can be seen even today. A deep gully (2 to 4 m deep) is cutting into the flood plain, progressing at a rate of tens of meters per year and so changing the base level of the area.

47. The cultivation of shallow depressions (Stage 1) probably also occurred in the small watersheds, and the simple terraces and cultivation of wadis found in these areas may relate to this stage. But we have not yet studied in detail either the hydrology or the historic development of this type of desert agriculture. Moreover, we have only begun an investigation of the role the numerous water cisterns played in collecting runoff from small watersheds for domestic purposes. It should also be pointed out that, although the final use of the diversion areas was as runoff farms, the runoff farms occur mainly in areas which were never related to diversion projects.

48. See, for example, M. Cressey [*Geograph. Rev.*, **48**, 27 (1958)], A. Smith [*Blind White Fish in Persia* (Allen and Unwin, London, 1953)], and A. Reifenberg (note 32).

49. M. G. Jonides, *op cit.*

50. B. Aisenstein, *J. Assoc. Engrs. and Architects in Palestine*, **8**, 5 (1947).

51. M. Evenari, L. Shanan, N. Tadmor, *Ktavim*, **9**, 223 (1959).

52. Y. Aharoni, *Eretz-Israel* (1953), vol. 2, p. 112 (in Hebrew); F. Franck, *Z. deut. Pal. Ver.*, **57**, 191 (1934).

53. N. Glueck, in many of his papers cited in these notes; Y. Kedar, *Econ. Quart.*, **5**, 444 (1958) (in Hebrew); Carton (see note 12), who writes: "Il peut être . . . intéressant de montrer que les études archéologiques méritent d'être favorisées *en raison des enseignements utiles* qu'elles puvent donner" (italics ours); Woolley and Lawrence [*Palestine Exploration Fund Annual* (1914–15)], who write: "We believe that today . . . the Negev could be made as fertile as it ever was in Byzantine times."

54. C. J. Kraemer, Jr., *Excavations at Nessanah* (Princeton Univ. Press, Princeton, 1958), vol. 3.

55. Jossi Feldmann, then a member of *kibbutz* Revivim, carried out the reconstruction work at the Shivtah farm. The agricultural planning for both farms is done by an agricultural committee headed by Dr. J. Carmon. Its members are Dr. Samish (fruit trees) and Dr. R. Fraenkel (field crops), both from the National and University Institute of Agriculture; M. Hilb, Government Minister of Agriculture; J. Dekel, Jewish Agency; and M. Eshel, Government Department of Soil Conservation. Joel de Angeles, from Revivim, is responsible for carrying out the agricultural planning.

56. The farmhouse is called "The Lauterman Negev House" and is a gift of Rose Annie Lauterman, Montreal, Canada. We hope that scientists interested in animal or plant ecology of deserts will use this opportunity to study desert fauna and flora *in situ*. We ourselves, in addition to pursuing the hydrological and agricultural aims of the project, are carrying out an ecological-physiological investigation of the main desert plants, based on fixed observation plots around the two farms.

57. This investigation was supported by the Ford Foundation, the Rockefeller Foundation, and the Israeli Government. Our studies of ancient agriculture were supported by a grant from the Ford Foundation; the reconstruction and agricultural work was and is being financed by the Rockefeller Foundation and the office of the Israeli prime minister. Our thanks are due to the Israeli Air Force for the aerial photographs and to Mrs. L. Evenari for the ground photographs.

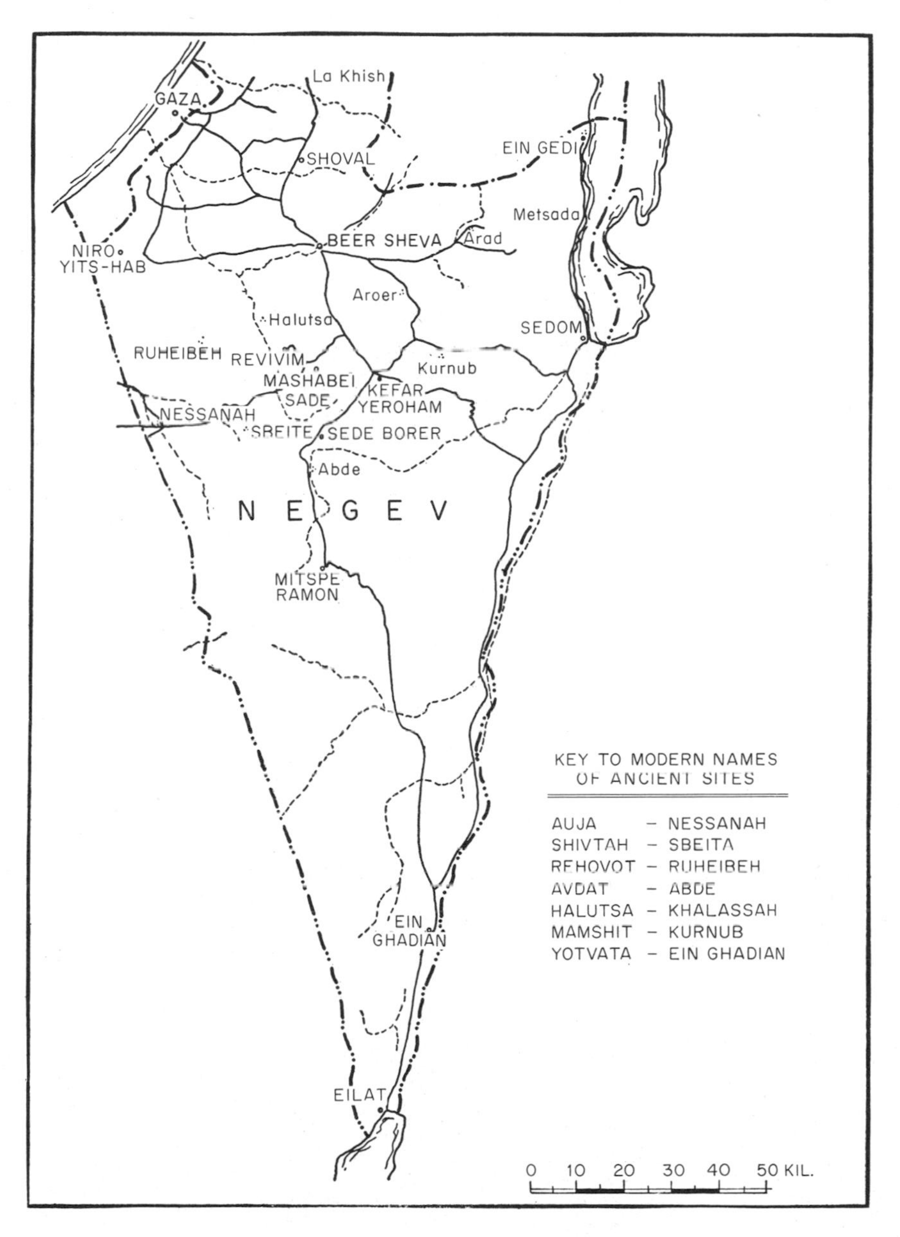

Fig. 19–1. Map of the Negev. The key gives modern names at left, ancient names at right.

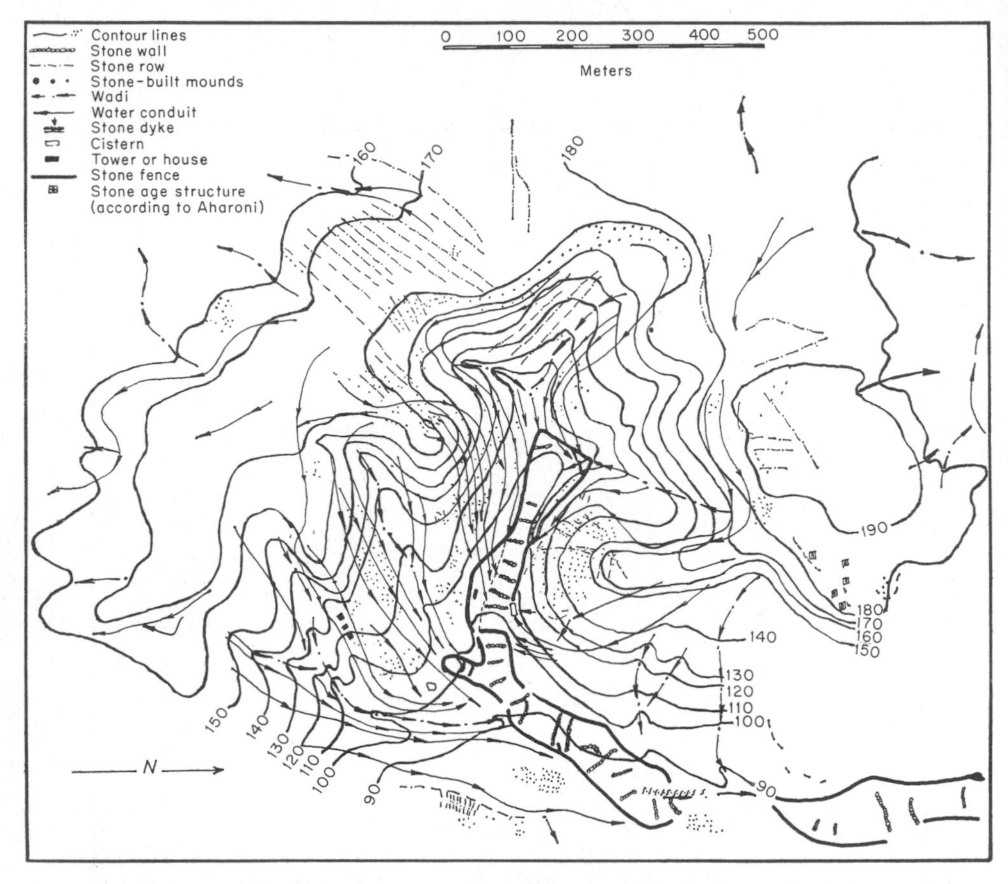

Fig. 19–2. Map of a runoff farm area near Avdat. Note the conduits and stone mounds.

Fig. 19–3. An oblique aerial photograph showing a number of ancient runoff farms near **Shivtah.**

Fig. 19–4. A vertical aerial photograph of a gravel mound and strip area near Shivtah.

Fig. 19–5. A field of gravel mounds near Shivtah.

Fig. 19–6. A field of gravel strips near Shivtah.

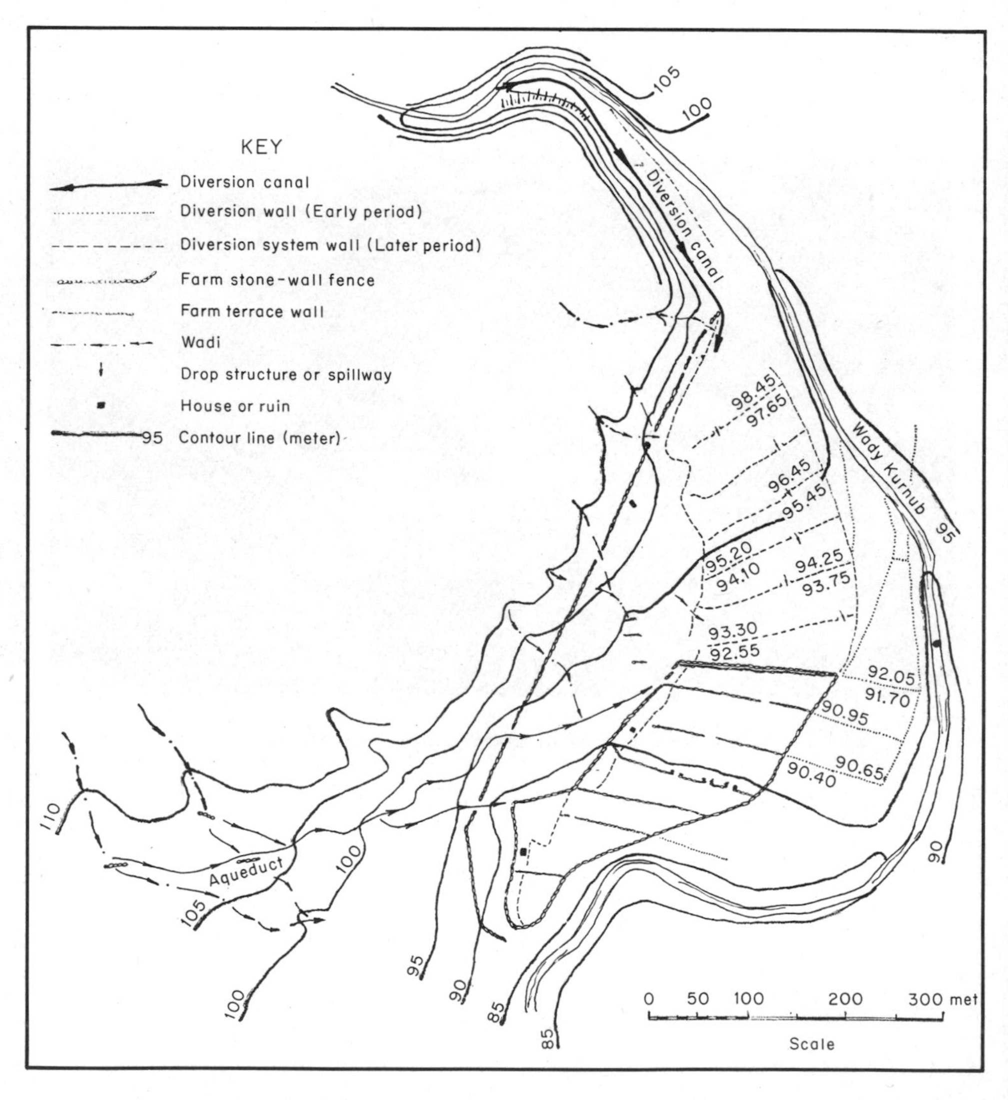

Fig. 19–7. A map of the Mamshit system, which exploited runoff from a large watershed, showing the various periods of development.

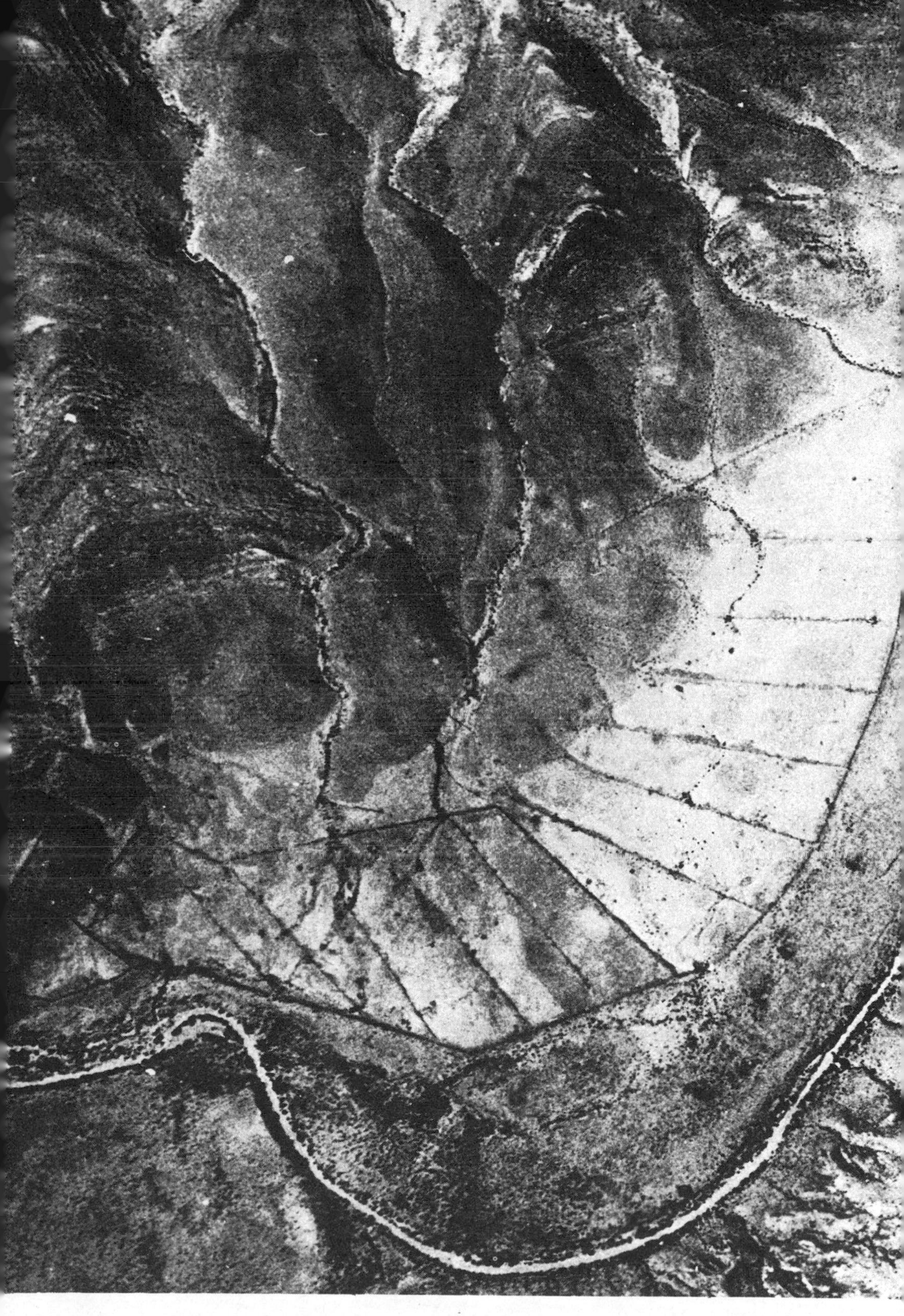

Fig. 19–8. A vertical aerial photograph of the Mamshit system.

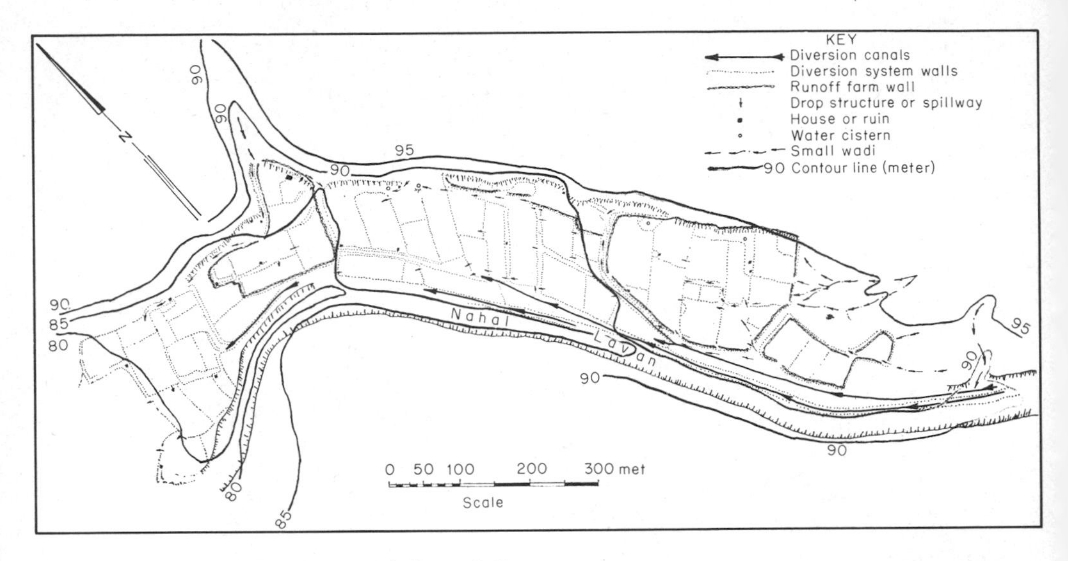

Fig. 19–9. Map of a section of the Nahal Lavan system.

Fig. 19–11. A large spillway with a crest length sufficient to allow passage of large floods.

Fig. 19–10. A diversion canal wall of the Nahal Lavan system. Note the different stages of construction. The foundations are about 2 meters below the present soil surface.

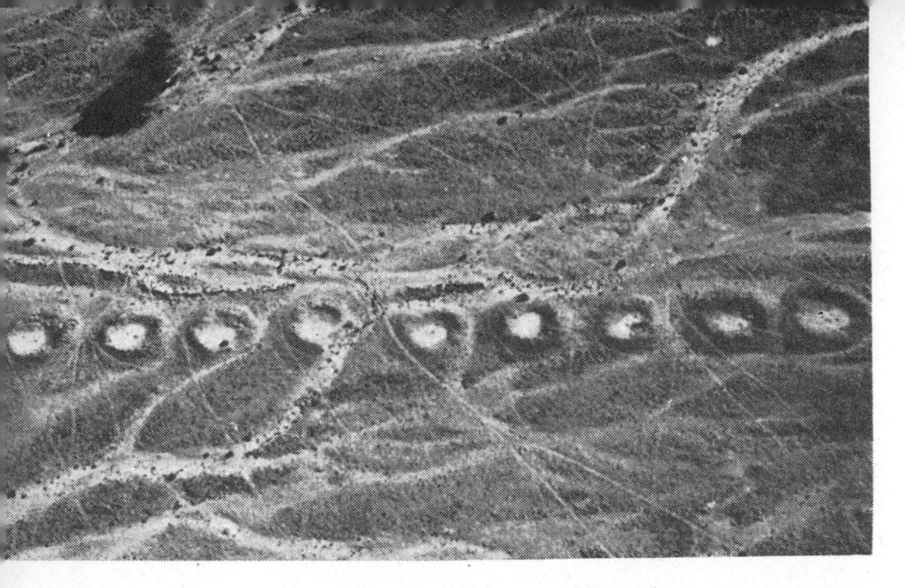

Fig. 19–12. An oblique aerial photograph of a chain-well system near Ein Ghadian.

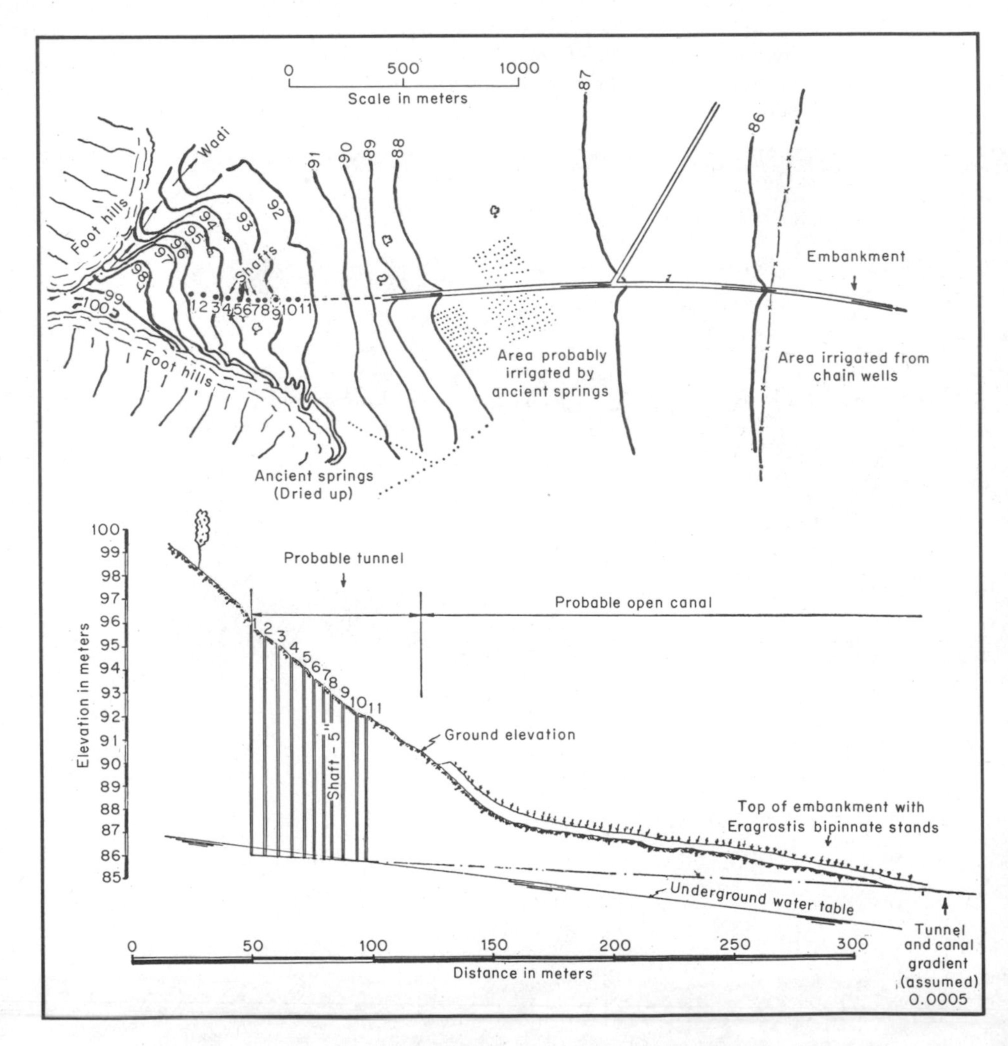

Fig. 19–13. Details of a chain-well system near Ein Ghadian.

Fig. 19–14. An aerial photograph of the reconstructed Shivtah farm. Note one branch of one runoff conduit entering the farmhouse, bringing water to an underground cistern.

Fig. 19–15. The reconstructed ancient farm near Avdat. Note the farmhouse on the hill. The reconstructed conduits leading runoff from the small watersheds may be seen in the background.

Fig. 19–16. Harvesting barley on the Avdat farm in May 1960.

Part IV

SCIENCE IN ARCHAEOLOGY

Chapter 20

Patination of Cultural Flints

Vernon J. Hurst
A. R. Kelly

Certain flint artifacts have undergone cortical decomposition since they were fashioned several hundred to several thousand years ago and are now invested with a rind or patina. A rough correlation between the thickness of the patina and the age of the artifact has been noted.[1]

However, many artifacts that are known to be old exhibit little or no cortical change. It is apparent, therefore, that age is not the sole variable on which patination depends. Before patination can be utilized chronometrically the other variables must be determined and their relative effects must be evaluated.

For obsidian, I. Friedman and R. L. Smith[2] have shown that only two main variables must be evaluated: composition and mean temperature.

Flint is a more complex material. Its microstructure is more variable and is an important patina-controlling factor. The kind, size, and manner of distribution of impurities profoundly influence flint's susceptibility to attack and the rate of penetration by water and other weathering agents. Even limited study suffices to show that the patination rate may vary more with microstructure and mineralogical impurities than with age. The number of variables is greater for flint than for obsidian, but with careful use of evaluation techniques and a fuller understanding of the processes involved, flint patination can be made to yield useful chronometric data.

Two general types of flint patination can be distinguished. In one type the patina is some shade of brown and is darker than the core. In the other type the patina is a chalky white and is lighter in color than the core. The patina is visible primarily as a color difference in both cases. Before considering in detail the origin of the patina, let us review the properties of flint, such as composition and texture, which are modified during patination and on which the color differences depend.

Composition and Texture of Flint

Flint is composed mainly of small crystallites of silica. The individual crystallites may be so small that their boundaries are invisible even at high magnification, or they may be large enough to be seen by the unaided eye. Their shapes range from needles or fibers to equant (nearly equidimensional) grains. The size and shape of the grains may be regular within a given specimen, or they may vary greatly (Fig. 20–1). Radiating spherulitic and oölitic textures are common (Fig. 20–2). Many flints originated as siliceous replacements of fossiliferous carbonate rocks, and outlines of fossils or fossil fragments are commonly preserved (Fig. 20–3). The mineral impurities in flint include carbonates, iron and manganese oxides, clay minerals, carbonaceous matter, and iron sulfide. The carbonates are typically in the form of rhombs 0.01 to 0.1 millimeter across, distributed unevenly through the flint and often imparting to it a mottling or banding. Less frequently the carbonate is in the form of granular patches or partially replaced fossil fragments. The other mineral impurities are generally distributed unevenly as "dust" or clay-size inclusions, which are not individually visible to the eye, but which impart a haziness, milkiness, or other color change.

Color of Flint

The color of flint is determined by the grain size, the texture, and the kind and number of impurities.

Grain size and texture largely determine how much of the incident light enters the specimen and passes through and how much is reflected back to the observer by a series of refractions and reflections at grain intersurfaces. The ratio of transmitted light to diffused and reflected light fixes the diaphaneity of the specimen and the *value* of its color (according to the Munsell color system). A flint composed of coarse even-grained quartz, for example, has relatively few refracting and reflecting surfaces, so that most of the incident light enters and passes through (unless it is absorbed by impurities) and the specimen appears colorless-to-light-gray and transparent-to-translucent. A flint composed of fine crystallites, on the other hand, particularly if they vary in size and are irregular in shape, has numerous intersurfaces. Most of the light incident on such a specimen is diffused and reflected out. Consequently the specimen is white and opaque.

The impurities in flint affect the color in two ways: (i) by their influence on the ratio of transmitted light to reflected light, and (ii) by their absorption of light. The surfaces between impurities and the enclosing quartz are more effective in refracting and reflecting light than the surfaces between the quartz crystallites themselves. A fractional percentage of impurities may reflect or absorb, or reflect and absorb all the incident light. Whether the flint appears white, gray, black, or some other color depends on how much of the light is reflected and how much is absorbed, and on whether all wavelengths are absorbed equally. Specimens that reflect all the incident light are white. Those that absorb all the light are black. Those that absorb a part of the light are some shade of gray, if all wavelengths are absorbed equally, and are red, brown, or some other color if certain wavelengths are absorbed more than others. The grain size and the texture strongly influence the ratio of reflected light to transmitted light and therefore affect the property of lightness—the *value* of the color. The kinds of impurities determine which wavelengths are absorbed—that is, they determine the *hue* of the color. The amount of the impurity affects the degree of saturation of the color, or its *chroma*, as well as its *value*.

These generalizations are illustrated by figs. 20–4, 20–5, and 20–6, which are photomicrographs, made with transmitted light, of thin sections of flints. The light gray areas are silica. The moderate gray areas are silica clouded by clay inclusions. The gray rhombic shapes are carbonate minerals. The black areas are opaque impurities in the flint. The proportion of opaque matter is not so great as it appears, inasmuch as the photograph is a projection onto a single plane of all the opaque matter in the body of the section, which is 0.035 millimeter thick. Most of the opaque matter is white—that is, it reflects all visible wavelengths of light. Some of it absorbs all visible wavelengths equally, and is black. Light can penetrate these three flints for only a fraction of a millimeter before a part of it is reflected out and the remainder is absorbed.

The flint shown in Fig. 20–4 contains abundant white opaque matter that reflects out the incident light before much of it has been absorbed; consequently the flint is light gray. The flint shown in Fig. 20–6 contains much less opaque matter than that of Fig. 20–4 and a greater proportion of the opaque matter is black. Incident light penetrates deeper and is largely absorbed; consequently the upper half of this flint is black. A flint with the same impurity content but with a finer, uneven texture would be light gray because light would be scattered more at the surface and a greater proportion would be reflected. The opaque-impurity content of the moderately gray flint (Fig. 20–5) is greater than that of the light gray flint (Fig. 20–4) and still greater than that of the black flint (upper half of Fig. 20–6). The thin section represented by Fig. 20–6 was cut from a black arrowhead mottled with light reddish brown. The black portion (upper half) contains much less opaque matter than the light-colored portion (lower half), but in the lower half the opaque matter is iron stained and reflects out more light, even while absorbing the shorter wavelengths; hence this portion appears light-colored and reddish brown.

The absorption of light by color centers may be a secondary cause of color in a few flints. The color of most, however, is attributable to the refraction and reflection of light along intergranular surfaces and to the pigmenting effects of a few included impurities.

The Patination Process

Obsidian is volcanic glass. Patination in obsidian is essentially a single process, hydration, the rate of which is controlled by the rate of diffusion of water into the glass. For this reason the patination rate is strongly influenced by temperature.

Flint, on the other hand, is mainly silica, which is only slightly susceptible to hydration. Patination in flint proceeds by the interaction of at least three different processes operating, for the most part, on impurities in the flint: (i) oxidation and hydration, (ii) dissolution and leaching, and (iii) chemical and mechanical disaggregation.

Flints that are pure silica, or that contain only chemically stable impurities, cannot patinate, regardless of grain size or texture. To this class belong many of the light gray, translucent flints and the dark flints that owe their color to organic matter.

The flints that contain chemically unstable impurities are all susceptible to patination, but the rate varies greatly depending on what is the predominant mechanism by which water and other weathering agents penetrate.

For the very fine-grained, even-textured flints whose unstable impurities are thinly dispersed, the principal mechanism is diffusion along intergranular boundaries. Flints of this class are most useful for chronometric purposes, because their patination is controlled by only a few variables, foremost of which are (i) mean temperature and (ii) composition, as in the case of obsidian.

Most flints have grains of various sizes, an uneven texture, and inhomogeneously distributed impurities, and their patination is influenced by several factors in addition to temperature and composition. When a new surface is formed on one of these flints, any exposed carbonates begin to slowly dissolve and leach. As they are removed, new surfaces are exposed, the flint becomes more porous, and quartz crystallites are somewhat loosened. Where the weathering agents gain access to inclusions of iron sulfide, the ensuing oxidation and hydration cause local volume changes and create dilute acids that, in turn, attack some of the clays. The iron oxides hydrate and disperse along grain boundaries and in voids. The aggregate effect of the various oxidation, hydration, and dissolution processes is a loosening of quartz crystallites and a greatly increased porosity. This over-all textural change, which greatly increases the reflectivity of the specimen, is seen as a chalky white patina. Where iron oxide or other pigment is dispersed near the surface, the whiteness is correspondingly modified to cream, pink, brown, and so on.

Figure 20–7 illustrates the characteristic development of a chalky white patina. The core (Zone *1*) shows relatively little change. The boundary between zones *1* and *2* is the depth to which patination has progressed. In Zone *2* the ferruginous masses have been attacked and partly dispersed. In Zone *3* dispersal is pronounced and leaching of ferruginous and calcareous matter is perceptible. In Zone *4* leaching is pronounced and the flint is quite porous. The whiteness of the patina is largely due to textural changes and to the increase in reflectivity occasioned by dispersal of the opaque matter, and partly to leaching of carbonates and reddish brown pigment. The formation of the chalky white patina is favored by the high impurity content of clay and carbonate minerals.

Where the water in contact with a ferruginous flint is acidic, a large part of the iron in the patinating layer may be dissolved and leached away from the flint or, instead, may be dispersed along intergranular spaces and only slightly concentrated at the flint's surface. In the first case a chalky white patina develops; in the second case, a brown goethitic rind develops, as shown in Fig. 20–8. Different sets of local conditions can produce one or the other type of patina from the same kind of flint. Where the flint is exposed to circulating acidic ground water, leaching of iron and development of the chalky patina are favored. Where the moisture involved in patination gains its acidity from chemical reactions within the flint, or where the flint reposes in stagnant phreatic water, the development of the brown goethitic patina is favored.

Texture and microstructure strongly affect patination, as illustrated in the flint shown in Fig. 20–9. This flint was formed by the silicification of a porous fossiliferous rock. Silica has replaced foraminiferal tests and other fossil fragments in the dark gray areas (Fig. 20–9, bottom) and has filled voids (the light gray, granular areas). The void-filling silica can be penetrated only through intergranular diffusion, and it contains no visible impurities; consequently it acts as a barrier to patinating agents and cannot, itself, develop a patina. On the right side, the impurity-laden flint in Fig. 20–9 has patinated normally. At the top, patination has proceeded normally down to the void-filling silica, which has a different texture and has acted as a barrier. The two outer zones of the typical chalky patina—the zone of pronounced leaching (Zone *4* in Fig. 20–7), and the zone of homogenization of impurities (Zone *3* in Fig. 20–7)—continue around the specimen from the right side to the top, as shown by the dashed lines, but the thickness of the next inner zone (Zone *2* in Fig. 20–7) is much reduced. Continued patination of this specimen would thicken the patina on the right side by the inward migration of each zone boundary, but at the top only Zone *4* would thicken, at the expense of zones *3* and *2*, without any over-all increase in the thickness of the patina.

Summary

All flints containing unstable impurities are susceptible to patination. The rate of patination varies with many factors: (i) the texture and microstructure of the flint; (ii) its permeability; (iii) the kind, proportion, and distribution of impurities; and (iv) environmental factors, such as temperature and soil chemistry. The thickness of the patina varies also with time.

Two contrasting types of patina can develop: a chalky white patina and a ferruginous brown patina. Both types are observable primarily as a color change, and study of these types is facilitated by a clear understanding of the causes of color in flint.

The color of most flints is the result of repeated refraction and reflection of light at numerous intergranular surfaces, whereby part of the light is internally absorbed and part is reflected back to the observer. The ratio of reflected to absorbed light governs the lightness of the color, or its value. The preferential absorption of certain wavelengths by natural pigments (such as iron oxide and hydrous iron oxide) disseminated through the flint determines the hue of the color.

The color changes produced during patination relate to changes in texture and impurity content occasioned by the attack of weathering agents on impurities in the flint. The creation of voids by the dissolution and leaching of carbonates, the loosening of quartz crystallites, and the dispersal of clays all modify the reflectivity of the flint. Chemical changes involving the pigments, their dispersal along intergranular surfaces, or removal by leaching modify both reflectivity and capacity to preferentially absorb.

Attempts to correlate patina thickness with age, and thus to use flint patinae chronometrically, have proven unsatisfactory because other factors, whose importance in some cases exceeds that of age, have not been taken into acount. The texture and microstructure of flint, its permeability, and the kind, proportion, and distribution of impurities can be evaluated by regular petrographic techniques. Environmental factors can be assumed constant for artifacts from the same types of soil in a given climatic region. Only after allowances have been made for these additional variables does the age-dependence of flint patination become clear.

NOTES

1. A. R. Kelly, "Age measurements in decomposed flint," *Georgia Dept. of Mines, Mining and Geol., Geol. Survey Bull. No. 60* (1953), pp. 321–30; A. R. Kelly and V. J. Hurst, *Am. Antiquity*, **22**, No. 2 (1956).
2. I. Friedman and R. L. Smith, *Am. Antiquity*, **25**, 476 (1960).

Fig. 20–1. Photomicrograph of flint, in thin section, showing the variation in grain size and the radiating texture.

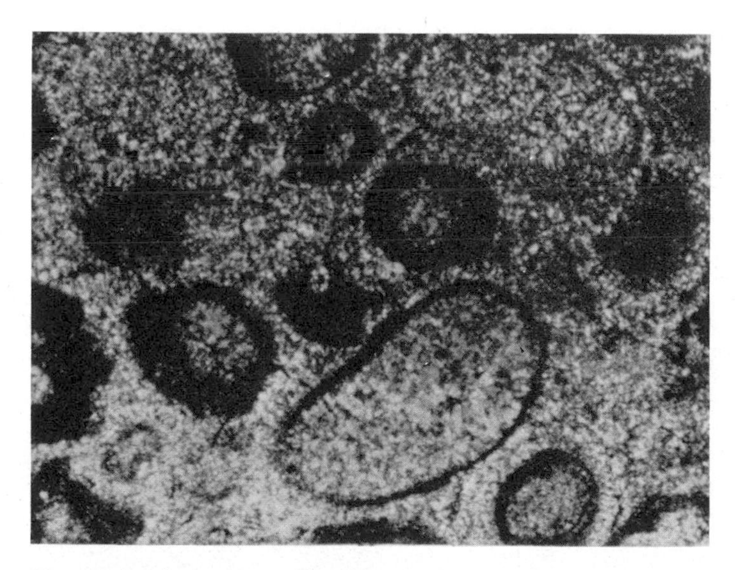

Fig. 20–2. Photomicrograph of flint with relic oölitic texture thin section. (Polarized light; about × 73)

Fig. 20–3. Photomicrograph of a portion of a brown flint projectile, in thin section. The flint is composed of silicified foraminiferal tests and is colored red brown by unevenly distributed ferruginous matter. Patination processes have modified the exterior distribution of the ferruginous matter, dispersing it more or less evenly and investing the projectile with a continuous smooth rind of hydrous iron oxide. (Transmitted light; about × 9.3.)

Fig. 20–4. Photomicrograph of light gray flint, in thin section (transmitted light; × 147).

Fig. 20–5. Photomicrograph of moderate gray flint, in thin section (transmitted light; × 147).

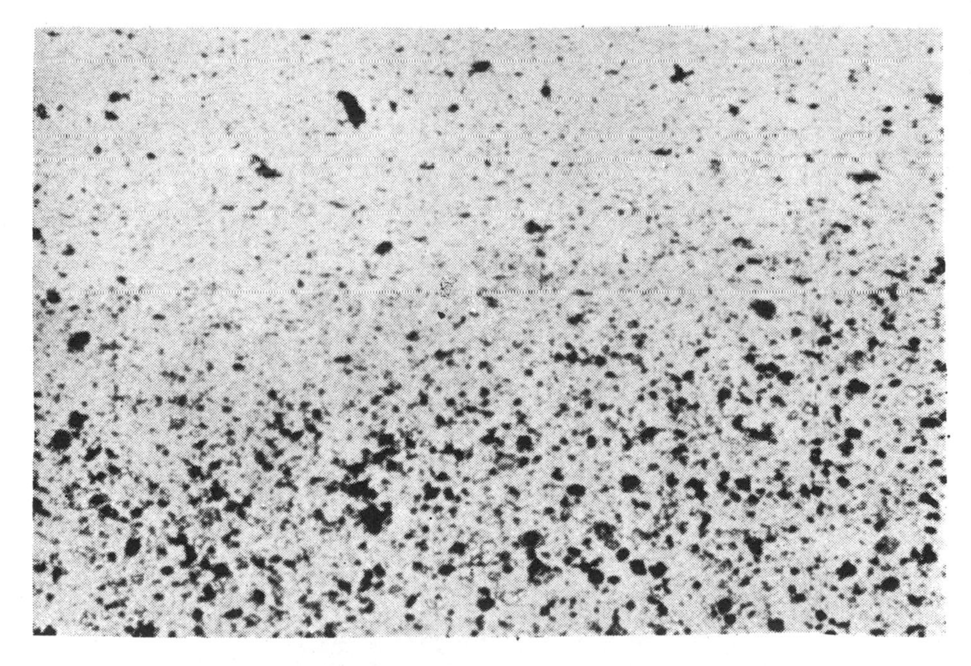

Fig. 20–6. Photomicrograph of black flint (upper half), mottled by pale reddish brown flint (lower half), in thin section (transmitted light; × 175).

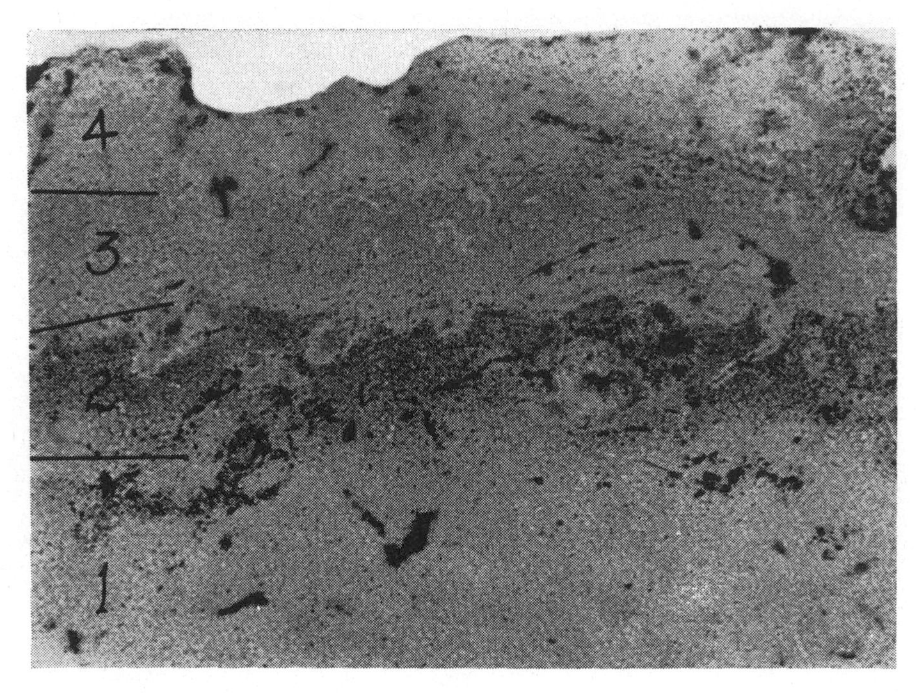

Fig. 20–7. Photomicrograph of a thin section, showing the characteristic development of the chalky white patina (see text). (Transmitted light; × 96.)

Fig. 20–8. Photomicrograph of a thin section, showing patina of the brown goethitic type. The patinating processes have caused the ferruginous matter (black areas), which was unevenly distributed originally and is still unevenly distributed in the core, to be dispersed along intergranular spaces in the patinating layer and concentrated on the exterior of the specimen to form a smooth brown goethitic rind. (Transmitted light; × 70.)

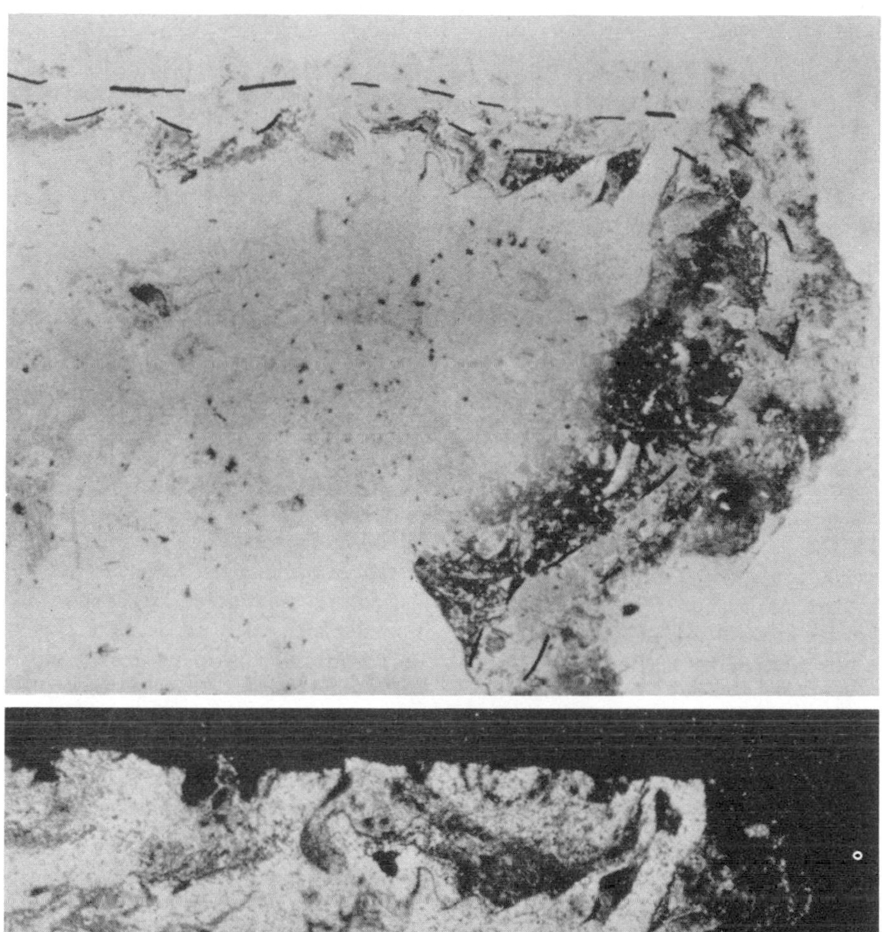

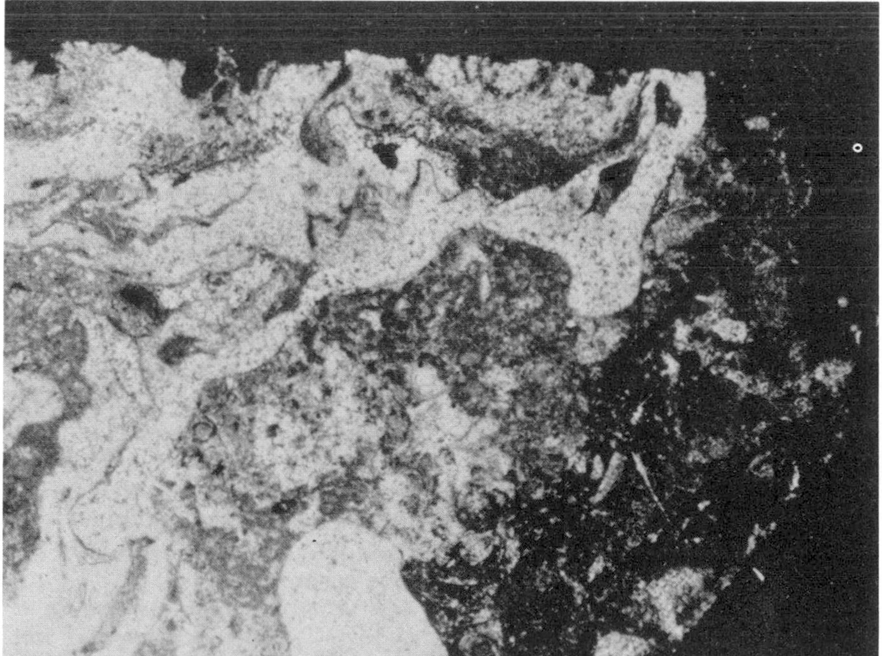

Fig. 20–9. Photomicrographs of a flint formed by the silicification of a porous fossiliferous rock (see text). (Top, plain light; bottom, polarized light; × 10.)

Index